Fundamentals of Aeronomy

SPACE SCIENCE TEXT SERIES

A. J. Dessler and F. C. Michel, Editors

Rice University, Houston, Texas

Associate Editors

D. L. Anderson

California Institute of Technology, Pasadena, California

Carl-Gunne Fälthammar

Royal Institute of Technology, Stockholm, Sweden

William M. Kaula

University of California, Los Angeles, California

K. G. McCracken

University of Adelaide, Adelaide, South Australia

I. A. Zhulin

Izmiran, Moscow. U.S.S.R.

William M. Kaula

AN INTRODUCTION TO PLANETARY PHYSICS

Thomas L. Swihart

ASTROPHYSICS AND STELLAR ASTRONOMY

Frank D. Stacey

PHYSICS OF THE EARTH

R. C. Whitten and I. G. Poppoff

FUNDAMENTALS OF AERONOMY

Robert C. Haymes

INTRODUCTION TO SPACE SCIENCE

A. J. Dessler and F. C. Michel

PARTICLES AND FIELDS IN SPACE

S. Matsushita and W. H. Campbell

INTRODUCTION TO THE MAGNETISM OF THE EARTH IN SPACE

Fundamentals of Aeronomy

R. C. WHITTEN, Chief, Planetary
Environments Branch
I. G. POPPOFF, Assistant Chief,
Space Science Division
Ames Research Center
National Aeronautics and Space Administration

John Wiley & Sons, Inc.
New York · London · Sydney · Toronto

Library of Congress Catalog Card Number: 71-130435

ISBN 0 471 94120 4

Printed in the United States of America

10 9 8 7 6 5 4 3 2 1

Reprinted from Nature, June 17, 1933

Fine-Structure of the Ionosphere

I WAS much interested in the letter by Messrs. Schafer and Goodall in NATURE of June 3 dealing with the results of their experiments on the radio exploration of the ionosphere in the United States, since an independent set of observations in Great Britain, which are dealt with at length in a paper now awaiting publication, have led to very similar conclusions.

The method[1] used in the British series of observations has been to measure the maximum ionisation content of the various upper atmospheric regions by finding the critical penetration frequencies.

The British series of observations suggests therefore that there are four main components in the ionosphere caused by the influence of ultra-violet light from the sun. Such a composite structure is not considered unlikely when it is remembered that Pannekoek[3] has shown that the level of maximum ionisation caused by ultra-violet radiation depends on the ionisation potential of the gaseous constituent. It is tempting to associate the four components with the four ionisation potentials of oxygen and nitrogen atoms and molecules, and the suggestion that F^{II} is due in this way to oxygen atoms and F^{I} to oxygen

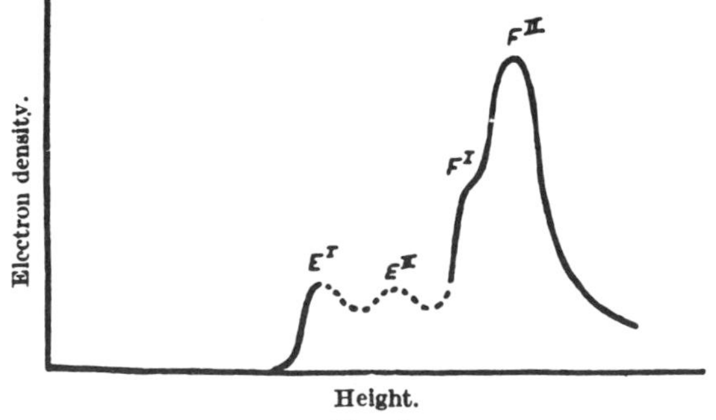

FIG. 2.

molecules has, indeed, already been made by T. L. Eckersley[4], who independently obtained evidence indicating a dual structure for region F.

E. V. APPLETON.

Halley Stewart Laboratory,
King's College,
London.
June 3.

Preface

Atmospheric phenomena have been the objects of curiosity for millenia and of more serious scientific study for centuries. Unlike the situation in areas of physics which have developed more recently such as elementary particles, the basic physical principles of atmospheric processes are well established. Nevertheless, the manner in which these principles are involved in many geophysical phenomena is poorly understood. This is particularly true of the science of the upper atmosphere, first called *aeronomy* by Nicolet some years ago. Despite the lack of new principles to be discovered, aeronomy has been a very active research field in recent years. This activity has been due mainly to great strides in observational techniques (i.e., by use of rockets and spacecraft), and to a lesser extent to developments in the ancillary sciences such as plasma physics and atomic physics. In fact, much of the impetus behind the recent revival of interest in the latter is due to the demands of aeronomy.

Although the field of aeronomy is now quite well developed, no textbook has previously been published. It is true that a number of excellent monographs, which are cited at appropriate places in the book, have appeared in recent years. However, for a number of reasons, they are not completely satisfactory for classroom use. It is our aim to remedy this deficiency with the present book by developing the principal themes of aeronomic science from basic physical principles. We try to emphasize the fundamental physics throughout rather than concentrate on the results of the most recent work in each area. This has the advantage of saving the student from a mass of confusing detail which may be obsolete tomorrow. Of course, a sizable portion of the text is descriptive in nature, particularly the chapter on optical phenomena. This approach is unavoidable in a text on geophysics. Although the earth is understandably given most emphasis, the upper atmospheres of other planets are also discussed to the extent that present knowledge permits. In any case, the methods developed for dealing with the earth's upper atmosphere are directly applicable to the atmospheres of other planets.

We introduce and develop those topics which seem to us to be of greatest significance to aeronomy and which should, in our opinion as practitioners of the science, be included in a course on the subject. The book is *not* based on classroom notes, and although we make no apology for this, we do wish that there might have been more opportunities for classroom trial. We were fortunate in having the advice and criticism of students of Rice University and the U.S. Naval Post Graduate School and their suggestions have been followed in many instances.

Fundamentals of Aeronomy is intended for use in a two-semester course at the senior or first year graduate level. The student is assumed to have a basic knowledge of mechanics, electrodynamics, kinetic theory of gases, thermodynamics, and atomic physics, as well as undergraduate mathematics through differential equations and vector analysis. As a refresher, the basic physics is reviewed at some length in Chapter 2. Subsequently we discuss the structure of the neutral upper atmosphere (Chapters 3 to 5), aurora and airglow (Chapter 6), ionospheric phenomena (Chapters 7 to 10), and electromagnetic waves in the ionosphere (Chapter 11). General as well as specific references and a set of appropriate problems are given at the end of each chapter. The specific references are by no means exhaustive; they have not been offered to support particular statements but to aid those students who wish to consult original sources.

We are indebted to the aurhors of all the books and papers we have read, to all the lecturers we have heard, and to all the colleagues we have known. Specifically we wish to thank Dr. Donald M. Hunten of Kitt Peak National Observatory for many helpful comments and much useful advice, as well as Professors A. J. Dessler and R. F. Stebbings of the Department of Space Science at Rice University, Professor Otto Heinz of the Department of Physics at the U.S. Naval Post Graduate School, Dr. C. A. Riegel of the Department of Meteorology at San Jose State College, and Professor R. Parthasarathy of the Geophysical Institute at the University of Alaska for helpful criticism. We also wish to express our thanks to Mrs. Jane Gutter for typing the final manuscript. Lastly, we thank our families for their encouragement and their forbearance.

R. C. WHITTEN
I. G. POPPOFF

September, 1970

Contents

1. Introduction 1

1.1 Historical Notes 2
1.1.1 Aurora *2*
1.1.2 Airglow *3*
1.1.3 Ionosphere *4*
1.2 Overview of Aeronomy 5
1.2.1 Atmospheres *6*
1.2.2 Radiations *7*
1.2.3 Interactions *14*
1.2.4 Layers and Regions *15*
1.3 Limitations of Knowledge 19
1.4 Outlook 20

2. Fundamental Physical Principles 21

2.1 Electrodynamics 22
2.1.1 Fundamentals *22*
2.1.2 Electromagnetic Wave Equations *23*
2.1.3 Electrical Properties of a Plasma *24*
2.1.4 Energy Flow *25*
2.2 Thermodynamics 26
2.2.1 The First and Second Laws *26*
2.2.2 Thermodynamics of Ideal Gases *28*
2.2.3 Miscellaneous Thermodynamic Equations *29*
2.3 Kinetic Theory of Gases 30
2.3.1 The Distribution Function *30*
2.3.2 Equation of State of an Ideal Gas *31*
2.3.3 The Boltzmann Equation *32*
2.3.4 The Boltzmann Transport Equations and Applications *33*
2.3.5 DeBye Shielding in a Plasma *36*
2.4 Atomic and Molecular Structure and Spectra 37
2.4.1 The Black Body Spectrum and Einstein Coefficients *37*
2.4.2 Atomic Structure *39*

2.4.3 Molecular Structure *45*
2.4.4 Atomic Spectra *48*
2.4.5 Molecular Spectra *48*
2.5 Collision Processes 49
2.5.1 Collision Theory *50*
2.5.2 Photo Processes *54*
2.5.3 Passage of Charged Particles Through Matter *56*
2.5.4 Collisions of Low Energy Electrons *59*
2.6 Reactions 60
2.6.1 Kinetics *60*
2.6.2 Ionic Recombination *62*
2.6.3 Negative Ions and Their Reactions *62*
2.6.4 Ion-molecule Reactions *63*
2.6.5 Chemical Reaction Rates *64*

3. Physical Aeronomy 69

3.1 Hydrostatics 70
3.2 Heating of the Upper Atmosphere 72
3.2.1 Energy Deposition *72*
3.2.2 Heat Transfer by Conduction *75*
3.2.3 Heat Transfer by Radiation *77*
3.3 Mechanical and Chemical Equilibrium in the Earth's Upper Atmosphere 79
3.4 Molecular Diffusion 80
3.5 Variations in the Earth's Atmosphere 83
3.6 Measurements of Atmospheric Properties 87
3.6.1 Density *87*
3.6.2 Temperature *92*
3.6.3 Diffusion *94*
3.6.4 Observation of Planetary Atmospheres by Spacecraft *96*
3.7 Model Atmospheres 98
3.7.1 Model of Harris and Priester *99*
3.7.2 Empirical Models *101*
3.7.3 Atmospheric Models of Venus and Mars *105*
3.7.4 Atmospheric Models of Jupiter *106*
3.8 Planetary Exospheres 107
3.8.1 Theory of the Exosphere *108*
3.8.2 The Earth's Exosphere (Geocorona) *110*
3.8.3 The Exosphere of Venus *111*

4. Chemical Aeronomy 117

4.1 Dissociation and Recombination 118
4.1.1 Photodissociation Reactions 121
4.1.2 Recombination Reactions 132
4.1.3 Deactivation of Excited Oxygen Atoms 135
4.2 Ionization, Recombination, and Interchange 136
4.3 Photochemical Equilibrium 137
4.3.1 Terrestrial Atmosphere 138
4.3.2 The Atmospheres of Mars and Venus 145

5. Fluid Aeronomy 153

5.1 Fundamentals of Fluid Dynamics 153
5.1.1 Equations of Motion 154
5.1.2 Perturbation Theory 156
5.1.3 The Significance of N 158
5.1.4 The Geostrophic Approximation and the Thermal Wind 160
5.1.5 Vorticity and Cyclogenesis 161
5.1.6 Atmospheric Oscillations 162
5.2 Internal Gravity Waves 162
5.3 Tidal Oscillations 165
5.4 Winds and Circulation 170
5.4.1 Mesosphere 170
5.4.2 Thermosphere 173
5.4.3 Wind Observations 174
5.5 Turbulence in the Upper Atmosphere 175
5.5.1 Theory of Turbulence 175
5.5.2 Mass and Heat Transport 180

6. Optical Phenomena 185

6.1 Emission Mechanisms 186
6.1.1 Resonance and Fluorescence 186
6.1.2 Chemiluminescence 186
6.1.3 Excitation by Charged Particles 187
6.1.4 Energy Transfer 187
6.1.5 Emission Rate 188
6.2 Airglow 189
6.2.1 Dayglow 189
6.2.2 Twilight Glow 195

6.2.3 Nightglow *197*
6.2.4 Venus *200*
6.2.5 Mars *201*
6.3 Aurora 202
6.3.1 Morphology *203*
6.3.2 Excitation Mechanisms *207*
6.3.3 Auroral Spectra *208*
6.3.4 Red Auroras *213*
6.3.5 Polar Glow Auroras *215*
6.3.6 Other Planets *215*

7. Electric Currents in the Upper Atmosphere 219

7.1 Ionospheric Conductivity 221
7.1.1 Motion of Charged Particles in a Magneto Plasma *222*
7.1.2 The Conductivity Tensor for a Magneto Plasma *223*
7.1.3 Electrical Conductivity of the Middle Ionosphere *226*
7.1.4 Ionospheric Conductivity at Very High Altitudes *227*
7.2 The Dynamo Theory 229
7.3 Magnetic Disturbances 234
7.4 *F*-Region Drifts 234
7.5 Currents in the Interface between the Solar Wind and Ionosphere for a Planet with no Magnetic Field 235

8. Structure of the Lower and Middle Ionosphere 240

8.1 Ion-Electron Pair Production 240
8.1.1 Photoionization *242*
8.1.2 Corpuscular Ionization *243*
8.2 Ion-Kinetics 245
8.3 Equilibrium 247
8.4 Ionospheric Regions 252
8.4.1 D-region *253*
8.4.2 E-region *259*
8.4.3 F_1*-Region* *271*
8.5 Variations 273
8.5.1 Diurnal Variations *273*
8.5.2 Seasonal Variations *277*
8.5.3 Solar Cyclic Variations *278*

9. The Upper Ionosphere 282

9.1 Formation of the F_2 Layer 283
9.1.1 Photoionization 283
9.1.2 Electron Loss 284
9.1.3 Ambipolar Diffusion 285
9.2 The Continuity Equation 287
9.2.1 Solutions for Simple Models of the F_2-Region 287
9.2.2 A Multicomponent Topside Ionosphere 291
9.2.3 The F_2 Layer at Equatorial Latitudes 293
9.2.4 Time-Dependent Model of the F_2-Region 293
9.3 Anomalies 294
9.3.1 Electrodynamic Drifts 294
9.3.2 Nocturnal Ionization 296
9.3.3 Geographical Anomalies 297
9.3.4 Temporal Anomalies 299
9.4 Thermal properties of the F_2-Region 300
9.4.1 Energy Spectrum of the Photoelectrons 300
9.4.2 Energy Loss Mechanisms 301
9.4.3 Cooling of the Electron and Ion Gases; Heat Conduction 303
9.4.4 Nonlocal Energy Deposition 308
9.4.5 The Heat Equations for the Ionosphere 308
9.4.6 Observations of Electron and Ion Temperature 310
9.5 The Protonosphere 314
9.6 Spread-F and Other Irregularities 315
9.7 Traveling Ionospheric Disturbances 320
9.8 Upper Ionospheres of Other Planets 320
9.8.1 The Daytime Upper Ionosphere of Venus 320
9.8.2 Thermal Structure of the Ionosphere of Venus 322

10. Disturbances in the Ionosphere 327

10.1 Perturbations by Electromagnetic Radiation 328
10.1.1 Solar x-ray Flare 328
10.1.2 Observed Effects 329
10.1.3 Balance Equations 333
10.1.4 Illustrative Examples 335
10.2 Perturbations by Corpuscular Radiation 335
10.2.1 Solar Proton Flares 335
10.2.2 Auroral Electrons 341
10.2.3 Ionospheric Effects 344
10.2.4 Balance Equations 347

10.3 Ionospheric Storms 364
10.4 Sporadic-*E* 365
10.5 Solar Eclipses 367

11. Electromagnetic Waves in the Ionosphere 373

11.1 Properties of a Cold Magnetoplasma 373
11.1.1 The Fundamental Electrodynamics 374
11.1.2 The A.C. Conductivity of a Plasma 375
11.1.3 A Plasma as a Polarizable Medium 377
11.1.4 The Dispersion Relation for a Cold Magnetoplasma 379
11.2 Electromagnetic Waves in a Warm Magnetoplasma 382
11.2.1 Dispersion Relations for a Warm Plasma 382
11.2.2 Modes of Propagation in a Hot Magnetoplasma 384
11.2.3 Energy Flow in the Plasma 391
11.3 Faraday Rotation in a Plasma 393
11.4 The Hydromagnetic Mode 393
11.5 The Propagation of Radio Waves 397
11.5.1 HF Wave Propagation 398
11.5.2 LF and VLF Wave Propagation 403
11.5.3 Scattering of HF and VHF Waves 407
11.5.4 Radio Wave Interaction (Luxembourg Effect) 410
11.5.5 Partial Reflection of Radio Waves in the Lower Ionosphere 412
11.6 The Two-Stream Instability 413
11.7 Techniques for Ionospheric Measurements 414
11.7.1 Radio Wave Sounding 414
11.7.2 Cosmic Noise Absorption 417
11.7.3 Thomson Scatter Radar 419
11.7.4 Cross Modulation 419
11.7.5 Partial Reflection 420
11.7.6 Faraday Rotation 421
11.7.7 Differential Doppler-Effect 422
11.7.8 Bistatic Radar 423
11.7.9 Differential Absorption 423
11.7.10 Langmuir Probe 424

Appendix 428

List of Symbols 433

Index 439

Fundamentals of Aeronomy

1

Introduction

> *"There are two things which I am confident I can do very well: one is an introduction to any literary work, stating what it is to contain, and how it should be executed in the most perfect manner; the other is a conclusion, shewing from various causes why the execution has not been equal to what the author promised to himself and to the public."*
>
> —Samuel Johnson

We cannot help but be interested in the atmosphere: It affects our attitudes, it limits our endeavors, it causes tragedies, it provides spectacular scenery. The atmosphere cannot be ignored, it is our environment. We are curious and want to know the how's and why's of atmospheric processes. We are practical and want to control, or at least make good use of, the vital forces we observe. We are also artistic, poetic, philosophical and religious; we are awed by phenomena that affect all our senses and constantly remind us of the crudeness of our knowledge, the vulnerabilities of our lives and endeavors, the infinite complexities of nature, and the sheer beauty of our world. It is no wonder, then, that the study of the atmosphere should be one of the oldest intellectual pursuits and, at the same time, perhaps the most disorganized.

The lower atmosphere of the earth has been studied intensively and formally for many generations as an area of knowledge known as *meteorology*. Meteorology deals with the lower regions of the atmosphere, regions that are accessible by ship, automobile, shank's mare, ladder, kite, aircraft and balloon. Meteorology is mainly concerned with dynamics, the heating and transport of dense gases, interactions with the surface of the earth, and to a smaller extent, the physical chemistry involved in the formation of rain and snow. Many volumes have been written and many scholars and practitioners have been trained in meteorology.

This book, on the other hand, is concerned with a much newer study, at least newer as a formal endeavor. The study is called *aeronomy* and encompasses the upper atmosphere, a region that is only accessible with rockets, satellites, and radiowaves and is studied mainly on the atomic and molecular scale. Inásmuch as the research techniques have required the use of very recently developed technology, knowledge of the upper atmosphere is very recent. Consequently, few textbooks have been written and few classes have been taught. Academically, then, aeronomy is comparatively new, even the name is not universally accepted; however, as an intellectual pursuit, studies of the upper atmosphere may well be as old as mankind.

There is a natural division between the lower and upper atmosphere (and, hence, between meteorology and aeronomy). This boundary is called the tropopause (see Figure 1.5). It is distinguished by marked changes in temperature, structure, dynamics, and constituents, and until recently, it marked the upper limit of accessibility. This boundary also marks a rather distinct change in the kinds of processes that are emphasized in research. Meteorology is principally concerned with hydrodynamics and thermodynamics, whereas aeronomy is concerned with chemical reactions, interactions between photons, protons, and electrons and atoms and molecules, and with the behavior of plasmas.

However, there appear to be important interactions between the macroscopic world of meteorology and the microscopic world of aeronomy. They are not emphasized in this book because they are not yet well delineated. The omission reflects our ignorance rather than a desire to oversimplify the distinction between meteorology and aeronomy.

The basics of aeronomy (or of meteorology) can be applied to any planetary atmosphere, and this will be illustrated throughout the book. The limited successes that have been achieved in applying the principles of aeronomy to the atmospheres of Mars and Venus have strengthened our confidence in the basics of aeronomy, have increased our knowledge of terrestrial processes, and have stimulated the entire field of atmospheric studies.

1.1 Historical Notes

1.1.1 Aurora†

The mysteries of the upper atmosphere have fascinated mankind since the first human observed an aurora. Greek writers described auroras (which are rare occurrences in the Mediterranean area) as early as the sixth century B.C.

† The historical notes on Aurora and Airglow are based largely on the writings of an illustrious pioneer in the field of aeronomy, the late Professor Sydney Chapman (*see* General References).

Aristotle discussed auroras in his work, *Meteorologia*, in the fourth century B.C.; he called them *chasmata*, which may have implied that he thought they were chasms or cracks in the sky.

A French mathematician and astronomer, P. Gassendi, described an outstanding display observed in sourthern France on September 12, 1621, and named the phenomenon *aurora borealis* (northern dawn). About a century later, on March 16, 1716, the British astronomer, Halley, observed a great aurora in London. He formulated a theory that involved magnetic particles flowing along magnetic lines of force and exciting luminescence of the atmosphere; his theory included field lines around a uniformly magnetized sphere. Considering the data he worked with, Halley's theory showed a remarkable insight. The first work devoted entirely to the aurora was written by a member of the French Academy of Sciences, J. J. de Mairan, in 1733. He debunked the popular notion that aurora was simply the reflection of sunlight on snow and ice. He also criticized the theory proposed by Halley as well as a theory proposed in 1724 by the Swiss mathematician, Leonard Euler. De Mairan believed that aurora was related to the sun's atmosphere—which he thought extended as far as the earth. Obviously, both Halley and de Mairan had glimmers of what we now believe to be correct, although it should be added that the full story has not yet been revealed.

The history of auroral research is dotted with scientific notables. The association of magnetic disturbance with auroras was discovered in 1741 by Celsius and Hiorter. Benjamin Franklin proposed an auroral theory to the French Academy in 1779; he suggested that hot air from the tropics rises and travels at a high level to the polar regions where it descends to produce "auroral lightning." The first sighting of the southern aurora was made by the famous explorer Captain James Cook, who named it *aurora australis*. The aristocrat-scientist Henry Cavendish used accounts of auroral observations to deduce the altitude of occurrence of the phenomena; his estimate of 84 to 114 km is reasonable. The atomic chemist John Dalton was among many who compiled auroral catalogs. The prominent yellow-green line emission was found by Ångstrom, though it was not identified until much later.

1.1.2 Airglow

The airglow was discovered in 1901 by Newcomb, who erroneously explained it as light from stars too faint to be seen individually. Much later (1933), Dufay showed that Newcomb's explanation could not be correct and that the source of the "light of the night sky" must be zodiacal light and atmospheric luminescence. Among the many prominent workers who predated Dufay were Van Rhijn, McLennan, and Babcock; the fourth Lord Rayleigh, who also studied auroral spectra intensively, named the airglow phenomena "nonpolar aurora." The theory of the airglow mechanism that is now accepted (photons emitted by the recombination of ionized and/or dissociated atmospheric species) was proposed in 1931 by Professor Sydney Chapman.

1.1.3 Ionosphere

Early ideas about the ionosphere were propounded in 1878 by Balfour Stewart in this theory of diurnal variations in the geomagnetic field. Although Stewart is generally credited with the first suggestion of the existence of an ionosphere, Lord Kelvin (in 1860), Gauss (in 1839), and Faraday (in 1832) all proposed similar ideas independently. Gauss published a "General Theory of Terrestrial Magnetism" which, like Stewart's work, was concerned with electrical current systems in the upper atmosphere. Faraday's proposal was contained in a letter to the Royal Institution which was not opened until 1937!

Experimental evidence for an ionosphere was produced in 1901 when Marconi received a signal in Newfoundland that was transmitted from Cornwall (to the surprise of many scientists who had predicted failure). The reason for the success of the Marconi demonstration was deduced in 1902 by two scientists, Kennelly and Heaviside, working independently; they suggested that the radio signals must have been deviated by a conducting layer of ions at approximately 80 km altitude. The layer was known as the Kennelly-Heaviside layer for many years. However, Lord Rayleigh, Poincaré, and Sommerfeld objected to the deviation idea; instead, they favored a diffraction theory, which was shown later to be inadequate.

In 1912, Eccles investigated the optical index of refraction of ionized gas and found it to be less than unity; this means that rays are bent away from the normal, back towards the source. Larmor, in 1924, applied Eccles' theory to radiowaves and showed that refraction phenomena must occur at high altitudes where the electron-neutral particle collision frequency is small, if severe attenuation of the signal is to be avoided. Appleton (1925) and, independently, Hartree (1929) extended the Larmor theory to the case of a plasma with a superposed magnetic field.

The definitive experiments that demonstrated the existence of an ionosphere were performed in England (1925) by Appleton and Barnett, who employed a wave interference technique, and in the U.S. (1926) by Breit and Tuve, who used a pulse method. The latter technique (in which the time delay is measured between transmission of a pulse and the receipt of an echo from the reflecting layer) is used today for the exploration and monitoring of ionospheres from ground stations. Appleton is credited with naming the reflecting layer; he called it the *E*-layer after the electric field strength, **E**, recorded in his notebook.

Although the idea was widely held that the ionosphere consisted of gas ionized by solar radiation, it was Chapman who first (1931) formulated a quantitative theory of the formation of ionospheric layers. His work is still used as the basis of many contemporary ionospheric calculations.

1.2 Overview of Aeronomy

Aeronomy† is the study of chemical and physical processes of the upper atmosphere. This would seem to be all-inclusive; however, some writers restrict the scope of the term *aeronomy* to include only such processes as heating, and/or chemiluminescence, and/or photolysis, and/or diffusion. Most writers will specifically exclude ionospheric processes, especially processes involving interaction with radiowaves. A few writers will venture to apply the term to studies of atmospheres of other planets.

Much of the arbitrary classification of upper atmospheric processes is the residue of historical development. Electrical engineers were interested in the propagation of radiowaves; physicists in terrestrial magnetism and electricity, chemists in the dissociation and diffusion of atmospheric molecules and atoms; meteorologists in dynamics; and so on. As the body of knowledge grew, it became apparent that all aspects were important and that, in fact, all the processes are interdependent. In upper atmospheric research, as in no other field, the disciplines of electrical engineering, meteorology, chemistry, and physics are intertwined. Certainly, each discipline contributes to the understanding of atmospheric processes, but the importance of each contribution should be judged by how it helps to advance the entire field. It is much like a good pot of stew; it is necessary to have the separate ingredients, such as potatoes, onions, meat, water, wine, spices, etc., but it is also necessary for the ingredients to blend their flavors and consistencies during hours of simmering on the stove—with the result being more than the sum of all its parts.

Nowadays, in order to design meaningful experiments and interpret the results, not only must the aeronomer be expert in his particular specialty, he must also understand the entire field of upper atmospheric phenomena. Except for rare individuals, it is not possible to understand the entire field in detail; but it is possible to be aware of and to appreciate the subtle flavor that each specialty contributes. In this book we shall consider *aeronomy* to mean the combination of all the ingredients of upper atmospheric knowledge, hydrodynamic, ionic, electromagnetic, and chemical. Further, we shall use the term to include processes occurring in the atmospheres of other planets.

The unifying theme in aeronomy is the interaction between radiation and

† See *Webster's Third New International Dictionary* and/or *The Random House Dictionary of the English Language*.

matter. The radiation of interest is primarily from the sun (but includes a component from the galaxy); the matter of interest is the gaseous envelope that surrounds a planet. Let us look at the ingredients separately and then consider how they interact.

1.2.1 Atmospheres

Because it is much more accessible, we know far more about the atmosphere of the earth than we do of the atmospheres of other planets. We know that the major constituents are molecular nitrogen and oxygen in the ratio of approximately 4:1. Carbon dioxide and water vapor are known to be important minor constituents. Furthermore, we know the surface pressure of the atmosphere (1013 mb.) and the number density of each constituent. We have determined the temperature variations with altitude (see Figure 1.5) and can deduce the concentrations of molecules with altitude by applying the hydrostatic equation (see Chapter 3). It is known that other minor species occur in the upper atmosphere such as argon, ozone, atomic oxygen, nitric oxide, hydroxyl radicals, hydrogen, and helium, and that at even higher altitudes the atomic species become the major constituents; however, this occurs as a result of the radiation-atmosphere interaction and will be discussed later.

The next best known atmosphere is that of the planet Mars. This is because of the opportunities for observation with very high resolution spectroscopic techniques and with instrumentation carried on planetary probes of the Mariner series. From these observations, we know that carbon dioxide is *a* major constituent and that oxygen is a minor constituent, as are water vapor and carbon monoxide. It is also thought that the Martian atmosphere contains another major constituent because the surface pressures appear to be higher than can be accounted for by assuming pure carbon dioxide; both nitrogen and argon have been proposed as candidates for the missing component. A surface temperature around 200° K is usually accepted but temperatures over 300° K have been deduced. Surface pressures from 4.8 to 25 mb have been derived by various techniques.

Venus appears to have an almost pure carbon dioxide atmosphere; earth-based spectroscopic measurements and planetary probes appear to agree on this point. However, the identity and abundance of minor components are in doubt. Carbon monoxide and the vapors of hydrofluoric and hydrochloric acids have been detected spectroscopically. Water vapor, nitrogen, hydrogen, and oxygen were detected or deduced by planetary probes but the abundances reported have not been verified by spectroscopic observations. Spectroscopic techniques cannot penetrate the cloud layer of the Venus atmosphere and hence the composition and characteristics of the lower atmosphere cannot be determined except by planetary probes. Surface pressure and temperature measurements and deductions are presently unreconciled but it appears that the surface pressure is between 20 and 100 equivalent earth atmospheres and the temperature is between 545 and 800°K.

Evidence for an atmosphere on Mercury is virtually nonexistent.

Jupiter's atmosphere is clearly different from the pattern of oxygen, nitrogen, and carbon dioxide mixtures found on Mars, Venus, and Earth. A very thick atmosphere of hydrogen has been measured repeatedly by spectroscopic observations of Jupiter, although the results differ by factors as large as 25 or 30. Ammonia and methane are present as minor components. Helium, with an abundance perhaps as great as that of hydrogen, is theorized from stellar occultation measurements of density and from solar abundance and thermal balance considerations. The surface temperature appears to be low ($\sim$130°K) but the planet apparently has an internal heat source.

Hydrogen and methane have been detected on Saturn but little else is known about its atmosphere.

General information on the planets is summarized in Table 1.1.

1.2.2 Radiations

In the study of aeronomy, we are principally concerned with radiation from the sun; however, as we shall note later, galactic radiation may also be important.

On a warm, sunny day, we can easily sense the life-giving energy radiated from the glowing ball of hot gases we see in the sky. On a cold, foggy night, it may require a considerable amount of faith to believe that the sun is a source of energy; the evidence exists not only in the wood or coal burning cheerfully in the fireplace, but also in the fact that the air temperature is as warm as it is. The radiation we feel, or feel the lack of, is not, however, of great interest in aeronomy. The radiations that are most important in upper atmospheric processes are screened by the upper atmosphere and never, fortunately, reach the surface of the earth nor, presumably, the surfaces of any of the planets—with the probable exception of Mercury.

The sun radiates in the visible and near visible portions of the spectrum like a black body with a surface temperature of 6000°K (see Figure 1.1) and the energy is received at the top of the earth's atmosphere at the rate of 1.39×10^6 ergs cm^{-2} $sec.^{-1}$ Less than 10^3 ergs cm^{-2} sec^{-1} (or $<10^{-3}$ of the total radiation) is absorbed in the upper atmosphere.

The portion of the solar spectrum of principal interest to us is below 2900 Å. This contains the far ultraviolet, the extreme ultraviolet (EUV), and the X-ray regions; often, the EUV and X-ray regions are lumped together and called the XUV region. The spectrum of this region is represented by Table 1.2, Figure 1.2, as well as Figure 8.17 and Table 8.2.

TABLE 1.1. GENERAL INFORMATION ABOUT PLANETS

Planet	Mass (kg)	g (cm sec^{-2})	Mean Diameter (km)	Mean Solar Distance (A.U.)	Surface Temp (°K)	Surface Pressure	Magnetic Field	Atmospheric Constituents	
								Major	Minor
Mercury	3.33×10^{23}	392	4,840	0.39	250–400	?	?	?	?
Venus	4.87×10^{24}	882	12,228	0.72	545–800	20–100 Atmospheres	No	CO_2	O_2, N_2(?), H_2O, CO, HCl, HF
Earth	5.98×10^{24}	980	12,742	1.00	~300	1 Atmosphere (1013 mb)	Yes	N_2, O_2	CO_2, H_2O, O, H_2, H, He, A NO, CO, OH
Mars	6.64×10^{23}	392	6,770	1.52	195–310	5–25 mb	No	CO_2, N_2(?), A(?)	H_2O, CO, O_2, O
Jupiter	1.90×10^{27}	2646	140,720	5.20	130†	?	Yes	H_2, He(?)	H, CH_4, NH_3
Saturn	5.68×10^{26}	1176	116,820	9.54	95†	?	?	H_2	CH_4, NH_3(?)

† Cloud top temperature.

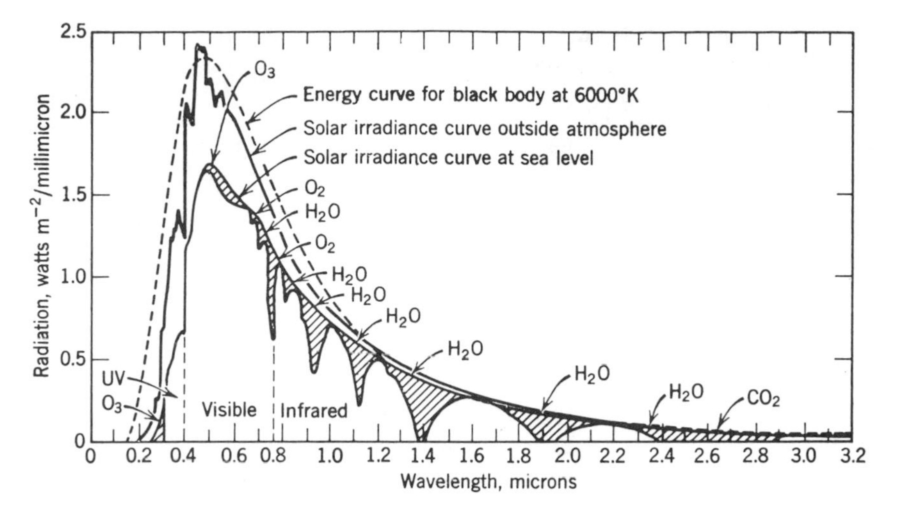

Figure 1.1 Visible and IR solar spectrum (from *Handbook of Geophysics*, reprinted by permission of The Macmillan Co.).

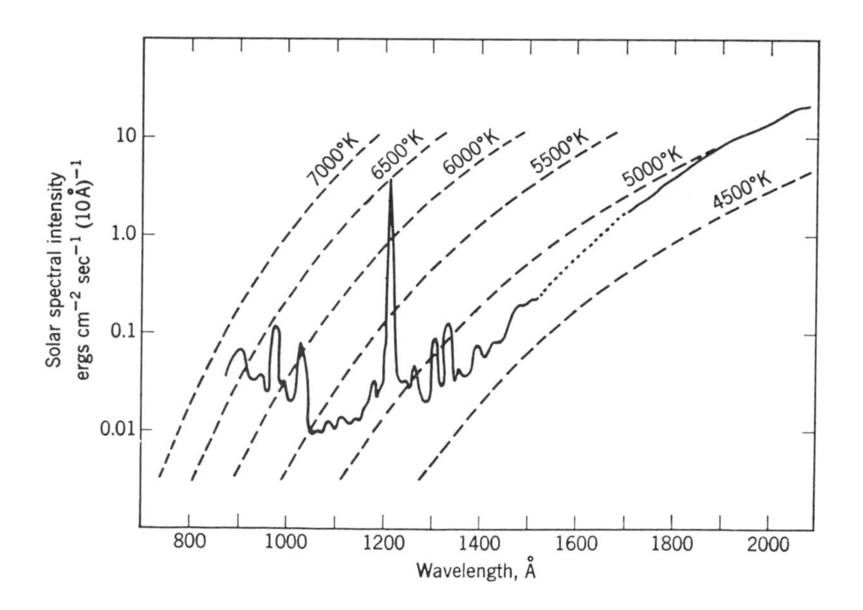

Figure 1.2 Intensity of solar spectrum 2600–850 Å. (After Detwiler, Garrett, Purcell, and Tousey, *Ann. de Géophysique*, **17**, 265 (1961); reprinted by permission of Service de Publications du C.N.R.S., Paris.)

TABLE 1.2. THE INTENSITY OF THE SOLAR SPECTRUM, (CONTINUUM AND LINES), IN 50-Å INTERVALS, AT A DISTANCE OF 1 A.U. (ASTRONOMICAL UNIT)†

Interval $\lambda \pm 25$ Å	Intensity erg cm^{-2} sec^{-1} (50 Å)	Interval $\lambda \pm 25$ Å	Intensity erg cm^{-2} sec^{-1} (50 Å)
2600	700	1700	8.2
2550	560	1650	5.0
2500	380	1600	3.2
2450	390	1550	1.7
2400	340	1500	0.95
2350	320	1450	0.50
2300	360	1400	0.26
2250	350	1350	0.26
2200	310	1300	0.18
2150	240	1250	0.15
2100	145	1200	5.7
2050	90	1150	0.08
2000	70	1100	0.06
1950	55	1050	0.10
1900	41	1000	0.18
1850	28	950	0.15
1800	19	900	0.25
1750	12	850	0.11

† From Detwiler, *et al.* Ann. de Geophysique **17,** 265 (1961). Reprinted by permission of Service des Publications du C.N.R.S., Paris.

The solar radiation with which we are most familiar, the radiation that warms the body, raises the foodstuffs and drives the winds, is emitted from the surface of the sun, called the photosphere; this is the 6000°K "black" body. The absorption spectrum superimposed on the black body radiation (the Fraunhofer lines) is produced by a cooler (4200°K) layer of gas between the photosphere and a warmer (6000 to 30,000°K) transition region that is approximately 12,000 km thick and called the chromosphere. The XUV radiation is produced by the very hot (>1,000,000°K) and extensive solar atmosphere or corona. If the solar disc is examined through a filter that transmits only X-ray or EUV energy (which can only be done outside the atmosphere) bright patches are seen on the disc. These patches, or areas of XUV emission, vary in size, shape, number, and location with time; they grow more numerous and larger as the sunspot number increases. If the solar disc is examined in white light (which can be done from the surface of the earth), it is seen that sunspots are found in the XUV patches; and if the disc is examined in the light of the hydrogen Balmer Hα line (which can also be done from the earth's surface), bright areas, called plages, are seen which correspond roughly with the XUV areas. (See Figure 1.3.) Obviously, sun

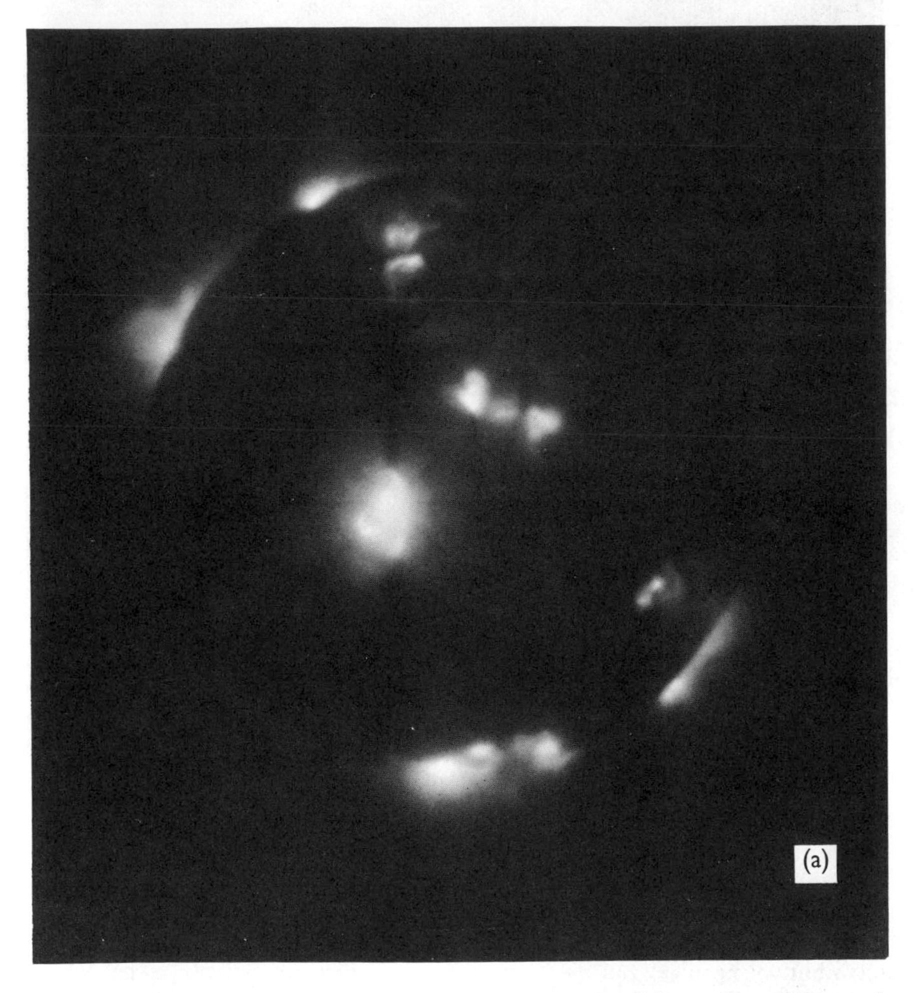

Figure 1.3 Photographs of the sun in X-ray and hydrogen Hα light. a. X-ray photograph of the sun taken on June 8, 1968. Note the bright flare slightly left and below center. b. Photograph of the sun taken at the same time as the X-ray photograph above. This photograph was taken through a filter that passes only the Balmer Hα line of hydrogen. c. Enlargement of the flare in X-ray light. d. Enlargement of the flare in Hα light. (After Vaiana and Giacconi, *Plasma Instabilities in Astrophysics* pp. 91–118, 1969. Reprinted by permission of Gordon and Breach, Publishers.)

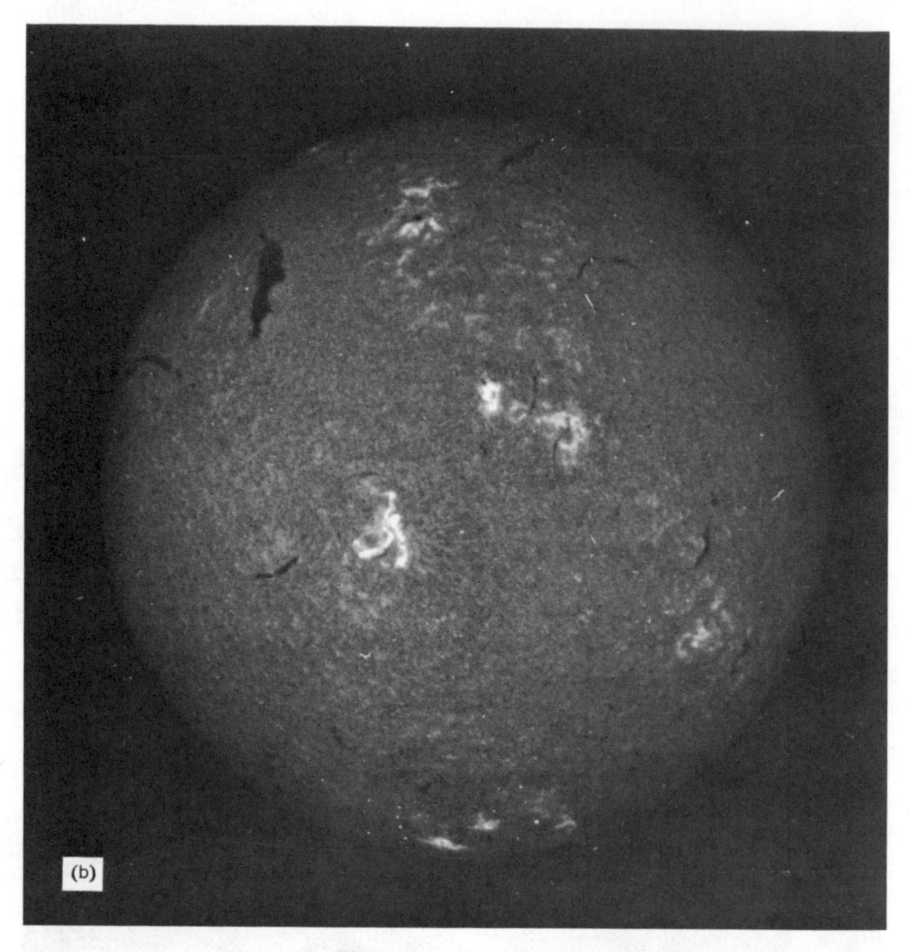

Figure 1.3 (*continued*)

spots, plages, and XUV emitting areas are associated phenomenologically, and some idea of the XUV radiation intensity can be obtained by monitoring the number of spots and the area of Hα plages on the solar disc. Unfortunately, quantitative correlations which might be used for aeronomy studies have not yet been established.

Occasionally (in Hα light), a portion of a plage area will be seen to brighten or flare. The flares emit additional X-radiation, especially in the shorter wavelength portion of the X-ray spectrum, for durations as long as several minutes. Some flares will also emit charged particles, which may bombard the planet for several hours or days; these exceptional events are known as "proton flares" and the radiation is known as solar cosmic radiation (SCR).

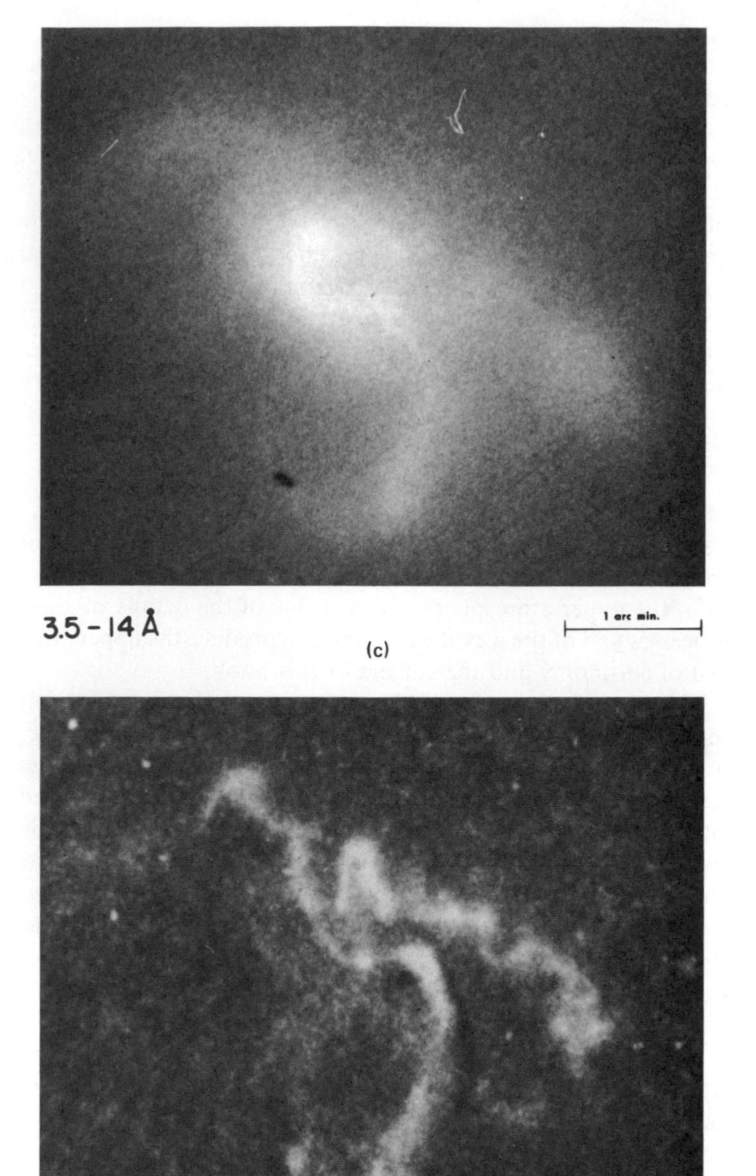

Hα

(d)

Figure 1.3 (*continued*)

Radiation from proton flares will be funneled into the magnetic polar regions of planets with strong magnetic fields (such as Earth and Jupiter), but could bombard the entire surface of planets without strong magnetic fields (such as Mars and Venus). Radiation from X-ray flares, on the other hand, does not interact with magnetic fields and will irradiate the daylight side of any planet. The spectra of flare radiations are illustrated in Figures 10.1, 3, 5.

Galactic cosmic radiation (GCR) is a very "hard" (>GeV) corpuscular radiation that penetrates planetary magnetic shielding and bombards the entire surface of a planet; the radiation originates outside the solar system but is apparently influenced by magnetic fields within it. Although GCR has a definite role in the formation of ionospheres, it is so energetic that it also penetrates atmospheres and irradiates planetary surfaces. GCR varies with the solar cycle, but not greatly.

1.2.3 Interactions

The interactions between radiations and atmospheric constituents are the processes responsible for the screening of planetary surfaces from solar radiations; the very same processes determine the unique composition and structure of the upper atmosphere. Knowledge of the details of these interaction processes and of the way they combine to produce the upper atmosphere is the goal of aeronomy and the subject of this book.

Types of primary interactions in the upper atmosphere are manifold. The most energetic radiations ionize the atoms and molecules that they encounter; less energetic radiations dissociate molecules and excite atoms and molecules to higher electronic and vibrational levels; the least energetic radiations excite rotational levels in molecules and detach weakly bound electrons from negative ions. The frequency and relative importance of these reactions depend on the spectral irradiance of solar and galactic emissions as well as on the distribution of atmospheric constituents and temperatures; these are all known to vary with the time of day, latitude, season, solar sunspot cycle, and irregular solar disturbances.

Primary interactions, however, are only the initial reactions of a long and complex web of events. Subsequent reactions include recombination of ions and electrons, recombination and rearrangement of atoms and molecules, radiation of energy, and absorption of energy. The results are optical displays, free electrons and ions, altered distributions and identities of molecular and atomic species, atmospheric heating, and mass motions—and the formation of layers or regions with distinctive physical and chemical characteristics. Several hundreds of known reactions occur simultaneously, and from the frustrating inconsistencies that are noted between theory and observations, we might suspect that several hundreds of *un*known reactions are also involved.

In the earth's atmosphere for example, solar X-radiations, solar cosmic radiations, and galactic radiations ionize all atmospheric species to some extent (depending on the energies of the radiations and the ionization cross sections of the constituents) to produce electrons, oxygen and nitrogen ions, and a number of minor ions. Ultraviolet radiations in the solar hydrogen Lyman α and β lines ionize nitric oxide and oxygen, respectively; atomic oxygen, molecular oxygen, and molecular nitrogen are ionized by X and EUV radiation. When the molecular ions recombine with electrons they dissociate to produce neutral, but probably excited, atoms. Ultraviolet radiation also dissociates molecular oxygen: At higher altitudes, oxygen persists in the atomic state; at lower altitudes atomic oxygen recombines to produce molecular oxygen; and at even lower altitudes the atomic oxygen combines with molecular oxygen to produce ozone. The ozone that is formed acts as a screen by absorbing other ultraviolet radiations; at the same time this causes the ozone to be dissociated back into atomic and molecular oxygen. Collisions of the excited ions, electrons, atoms, and molecules with the surrounding gas particles leads to "cooling" of the energetic species and heating of the ambient gas. Thus, the basic neutral molecular nitrogen and oxygen mixture of the earth's atmosphere is converted by interactions with radiation to a mixture of ions, electrons, atoms, and molecules—with a vertical temperature and concentration structure.

Similar but not as well documented reactions occur in the atmospheres of other planets. Carbon dioxide, the basic ingredient in the Mars and Venus atmospheres, undergoes ionization, dissociation, and excitation which should produce a particular mixture of ions, electrons, atoms, and molecules and a specific temperature and concentration profile. The fact that observations do not satisfactorily confirm calculations underlines the limitations of present theoretical, experimental, and observational results.

Details of ionization processes are covered in Chapters 8 through 10, chemical processes in Chapter 4, excitation and radiation processes in Chapter 6, and heating processes in Chapter 3.

1.2.4 Layers and Regions

It has not been possible to unravel the intricacies of the upper atmosphere by dealing simultaneously with all possible reactions. Simplifications have been sought to explain the basic phenomena, then elaborations have been made to the extent necessary to explain the complexities found in nature. Hence, it has been useful to define regions of the atmosphere in which the character is determined by only a few major processes. The notion that layers with unique properties exist in the upper atmosphere was inherent in the development of ionospheric studies, e.g., the Kennelly-Heaviside theory that radio signals were reflected by a conducting layer.

Let us take the concept of an ionized layer in the upper atmosphere as an illustration of the basic mechanism of layeı or region formation. We know (see Chapter 3) that the density of an atmosphere decreases exponentially with altitude (neglecting, for the moment, variations cau. ed by the temperature structure or the diffusive separation of species). We know also that monochromatic radiation is attenuated exponentially by an absorbing medium (see Chapter 2). With these two facts, we can conceive, qualitatively, how solar radiation produces an ionized layer. At the outer fringes of the atmosphere, the density is low and radiation is absorbed only slightly; but deeper in the atmosphere, the density increases and so does the absorption of radiation. Both the density of the atmosphere and the absorption of radiation increase exponentially with depth; and in a particular altitude region, this double exponential absorption process will produce a very rapid attenuation of a particular wavelength radiation; below this region, virtually none of that particular radiation will penetrate. If the absorption is caused by ionization processes, an ionized layer will result. This is represented schematically by Figure 1.4.

This layer formation theory was put in good quantitative form by Chapman in 1931 (see Chapter 8) and this type of layer is known as a "Chapman Layer." The basic idea is fundamental for ionospheric models, but the specific assumptions must be understood, viz , a specific absorbing species, a specific wavelength radiation, an isothermal atmosphere. In practice, there

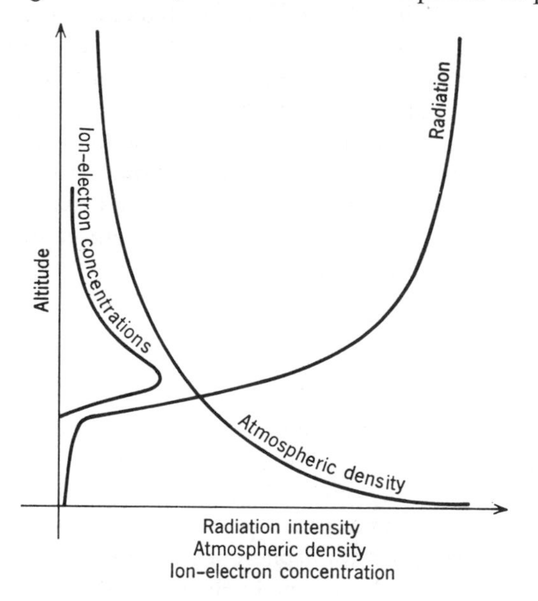

Figure 1.4 Schematic representation of layer formation.

are many species, a wide range of radiation energies, temperature structure, etc., and the fundamental theory must be elaborated to fit the specific conditions. In earlier days when radio sounding was the sole technique for exploring the ionosphere, layer formation theories appeared to be very explicit, relatively simple, and very relevant because radio sounding detected what appeared to be specific layers. Thus, the *E*-layer was very distinct and easily identified. As techniques improved, an *F*-layer was found above the *E*-layer; later, the *F*-layer was found to produce multiple echoes, presumably by reflection from what is now known as the F_1 and F_2 layers. When it was found that signals sometimes were not reflected, it was deduced that a lower layer was absorbing a part or all (depending on solar conditions) of the signal before it could reach the *E*-layer; this was known as the *D*-layer. More recently, some authors have introduced an even lower layer, the *C*-layer, which is formed by GCR; incidently, it was fortuitous that the letter *C* preceded *D* in the alphabet and was also the first letter of the name (cosmic) of the responsible radiations.

When the upper atmosphere was explored by rockets, however, it became evident that the layers were not at all distinct but were more accurately described as regions; at about the same time, better ionogram inversion techniques were developed which also led to similar evidence. There appears to be only one large ionospheric layer, the *F*-layer; the other "layers" are actually ledges or steep electron concentration gradients, which also reflect (or, more accurately, refract) radio signals. A schematic representation of ionospheric regions is shown in Figure 1.5. As is discussed in Chapter 8, each region can be characterized by dominant electron removal processes and these dominant types of processes can be used with some care as a common basis for classifying ionospheric regions in other planetary atmospheres.

The concept of regions with unique characteristics resulting from certain dominant processes has much wider application to the upper atmosphere. For example, the Chapman function used to describe a basic ionization layer is also used to describe heating processes (see Chapter 3) and the upper atmosphere can also be classified into regions according to temperature gradients. This is illustrated in Figure 1.5.

Similarly, the atmosphere can be classified according to the distribution of species. For example, the ozonosphere is the region (centered near 30 km) which is rich in ozone, the homosphere is the region where atmospheric species are well mixed and the composition remains constant with altitude. The heterosphere is the region just above the homosphere where photochemical reactions and diffusive separation combine to change the composition with altitude. The division between the homosphere and the heterosphere is near the turbopause ($\sim$105 km). The outermost portion of the atmosphere which is almost fully ionized and where helium and hydrogen

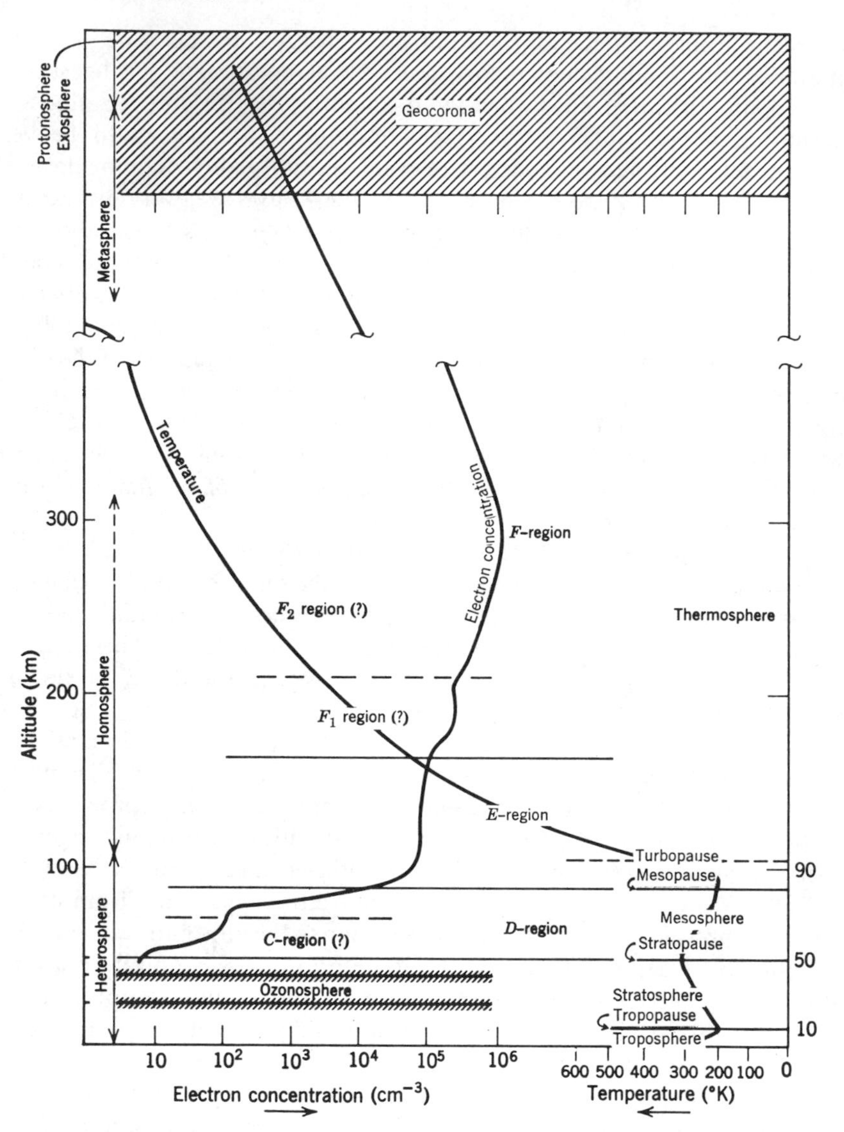

Figure 1.5 Schematic representation of atmospheric regions.

ions predominate is called the protonosphere by some and the exosphere by others, depending on whether they are emphasizing the ionic or neutral composition. The region just below the protonosphere, where laws derived from equations of state and hydrostatics appear to have limited application and the dominant species is neutral hydrogen, is sometimes called the metasphere. On some occasions the metasphere and protonosphere are known as the geocorona (see Figure 1.5).

It is evident that the classification schemes were developed historically, in accordance with the desires of researchers who were interested only in certain aspects of upper atmospheric science. Their justification at present is convenience in studying and describing the upper atmosphere. It will be seen as we proceed through this book, that there are many points in common between the three classification schemes. Future developments, based on aeronomy as an integrated study will perhaps lead to a unified system of classification.

1.3 Limitations of Knowledge

It cannot be stressed too strongly that present concepts and theories are built on a skeleton of experimental data (obtained in laboratories and observatories), stuffed with intuition and opinion, and somehow held together by a covering of basic scientific principles and accepted methodology. Although of paramount importance in the creative development of theory, intuition and opinion are often the least reliable components. The covering of principles and methodology is usually flexible enough to accommodate change. The basic limitation, generally, is in the quality and quantity of experimental data.

We depend on experimental data for our knowledge of the basic constituents of the atmosphere, the energy states of the constituents, the spectral and temporal character of important radiations, the rates and cross sections of specific reactions, and the temperature of the environment. It does not require a great deal of uncertainty in all these factors to change the dimensions and form of our concepts.

At present, some of the more important data limitations are the following: uncertainties about the concentrations and identities of many constituents, particularly ionic, atomic, and minor molecular species; uncertainties about the spectral irradiances of solar emissions, particularly their variations with time; uncertainties about the rates of reactions that are thought to be important, particularly their variations with temperature and energy state; and uncertainties about the temperatures and dynamics of most regions.

It is easy enough to understand the reasons for these uncertainties. Data from rockets and flyby planetary probes represent single points in time and space and we are not yet confident that we can fit them into a meaningful pattern. Satellite orbits and flyby trajectories are usually too high to furnish continuous data in the altitudes of interest. Solar measurements, though invaluable, have been designed to acquire a better understanding of the sun rather than of atmospheric phenomenon. No probes have been adequately instrumented to obtain all the necessary data about species, radiations, and physical environment simultaneously. It is not yet possible to completely describe laboratory measurements in terms of the energy states of all reactive species; nor can we explore in the laboratory all the conditions of temperature, density, and radiation environment that are thought to be valid in the atmosphere.

Hence, in summary, one should be cautious about using data, careful about the application of established principles and methods, and skeptical of interpretations.

1.4 Outlook

The period ahead appears to be very challenging and exciting for aeronomy. New data will continue to flow not only from terrestrial experiments but also from planetary probes to Mars, Venus, Mercury, Jupiter, and Saturn. The basic tenets of aeronomy will be tested, stressed and modified. Pioneering work still must be done not only with regard to the atmospheres of other planets but also with regard to the terrestrial upper atmosphere. The interests of astronomers, meteorologists, and aeronomers are moving closer together and may well merge within the next generation. Communications engineers, physical chemists, and physicists have already merged their upper atmosphere interests in *aeronomy*, which now promises to become a unified and sophisticated scientific discipline.

It is hoped that this book will serve as a stimulating and thorough introduction to aeronomy for those who will be responsible for the realization of the great enlightenment we can only imagine. To the physicists, chemists, astronomers, engineers, and meteorologists who are now striving to develop aeronomy as a science, we offer a quotation (out of context) from Emerson:

"Keep cool: it will all be one a hundred years hence."

GENERAL REFERENCES

J. A. Chalmers, The first suggestion of an ionosphere, *J. Atmos. Terr. Phys.* **24,** 219 (1962).

Sydney Chapman, Perspective in *Geophysics of Geomagnetic Phenomena* Vol. I, edited by S. Matsushita and W. H. Campbell, Academic Press, New York, 1967.

Sydney Chapman, History of Aurora and Airglow, in *Aurora and Airglow*, edited by B. M. McCormac, Reinhold, New York, 1967.

Sydney Chapman, Historical Introduction to Aurora and Magnetic Storms, *Annales de Geophysique* **24,** 1 (1968).

T. R. Kaiser, The first suggestion of an ionosphere, *J. Atmos. Terr. Phys.* **24,** 865 (1962).

2

Fundamental Physical Principles

Although the details of many of the mechanisms involved in aeronomic phenomena are poorly understood, the underlying physical principles are well known. This is true not only of the older classical ideas as embodied in electrodynamics, thermodynamics, and the kinetic theory of gases, but in the newer (but no longer "new!") quantum concepts of atomic structure. Most students with a thorough background in physics on the undergraduate level are probably quite familiar with the subject matter of this chapter. Nevertheless, many will find a review helpful and this we hope to provide in the following sections. In no way are they substitutes for a careful study in each area. Such a program would fill several volumes, not just the few pages allotted here.

TABLE 2.1. MAXWELL'S EQUATIONS

Name	Differential Form	Integral Form	
Gauss' Law (Magnetic field)	$\text{div}\,\mathbf{B} = 0$	$\iint_{\Sigma} \mathbf{B} \cdot \mathbf{dS} = 0$	(2.1)
Gauss' Law (Electric field)	$\text{div}\,\mathbf{D} = \rho$	$\iint_{\Sigma} \mathbf{D} \cdot d\mathbf{S} = q$	(2.2)
Faraday's Induction Law	$\text{curl}\,\mathbf{E} = -\dfrac{\partial \mathbf{B}}{\partial t}$	$\oint \mathbf{E} \cdot d\mathbf{r} = -\dfrac{\partial}{\partial t}\iint_{\sigma} \mathbf{B} \cdot \mathbf{dS}$	(2.3)
Ampere's Law	$\text{curl}\,\mathbf{H} = \mathbf{j} + \dfrac{\partial \mathbf{D}}{\partial t}$	$\oint \mathbf{H} \cdot d\mathbf{r} = I + \dfrac{\partial}{\partial t}\iint_{\sigma} \mathbf{D} \cdot \mathbf{dS}$	(2.4)

2.1 Electrodynamics

2.1.1 Fundamentals

The basis of electrodynamics lies in the equations of Clerk Maxwell which describe the interrelationships which exist between electric and magnetic fields under essentially all conditions (provided, of course, that quantum and electron self-energy effects can be neglected). They are summarized in Table 2.1 which presents both the differential and integral forms. In the table $\mathbf{B}$ = magnetic field induction; $\mathbf{H}$ = magnetic field intensity; $\mathbf{D}$ = electric field displacement; $\mathbf{E}$ = electric field intensity; ρ = charge density; $\mathbf{j}$ = current density; q = total charge; I = total current. The surface Σ is a closed one while σ is not. The boundary of σ is in fact the contour over which the line integrals of the form $\oint \mathbf{V} \cdot d\mathbf{r}$ are computed.

The field quantities $\mathbf{B}$, $\mathbf{H}$, $\mathbf{D}$, and $\mathbf{E}$ are related by the equalities

$$\mathbf{B} = \mu\mathbf{H} = \mu_0\mathbf{H} + \mathbf{M} \tag{2.5a}$$

$$\mathbf{D} = \epsilon\mathbf{E} = \epsilon_0\mathbf{E} + \mathbf{P} \tag{2.5b}$$

in which μ and ϵ are, respectively, the magnetic permeability and electric permittivity, $\mathbf{M}$ the magnetization, and $\mathbf{P}$ the polarization; μ_0 and ϵ_0 refer to *free space*. It is appropriate at this point to say a few words about units. In Table 2.1 and throughout the text the so-called mks rationalized system is used. We do not think this the best system, but it has gained almost universal acceptance in recent years. We defer to the majority.

Maxwell did not, of course, originate these equations which individually bear other people's names. They are collectively named for him because he incorporated them into a unified theory of electricity and magnetism. However, he did add one term to Eq. (2.4), the term involving the electric displacement $\mathbf{D}$. Only by doing this could the four equations be made consistent with the demands of charge conservation which is embodied in the continuity equation

$$\frac{\partial \rho}{\partial t} + \operatorname{div} \mathbf{j} = 0 \tag{2.6}$$

One can easily see the necessity for including the displacement current by taking the divergence of both sides of Eq. (2.4) and employing Eq. (2.2). For many types of electrical systems, including plasmas, the current density is related to the electric field by a generalized form of Ohm's law

$$\mathbf{j} = \underset{\sim}{\boldsymbol{\sigma}} \cdot \mathbf{E} \tag{2.7}$$

where $\underset{\sim}{\boldsymbol{\sigma}}$ is the *conductivity tensor*.

If one investigates the motion of a single charged particle in a uniform magnetic field, it is found that the particle moves in a circle or a helix about the lines of force. The mathematical expression for the force $\mathbf{F}$ acting on it is given by

$$\mathbf{F} = e\mathbf{v} \times \mathbf{B} \tag{2.8}$$

where $\mathbf{v}$ is the particle velocity and e is its charge. $\mathbf{F}$ is frequently called the *Lorentz force*. This force will be of great importance to our discussion in Chapters 7 and 11.

2.1.2 *Electromagnetic Wave Equations*

By taking the curl of both sides of Eqs. (2.3) and (2.4), and using the vector identity

$$\text{curl curl } \mathbf{A} = \boldsymbol{\nabla}(\text{div } \mathbf{A}) - \nabla^2\mathbf{A},$$

we obtain the Helmholtz or *wave* equations[1] for the electric and magnetic fields

$$\nabla^2\mathbf{E} - \mu\epsilon \frac{\partial^2\mathbf{E}}{\partial t} = \mu \frac{\partial \mathbf{j}}{\partial t} + \nabla(\epsilon\rho) \tag{2.9}$$

$$\nabla^2\mathbf{H} - \mu\epsilon \frac{\partial^2\mathbf{H}}{\partial t^2} = -\text{curl } \mathbf{j} \tag{2.10}$$

which describe the propagation of the familiar electromagnetic waves. The reader is cautioned that care must be exercised in applying the Laplacian operator to a vector written in component form: Only in cartesian coordinates are the unit vectors independent of the coordinates. In free space, the inhomogeneous terms vanish; μ and ϵ are replaced by μ_0 and ϵ_0, yielding

$$\nabla^2\mathbf{E} - \epsilon_0\mu_0 \frac{\partial^2\mathbf{E}}{\partial t^2} = 0 \tag{2.11}$$

$$\nabla^2\mathbf{H} - \epsilon_0\mu_0 \frac{\partial^2\mathbf{H}}{\partial t^2} = 0 \tag{2.12}$$

Comparison with a typical vector Helmholtz equation shows that $c = \dfrac{1}{\sqrt{\epsilon_0\mu_0}}$ is to be interpreted as the phase velocity of the wave, i.e., the velocity of a wave crest or trough of a wave of a single frequency. One of the striking confirmations of Maxwell's theory was the fact that substitution of experimentally determined values of ϵ_0 and μ_0 into the equation for c yielded the measured velocity of light to a high degree of precision.

As a very simple example, let us consider the case where $\mathbf{E}$ is a sinusoidal wave of the form

$$\mathbf{E} = \mathbf{E}_0 e^{i(\mathbf{k}\cdot\mathbf{r} - \omega t)} \tag{2.13}$$

Substitution into the equation describing the propagation of waves in a conducting medium yields

$$\left(\frac{\omega^2}{c^2} - k^2\right)\mathbf{E} = i\mu\omega\boldsymbol{\sigma}\cdot\mathbf{E} \tag{2.14}$$

which is a *dispersion relation* for the medium. We shall return to it in a subsequent chapter.

In the foregoing analysis it was assumed that the conductivity does not vary much in a distance of one wavelength. If it should vary appreciably, reflection will occur in the region of sharp change. In general, this case is difficult to handle mathematically, but there is one situation in which the treatment is fairly easy. This is the case of a sharp boundary. The following conditions must be satisfied at such an interface:

1. The tangential components of **E** and **H** must be continuous across the boundary.
2. The normal components of **D** and **B** must be continuous across the boundary, assuming that it contains no net free charge.

2.1.3 Electrical Properties of a Plasma: Plasma Frequency

If the medium is an ionized but electrically neutral fluid or *plasma* (see Section 2.3.5), the perturbing wave will cause time-varying displacements of the positive and negative charges amounting to a polarization. From the definition of the latter quantity, it can be written

$$\mathbf{P} = ne\mathbf{r} \tag{2.15}$$

where n is the number density of chaıged particles, e is the charge on each particle, and $\mathbf{r}$ is the relative displacement of positive and negative charges. The charge displacement produces an electric field

$$\mathbf{E}_r = -\frac{\mathbf{P}}{\epsilon_0} \tag{2.16}$$

which acts to restore the charges to their unperturbed positions. In accord with Newton's second law of motion, one can equate the inertial force on a particle of mass m, $\mathbf{F} = m(d^2\mathbf{r}/dt^2)$, to the electrical force $\mathbf{F} = e\mathbf{E} = -n(e^2/\epsilon_0)\mathbf{r}$:

$$\frac{d^2\mathbf{r}}{dt^2} + \frac{ne^2}{m\epsilon_0}\mathbf{r} = 0 \tag{2.17}$$

which is the equation of motion of an oscillator of resonant angular frequency

$$\omega_p = \left(\frac{ne^2}{m\epsilon_0}\right)^{1/2} \tag{2.18}$$

Hence we call it the *plasma frequency;* in general the plasma frequency of the positive component will differ from that of the negative component.

2.1.4 Energy Flow

The electric and magnetic field vectors **E** and **H** can be used to describe the rate of the energy flow and its direction. In free space **E**, **H**, and the wave vector **k** are mutually orthogonal, but this is not the case in a conducting medium such as a plasma. Because of the polarization charges which are present, Gauss' Law [Eq. (2.2)] requires that $\mathbf{k} \cdot \mathbf{E} \neq 0$. Nevertheless, the direction of energy flow is perpendicular to **E** and **H**; it is given by the Poynting vector

$$\mathbf{\Pi} = \mathbf{E} \times \mathbf{H} \tag{2.19}$$

whose average over a cycle is

$$\overline{\mathbf{\Pi}} = \tfrac{1}{2}\,\mathrm{Re}\,(\mathbf{E} \times \mathbf{H}^*) \tag{2.20}$$

where the asterisk denotes complex conjugation and Re means "take the real part of." From our previous statements we conclude that in general $\overline{\mathbf{\Pi}}$ and **k** are not parallel.

The propagation of monochromatic waves is described completely by the phase velocity; in the real world, however, we never deal with such waves. We are interested, rather, in superpositions of waves with a spread of frequencies, so-called "wave packets." Here we are usually interested in the motion of the packet "crest" which is formed by constructive interference of the component waves. The associated velocity is the "group velocity" of the system. Consider a wave packet which is written in Fourier component form as

$$\mathbf{A}(x, t) = \int_{-\infty}^{\infty} \mathbf{A}(\omega) e^{i(kx-\omega t)}\, d\omega \tag{2.21}$$

and represent the phase as $\psi(\omega) = k(\omega)x - \omega t$. Since $\mathbf{A}(x, t)$ is peaked at the mean frequency, ω_0, the phase $\psi(\omega)$, of the wave must be stationary there, i.e., $\left.\dfrac{\partial \psi}{\partial \omega}\right|_{\omega=\omega_0} = 0$, or

$$\left(\frac{\partial \psi}{\partial \omega}\right)_{\omega_0} = \left(\frac{\partial k}{\partial \omega}\right)_{\omega_0} x - t = 0 \tag{2.22}$$

The quantity x/t is evidently the velocity of the packet crest or

$$v_g = \frac{x}{t} = \left(\frac{\partial \omega}{\partial k}\right)_{\omega_0} \tag{2.23}$$

which we call the "group velocity." It should be obvious that these developments apply to any wave, not just those of electromagnetic character. In fact, we shall first use them in our discussion of acoustic-gravity waves in the atmosphere.

2.2 Thermodynamics

As we shall see later, the upper atmosphere is for many purposes sufficently close to local thermodynamic equilibrium that the methods of equilibrium thermodynamics can be applied. The basic principles are the first and second laws of thermodynamics. There is also a third law but it need not concern us here.

2.2.1 The First and Second Laws

The first law is a statement of the conservation of energy of all forms, including the *internal thermal energy* U of the system. It is expressed in mathematical form as

$$đQ = dU + đW \tag{2.24}$$

where Q is the heat energy absorbed and W is the work done by the system. For the applications presented in this text, work is limited to that done by an expanding gas

$$đW = p\, đ\alpha \tag{2.25}$$

where p is the pressure and α is the volume. The symbol "$đ$" means that the differential is *not* an *exact* one. In other words Q and W are not thermodynamic functions, although U is. One can relate the internal energy of a system to its temperature by defining the heat capacity C_v of the system:

$$dU = C_v\, dT \tag{2.26}$$

Although we shall not do so here, it is possible to show that U is a function of the temperature only. Substitution of Eqs. (2.25) and (2.26) into Eq. (2.24) yields

$$đQ = C_v\, dT + p\, đ\alpha \tag{2.27}$$

If we now differentiate with respect to T, holding α constant, we obtain

$$\left(\frac{đQ}{dT}\right)_\alpha = C_v, \tag{2.28}$$

showing that C_v is the heat capacity at *constant volume*. The symbol $(\ )_\alpha$ means that α is held constant during the differentiation.

The second law of thermodynamics is somewhat more difficult to express but it is associated with inability of a cyclical system or *heat engine* to be 100 percent efficient. In fact, one statement of the law is: "It is impossible to construct an engine which, acting in a cycle, extracts a given amount of heat from a source, and performs an equivalent amount of work." Although we shall not do so here,[2] it is possible to show that the most efficient cycle is the *Carnot cycle* shown in Figure 2.1. It consists of an isothermal compression stroke, an isothermal expansion stroke, and two others in which the engine is thermally isolated from its environment (adiabatic compression and expansion strokes). Such a system is *reversible* in that it returns to its initial thermodynamic state upon completion of a cycle. This fact can be used to define another thermodynamic function called the *entropy*. The net change of entropy after completion of a Carnot cycle is zero. An infinitesimal change in entropy dS is related to the heat input $đQ$ by

$$dS = \frac{đQ}{T} \tag{2.29}$$

Obviously the entropy change during an adiabatic process must be zero. The foregoing statements lead us to another statement of the second law: In any closed system the total entropy change is positive semidefinite.

$$\Delta S \geq 0 \tag{2.30}$$

Entropy is also related to the degree of disorder of a system. Nearly one hundred years ago Boltzmann derived an expression which showed that the greater the microscopic disorder, the greater the entropy.

With the aid of the definition expressed by Eq. (2.29), one can rewrite the first law entirely in terms of the thermodynamic variables S, T, U, P, and α:

$$T\,dS = dU + p\,d\alpha = C_v\,dT + p\,d\alpha \tag{2.31}$$

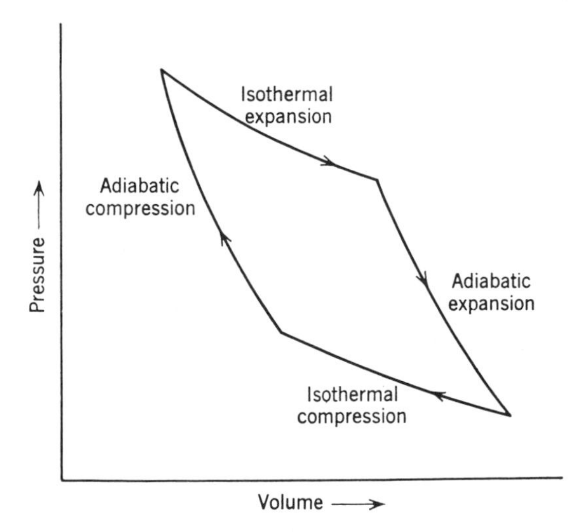

Figure 2.1 The Carnot cycle.

It is frequently useful to introduce another thermodynamic function, the enthalpy H defined by

$$H = U + p\alpha \tag{2.32}$$

When the internal energy U is eliminated from Eq. (2.31), the mathematical expression of the first law becomes

$$T\,dS = dH - \alpha\,dp \tag{2.33}$$

Differentiation with respect to T, holding p constant, leads to a definition of the heat capacity at constant pressure, C_p:

$$\left(\frac{dQ}{dT}\right)_p = \left(\frac{\partial H}{\partial T}\right)_p = C_p \tag{2.34}$$

The first law can be expressed in terms of C_p as

$$T\,dS = C_p\,dT - \alpha\,dp \tag{2.35}$$

2.2.2 *Thermodynamics of Ideal Gases*

At the pressures considered in this text all gases can be treated as ideal gases to a very good approximation. The reader will recall that an ideal gas is composed of point "molecules" which interact only when in actual contact. The equation of state, i.e., the relationship among temperature, pressure, and volume, is easily shown to be (see Section 2.3)

$$p\alpha = RT, \tag{2.36}$$

where R is the gas constant. One of the very useful relationships in dealing with an ideal gas is that between C_p and C_v. With the aid of Eqs. (2.31) and (2.35) it is easily shown that

$$(C_p - C_v)\,dT = p\,d\alpha + \alpha\,dp = R\,dT \tag{2.37}$$

Hence,

$$C_p - C_v = R \tag{2.38}$$

An interesting question which arises in aeronomy (and meteorology) is concerned with the behavior of pressure and volume in a rapidly rising cell of air. Under this condition little heat is lost by the air mass, and the process is very close to adiabatic. From our earlier discussion we have $T\,dS = 0$ and

$$C_p\,dT - \alpha\,dp = 0 \tag{2.39}$$

from which we derive the equation

$$p\alpha^\gamma = \text{constant} \tag{2.40}$$

where $\gamma = C_p/C_v$. The derivation is left as an exercise.

It will be necessary in later chapters to know the value of γ for atmospheric gases. One can show by the methods of statistical mechanics that the heat capacity of a gas at constant volume is

$$C_v = \frac{n}{2} R \tag{2.41a}$$

where n is the number of degrees of freedom of a molecule. In this text we shall be interested almost exclusively in monatomic and diatomic molecules. The former have only three degrees of freedom (translational) so that

$$C_v = \frac{3}{2} R \text{ (monatomic)} \tag{2.41b}$$

whereas the latter (except at very high temperatures) have 5 (3 translational, and 2 rotational about axes perpendicular to the internuclear axis and to each other); hence

$$C_v = \frac{5}{2} \text{ (diatomic)} \tag{2.41c}$$

Obviously $\gamma = \frac{5}{3}$ and $\frac{7}{5}$, respectively, for the two cases. If the temperature is high enough to populate the vibrational levels according to a Maxwell-Boltzmann distribution (see section 2.3), there are 6 degrees of freedom.

2.2.3 Miscellaneous Thermodynamic Equations

We conclude with a discussion of the interrelationships among variations δ in the thermodynamic variables. Let us expand δp, δT, and δS in the following manner

$$\delta p = \left(\frac{\partial p}{\partial \alpha}\right)_s \delta\alpha + \left(\frac{\partial p}{\partial S}\right)_\alpha \delta S \tag{2.42}$$

$$\delta T = \left(\frac{\partial T}{\partial \alpha}\right)_s \delta\alpha + \left(\frac{\partial T}{\partial S}\right)_\alpha \delta S \tag{2.43}$$

$$\delta S = \left(\frac{\partial S}{\partial T}\right)_\alpha \delta T + \left(\frac{\partial S}{\partial \alpha}\right)_T \delta\alpha \tag{2.44}$$

$$\delta S = \left(\frac{\partial S}{\partial T}\right)_p \delta T + \left(\frac{\partial S}{\partial p}\right)_T \delta p \tag{2.45}$$

and use some previously derived expressions to evaluate the partial derivatives. For example, the first law can be written

$$T\, dS = C_v\, dT + p\, d\alpha \tag{2.31}$$

Comparison with Eq. (2.44) shows that

$$C_v = \left(\frac{\partial S}{\partial T}\right)_\alpha \tag{2.46}$$

$$\left(\frac{\partial S}{\partial \alpha}\right)_T = p/T \tag{2.47}$$

We shall make use of such relations (see exercises for further examples) in Chapter 5. Thermodynamics is, of course, applicable to systems other than gases but they are of no interest in the present context.

2.3 Kinetic Theory of Gases

In the preceding section the kinetic theory was anticipated when an ideal gas was defined as being composed of point molecules with contact interaction only. Actually these assumptions can be relaxed within the framework of kinetic theory, leading to the inclusion of real gases with finite size molecules and long range interactions. In the upper atmosphere, however, all gases including the charged ones are essentially ideal in their behavior because of their low densities.

2.3.1 The Distribution Function

It is intuitively obvious that if somehow one is able to set the molecules in motion with a given set of initial velocities, collisions will soon alter them radically. The question that one would like to answer is: "What is the *distribution* of velocities at any given instant of time?" That is, how many have velocities between $\mathbf{v}_1$ and $\mathbf{v}_1 + d\mathbf{v}_1$, between $\mathbf{v}_2$ and $\mathbf{v}_2 + d\mathbf{v}_2$, etc. Of particular interest is the distribution corresponding to the equilibrium situation in which the distribution of velocities no longer changes with time. This is the case of maximum disorder in which the molecular motion is completely random.

In general, the velocity distribution function depends explicitly upon the time and position as well as velocity: $f = f(\mathbf{v}, \mathbf{x}, t)$. However, in a homogeneous gas in thermodynamic equilibrium, it is a function of velocity only. If the gas is also isotropic, only the molecular speeds appear in the distribution function. It is almost a universal convention that the latter be normalized to the total number N_T of molecules in the system when integrated over all possible speeds (velocity space) and all configuration space:

$$N = \int f(\mathbf{v}, \mathbf{x})\, d\mathbf{v}\, d\mathbf{x} \tag{2.48}$$

If the gas is homogeneous and isotropic, the normalization is in terms of the number density N:

$$N = 4\pi \int_0^\infty f(v)\, dv \tag{2.49}$$

Of course, no molecule can have an infinite velocity, but the error introduced by using infinity as the upper limit of integration is negligible in all cases of interest.

The derivation of the equilibrium distribution function, usually called the Maxwell-Boltzmann distribution after its discoverers, is given in most books on statistical mechanics and kinetic theory of gases.[3] We present it here without proof:

$$f(v) = 4\pi N \left(\frac{m}{2\pi kT}\right)^{3/2} v^2 \exp\left[-\frac{mv^2}{2kT}\right] \tag{2.50}$$

where m is the molecular mass, T is the absolute temperature, and k is the Boltzmann constant. Some quantities which pıove to be of considerable importance in kinetic theory are the *velocity moments*

$$\overline{v^m} = \int_0^\infty v^m f(v)\, dv \tag{2.51}$$

particularly $\overline{v^2}$. Their evaluation in terms of the macroscopic variables is left as an exercise. Incidentally, knowledge of *all* the velocity moments is equivalent to knowledge of the distribution function.

2.3.2 Equation of State of an Ideal Gas

An interesting and instructive application of the Maxwell-Boltzmann distribution function is the derivation of the equation of state of an ideal gas. Consider a wall which is being continually bombarded by gas molecules whose positions and velocities are distributed isotropically. When a molecule strikes the wall and rebounds as shown in Figure 2.2, it transfers momentum $\Delta\mathbf{P}$ to the wall. Because of isotropy, the components ΔP_x and ΔP_y vanish, but ΔP_z is given by

$$\Delta P_z = mv \cos\theta - (-mv\cos\theta) = 2mv\cos\theta \tag{2.52}$$

where θ is defined in the figure. The average or *stochastic* force p on a unit area of the wall (p is thus the pressure) exerted by the random motion of the molecules is equal to the rate at which momentum is transferred to it:

$$p = \sum_i 2mv_i \cos\theta_i \cdot v_i \cos\theta_i = \sum_i 2mv_i^2 \cos^2\theta_i \tag{2.53}$$

where the summation is over all particles bombarding the unit area. Rather than attempt the impossible task of summing over all particles, we approximate it by integrating over the Maxwell-Boltzmann distribution, obtaining

$$p = nkT \tag{2.54}$$

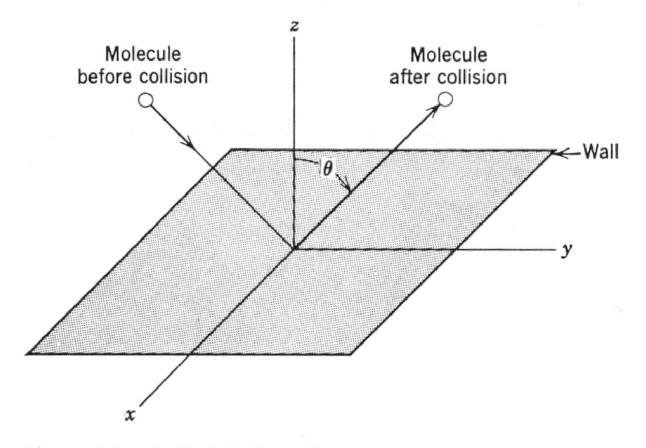

Figure 2.2 Collision of an ideal gas molecule with the container wall.

the equation of state of an ideal gas (compare with Eq. (2.36)). Actually, the complete agreement between the two is deceptive, because such a comparison is actually used to define the gas kinetic temperature T. As an exercise, write $f(v)$ in the form

$$f(v) = Av^2e^{-av^2} \tag{2.55}$$

and evaluate A and a by use of the normalization condition (2.49) and by comparison with the equation of state of an ideal gas.

2.3.3 *The Boltzmann Equation*

It frequently happens that the system under consideration is not in thermal equilibrium. How then do we find the corresponding distribution function? This question was answered in principle nearly a century ago by Boltzmann who derived an equation for $f(\mathbf{v}, \mathbf{x}, t)$. He did so by following the motion of an element of the gas as it moves through configuration *and* velocity space. The time rate of change of the number density of particles in the element (i.e., the distribution function) is equal to the difference between the rates at which the particles are scattered into and out of the element:

$$\frac{df}{dt} = \left(\frac{\partial f}{\partial t}\right)_{\text{coll}} \tag{2.56a}$$

or in terms of partial derivatives:

$$\frac{\partial f}{\partial t} + \mathbf{v} \cdot \nabla f + \mathbf{a} \cdot \nabla_v f = \left(\frac{\partial f}{\partial t}\right)_{\text{coll}} \tag{2.56b}$$

Here, ∇ is the configuration space gradient, ∇_v is the velocity space gradient, and **a** is the particle acceleration which is equal to the external force per unit mass exerted on the particles.

That part of Eq. (2.56b) which leads to the most difficulty in obtaining solutions is the collision integral $(\partial f/\partial t)_{\text{coll}}$. This term is nonlinear in the distribution function f; hence, attempts to find general solutions are fraught with all the difficulties usually associated with nonlinear integro-differential equations. In order to obtain an expression for $(\partial f/\partial t)_{\text{coll}}$, let us consider collisions between two species denoted by i and j. The corresponding collision integral for species i due to elastic collisions with species j is

$$\left(\frac{\partial f_i}{\partial t}\right)_{\text{coll}} = \int f_i(\mathbf{v}_i')f_j(\mathbf{v}_j')\,|\mathbf{v}_i' - \mathbf{v}_j'|\,Q(\theta)\,d\mathbf{v}_j'\,d\Omega' - \int f_i(\mathbf{v}_i)f_j(\mathbf{v}_j)\,|\mathbf{v}_i - \mathbf{v}_j|\,Q(\theta)\,d\mathbf{v}_j\,d\Omega \tag{2.57}$$

where $Q(\theta)$ is the differential scattering cross section (to be discussed in Section 2.5). The two terms on the right hand side represent the rates at which particles are scattered into and out of velocity space elements $d\mathbf{v}_i$ and $d\mathbf{v}_j$; $f_i(\mathbf{v}_i')$ and $f_j(\mathbf{v}_j')$ are the numbers of particles in position to be scattered into these velocity elements, whereas $f_i(\mathbf{v}_i)$ and $f_j(\mathbf{v}_j)$ represent the numbers about to be scattered out of it. The collision integral represents the rate at which molecules of species i undergo momentum transfer collisions with molecules of species j. It is related to the collision frequency for momentum transfer, ν_m, (also to be discussed in Section 2.5) by the equality

$$\left(\frac{\partial f}{\partial t}\right)_{\text{coll}} = f\nu_m \tag{2.58}$$

It is apparent that $\overline{f\nu_m} = 0$ in a homogeneous and isotropic gas because momentum cannot be transported by molecular motion from one point to another in that case.

Several approaches have been found for obtaining approximate solutions to the Boltzmann equation. Later we shall use the perturbation expansion of f in powers of the velocity; to the first order term, such an expansion is of the form

$$f \approx f_0 + \hat{\mathbf{v}} \cdot \mathbf{f}_1 \tag{2.59}$$

where

$$f_0 \gg |\mathbf{f}_1|$$

and $\hat{\mathbf{v}}$ is a unit vector in the direction of the velocity.

2.3.4 The Boltzmann Transport Equations and Applications

The Boltzmann equation deals with the microscopic properties of a gas, i.e., the velocity distribution of the molecules. One can relate it to macroscopic properties such as density, pressure, heat flux, etc., by multiplying each term by some function of the velocity denoted by $\phi(w)$ and integrating over all velocity space:

$$\int \phi \frac{\partial f}{\partial t} d\mathbf{v} + \int \phi \mathbf{v} \cdot \nabla f \, d\mathbf{v} + \int \phi \mathbf{a} \cdot \nabla_v f \, d\mathbf{v} = \int \phi \left(\frac{\partial f}{\partial t}\right)_{\text{coll}} d\mathbf{v} \tag{2.60a}$$

where the arguments of ϕ and f have been omitted for the sake of brevity. Upon integrating by parts all terms on the left hand side, and employing the condition that as v tends to infinity, f tends to zero faster than does $1/\phi$, Eq. (2.60a) can be recast as

$$\frac{\partial(N\bar{\phi})}{\partial t} + \nabla \cdot (N\overline{\phi \mathbf{v}}) - N\left(\overline{\frac{\partial \phi}{\partial t}} + \overline{\mathbf{v} \cdot \nabla \phi} + \overline{\mathbf{a} \cdot \nabla_v \phi}\right) = \overline{\phi \left(\frac{\partial f}{\partial t}\right)}_{\text{coll}} \tag{2.60b}$$

where the bar indicates average value. Equation (2.60b) is called the Boltzmann transport equation because it can be used to relate various transport properties by means of conservation equations. For example, if $\phi = 1$, we obtain the continuity equation

$$\frac{\partial N}{\partial t} + \nabla \cdot \overline{(N\mathbf{v})} = S, \tag{2.61}$$

where S represents the net rate at which particles are introduced into the system. Setting $\phi = m\mathbf{v}$ and $(m/2)v^2$, leads to the equations for conservation of momentum and energy ϵ, respectively

$$\frac{\partial \overline{Nm\mathbf{v}}}{\partial t} + \nabla \cdot (\overline{Nm\mathbf{v}\mathbf{v}}) = Nm\mathbf{a} + m \int \mathbf{v} \left(\frac{\partial f}{\partial t}\right)_{\text{coll}} d\mathbf{v} \tag{2.62}$$

$$\frac{\partial \epsilon}{\partial t} + \nabla \cdot (N\overline{\epsilon \mathbf{v}}) = N\overline{m\mathbf{a} \cdot \mathbf{v}} + \int \frac{m}{2} v^2 \left(\frac{\partial f}{\partial t}\right)_{\text{coll}} d\mathbf{v} \tag{2.63}$$

These equations are of particular importance because they provide a connection between kinetic theory and large scale fluid dynamics.

It will prove useful later in the text if we now apply the theory developed above to two transport phenomena: diffusion and heat flow. We begin by considering diffusion for which we need to evaluate the quantity $N\mathbf{v} = \int \mathbf{v} f \, d\mathbf{v}$. The perturbation expansion of Eq. (2.59) is convenient for this purpose. Upon substitution into the steady state field-free Boltzmann equation [Eq. (2.56b) with the first and third terms on the left hand side omitted], we obtain the first order (see the appendix to Chapter 11 for more details concerning the collision integral)

$$\nu_m f_1^i = -v \frac{\partial f_0}{\partial x_i} \tag{2.64}$$

where ν is collision frequency for momentum transfer. We now average over all velocity space, obtaining

$$\overline{N\mathbf{v}} = \int \mathbf{v} f_0 \, d\mathbf{v} + \int \mathbf{v}\hat{\mathbf{v}} \cdot \mathbf{f}_1 \, d\mathbf{v} = -\int \frac{\mathbf{v}}{\nu_m} \mathbf{v} \cdot \nabla f_0 \, d\mathbf{v} \tag{2.65}$$

where the first term averages to zero because of the isotropy of f_0 in velocity space. If f_0 is Maxwellian and the temperature is constant, the above average becomes

$$\overline{\mathrm{N}\mathbf{v}} = -D\nabla \mathrm{N} \tag{2.66}$$

where

$$D = \frac{kT}{m\nu_m} \tag{2.67}$$

is called the *diffusion coefficient*. The equation of continuity is thus

$$\frac{\partial N}{\partial t} + \nabla \cdot D\nabla N = S \tag{2.68}$$

The equation for heat flow due to conduction is obtained with the aid of Eq. (2.63); for convenience we neglect the first term on the right hand side. In Problem 2.10 we prove that the internal energy density of an ideal gas is related to the temperature T by

$$E = \frac{3}{2} RT \tag{2.69a}$$

or in terms of the Boltzmann constant k and the number density N,

$$E = \frac{3}{2} NkT \tag{2.69b}$$

Substitution into Eq. (2.63) yields

$$\frac{\partial(NkT)}{\partial t} + \nabla \cdot \overline{(N\epsilon\mathbf{v})} = \dot{Q} \tag{2.70}$$

where $\dot{Q} = m/2 \int v^2(\partial f/\partial t)_{\mathrm{coll}} \, d\mathbf{v}$ is the heat source term. It is left as an exercise to prove that

$$C_v \frac{\partial T}{\partial t} - \nabla \cdot K_T \nabla T = \dot{Q} \tag{2.71}$$

and to express K_T, the thermal conductivity, in terms of molecular mass, collision frequency, and number density.

2.3.5 *Debye Shielding in a Plasma*

As a final example of the application of the kinetic theory of gases, consider a gas of positive ions and electrons which is in thermal equilibrium. Each ion repels other ions in its neighborhood but attracts nearby electrons. A space charge

$$q_{\text{sp}} = e(N_+ - N_e) \tag{2.72}$$

is thus created about each ion. From Gauss' law [Eq. (2.2)] one can relate q_{sp} to the associated electric field $\mathbf{E}$ by

$$\nabla \cdot \mathbf{E} = \frac{q_{\text{sp}}}{\epsilon_0} = \frac{e}{\epsilon_0}(N_+ - N_e) \tag{2.73}$$

or in terms of the electrostatic potential ϕ defined by:

$$\mathbf{E} = -\nabla\phi, \tag{2.74}$$

$$\nabla^2\phi = -\frac{e}{\epsilon_0}(\bar{N}_+ - N_e) \tag{2.75}$$

Here N_+ is the *averaged* ion number density.

In order to proceed further, it is necessary to relate ϕ to the distribution function f for the electrons, by means of the time-independent collisionless Boltzmann equation. We therefore set $\partial f/\partial t = 0$, $(\partial f/\partial t)_{\text{coll}} = 0$, and $\mathbf{a} = e\mathbf{E}$ in Eq. (2.56b)

$$\mathbf{v} \cdot \nabla f + e\mathbf{E} \cdot \nabla_v f = \mathbf{v} \cdot \nabla f - e\nabla\phi \cdot \nabla_v f = 0 \tag{2.76}$$

The fact that $e\phi$ is the potential energy of an electron suggests that we try an f of the form

$$f = f_0 e^{-e\phi/kT} \tag{2.77}$$

where f_0 is Maxwellian. Substitution into Eq. (2.76) yields

$$\left(-\frac{e}{kT}\mathbf{v} \cdot \nabla\phi + e\nabla\phi \cdot \frac{\mathbf{v}}{kT}\right) f \equiv 0 \tag{2.78}$$

proving that our guess was a good one. The number density of the electrons can now be written (recall Eq. (2.49))

$$N_e = \int f \, d\mathbf{v} = \bar{N}_e e^{-e\phi/kT} \tag{2.79}$$

Because the gas is macroscopically neutral, $\bar{N}_e = \bar{N}_+$ and Eq. (2.75) becomes

$$\nabla^2\phi = -\frac{e}{\epsilon_0}\bar{N}_+(1 - e^{-e\phi/kT}), \tag{2.80}$$

a nonlinear differential equation for ϕ. Fortunately, we usually find that $e\phi \ll kT$, permitting the equation to be linearized to

$$\nabla^2\phi \approx \frac{e^2}{\epsilon_0} \bar{N}_+ \frac{\phi}{kT} \tag{2.81}$$

Assuming spherical symmetry (which is well justified), a solution is easily found to be

$$\phi = \frac{e}{4\pi\epsilon_0 r} \exp\left(-\frac{r}{h}\right) \tag{2.82}$$

where

$$h = \sqrt{\frac{\epsilon_0 kT}{\bar{N}_+ e^2}}$$

is called the Debye shielding length. Note that ϕ differs from the Coulomb potential only by the factor $e^{-r/h}$ which has the following physical significance. It is a cut-off factor which arises from the screening effect of the electron cloud about each ion. Inside the *Debye sphere* of radius h, the particles interact with each other *individually*, but they interact with any particles located outside the sphere *collectively*. This brings us to the criterion for the existence of a true plasma: The dimensions of the plasma container must be much larger than the Debye shielding length. If this is not true, collective motions cannot be supported. The other criterion is that the number of particles within a Debye sphere must be $\gg 1$.

2.4 Atomic and Molecular Structure and Spectra

Many aeronomic phenomena such as the aurora, airglow, and radio wave propagation are directly or indirectly dependent upon atomic and molecular structure and collision processes. We shall, therefore, review pertinent aspects of atomic physics in this and the following two sections.

2.4.1 The Black Body Spectrum and Einstein Coefficients

One of the few riddles left in physics at the close of the nineteenth century was the nature of the "black body spectrum," i.e., the radiation spectrum inside an enclosure which is thermally isolated from its surroundings. Planck solved the problem by introducing the quantum hypothesis and the associated "quantum of action," usually denoted by h (Planck's constant). Stated very briefly, Planck assumed that the electromagnetic oscillators in the walls of the enclosure could not have any energy, but only those energies ϵ related to the fundamental frequency of oscillation ν_0 by

$$\epsilon = nh\nu_0 \tag{2.82}$$

where n is an integer. Furthermore, the radiation could be emitted only in discrete quanta of energy $h\nu_0$. From these assumptions and the condition of thermal equilibrium, Planck was able to deduce the black body spectrum (i.e., energy density per unit frequency)

$$\rho(\nu) = 8\pi \frac{h\nu^3}{c^3} \left(\frac{e^{h\nu}}{kT} - 1 \right)^{-1} \quad (2.83)$$

Here c is the velocity of light, k is the Boltzmann constant, and T is the absolute temperature of the enclosure.

Einstein carried the quantum hypothesis a step further by treating the assemply of radiation quanta as a gas. Each quantum or "photon" was entirely analogous to a gas molecule except that it could be destroyed or created by the oscillators in the enclosure walls. Einstein assumed that the radiative transitions are governed by the laws of probability, i.e., that one could not tell exactly when a given oscillator would radiate but only the probability of radiation per unit time. Hence the probability that an oscillator will spontaneously radiate in time dt is given by $A_{n' \to n''}\, dt$ where $A_{n' \to n''}$ is the Einstein coefficient for spontaneous emission from state n' to state n''. The total rate of emission is thus $N_{n'} A_{n' \to n''}$; $N_{n'}$ is the number of oscillators in state n'.

The oscillators can absorb as well as emit radiation and it is necessary to include an expression for the corresponding absorption rate. For a transition from state n'' to state n' the rate is $N_{n''} B_{n'' \to n'} \rho(\nu_{n'n''})$ where $\rho(\nu_{n'n''})$ is the energy density of the radiation pei frequency interval at frequency $\nu_{n'n''}$, and $B_{n' \to n''}$ is the Einstein absorption coefficient.

Finally, there is the possibility of emission which is stimulated by the radiation field. The associated probability is $N_{n'} B_{n' \to n''} \rho(\nu_{n'n''})$. At equilibrium the sum of the two emission rates is equal to the absorption rate:

$$N_{n'}[A_{n' \to n''} + B_{n' \to n''} \rho(\nu_{n'n''})] = N_{n''} B_{n'' \to n'} \rho(\nu_{n'n''}) \quad (2.84)$$

Furthermore, thermal equilibrium requires that

$$\frac{N_{n''}}{N_{n'}} = \frac{g_{n''}}{g_{n'}} \exp \frac{h\nu_{n'n''}}{kT} \quad (2.85)$$

in which $g_{n'}$ and $g_{n''}$ are the *statistical weights* of states n' and n''. Upon solving Eq. (2.84) for $\rho(\nu)$, one obtains

$$\rho(\nu_{n'n''}) = \frac{A_{n' \to n''}}{\dfrac{g_{n''}}{g_{n'}} B_{n'' \to n'} \exp \dfrac{h\nu_{n'n''}}{kT} - B_{n' \to n''}} \quad (2.86)$$

In order to deduce the Planck formula it is only necessary to assume the relationships

$$g_{n''} B_{n'' \to n'} = g_{n'} B_{n' \to n''} \quad (2.87)$$

and

$$A_{n'\to n''} = \frac{8\pi h\nu_{n'n''}^3}{c^3} B_{n'\to n''} \tag{2.88}$$

By analogy with classical electrodynamics one can show that the coefficient $A_{n'\to n''}$ is related to the *dipole matrix element* $\langle n''|\mathbf{r}|n'\rangle$ for the transition by

$$A_{n'\to n''} = \frac{64\pi^4\nu_{n'n''}^3}{3hc^3} |\langle n''|\,\mathbf{r}\,|n'\rangle|^2 \tag{2.89}$$

with the aid of Eq. (2.88), one can readily show that

$$B_{n'\to n''} = \frac{8\pi^3}{h^2} |\langle n''|\,\mathbf{r}\,|n'\rangle|^2 \tag{2.90}$$

The dipole matrix elements will be discussed further in Section 2.4.3.

2.4.2 *Atomic Structure*

The modern quantum theory of atomic structure has its origin in the Bohr model of the hydrogen atom. Bohr employed two fundamental postulates: (1) that the angular momentum of the electron bound in the atom must be an integer times Planck's quantum of action divided by 2π, i.e., $h/2\pi$ or $\hbar$, and (2) that when an electron jumps from one allowed state to another, the frequency ν of the radiation emitted or absorbed is related to the energy difference $\Delta\epsilon$ between the two states by

$$\Delta\epsilon = h\nu \tag{2.91}$$

As a consequence of the first assumption and the radial dependence of the Coulomb force, he was able to show that

$$\Delta\epsilon = \frac{me^4}{2\hbar^2}\left(\frac{1}{n_1^2} - \frac{1}{n_2^2}\right) \tag{2.92}$$

where e and m are the electronic charge and mass, respectively, and n_1 and n_2 are integers ($n_1 < n_2$). Another consequence was the restriction of the electron to allowed orbits, each of radius

$$r_n = \frac{n^2\hbar^2}{me^2} \tag{2.93}$$

However, the Bohr model did not prove to be satisfactory when extended to multi-electron atoms and to molecules. Ultimately, Schrödinger and Heisenberg were led to abandon Newtonian mechanics completely by replacing the dynamical variables such as momentum and energy with linear operators. In Schrödinger's scheme momentum was replaced by a differential operator in coordinate space; e.g., the x component is

$$p_x = -i\hbar \frac{\partial}{\partial x} \tag{2.94}$$

and the energy or "Hamiltonian" operator H is

$$H = i\hbar \frac{\partial}{\partial t} \tag{2.95}$$

Use of such linear operators gave rise to the Schrödinger "wave equation"

$$\left(-\frac{\hbar^2}{2m}\nabla^2 + V\right)\psi = i\hbar \frac{\partial \psi}{\partial t} \tag{2.96}$$

in which ∇^2 is the Laplacian operator,[3] V is the potential energy operator (e^2/r for the Coulomb potential), and ψ is a "state function" (sometimes called a "wave function") which completely describes the system. In the Schrödinger scheme the expectation value (mean of a set of measured values) corresponding to any dynamical variable $\mathcal{O}$ is computed as

$$\bar{\mathcal{O}} = \int \psi^* \mathcal{O} \psi \, d\tau \tag{2.97}$$

where ψ^* is the complex conjugate of ψ and the integral is to be taken over all space. The Schrödinger-Heisenberg "quantum mechanics" can be used, at least in principle, to account for nearly all atomic phenomena including the Bohr energy levels in the hydrogen atom. Unfortunately, the mathematical expressions usually are extremely difficult to solve. Even then, the solutions can be obtained in approximate form only.

One of the consequences of the quantum theory is a general rule, the Heisenberg uncertainty principle: Certain physical quantities (those which are "canonically conjugate") cannot be measured precisely even with perfect measuring devices. Among these dynamical variables are position and momentum. If we wish to measure their "x" components, the uncertainty principle says that the products of the deviations Δx and Δp_x in a series of measurements must satisfy†

$$\Delta x \, \Delta p_x \geq \hbar \tag{2.98}$$

The principle itself stems directly from the noncommutativity of the operators which represent these dynamical variables (see Problem 2.13).

Although the use of the Schrödinger-Heisenberg quantum mechanics is necessary for the computation of atomic structures, it is not very useful for illustrative purposes. Instead we shall employ a sort of hybrid of the Bohr theory and the more modern theory. This is the vector model of the atom,

† The exact form of this inequality depends upon the way in which the deviation is defined.

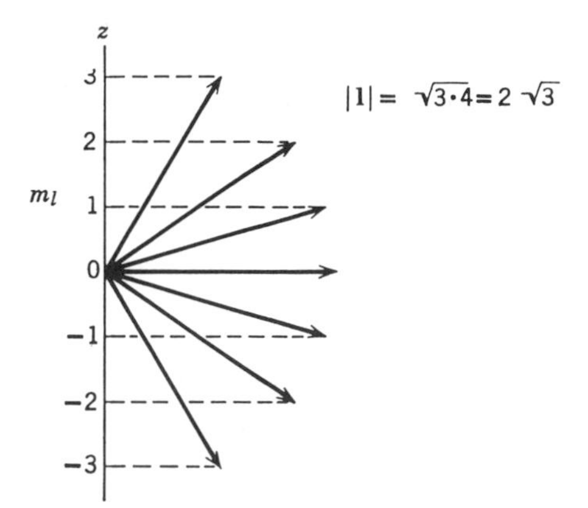

Figure 2.3 Allowed angular momentum states for $l = 3$.

sometimes called the semiclassical model. It is concerned mainly with the angular momentum states of the electrons and the way in which the angular momenta of different electrons couple together.

With the aid of quantum mechanics we find that the only allowed values of the angular momentum magnitude are

$$|\mathbf{L}| = \sqrt{l(l + 1)}\hbar \tag{2.99}$$

where l is an integer equal to or greater than zero. Furthermore, we can know only one component of the vector **L**. Denoting the known component by L_z, we find that it is restricted to certain values given by

$$L_z = m_l \hbar \tag{2.100}$$

where $|m_l| \leq l$ is a positive or negative integer, including zero. The allowed values of these parameters are shown in Figure 2.3. Because only one component of **L** is known, the orbital angular momentum may be regarded as precessing about the z-axis as shown in Figure 2.4.

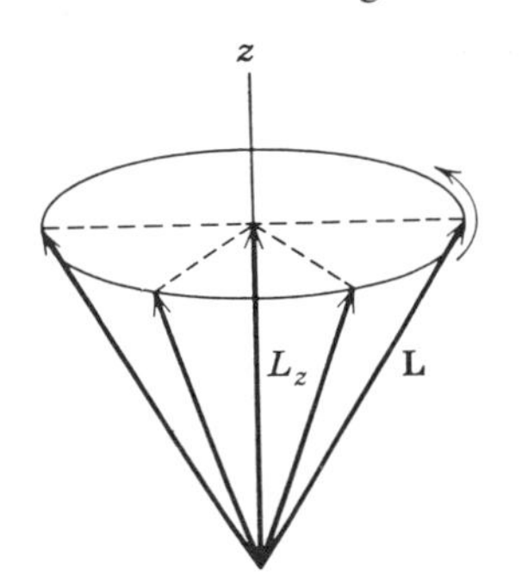

Figure 2.4 Precession of the angular momentum vector **L**.

In addition to the orbital angular momentum of an electron owing to its motion about the nucleus, the former possesses an intrinsic angular momentum or "spin." The spin quantum number is restricted to a value $s = \frac{1}{2}$; the magnitude of the spin is, therefore

$$|\mathbf{s}| = \sqrt{\tfrac{1}{2}(\tfrac{1}{2} + 1)}\hbar = \sqrt{\tfrac{3}{4}}\,\hbar \tag{2.101}$$

The z-component can assume only two values:

$$s_z = m_s\hbar = \pm\tfrac{1}{2}\hbar \tag{2.102}$$

The spin was first inferred from observations made before the discovery of quantum mechanics but was not satisfactorily explained until the development of relativistic quantum theory by Dirac.

There is another quantum number which is very important in atomic structure, but it has no classical analogue. It is called the parity, and it describes the behavior of the state function of the system under a space inversion†. If it changes sign, the parity is said to be odd; if it does not, the parity is even. Parity changes play an important role in the radiation selection rules discussed in later parts of this section.

The orbital and spin angular momenta, $\mathbf{l}$ and $\mathbf{s}$, of an electron are added vectorially, as in classical mechanics, to yield the total angular momentum

$$\mathbf{j} = \mathbf{l} + \mathbf{s} \tag{2.103}$$

of magnitude

$$|\mathbf{J}| = \sqrt{j(j+1)}\,\hbar = (l(l+1)\hbar^2 + \tfrac{3}{4}\hbar^2 + 2\mathbf{s}\cdot\mathbf{l})^{1/2} \tag{2.104}$$

In developing his relativistic theory, Dirac showed that l and s are not good quantum numbers even for a single electron in a central field because of the appearance of a *spin-orbit coupling term* $\xi(r)\mathbf{s}\cdot\mathbf{l}$ in the energy operator for the atom [$\xi(r)$ is a function of the orbital radius which we leave unspecified]. Even nonrelativistically, there is an interaction between the magnetic field due to orbital motion and the intrinsic magnetic moment arising from spin; but the resultant interaction energy is only one-half the correct value. Hence, we need to evaluate the scalar product $\mathbf{s}\cdot\mathbf{l}$ when computing electron energy levels in the atom.

$$\mathbf{s}\cdot\mathbf{l} = \tfrac{1}{2}(j(j+1) + l(l+1) - \tfrac{3}{4})\hbar^2 \tag{2.105}$$

Because of the vector addition rule, l is restricted to the two values $j + \frac{1}{2}$ and $j - \frac{1}{2}$. Thus,

$$\begin{aligned} \mathbf{s}\cdot\mathbf{l} &= j(2j+3)\hbar^2, \quad \text{if} \quad l = j + \tfrac{1}{2} \quad \text{or} \\ &= j(2j-1)\hbar^2, \quad \text{if} \quad l = j - \tfrac{1}{2} \end{aligned} \tag{2.106}$$

Since all atoms except hydrogen contain more than one electron, it is important to know how to add the angular momenta of the latter in order to

† In space inversion, the x, y, and z axes are reversed.

obtain the total angular momentum of the atom. Of course, not all of the states predicted by simple compounding of orbital and spin angular momenta are realizable. The Pauli exclusion principle rather severely restricts them. This law which is fundamental to atomic, molecular, and nuclear structure states that no two electrons in the atom can have the same set of quantum numbers, or more generally, in the language of quantum mechanics, that the state function which represents the atom must be antisymmetric under the interchange of any two electrons. The exclusion principle plays a central role in the building-up of atoms by adding electrons. For example, when the principal quantum number n is 1, $l = 0$, and $m_s = \pm\frac{1}{2}$ are the only possible states; a *full shell* is attained with only two electrons. When $n = 2$, we can have $l = 0$ (S shell) or $l = 1$ (P shell). In the first case two electrons fill the shell, but in the second there are six possibilities, $m_l = 0, \pm 1$, $m_s = \pm\frac{1}{2}$, giving a total of six electrons in a full shell. Actually, an "individual particle" model of this type does not accurately describe the electronic structure of an atom, but it is adequate for a qualitative treatment of the building-up, or "aufbau," principle.

If there were no spin-orbit coupling term in the Hamiltonian operator (i.e., $\mathbf{s} \cdot \mathbf{l} = 0$), the $\mathbf{l}$ and $\mathbf{s}$ vectors for the various electrons would add separately to yield total orbital and spin angular momenta:

$$\mathbf{L} = \sum_i \mathbf{l}_i \tag{2.107}$$

$$\mathbf{S} = \sum_i \mathbf{s}_i \tag{2.108}$$

where the sum is to be taken over all electrons. The total angular momentum for the atom is the sum of $\mathbf{L}$ and $\mathbf{S}$:

$$\mathbf{J} = \mathbf{L} + \mathbf{S} \tag{2.109}$$

Angular momentum addition is illustrated in Figure 2.5. The quantization rules apply as in the case of single electrons except that the total spin quantum numbers S and M_s are integers or half odd integers according to whether the number of electrons is even or odd. Thus,

$$|\mathbf{L}| = \sqrt{L(L+1)}\hbar, \; L = 0, 1, 2, \ldots$$

$$|\mathbf{S}| = \sqrt{S(S+1)}\hbar, \; S = 0, 1, 2, \ldots \quad \text{or} \quad S = \tfrac{1}{2}, \tfrac{3}{2}, \tfrac{5}{2}, \ldots \tag{2.110}$$

Hence,

$$L_z = M_l\hbar; \;\; M_l = 0, +1, +2, \ldots (-L \leq M \leq L)$$

$$S_z = M_s\hbar; \;\; M_s = 0, \pm 1, \pm 2, \pm 3, \ldots \quad \text{or}$$

$$M_s = \pm\tfrac{1}{2}, \pm\tfrac{3}{2}, \pm\tfrac{5}{2}, \ldots (-S \leq M_s \leq S)$$

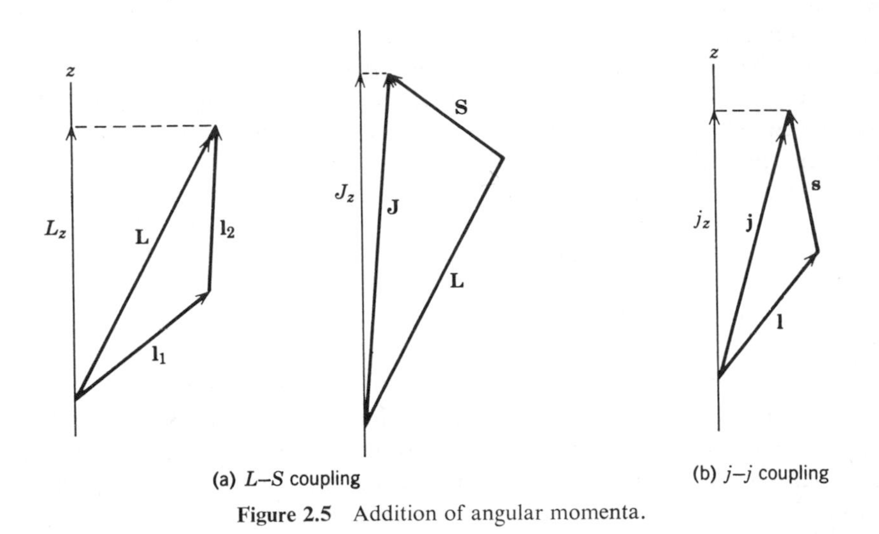

(a) L–S coupling (b) j–j coupling

Figure 2.5 Addition of angular momenta.

Furthermore, the total angular momentum must obey the rules

$$|\mathbf{J}| = \sqrt{J(J+1)}\,\hbar, \qquad |L - S| \leq J \leq L + S \tag{2.111}$$

where $J = 0, 1, 2, \ldots$ if S is an integer, or $J = \frac{1}{2}, \frac{3}{2}, \frac{5}{2}, \ldots$ if S is half odd integer. Since the spin-orbit interaction is present in every atom, the electron energy levels corresponding to a given L and S are split into several levels, each of which corresponds to a particular value of J. For the lighter atoms the L-S coupling approximation is quite a good one. As an example, consider the element calcium which has two electrons outside the "closed shells."† The lowest value of L is 0 (called an S state), the next is $L = 1$ (called a P state), followed by $L = 2$ (D state) and so on. S can have two values, 0 and 1 (i.e., $\frac{1}{2} + \frac{1}{2}$ or $\frac{1}{2} - \frac{1}{2}$) since there are two electrons. Some of these states are labeled ${}^1S_0$, ${}^3S_1$, ${}^1S_1$, ${}^3P_{0,1,2}$, ${}^1D_2$, etc. The superscript μ (called the multiplicity) labels the spin according to the formula $\mu = 2S + 1$ ($\mu = 1$ is called singlet state, $\mu = 3$, a triplet state, etc.). It is equal to the number of J states produced by the splitting. The total angular momentum quantum number J is given by the subscript. An energy level diagram showing some of the levels of calcium is shown in Figure 2.6. The lowest lying state (${}^1S_0$ state in this case) is called the ground state.

If the spin-orbit interaction should be large compared to the coulomb potential between electrons and between nucleus and electrons, the L-S coupling model is not valid and is replaced by one in which the **l** and **s** for each electron are combined into a resultant **j**. The **j**'s for all the electrons are

† A closed shell has no angular momentum.

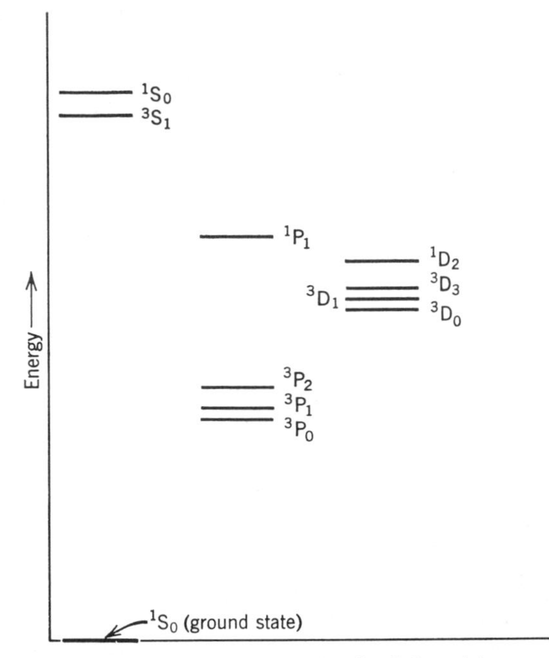

Figure 2.6 Some energy levels of the calcium atom.

then combined into a resultant **J** for the atom:

$$\mathbf{j}_i = \mathbf{l}_i + \mathbf{s}_i$$
$$\mathbf{J} = \sum_i \mathbf{j}_i \tag{2.112}$$

This scheme is usually referred to as the "$j - j$ coupling model." The foregoing is only a cursory outline of the rudiments of atomic theory; readers who desire a more complete account will find it in one of the references listed at the end of the chapter.

2.4.3 Molecular Structure

The underlying principles of molecular structure are identical to those of atomic structure, i.e., quantum mechanics and the Coulomb force law. The computational difficulties are much greater, however, because of the presence of more than one nucleus. This complication destroys the spherical symmetry which makes the mathematics of atomic structure relatively simple. Nevertheless, recent attempts to carry out programs for calculating the structure of molecules has met with considerable success.

Because of symmetry about the internuclear axis, diatomic molecules are much simpler than polyatomic ones. The magnitude of the electronic angular momentum is, of course, not conserved because the attractive force between electrons and the nuclei is no longer a central one. However, the projection of the total orbital angular momentum on the internuclear axis, Λ (the analogue of L_z) is a good quantum number; this offers a way in which to classify the rotational symmetry of the molecule.† By analogy with the nomenclature of atomic spectra, the Greek characters Σ ($\Lambda = 0$), Π ($\Lambda = 1$), $\Delta(\Lambda = 2)$, etc., are employed. Under certain conditions reflection symmetries may also be present. There are two possible reflection planes; one bisects the internuclear axis and is perpendicular to it; the other contains the internuclear axis. Reflection symmetry exists for the first plane only if the molecule is homonuclear. If so, the wave function can either change sign or remain the same when reflected; the corresponding labels are *u* (*ungerade*) and *g* (*gerade*). In the case of the second plane, reflection symmetry can exist only for the cylindrically symmetric Σ states. The two possibilities are again a change in the sign of the wave function (−) or no change (+). Finally, the projection on the internuclear axis of the total electronic spin Σ is a conserved quantity. We now have all the information necessary to label the electronic symmetries of the molecule. An example is ${}^3\Sigma_g^+$ which indicates that $\Sigma = 1$, $\Lambda = 0$, and the wave function has symmetries *g* (*gerade*) and (+) under the two reflections mentioned earlier. It is also customary to indicate something about the electronic energy state. This is done by means of a prefix-X if the molecule is in the electronic ground state, a letter near the beginning of the alphabet if it is excited. Examples for the N_2 molecule are $X^1\Sigma_g^+$ and $C^3\Pi_u$. The projection on the internuclear axis of the total angular momentum Ω is also a conserved quantity; it is obtained by simple algebraic addition of Σ and Λ:

$$\Omega = |\Sigma + \Lambda| \tag{2.113}$$

In addition to electronic excitation, molecules can experience vibrational and rotational excitation. Most diatomic molecules can be approximated by a simple harmonic oscillator. The energies E_v of the vibrational levels are thus given approximately by

$$E_v = h\nu_0(\mathrm{v} + \tfrac{1}{2}) \tag{2.114}$$

where ν_0 is the fundamental frequency of the oscillator and v is an integer (or 0 for the vibrational ground state). In practice the interatomic potential is anharmonic so that Eq. (2.114) also contains terms which are nonlinear in v. The nonlinear terms are particularly important for the highly excited vibrational states. This aspect is illustrated in Figure 2.7 which shows two potential energy curves for a hypothetical diatomic molecule. When the two atoms come together, their potential energy follows one of these curves.

† It is assumed that the molecular analogue of L-S coupling is a valid approximation.

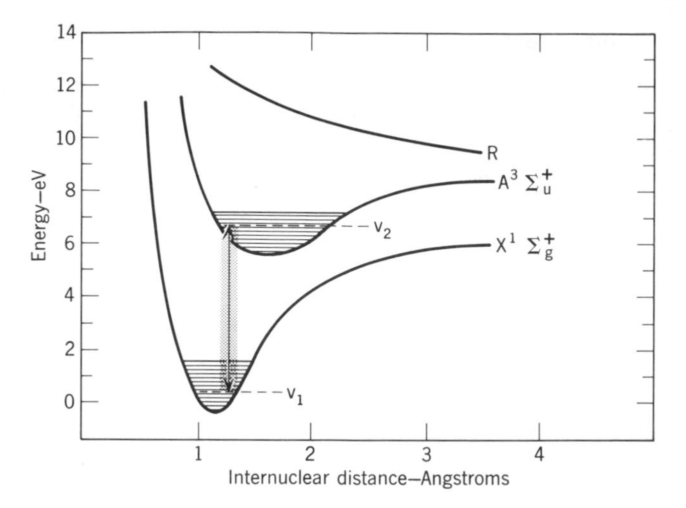

Figure 2.7 Potential energy curves of a hypothetical diatomic molecule. The curve R is repulsive, the others attractive. The electronic transition from the v_2 vibrational state of $A^3\Sigma$ to the v_1 state of $X^1\Sigma$ is shown as a vertical arrow. The region of overlap of the two vibrational levels is denoted by the shaded band.

Near the bottom, the curves have shapes very close to parabolic (pure harmonic potential), but at higher energies the deviation therefrom is quite marked. Electronic transitions can occur as shown in the diagram; when they do, the nuclei, because of their large mass, behave during the transition as if they had no relative motion. Consider a transition from vibrational level v_2 of electronic state A to level v_1 of electronic state X. In order that it may occur, at least portions of the vibrational levels must overlap as indicated by a vertical band. This rule is related to the Frank-Condon principle which states that the internuclear distance is essentially the same after the electronic transition as before. Those transitions are most favored in which the internuclear distance in the final state is nearly equal to that in the initial state.

Finally, we mention the rotational states of diatomic molecules. These molecules can rotate freely about axes perpendicular to the internuclear axis with energies given by

$$E_r = \frac{J(J + 1)\hbar^2}{2I} \tag{2.115}$$

where J is the angular momentum quantum number of the free rotor and I is the moment of inertia about the rotational axis. Typical spacings of rotational energy levels of diatomic molecules are of the order of a few ten thousandths of an eV.

2.4.4 *Atomic Spectra*

The radiation emitted by atoms is characterized by spectral *lines* at frequencies related to the energy difference between the associated electron states by Eq. (2.91). Whether or not such a transition is possible is dependent upon certain *selection rules.* One computes the intensity I of the lines from the so-called transition matrix elements $\int \psi_f \mathbf{R} \psi_i \, d\tau$ in which ψ_i and ψ_f are the initial and final state functions of the atom and $\mathbf{R}$ is an operator which induces the transition:

$$I \propto \left| \int \psi_f \mathbf{R} \psi_i \, d\tau \right|^2 \equiv |\langle f | \, \mathbf{R} \, | i \rangle|^2 \tag{2.116}$$

If the matrix element vanishes, the transition is not possible. The simplest (and most rapid) type of transition is due to the *electric dipole interaction* in which $\mathbf{R}$ is just the position operator $\mathbf{r}$ for an electron multiplied by the electronic charge e. The next simplest (and much slower) are the magnetic dipole and electric quadrupole interactions. In the first, $\mathbf{R}$ is the magnetic dipole operator

$$\mathbf{R} = \frac{e\hbar}{2m}(\mathbf{L} + 2\mathbf{S}) \tag{2.117}$$

in which e and m are the charge and mass of the electron, and $\mathbf{L}$ and $\mathbf{S}$ are the orbital and spin angular momentum operators, respectively. In the second, $\mathbf{R}$ is the electric quadrupole operator

$$R_m = er^2 Y_{2,m}(\theta, \phi) \tag{2.118}$$

in which $Y_{2,m}$ is the spherical harmonic of order $(2, m)$ and (θ, ϕ) are the angular coordinates of the electron undergoing the transition. One can easily prove the selection rules presented in Table 2.2 from quantum mechanical principles. An example of each is presented at the bottom of the appropriate column.

As we mentioned earlier, electric dipole transitions are much faster than the electric quadrupole and magnetic dipole emissions. Typically the mean life associated with the former are of the order 10^{-8} second whereas atoms and molecules which emit via the latter mechanisms may have lifetimes ranging from 10^{-5} to one second. For example, the ^1S state of atomic oxygen has a mean life of about 0.77 sec. Such states are said to be *metastable.*

2.4.5 *Molecular Spectra*

The emission of radiation by molecules also obeys the same general quantum principles set forth in the preceding subsection. However, because

TABLE 2.2. SELECTION RULES FOR L-S COUPLING

	Electric Dipole	Electric Quadrupole	Magnetic Dipole
S	$\Delta S = 0$	$\Delta S = 0$	$\Delta S = 0$
L	$\Delta L = 0, \pm 1$ (no $0 \to 0$)	$\Delta L = 0, \pm 1, \pm 2$ (no $0 \to 0$ or $0 \leftrightarrow 1$)	$\Delta L = 0$
J	$\Delta J = 0, \pm 1$ (no $0 \to 0$)	$\Delta J = 0, \pm 1, \pm 2$ (no $0 \to 0$ or $0 \leftrightarrow 1$)	$\Delta J = 0, \pm 1$
parity (π)	$\Delta \pi$ = Yes	$\Delta \pi$ = No	$\Delta \pi$ = No
Example	$Na(^2P_{0,1}) \to Na(^2S_1)$	$O(^1S_0) \to O(^1D_2)$	$O(^1D_2) \to O(^3P_{2,1})$
	$\Delta S = 0$	$\Delta S = 0$	$\Delta S = 1$†
	$\Delta L = -1$	$\Delta L = 2$	$\Delta L = -1$
	$\Delta J = 0$ or 1	$\Delta J = 2$	$\Delta J = 0, -1$
	$\Delta \pi$ = Yes	$\Delta \pi$ = No	$\Delta \pi$ = No

† Obviously, the requirement that $\Delta S = \Delta L = 0$ is not satisfied by this transition; it is an example of an *intercombinational* line and results from the approximate nature of the L-S coupling model. Because of spin-orbit interaction the spin S and orbital angular momentum L are not, strictly speaking, good quantum numbers.

of the somewhat different symmetries involved, the selection rules are not quite the same. Each electronic state has numerous vibrational levels embedded in it. Transitions can occur between a given vibrational state and any vibrational level in the final state. The result is a series of closely spaced spectral lines called a *band*. There is also band structure associated with pure vibrational transitions. These *vibration-rotation bands* occur because the rotational energy states of a molecule are not equally spaced [see Eq. (2.115)].

The electric dipole selection rules for electronic transitions in molecules are

$$\Delta \Sigma = 0$$
$$\Delta \Lambda = 0, \pm 1$$
$$\Delta \Omega = 0, \pm 1 \text{ (no } 0 \to 0)$$
$$u \leftrightarrow g$$
$$+ \leftrightarrow - \quad (\Sigma \text{ states only})$$

2.5 Collision Processes

Collision processes of various types are particularly important in the physics of the upper atmosphere. They include collisions which lead to chemical and ionic reactions; momentum transfer (elastic) collisions between electrons and atoms or molecules; ionization of neutral particles by fast charged particles; rotational, vibrational, and electronic excitation by charged particle impact; etc. We shall also discuss various photo processes such as photoionization in the present section although, strictly interpreted, they are not collision phenomena. Reactions, on the other hand, will be deferred to Section 2.6.

2.5.1 Collision Theory

Collisions between particles in the atomic domain, e.g., between atoms and electrons, can be classed according to the change in their kinetic energy and to their identity. If the kinetic energy and identity are unchanged, the collision is said to be *elastic*. If the kinetic energy does change so that one or more of the particles is excited, the collision is called inelastic. If the identity of the particles changes, we speak of a *rearrangement* collision. Actually the definition of *rearrangement* collisions is somewhat more complicated than this implies, but our definition is adequate for present purposes. It is convenient to treat collisions theoretically in the center of mass coordinate system (vanishing total momentum) and this convention is adopted in the following.

The physical quantity which is fundamental to all collision processes is the *cross section*. The *differential* cross section $dQ/d\Omega$ is defined as

$$\frac{dQ}{d\Omega} = \frac{\text{number of particles scattered into a unit solid angle in unit time}}{\text{number of incident particles per unit area per unit time}} \tag{2.119}$$

It is apparent that the dimensions are area per unit solid angle. One can show that $dQ/d\Omega$ can be expressed in terms of a scattering angle Θ and an impact parameter b as

$$\frac{dQ}{d\Omega} = \left| \frac{b}{\sin \Theta} \frac{db}{d\Theta} \right| \tag{2.120}$$

The meaning of Θ and b is evident from Figure 2.8. Loosely speaking, b would be the closest distance of approach of the incident particle if the scattering force were "turned off." It is frequently convenient to use quantum theory to compute a probability $\mathscr{P}$ of scattering in terms of the impact parameter. For a spherically symmetric scattering force the *total cross section* is

$$Q = \pi \int \mathscr{P}(b) b \, db \tag{2.121}$$

The latter is just the integral of $dQ/d\Omega$ over a complete solid angle

$$Q = \int \left(\frac{dQ}{d\Omega} \right) d\Omega \tag{2.122}$$

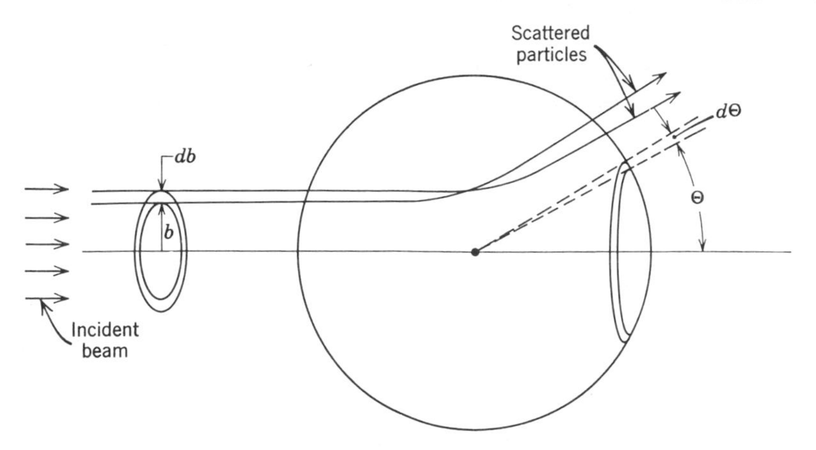

Figure 2.8 Scattering of a beam of particles by a spherically symmetric center of force.

It also proves useful to define a *momentum transfer* cross section

$$Q_m = \int \frac{dQ}{d\Omega}(1 - \cos \Theta)\, d\Omega \tag{2.123}$$

which weights most heavily those collisions in which momentum is to be exchanged (i.e., ones in which the incident particle is scattered backward).

As a simple but interesting and useful application of Eq. (2.121) consider the case of two particles, one neutral and the other ionic, which are in a quasi-stationary orbit[1]. The potential energy due to the attractive polarization force is

$$V(r) = -\frac{\alpha e^2}{2r^4} \tag{2.124}$$

where α is the polarizability of the neutral particle, e is the electronic charge, and r is the distance between the particles. Their kinetic energy of orbital motion is

$$K = \frac{J^2}{2\mu r^2} \tag{2.125}$$

where J is the angular momentum, and μ is the "reduced mass"† of the system. K is equivalent to a potential barrier which must be surmounted if the reaction is to occur. Thus the distance of closest approach for which reaction occurs is equal to the separation r_0 at which the barrier energy is maximum and is given by the condition

$$\frac{\partial E}{\partial r} = 0 \tag{2.126}$$

† If m_1 and m_2 are the masses of the two particles, $\mu^{-1} = m_1^{-1} + m_2^{-1}$.

where

$$E = K + V \tag{2.127}$$

Since J is related to the impact parameter b and initial relative velocity v by

$$J = \mu v b, \tag{2.128}$$

it is easy to show that

$$b_0 = \sqrt{2}\, r_0 \tag{2.129}$$

Using the result of Eq. (2.126) we find that the impact parameter at the top of the barrier b_0 is related to v by the equation

$$b_0 = \left(\frac{4e^2\alpha}{\mu v^2}\right)^{1/4} \tag{2.130}$$

Using Eq. (2.121) and assuming that $\mathscr{P} = 1$ if $b \leq b_0$ and $\mathscr{P} = 0$ if $b > b_0$, we obtain the reaction (i.e., rearrangement) cross section

$$Q = \pi b_0^2 = \frac{2\pi}{v}\frac{\sqrt{\alpha e^2}}{\mu} \tag{2.131}$$

This approach has been applied with some success to ion-molecule reactions.

The absorption as well as scattering of a beam of incident particles (or photons) can be described quantitatively by a cross section. Hence, we speak of an *absorption cross section* which is defined as

$$Q_a = \frac{\text{number of particles removed from beam in unit time}}{\text{number of incident particles per unit area per unit time}} \tag{2.132}$$

Q_a is related to the flux density of particles (or photons) which enter and emerge, I_0 and I, respectively, from an absorbing medium of thickness x and particle number density n by the equation†

$$I = I_0 \exp(-Q_a n x) \tag{2.133}$$

We shall make considerable use of this equation in Chapters 3, 8, 9, and 10 when we discuss photoabsorption in the upper atmosphere.

Collision dynamics can be approached via classical or quantum mechanics. Of course, the latter is really the correct way in which to describe atomic processes, but very frequently (i.e., in the large quantum number limit) classical mechanics is adequate. In the case of a spherically symmetric potential $V(\rho)$ it can be shown[5] that the scattering angle is related to the asymptotic velocity v_0, the ratio $\rho_0 = b/r_0$ of impact parameter b to distance of closest approach r_0, and the reduced mass μ by the equation

$$\Theta = \pi - 2\int_0^{\rho_0} \frac{d\rho}{\left[1 - \rho^2 - \dfrac{2V(\rho)}{\mu v_0^2}\right]^{1/2}} \tag{2.134}$$

† Here we ignore the effect of spectral line shape.

An interesting and instructive application of Eq. (2.134) is the collision of two hard spheres (billiard balls). Computation of the corresponding cross section is left as an exercise.

As we discussed in the preceding section, the behavior of an atomic or molecular system is completely described by its state function ψ. In the case of two colliding particles we are interested principally in the asymptotic state functions, i.e., the state functions long before and long after collision. The condition of the system during close encounter is of little interest. For the collision of two structureless particles we write ψ as a sum of two terms

$$\psi = e^{i\mathbf{k}\cdot\mathbf{r}} + \frac{e^{ikr}}{r} f(\Theta), \qquad r = |\mathbf{r}|, \qquad k = |\mathbf{k}| \tag{2.135}$$

in which the first represents the system initially and the second represents it long after the interaction has occurred. Here $\mathbf{k}$ is the wave vector of the particles before collision, $\mathbf{r}$ is their relative position vector, and $f(\Theta)$ is the *scattering amplitude* for scattering angle Θ. It is usually convenient in scattering problems to express the Schrödinger equation for ψ as an integral equation

$$\psi = e^{i\mathbf{k}\cdot\mathbf{r}} - \frac{\sqrt{2\pi}}{\hbar^2} \mu \frac{e^{ikr}}{r} \int e^{-i\mathbf{k}'\cdot\mathbf{r}'} V(\mathbf{r}')\psi(\mathbf{r}')\, d\mathbf{r}' \tag{2.136}$$

which, by comparison with Eq. (2.135), yields an expression for $f(\Theta)$

$$f(\Theta) = -\frac{\sqrt{2\pi}}{\hbar^2} \mu \int e^{-i\mathbf{k}'\cdot\mathbf{r}'} V(\mathbf{r}')\psi(\mathbf{r}')\, d\mathbf{r}' \tag{2.137}$$

A particularly useful approximate solution to this equation is obtained if we assume that ψ is affected only slightly by the scattering process and can be replaced by a plane wave $e^{i\mathbf{k}\cdot\mathbf{r}''}$. It is obvious that this *Born approximation* to the scattering amplitude is proportional to the Fourier transform of the potential. Use of the Born approximation in atomic physics has been quite successful in cases where the kinetic energy is not too small. The flux F_i per unit area of the incoming particles is

$$F_i = \frac{|e^{i\mathbf{k}\cdot\mathbf{r}}|^2}{v} = \frac{\mu}{\hbar k} \tag{2.138}$$

where v is the relative speed and μ is the reduced mass. The outgoing flux dF_0 in an element of solid angle $d\Omega$ is

$$dF_0 = \left| \frac{e^{ikr}}{r} f(\Theta) \right|^2 r^2 \frac{\mu}{\hbar k}\, d\Omega = |f(\Theta)|^2 \frac{\mu}{\hbar k}\, d\Omega \tag{2.139}$$

From the definition of differential cross section given in Eq. (2.119), one readily arrives at the relation

$$\frac{dQ}{d\Omega} = |f(\Theta)|^2 \tag{2.140}$$

Should the collision be an inelastic one, Eq. (2.13)6 must be modified slightly:

$$\frac{dQ}{d\Omega} = \frac{k_i}{k_f} |f(\Theta)|^2 \tag{2.141}$$

where k_i and k_f are respectively the incoming and outgoing wave numbers. The computation of cross sections, or equivalently, scattering amplitudes by quantum mechanical means is fraught with difficulties in cases in which the Born approximation is not valid (low energies). Nevertheless, considerable success has been attained in many cases such as collisions of electrons with atoms.

2.5.2 *Photo Processes*

Photo processes include the excitation and ionization of atoms and molecules as well as molecular dissociation by the absorption of photons. In principle all of these processes can be computed by use of quantum mechanics. In practice, only the simpler cases, i.e., photoabsorption by atoms, are amenable to such an approach. Photo processes involving molecules must be measured experimentally. In the electric dipole approximation the cross section is given by the equation

$$Q = \frac{8\pi^3\nu}{3c} \sum_{i,f} \frac{1}{g_i} \left| \int \psi_f^* e\mathbf{r}\psi_i \, d\tau \right|^2 \tag{2.142}$$

Here ν is the frequency of the emitted radiation, c is the velocity of light, and g_i is the weight function of the initial state i; the final state is labeled f. In the case of ionization, the energy E_p of the photo electron is related to the photon energy $h\nu$, the ionization potential IP and the excitation energy E_e of the atom or molecule in its final state by

$$E_p = h\nu - IP - E_e \tag{2.143}$$

In most collision processes, there is a fairly large probability that the atom or molecule will be left in an excited state if there is enough energy available. The photoionization cross section for atomic oxygen is shown in Figure 2.9 as a function of wavelength of the incident radiation.

The photodissociation of molecules has never been successfully calculated because of the great complexity of the molecular wave function. In order to be dissociated, the molecule must transit from a bound state characterized by the lower two potential energy curves shown in Figure 2.7 to a repulsive

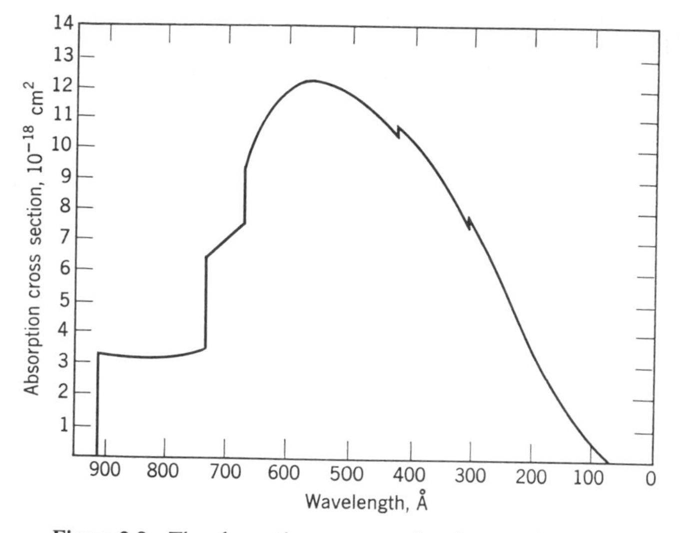

Figure 2.9 The absorption cross section for atomic oxygen in the "dipole length" approximation [Dalgarno, Henry and Stewart, *Planet. Space Sci.*, **12**, 235 (1964)].

potential energy curve such as R in the same figure. In general, one of the emerging atoms is in an excited state. For example, the oxygen molecule which has a very large dissociation cross section in the range 1300–1759 Å (Schumann-Runge continuum) commonly yields an atom each of $O(^3P)$ and $O(^1D)$ if the incident photon is of energy >7.08 eV. However, if $5.11 < h\nu < 7.08$ eV both atoms are in the ^3P state. It is also possible for the oxygen molecule to be raised to the excited state $^3\Sigma_u^+$ by absorption of radiation in the Schumann-Runge absorption bands ($1930 > \lambda > 1759$ Å); the corresponding potential energy curve crosses that corresponding to the repulsive state $^3\Pi_u$. The $^3\Sigma_u^+$ state can undergo a radiationless transition to the $^3\Pi_u$ state at the point where the potential energy curves cross. If it does so, the molecule dissociates into two $O(^3P)$ atoms. We call this process predissociation. The N_2 molecule is not so readily dissociated by absorption of radiation because of the great strength of the chemical bonding. Photoabsorption coefficients for O_2 and N_2 are shown in Figures 2.10 and 2.11 as functions of the wavelength of the incident radiation; the absorption coefficient is defined as the absorption cross section per unit concentration of the absorber.

Finally, we mention the production of *bremsstrahlung* (braking radiation) by electron bombardment of atmospheric species. When energetic electrons (i.e., energy above a few keV) pass close to the nucleus of the target atom, they are deflected by the electric field. The associated acceleration produces a photon, usually in the X-ray range. We shall not discuss the production of

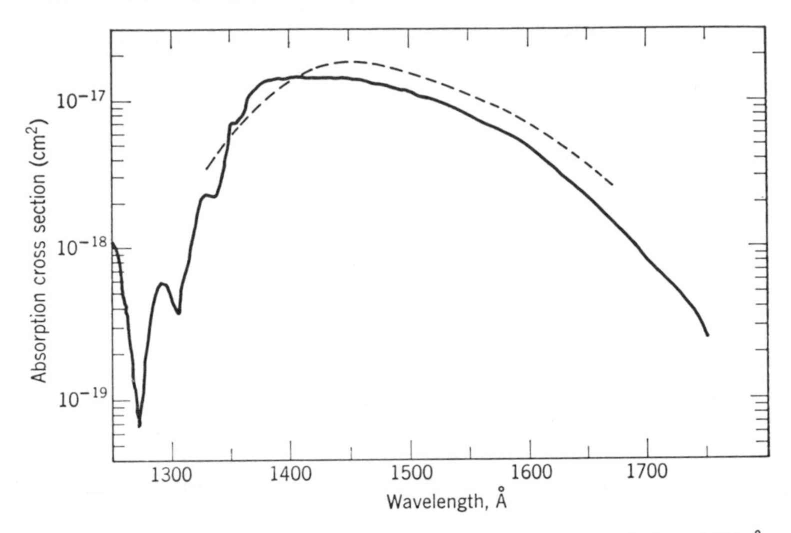

Figure 2.10 Absorption cross section of O_2 in the region 1250 to 1750 Å. Broken curve by Ladenburg and Van Voorhis [*Phys. Rev.* **43,** 315 (1933)]. Solid curve by Watanabe, Inn, and Zelikoff [*J. Chem. Phys.* **21,** 1026 (1953)]. [After K. Watanabe, *Adv. Geophys.* **5,** 153 (1958); reprinted by permission of Academic Press, Inc.]

bremsstrahlung further, but refer the reader to Chapter 10 and one of the references listed at the end of this chapter.

2.5.3 Passage of Charged Particles Through Matter

When charged particles such as electrons, protons, and alpha particles pass through matter, they lose energy through excitation and ionization of the atomic and molecular constituents. The details of the various mechanisms could be studied by evaluating the various cross sections as functions of the bombarding energy. However, there are numerous complicating factors which make this approach rather impractical. Such effects include large angle scattering in the case of electrons, and charge capture and loss processes in the case of protons and alpha particles. Typical capture and loss processes are

$$H^+ + N_2 \rightarrow H + N_2^+ \tag{2.144}$$

and

$$H + N_2 \rightarrow H^+ + N_2 + e \tag{2.145}$$

It should be remembered that the neutral hydrogen atoms in the above reactions have velocities of the same order as the incident protons.

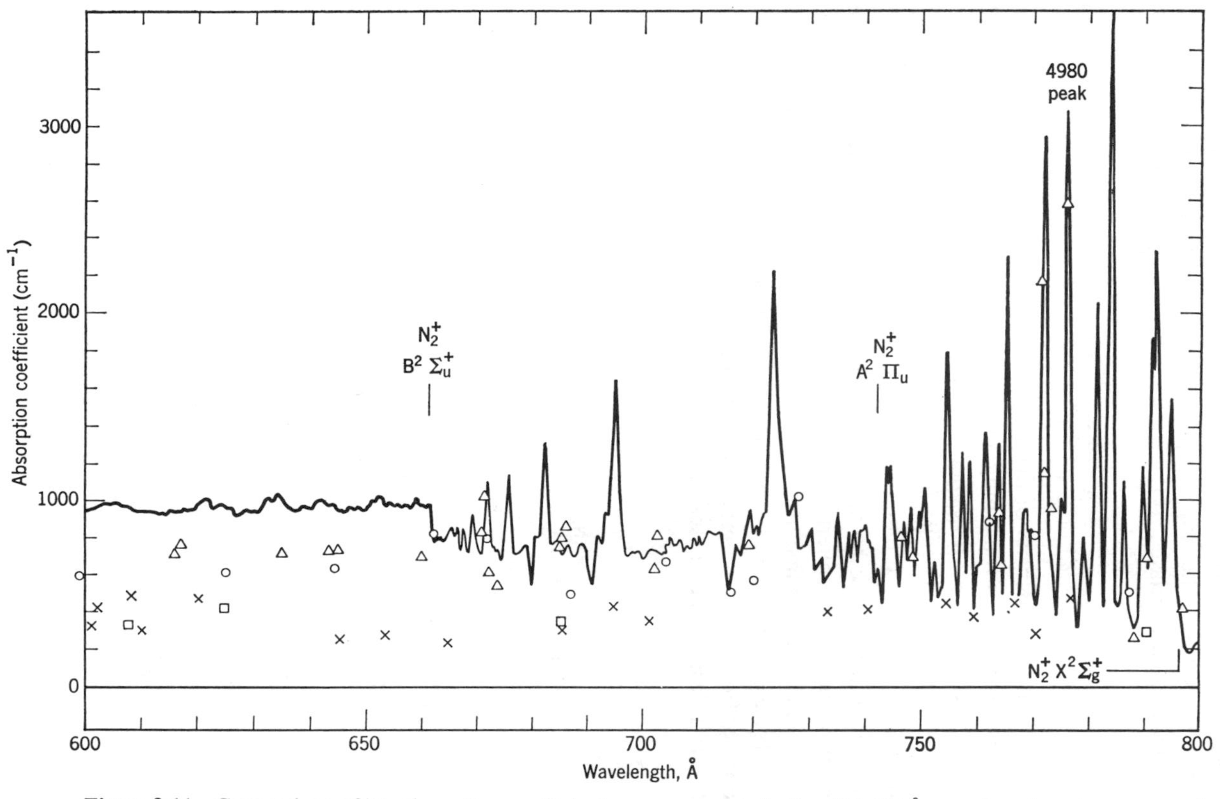

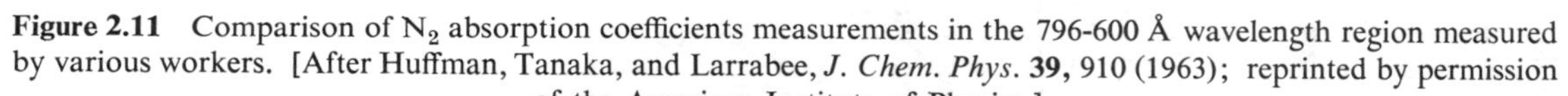

Figure 2.11 Comparison of N_2 absorption coefficients measurements in the 796-600 Å wavelength region measured by various workers. [After Huffman, Tanaka, and Larrabee, *J. Chem. Phys.* **39**, 910 (1963); reprinted by permission of the American Institute of Physics.]

As a result of such complications, the slowing down of charged particles is usually given in terms of the *stopping power* or energy loss per unit path length in the medium. The stopping power for heavy particles (e.g., protons and alpha particles) is related to the speed v and charge $Z'e$ by

$$-\frac{dE}{dx} = N\frac{4\pi(Z'e^2)^2}{mv^2} Z \ln \frac{2mv^2}{I} \tag{2.146}$$

where N is the number density of the absorber atoms and molecules, m is the mass of the electron, Z is the atomic number of the absorber material, and I is a sort of "average excitation energy." The stopping power for electrons in the nonrelativistic limit (valid for energies <100 keV) is given by a similar equation if we set $Z' = 1$ and replace the argument of the logarithm by $(mv^2/2)\sqrt{e/2}$ (e is the base of the natural logarithms). Some representative stopping powers of atmospheric interest are shown in Figures 2.12 and 2.13.

In considering bombardment of the atmosphere by cosmic rays and auroral electrons, for example, one of the physical quantities of great interest is the specific ionization rate q (number of electron-ion pairs per unit volume per second). It is easily shown that for a beam of monoenergetic protons or

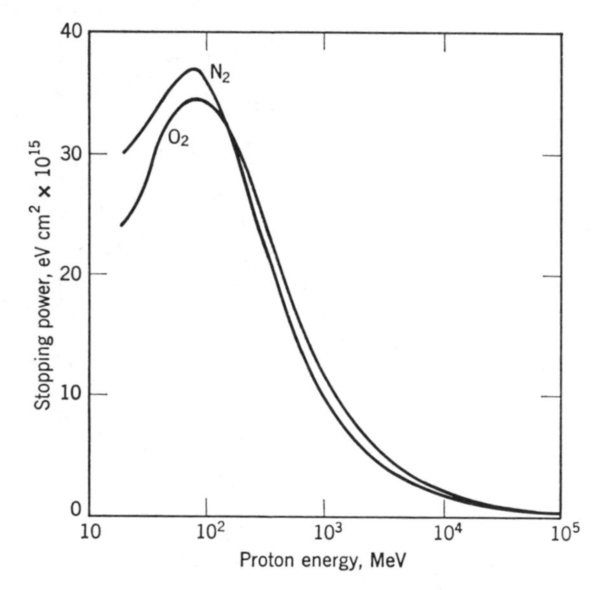

Figure 2.12 Stopping power of molecular nitrogen and oxygen for protons. (After Whitten and Poppoff, *Physics of the Lower Ionosphere*, © 1965; reprinted by permission of Prentice-Hall, Inc., Englewood Cliffs, N.J.)

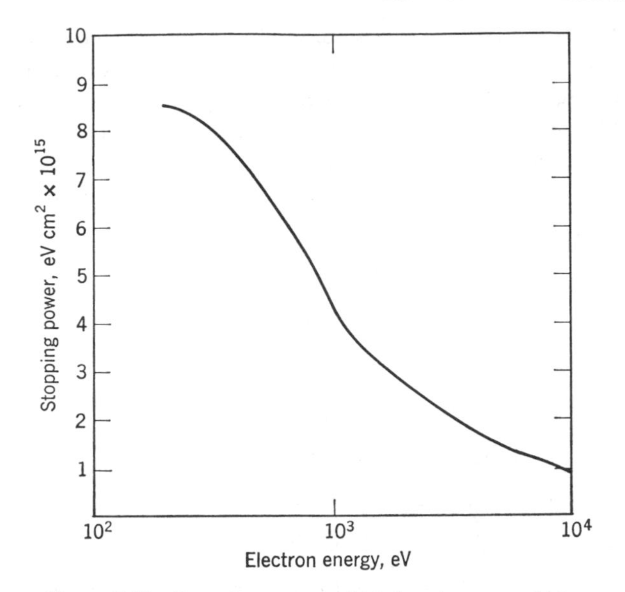

Figure 2.13 Stopping power of air for electrons. (After Whitten and Poppoff, *Physics of the Lower Ionosphere*, © 1965; reprinted by permission of Prentice-Hall, Inc.)

electrons, q is given by

$$q = \left(-\frac{dE}{dx}\right)\frac{F}{W_0} \tag{2.147}$$

where F is the beam intensity (number of charged particles per unit area per unit time) and W_0 is very close to 35 eV for proton and electron energies greater than 100 eV.

2.5.4 Collisions of Low Energy Electrons

The collisions of low energy electrons with atoms and molecules which occur in planetary atmospheres are important for several reasons. First, they control the cooling rate of the electron gas in regions where the neutral particle density is much greater than the ion number density. Secondly, they control the rate at which the ambient electron gas in an ionosphere extracts energy from an electromagnetic wave in the radio frequency range and transforms it into the kinetic energy of random molecular motion. Thirdly, the vibrational modes of ground state nitrogen molecules are quite efficiently excited by impact of electrons with energies of a few eV. The rate constant for the reaction

$$O^+ + N_2 \rightarrow NO^+ + N \tag{2.148}$$

is much larger if the N_2 molecules are vibrationally excited. This reaction is very important in the earth's F_2 layer. The first and third are due to inelastic collisions, the second to elastic (momentum transfer effects) collisions.

As we shall see in Chapter 11, the principal quantity of interest in the absorption of energy from radio waves is the momentum transfer collision frequency ν_m. The collision frequency is related to the momentum transfer cross section Q_m by the equality

$$\nu_m = \overline{NvQ_m} \qquad (2.149)$$

where N is the molecular or atomic number density and v is the electron speed. Phelps and coworkers[6,7] have carried out extensive investigations of the collision frequency of electrons in N_2, O_2, and CO_2 by use of electron "swarm" methods. For example, they obtained

$$\nu_m(N_2) = N \times 1.4 \times 10^{-7}\, E\ \text{cm}^{-3}\ \text{sec}^{-1} \qquad (2.150a)$$

$$\nu_m(O_2) = N \times 5 \times 10^{-9}\, E^{1/2}\text{cm}^{-3}\ \text{sec}^{-1} \qquad (2.150b)$$

$$\nu_m(CO_2) = N \times 8 \times 10^{-8}\ \text{cm}^{-3}\ \text{sec}^{-1} \qquad (2.150c)$$

for energies less than $\sim$1 eV; E is the electron energy expressed in eV.

The cooling rate of the electron gas at D-layer heights (70–90 km) plays a very important role in the phenomenon called cross-modulation (see Section 11.5.4). The energy lost by the electrons, which are actually quite slow ($<$1 eV), goes into the rotational excitation of nitrogen molecules. The cross section for this has been computed[8] using the Born approximation and the electric quadrupole interaction between the electron and the molecule. It was found to be about $2\frac{1}{2}$ orders of magnitude less than the momentum transfer cross section.

Finally we mention the vibrational excitation of N_2 by electron impact. This process has an unusually large cross section at about 2 eV because of the formation of a virtual negative ion.[9] The lifetime of this state against autodetachment is of the order 5×10^{-16} seconds which corresponds to a resonance width of about 1 eV.

2.6 Reactions

In this section we study collision phenomena among large numbers of particles, neutral or charged, which are in or close to thermal equilibrium. It is impossible to understand the chemistry of the upper air and ionosphere without a knowledge of the rate equations and rate constants discussed in the following subsections.

2.6.1 Kinetics

The term "kinetics" refers to the study of the change in time of the number densities of components of a chemical system. For two-body (bimolecular) reactions exemplified by

$$X + Y \rightarrow \text{products} \tag{2.151}$$

The rate equation for X is

$$\frac{d[X]}{dt} = -k_2[X][Y] \tag{2.152}$$

since the rate $d[X]/dt$ is proportional to the concentration of each of the reactants. The square brackets refer to number density or concentration and k_2 is the *rate constant* (in $cm^3\ sec^{-1}$) for the process. Since the rate constant is equal to the product of reaction cross section and relative molecular speed averaged over the speed distribution,

$$k = \overline{vQ_k} \tag{2.153}$$

it is in effect a collision frequency per unit concentration. If there were a source q of X molecules per unit volume per unit time, Eq. (2.152) would become

$$\frac{d[X]}{dt} = q - k_2[X][Y] \tag{2.154}$$

By similar reasoning the rate equation for a three-body or trimolecular reaction such as

$$X + Y + Z \rightarrow \text{products} \tag{2.155}$$

is written

$$\frac{d[X]}{dt} = -k_3[X][Y][Z] \tag{2.156}$$

where k_3 is given in units like $cm^6\ sec^{-1}$. The derivation of a rate equation for a component X which participates in a sequence of reactions such as

$$A + B \rightarrow X + Y \tag{2.157a}$$

followed by

$$X + Z \rightarrow \text{products} \tag{2.157b}$$

is left as an exercise.

One frequently wishes to know how long it takes the concentration of a reactant such as X to decrease by some factor, e.g., $\frac{1}{2}$, $1/e$, etc. We call such a characteristic time the *lifetime* or the *relaxation time* of species X. Only in the case of a unimolecular reaction of the type

$$X \rightarrow Y + Z \tag{2.158}$$

is the lifetime independent of the concentrations of the reactants. In this case

$$\frac{d[X]}{dt} = -\frac{1}{\tau}[X] \tag{2.159}$$

for which the solution is

$$[X] = [X]_{t=0} \exp\left(-\frac{t}{\tau}\right) \tag{2.160}$$

When $t = \tau$, $[X]/[X]_{t=0} = e^{-1}$; thus the lifetime τ of the species is a well-defined quantity. In the case of reaction (2.154) we *define* τ as

$$\tau = (k_2[Y])^{-1} \tag{2.161}$$

Unless $[Y]$ is held fixed during the reaction, τ is a continuously varying parameter.

2.6.2 *Ionic Recombination*

Positive ions can be removed from a plasma by recombining with electrons or with negative ions. In the first mechanism, the energy which must be lost by the electron when it recombines can be carried off by the emission of a photon. If so, we call the process radiative recombination. At ionospheric temperatures the radiative mechanism is quite slow; the rate constant is only of the order 10^{-12} cm^3 sec^{-1}. As a result, it is of little importance in aeronomy. This is the only mode of electronic recombination for atomic ions, but for molecular ions the energy can be carried off by dissociation; for example,

$$e + XY^+ \rightarrow X + Y \tag{2.162}$$

The associated rate constant α_D, usually called the dissociative recombination coefficient, is quite large; typically it is of the order 10^{-7} to 10^{-6} cm^3 sec^{-1} for diatomic or triatomic ions at 300°K.

The second type of positive ion removal mechanism is exemplified by

$$O_2^+ + O_2^- \rightarrow O_2 + O_2; \qquad X^+ + Y^- \rightarrow X + Y \tag{2.163}$$

We thus call the process ion-ion recombination or mutual neutralization. Although very little experimental and theoretical work has been done to evaluate ion-ion recombination coefficients, present indications are that they are of the same order of magnitude as α_D.

If both electrons and negative ions are present, one can write the rate equation for the positive ions as

$$\frac{d[X^+]}{dt} = q - \alpha_D[e][X^+] - \alpha_i[Y^-][X^+] \tag{2.164}$$

where q represents the specific rate of formation of X^+ ions.

2.6.3 *Negative Ions and Their Reactions*

Most neutral atoms and molecules do not attach electrons but a few, particularly the halogens, have large electron affinities. There are a number of modes of negative ion formation of which *radiative attachment* is the simplest, e.g.,

$$X + e \rightarrow X^- + h\nu \tag{2.165}$$

Typical rate constants are of the order 10^{-15} cm^3 sec^{-1}. Other mechanisms are the three-body collision

$$e + O_2 + M \rightarrow O_2^- + M \tag{2.166}$$

rearrangement collisions, such as

$$O_2^- + CO_2 \rightarrow CO_3^- + O \tag{2.167}$$

and charge transfer reactions like

$$O^- + NO_2 \rightarrow O + NO_2^- \tag{2.168}$$

Electrons can also be detached from negative ions. For example, the inverse of reaction (2.165), called photodetachment, is quite rapid in the presence of visible range sunlight for species with small electron affinities like O^- and O_2^-. Physically it is very similar to photoionization. Other detachment processes of possible importance in the upper atmosphere are

$$O_2^- + O \rightarrow O_3 + e \tag{2.169}$$

(associative detachment), and

$$N + NO_2^- \rightarrow N_2 + O_2 + e \tag{2.170}$$

2.6.4 *Ion-Molecule Reactions*

These processes are conveniently classified according to whether a molecular rearrangement occurs, or an electron is merely transferred from one molecule or atom to another, with both particles retaining their molecular identity:

$$O_2 + N_2^+ \rightarrow O_2^+ + N_2 \qquad X^+ + Y \rightarrow X + Y^+ \tag{2.171}$$

Rate constants for such processes have an upper limit of about 10^{-9} cm^3 sec^{-1} but may be much smaller.

A number of charge rearrangement collisions of the type

$$XY^+ + Z \rightarrow X + YZ^+ \tag{2.172}$$

have also been investigated in the laboratory. Attempts to compute them have been made but only a few very simple approaches such as that outlined in Section 2.5.1 have met with any degree of success. In Eq. (2.131) we presented a formula for an ion-molecule reaction cross section

$$Q = \frac{2\pi}{v}\left(\frac{\alpha e^2}{\mu}\right)^{1/2} \tag{2.131}$$

under the assumption that a reaction occurred for every collision in which the centrifugal barrier was surmounted. If the cross section Q is multiplied by the relative velocity v, the corresponding rate constant is obtained

$$k = \overline{vQ} = 2\pi\left(\frac{\alpha e^2}{\mu}\right)^{1/2} \tag{2.173}$$

(note that k is independent of the velocity distribution). This formula may be regarded as an upper limit on the actual value. In some cases such as

$$D_2 + D_2^+ \rightarrow D_3^+ + D \tag{2.174}$$

(where D denotes deuterium), the simple theory yields excellent results, but for most charge rearrangement reactions, the rate constant is greatly overestimated.

2.6.5 *Chemical Reaction Rates*

The theory of chemical reaction rates is quite complicated and we can do little justice to it in the space available. As we have already seen, any two-body rate constant k is really a collision frequency per unit concentration. In fact, it is the gas kinetic collision frequency multiplied by some probability factor that a reaction will occur. It can be shown[10] that k is proportional to the temperature and to another factor which is an exponential function of temperature

$$k = KT \exp(-F/RT) \tag{2.175}$$

This equation was first suggested in the last century by Arrhenius on purely empirical grounds. Here K is a constant which depends upon the nature of the reactants, R is the universal gas constant, and F is called the *free energy of activation*

$$F = H - TS \tag{2.176}$$

where H is enthalpy, and S is entropy (see Section 2.2). It represents a sort of "thermodynamic barrier" over which the molecules must pass in order to react.

As we shall see in Chapter 4, there are many chemical reactions of aeronomic importance, but they cannot all be discussed here because of lack of space. Instead we will merely mention as an example the reaction

$$O + O_3 \rightarrow O_2 + O_2 \tag{2.177}$$

which has the temperature-dependent rate constant

$$k = 5.0 \times 10^{-11} e^{-3000R/RT} \text{ cm}^3 \text{ sec}^{-1} \tag{2.178}$$

The quantity $F = 3000R$ may be regarded as the activation energy of the reaction; here R = Avogadro's number times the Boltzmann's constant.

In addition to the energy requirements of chemical reactions, there is one selection rule worth mentioning. It is called the Wigner spin conservation rule and states that the reaction is forbidden unless the spin of the reacting products can be formed by vector addition of the spins of the reactants. The rule applies only to light atoms and molecules (where the L-S coupling scheme is valid) and even then is violated quite often.

PROBLEMS

2.1. Use Ampère's Law [Eq. (2.4)] and Eq. (2.6) for the force exerted by a magnetic field on a moving charge to derive an expression for the force per unit length acting between two very long current-carrying wires.

2.2. Prove that **E** and **H** are perpendicular to the wave propagation vector **k** in free space. Show that, in general, **E** is not perpendicular to **k** in a plasma.

2.3. Derive Eq. (2.20) for the time-averaged Poynting vector.

2.4. Find the relationship between phase and group velocity in a dispersive medium [i.e., one in which the former is a function of wave number: $c = c(k)$].

2.5. Derive Eq. (2.40) for the pressure-volume relationship in an adiabatic process. Obtain the corresponding relation for temperature and volume.

2.6. Derive an expression for the entropy change in the isothermal expansion of an ideal gas.

2.7. Prove the following thermodynamic relationships

$$\left(\frac{\partial S}{\partial \alpha}\right)_T = \frac{p}{T}$$

$$\left(\frac{\partial S}{\partial p}\right)_\alpha = \frac{C_v}{p}$$

$$\left(\frac{\partial p}{\partial \alpha}\right)_S = -\frac{C_p p}{C_v \alpha}$$

$$\left(\frac{\partial T}{\partial S}\right)_\alpha = \frac{T}{C_v}$$

2.8. Find expressions for the "root mean square" (i.e., $\sqrt{\overline{v^2}}$), mean ($\bar{v}$), and most probable (v_m) molecular speeds for a Maxwellian distribution. *Hint* for v_m: The distribution of molecular speeds is a maximum at v_m.

2.9. Assume a molecular speed distribution function of the form

$$f(v) = Av^2 e^{-av^2}$$

Evaluate the constants a and A in terms of physical properties of the gas. *Hint:* Use the ideal gas law and the law of conservation of matter, Eq. (2.49).

2.10. Prove that the energy density of an ideal gas is

$$E = \tfrac{3}{2}nkT$$

2.11. Derive Eq. (2.71) and express the thermal conductivity K_T in terms of molecular mass, number density, and collision frequency.

2.12. Express the Debye shielding length h in terms of the root mean square (rms) electron speed and the plasma frequency.

2.13. Find the commutator of the position operator x and the operator corresponding to the x-component of momentum P_x, i.e., $xP_x - P_x x$.

2.14. Find the ratio of the spacing between $^3P_2$ and $^3P_1$ to that between $^3P_1$ and $^3P_0$ in calcium.

2.15. Verify that typical rotational energies of diatomic molecules are of the order 10^{-4} eV.

2.16. How would the *L-S* selection rules be affected if **R** in Eq. (2.117) did not contain the spin operator **S**?

2.17. Indicate in Figure 2.5 some possible dipole transitions for calcium.

2.18. By analogy with the atomic case find some selection rules for electric quadrupole and magnetic dipole transitions in diatomic molecules.

2.19. Prove Eq. (2.129).

2.20. Compute the differential and total cross sections in the center of mass system for the collision of two identical hard spheres of radius r_0. *Hint:* Use Eq. (2.134) and the fact that the distance between the centers cannot be less than $2r_0$.

2.21. Find the scattering amplitude in Born approximation when V is a spherically symmetric Yukawa potential of the form

$$V(r) = \frac{g}{r} e^{-\mu r}$$

2.22. Why do the "saw teeth" appear in the ionization cross section curves for atomic oxygen given in Figure 2.9?

2.23. The paths of heavy particles traversing a slab of matter are nearly straight lines and can be characterized by a length (range) in the material which is a function of the incident energy. Write an expression for the range in terms of stopping power.

2.24. In a sequence of chemical reactions, species X is formed by $A + B \rightarrow X + C + D$ (associated rate constant k_1) and removed by $X + Y \rightarrow W + Z$ (associated rate constant k_2). Write a rate equation for $[X]$ in terms of concentrations and the rate constants k_1 and k_2. Write an equation for $[X]$ when the system is in equilibrium.

2.25. Using Eq. (2.131), compute k for reactions $O^+ + N_2 \rightarrow NO^+ + N$ and $O + N_2^+ \rightarrow NO^+ + N$. Compare with the experimental results (4.6×10^{-12} and 2.5×10^{-10} cm^3 sec^{-1}, respectively). Look up the appropriate values of α (polarizability) in the *Handbook of Chemistry and Physics.*

REFERENCES

1. P. M. Morse and H. Feshbach, *Methods of Theoretical Physics*, McGraw-Hill, New York, 1953, p. 125, 1360–1582.
2. *See*, for example, M. W. Zemansky, *Heat and Thermodynamics*, McGraw-Hill, New York, 1951, p. 158ff.
3. Morse and Feshbach, *op. cit.*, p. 6, 115–117.
4. G. Gioumousis and D. P. Stevenson, *J. Chem. Phys.* **29**, 294 (1958).
5. For example, E. W. McDaniel, *Collision Phenomena in Ionized Gases*, Wiley, New York, 1965, p. 66.
6. A. V. Phelps and J. L. Pack, *Phys. Rev. Lett.* **3**, 340 (1959).
7. R. D. Hake and A. V. Phelps, *Phys. Rev.* **158,** 70 (1967).
8. E. Gerjuoy and S. Stein, *Phys. Rev.* **97,** 1671 (1955).
9. G. J. Schulz, *Phys. Rev.* **125,** 229 (1962).
10. Henry Eyring, John Walter, and George E. Kimball, *Quantum Chemistry*, Wiley, 1944, Chapter 16.

GENERAL REFERENCES

Electrodynamics

J. D. Jackson, *Classical Electrodynamics*, Wiley, New York, 1962, and

W. K. H. Panofsky and Melba Phillips, *Classical Electricity and Magnetism*, Addison-Wesley, Reading, Mass., 1955, are the two standard works at the graduate level.

E. H. Holt and R. E. Haskell, *Foundations of Plasma Dynamics*, Macmillan, New York, 1965, is an excellent elementary treatise on the electrodynamics of plasmas.

Thermodynamics

Mark W. Zemansky, *Heat and Thermodynamics*, McGraw-Hill, New York, 1951, and

Francis W. Sears, *An Introduction to Thermodynamics, The Kinetic Theory of Gases and Statistical Mechanics*, Addison-Wesley, Reading, Mass., 1953, are two excellent elementary texts on the subject.

Kinetic Theory of Gases

E. H. Kennard, *Kinetic Theory of Gases*, McGraw-Hill, New York, 1938,

J. H. Jeans, *An Introduction to the Kinetic Theory of Gases*, Cambridge University Press, Cambridge, 1960,

Earl W. McDaniel, *Collision Phenomena in Ionized Gases*, Wiley, New York, 1964, Chapter 2; and the book by Sears listed previously are excellent elementary accounts;

S. Chapman and T. G. Cowling, *The Mathematical Theory of Non-Uniform Gases*, University of Cambridge Press, Cambridge, 1970, is the standard advanced work on the subject.

Atomic and Molecular Structure and Spectra

Harvey White, *Introduction to Atomic Spectra*, McGraw-Hill, New York, 1934,

Gerhard Herzberg, *Atomic Spectra and Atomic Structure*, Dover, New York, 1944, and

Henry Eyring, John Walter, and George E. Kimball, *Quantum Chemistry*, Wiley, New York, 1944, are standard more elementary accounts, while

E. U. Condon and G. H. Shortley, *The Theory of Atomic Spectra*, Cambridge University Press, Cambridge, 1959,

Gerhard Herzberg, *Molecular Spectra and Molecular Structure*, Van Nostrand, New York, 1950,

J. C. Slater, *Quantum Theory of Atomic Structure*, 2 volumes, McGraw-Hill, New York, 1960, and

———, *Quantum Theory of Molecules and Solids*, Vol. 1, McGraw-Hill, New York, 1963, are more advanced works.

Collisions and Reactions

N. F. Mott and H. S. W. Massey, *The Theory of Atomic Collisions*, 3rd Ed, Oxford University Press, Oxford, 1965, is the standard work on collision theory.

J. B. Hasted, *Physics of Atomic Collisions*, Butterworths, London, 1964, and McDaniel's book mentioned previously are excellent works on experimental methods and results.

Chapter 16 of the book by Eyring, Walter, and Kimball gives a good elementary account of the theory of chemical reaction rates.

3

Physical Aeronomy

Although the earth's upper atmosphere is a region of extremely low density and pressure—in effect, a very good vacuum—it is nonetheless a region of great interest scientifically and technologically. The neutral gases present at these high altitudes constitute a screen which removes radiations harmful to life, and the partially ionized gas or plasma is closely associated with long distance radio communication and geomagnetic variations. Because of the compressibility of gases, atmospheric pressure decreases with increasing altitude in a roughly exponential way. However, departures from exponential behavior occur as a result of both atmospheric heating and variations of composition with altitude.

Before launching into a more detailed discussion of the physical processes involved in the structure of the earth's upper atmosphere, we shall briefly consider the several regions into which it is ordinarily divided. This scheme of characterizing upper atmospheric structure is based entirely on thermal properties, i.e., the temperature profile. Figure 3.1 shows on an expanded scale the temperature profile which was presented earlier in Figure 1.5. The lowest region, that which is actually in contact with the solid earth, is called the *troposphere* and is characterized by a positive lapse rate (temperature decreases with increasing altitude). This is, of course, the region of weather as we experience it, although meteorological effects may indeed extend to much greater altitudes. Above the temperature minimum (*tropopause*) at about 10 to 17 km (depending upon latitude) lies the *stratosphere* where the thermal lapse rate is negative, presumably because heat is somehow continuously deposited in and radiated from this region. As we shall see later in the chapter, the source of this heat is solar ultraviolet radiation which is absorbed and re-radiated by ozone. If there were very little or no ozone, as on Mars and Venus, there would be no stratopause. The *mesosphere* begins at the *stratopause* (about 50 km altitude) and is characterized by a positive lapse rate which extends to the *mesopause* at about 85 km. Above this altitude, heating occurs mainly through absorption of solar extreme ultraviolet radiation although charged particle influx as well as chemical reactions may produce detectable

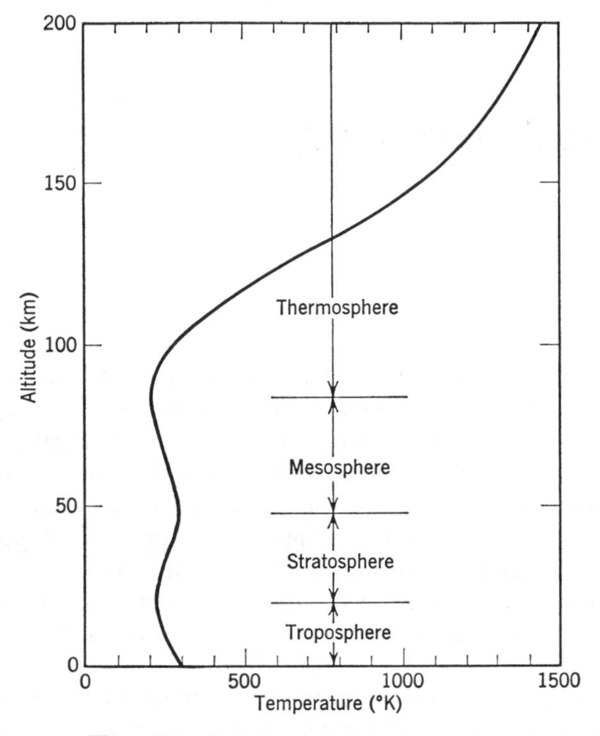

Figure 3.1 The atmospheric regions and a typical temperature profile.

effects. As is evident in Figure 3.1 the temperature increases monotonically in the *thermosphere*. This occurs up to the altitude where the mean free path of the molecules is so large that one can no longer speak of a gas in the normal usage of the term. Here in the *exosphere*, molecules move on ballistic trajectories and only very occasionally collide with each other.

The foregoing is a convenient classification scheme for the earth's upper atmosphere but really tells us nothing of such important properties as pressure, density, and composition. The purpose of this chapter is to investigate the principles which govern the behavior of these parameters and briefly to consider several models which seem to adequately describe aeronomic structure not only for the earth, but for other planets as well. Chemical aeronomy, the study of chemical reactions which affect upper atmospheric composition is deferred until Chapter 4.

3.1 Hydrostatics

The hydrostatic properties of the upper atmosphere are conveniently approached from the barometric equation which relates the pressure gradient

dp/dz to the mass density ρ

$$\frac{dp}{dz} = -\rho g \tag{3.1}$$

where the density can, in turn, be expressed in terms of the particle masses m_i and number densities n_i of the various constituents

$$\rho = \sum_i m_i n_i \tag{3.2}$$

Because of the very long mean free paths of the neutral particles at high altitudes, one can treat the gases as ideal. Dividing Eq. (3.1) by the equation of state of an ideal gas

$$p = nkT \tag{3.3}$$

yields

$$\frac{dp}{p} = -\frac{\rho g}{nkT}\,dz = -\frac{\bar{m}g}{kT}\,dz \tag{3.4}$$

where n is the total particle number density, T is the absolute temperature, k is Boltzman's constant, and $\bar{m} = \rho/n$ is the average particle mass.

The quantity $H = kT/\bar{m}g$ is customarily referred to as the scale height, the distance in which the pressure of an isothermal atmosphere of constant composition drops by a factor e^{-1}. It also has another rather interesting interpretation; it is twice the distance through which a single particle having the mean energy of a member of an ensemble of noninteracting particles will rise against the force of gravity. The proof of this statement is left as an exercise for the student. Of course, the scale height is not constant in practice and the integration of Eq. (3.4) becomes much more difficult; usually it must be performed numerically.

Within a sufficiently short range of altitude, however, it is always a good approximation to express the scale height as a linear function of altitude:

$$H = H_0 + \alpha z \tag{3.5}$$

Inserting this form of H into Eq. (3.4) and integrating, we obtain

$$p = p_0\left(1 + \frac{\alpha}{H_0}z\right)^{-1/\alpha} \tag{3.6}$$

and

$$n = n_0\left(1 + \frac{\alpha}{H_0}z\right)^{-(1+1/\alpha)} \tag{3.7}$$

If we now take the limit $\alpha \rightarrow 0$, we obtain (from the definition of the exponential function)

$$p = p_0 e^{-z/H} \tag{3.8}$$

as we would expect for an isothermal atmosphere.

The arguments advanced in the foregoing will certainly enable us to calculate the pressure and density profiles of any atmosphere provided we know the gravitational acceleration, the temperature, and the composition. For the determination of the last two quantities, however, one must investigate the heat budget and chemistry, both of which are quite complex. We shall consider the former in the following section and the latter in Chapter 4.

3.2 Heating of the Upper Atmosphere

Because of the law of conservation of energy, the thermal structure of an upper atmosphere involves budgetary considerations in which gains must be balanced against losses. Heat transfer by radiation and conduction must be included in the treatment; there is evidence that convection and advection (i.e., winds) should also be included, although the introduction of such complexity has only been attempted very recently. Bearing this in mind, one can write a relation which expresses the overall process quantitatively by equating energy gains to energy losses:

$$\frac{dQ}{dt} = q_T - R - \frac{d\Phi}{dz} \tag{3.9}$$

where dQ/dt is rate of energy gain (i.e., power gain) per unit volume at a given height in the atmosphere, q_T is the source function or rate of energy input, R is the rate of loss by radiation, and Φ is the heat energy flux. The thermal property of principal interest in an upper atmosphere is the absolute temperature T which is related to Q by

$$Q = \rho C_v T \tag{3.10}$$

where ρ is the gas density and C_v is the specific heat capacity at constant pressure.

We ought also to include a term describing heat transfer by eddy (turbulent) diffusion. However, turbulent transport processes will be discussed in some detail in Chapter 5 and we defer further consideration until then.

3.2.1 Energy Deposition

Although there are several sources of thermal energy in an upper atmosphere, photoabsorption of radiation from the sun is the principal one. We shall discuss it in some detail. The introduction of this energy into the atmosphere takes several forms such as electronic excitation of atomic and molecular species, ionization, and dissociation of polyatomic molecules.

In each case the energy is not initially thermal in nature, but collisions of the energetic species with molecules of the surrounding gas rapidly leads to thermalization. Although we have not done so, one can readily prove this by solving the appropriate form of the Boltzmann equation which was briefly discussed in Chapter 2. For a more complete presentation the interested reader is referred to Chapman and Cowling. Under some conditions the thermal equilibration time may be much longer than we have indicated. However, this does not lead to effects of great importance and so we shall disregard it. In any case, one can write the following equation for the thermal energy input:

$$q_T(z) = \sum_i \int F_\lambda(\lambda, z)\epsilon_i Q_i(\lambda) n_i(z)\, d\lambda \tag{3.11}$$

where $F_\lambda(\lambda, z)$ is the radiation flux per unit wavelength (or spectral irradiance) of photons in units of cm^{-2} sec^{-1} Å^{-1} at height z; n_i is the number density of the ith species at height z; $Q_i(\lambda)$ is the absorption cross section (not to be confused with thermal energy Q); and ϵ_i is the amount of heat energy liberated per unit radiation energy absorbed (i.e., thermal "efficiency").

If the intensity I were constant for all altitudes the energy input to the atmosphere would increase monotonically with descending altitude. Because of absorption at higher altitudes, it is not, and this fact must be carefully considered in evaluating $q_T(z)$. At a given wavelength λ the spectral irradiance at height z can be expressed in terms of the spectral irradiance at the top of the atmosphere, $F_\lambda(\lambda, \infty)$ and of the optical depth τ.

$$F_\lambda(\lambda, z) = F_\lambda(\lambda, \infty)e^{-\tau(\lambda, z)} \tag{3.12}$$

where

$$\tau = \sec\chi\, Q \int_z^\infty n(z')\, dz' \tag{3.13}$$

and χ is the solar zenith angle.

It is both interesting and instructive to derive the function q for monochromatic radiation of wavelength λ_0 and a single-component exponential atmosphere. From Eqs. (3.11)–(3.13) we can write (see Problem 3.3)

$$q_T(z) = F(\lambda_0, \infty)n(z_0)\epsilon Q \exp\left[-\frac{z}{H} - n(z_0)\sec\chi Q H e^{-z/H}\right] \tag{3.14}$$

where z_0 is the zero point for altitude measurements. It is quite obvious that q should show a maximum at some height due to the combined effects of increased heating rate and increased absorption of radiation with descending altitude. Eventually an altitude is reached where the attenuative effects in the atmosphere above offset the effect of higher local atmospheric density. One can easily determine this height by differentiating $q(z)$ and equating the result to zero. The altitude of maximum absorption z_m is related to the other parameters by

$$z_m = H \ln [n(z_0)QH \sec \chi] \tag{3.15}$$

Inserting this result into Eq. (3.14) yields

$$q(z) = q(z_m) \exp\left[1 - \frac{z - z_m}{H} - e^{-(z-z_m)/H}\right] \tag{3.16}$$

where

$$q(z_m) = \frac{F(\lambda_0)\epsilon \cos \chi}{eH}$$

This important result was obtained many years ago by Chapman[1] and for that reason the coefficient of $q(z_m)$ is frequently called a Chapman function Ch (z, χ). It will be employed in a different context in Chapter 8. The Chapman function which represents normalized $q(z)$ $[q/q(y_m)]$ corresponding to several zenith angles is shown as a function of normalized height ($y = -(z - z_m)/H$) in Figure 3.2.

In addition to absorption of ultraviolet radiation, an upper atmosphere can be heated by a number of other mechanisms such as chemical reactions,

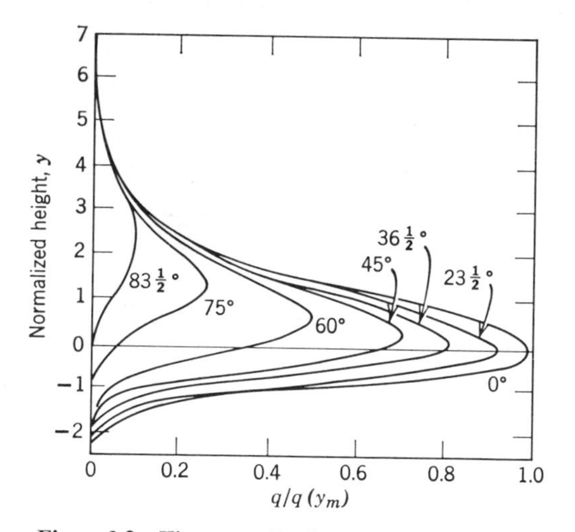

Figure 3.2 The normalized rate of energy deposition $q/q(z_m)$ plotted against the normalized height $y = \exp[z - z_m/H]$ for different values of the solar zenith angle. (After Ratcliffe and Weeks in *Physics of the Upper Atmosphere*, Ed. by J. A. Ratcliffe; reprinted by permission of Academic Press, Inc.)

absorption of energy from energetic charged particles, hydromagnetic waves, joule heating by electric currents, and atmospheric absorption of gravity waves. Although the gross effect of these is indeed usually negligible, they may be important under some circumstances. Heating by absorption of charged particles, which is of some importance in the earth's aurora, is very similar to that discussed above for photoprocesses. However, it is usually more convenient to employ the stopping power $-(dE/dz)$ or energy loss per unit path length rather than the cross section. Then the heating rate function q is written

$$q(z) = \sum_i n_i \int \epsilon_i \left(-\frac{dE}{dz}\right)_i \frac{d^2F}{dE\,d\Omega}\bigg|_z dE\,d\Omega \tag{3.17}$$

where n_i is the particle number density of the ith species, and $d^2F/dE\,d\Omega\,|_z$ is the charged particle flux per unit energy and solid angle Ω evaluated at height z. The other modes of thermal energy deposition in the neutral atmosphere will be considered in later chapters.

3.2.2 Heat Transfer by Conduction

Heat transfer in the earth's upper atmosphere occurs principally through conduction above 150 km where the atmosphere is optically thin, and through conduction and radiation at lower altitudes. While all modes of heat transfer in a gas are due to transport mechanisms, conductive processes are the result of gas kinetics (molecular motion). For this reason it has many of the characteristics of diffusion and, in fact, is described quantitatively by a diffusion-type equation (see Section 3.3 for a discussion of diffusion).

We introduced heat conduction from the standpoint of kinetic theory in Section 2.3.4, and obtained an expression for the heat flux $\mathbf{\Phi}$ due to conduction

$$\mathbf{\Phi} = n\overline{\epsilon\mathbf{v}} \tag{3.18}$$

where ϵ and $\mathbf{v}$ are, respectively, molecular energy and velocity. $\mathbf{\Phi}$ is just the operand of the second term on the left hand side of Eq. (2.63) and is related to the velocity distribution function $f(v)$ by

$$\mathbf{\Phi} = \tfrac{1}{2} \int f m v^2 \mathbf{v}\, d\mathbf{v} \tag{3.18a}$$

If f is isotropic, $\mathbf{\Phi}$ vanishes, but if f is of the form $f = f_0 + \mathbf{v} \cdot \mathbf{f}_1$ with $\mathbf{f}_1$ given by Eq. (2.64) and f_0 Maxwellian, we have the result that

$$\mathbf{\Phi} = \left[\frac{-1}{kT^2} \int \left(\frac{mv^2}{2}\right)^2 \mathbf{vv} \frac{f_0}{\nu_m}\, d\mathbf{v}\right] \cdot \nabla T = -K\,\nabla T \tag{3.18b}$$

where K, the *thermal conductivity*, is defined as the quantity in square brackets in Eq. (3.18b).

In a vertically stratified atmosphere the flux Φ of heat energy at a given point is proportional only to the z component of the temperature gradient:

$$\Phi = -K \frac{\partial T}{\partial z} \tag{3.19}$$

In a real gas, K can be approximated quite well by the expression

$$K = AT^{1/2} \tag{3.20}$$

For example, the constant A has the values[2] in joules m^{-1} sec^{-1} $°K^{-3/2}$ of 0.21 for atomic hydrogen, 0.036 for atomic oxygen, 0.018 for O_2 and N_2, and 0.016 for CO.

Referring to Eq. (3.9), we can express the heat budget above 150 km by the equation

$$\rho C_v \frac{dT}{dt} = q(z, t) + A \frac{\partial}{\partial z}\left(T^{1/2} \frac{\partial T}{\partial z}\right) - R \tag{3.21}$$

where the time dependence of q is explicitly stated. Because of the temperature dependence of the thermal conductivity, the equation is nonlinear and thus quite intractable to standard mathematical techniques for solution. If one approaches the problem by dividing the atmosphere into slabs in which T does not vary appreciably and then connecting the slabs thermally by suitable boundary conditions (temperature and thermal flux continuity), the equation can be linearized and solved by standard methods subject to boundary conditions on the thermosphere as a whole. If one is interested only in a crude description of the thermal structure, a value of K corresponding to the mean thermospheric temperature should be satisfactory. It is a reasonably valid approximation to assume isothermal conditions at very great heights:

$$\lim_{z \to \infty} \frac{\partial T}{\partial z} = 0$$

At the lower extremity of the thermosphere the situation is more complicated and it is difficult to treat the problem in this manner, principally because radiative transfer cannot be neglected. For the upper thermosphere though, it should yield fairly reliable results.

In the q which appears in Eq. (3.21), time is one of the arguments (this accounts for solar control). Inclusion of time variations means that the properties of the atmosphere at a given height will vary because of expansion and contraction. For this reason it is usually more convenient to formulate the problem in terms of pressure than of the actual height z. This was done by Harris and Priester in a model which we shall discuss briefly later in this chapter.

3.2.3 Heat Transfer by Radiation

In radiative heat transfer, electromagnetic energy is emitted by atoms or molecules of various atmospheric components at one point in space and absorbed or scattered, usually by the same components, at another point. If it is not absorbed at all, we have radiative cooling. Like other modes of thermal transport, radiative transfer is significant only in certain altitude regimes, but it is nevertheless important at some place in virtually every atmosphere.

The absorption of radiation by an atmosphere was discussed in some detail in Section 3.2.1 but we did not discuss the emission of radiation by atmospheric constituents. Because energy transfer is of considerable significance to planetary atmospheres, we shall present an introduction to the topic but refer the reader to the list of general references: Chapter 7 of Craig, Section 19 of Swihart, Chapter 4 of Brandt and Hodge, or (for more advanced treatment) Chapter 4 of Aller or Chapter 2 of Goody.

If there are no radiation sources within the atmosphere, one can express the spectral intensity dI_λ absorbed in a slab of optical depth $d\tau$ by the relation

$$\frac{dI_\lambda}{d\tau} = -I_\lambda \tag{3.22}$$

where I_λ is the intensity at the top of the slab and $d\tau$ is defined by Eq. (3.13). Spectral intensity is related to spectral irradiance introduced in Section 3.2.1 by the equality

$$I_\lambda = e_\lambda(dF_\lambda/d\Omega) \tag{3.23}$$

where e_λ is the energy of a photon of wavelength λ, and $d\Omega$ is the element of solid angle. We include the effect of emission by adding a source term J to the right hand side of Eq. (3.22):

$$\mu\frac{dI_\lambda(\mu, \tau)}{d\tau} = I_\lambda - J_\lambda \tag{3.22a}$$

which is usually called the *radiative transfer equation.* In accordance with common practice in radiative transport theory we have also removed the factor sec χ from the definition τ and included χ explicitly by means of the factor

$$\mu = -\cos\chi \tag{3.24}$$

Equation (3.22a) is a linear first order differential equation which has the solution

$$I_\lambda(\tau, \mu) = \int_\tau^\infty J_\lambda(\tau', \mu) \exp\left(-\frac{\tau'}{\mu}\right)\frac{d\tau'}{\mu} \tag{3.25}$$

Obviously, if we know the source function J_λ, we have a complete solution to the problem. A simple, but very useful case is one in which the atmosphere is in local thermodynamic equilibrium. The source function J is then given by Planck's relation

$$J_\lambda = B_\lambda = \frac{2hc}{\lambda^3} \frac{1}{[e^{hc/\lambda kT} - 1]} \tag{3.26}$$

In a real atmosphere, one usually finds that the black body approximation is not a satisfactory one. However, if departure from local thermodynamic equilibrium is not too great, one can use the *grey body* approximation

$$J_\lambda = \epsilon_\lambda B_\lambda \tag{3.27}$$

where ϵ_λ is the *emissivity*.

Although we have in the present context glossed over the optical depth, it usually is a rather complicated parameter owing to the effects of *line broadening*. There are two types of broadening which are important: Doppler and Lorentz broadening. The first is dependent only upon temperature (it results from Doppler shift of wavelengths) while the second is pressure-dependent (it results from shifts of energy levels during the collision process). Hence, one must include the spectral line shape when computing the absorption coefficient

$$K_\lambda = \frac{d\tau_\lambda}{ds} = Q(\lambda)n \tag{3.28}$$

[see Eq. (3.13)]. For a pure Lorentz profile, which, incidentally, is of great importance in studies of infrared radiation transfer, we have

$$K_\lambda = \frac{S\alpha}{\pi} \frac{1}{(\nu - \nu_0)^2 + \alpha^2} \tag{3.29}$$

where ν is the frequency ($\lambda = c/\nu$), α is the *half-width* of the line (where $K_\lambda = \frac{1}{2}K_{\lambda_0}$) and,

$$S = \int_{-\infty}^{\infty} K_\lambda \, d\lambda \tag{3.30}$$

Due to lack of space we cannot pursue the subject of line profiles at greater length here but refer the reader to Section 7.1.3 of Craig, Section 20 of Swihart, Chapter 4 of Aller, or Chapter 3 of Goody.

The radiative transfer equation could also be applied to scattering atmospheres. However, since we are interested principally in the transport of infrared radiation in upper atmospheres, we shall not pursue the matter.

In the case of the earth, radiative transfer is important only below about 150 km altitude because of the large thermal conductivity and optical thinness of the earth's atmosphere at greater heights. In the mesosphere, radiation in both the 15μ rotational vibrational band of carbon dioxide and the 9.6 μ ozone band accounts for significant cooling. At mid-latitudes near the stratopause the calculated cooling rate due to CO_2 is about 4 to 6°C day^{-1} while that for O_3 is about 2° day^{-1}. In regions of high latitude the role of mesospheric ozone changes in summer due to the very long days and consequently the time during which absorption occurs. Under these conditions ozone absorption leads to a heating rather than a cooling effect. It is easy to see why the temperature peak in the mesosphere is a consequence of infrared radiation and absorption by CO_2 and O_3. The rest of the atmosphere is quite transparent to these wavelengths, the radiation escaping to ground or to space. Since heat transfer by conduction and convection is quite slow, the mesosphere as well as the stratosphere is effectively insulated from the troposphere and thermosphere. Any heat energy absorbed by mesospheric gases will thus tend to maintain an elevated temperature there.

As we shall see in Section 3.6.4, the atmospheres of Mars and Venus are composed mainly of carbon dioxide. This gas is apparently not dissociated to any appreciable extent even at very high altitudes. As a consequence, radiative transport can be important as a means of radiation entrapment at low altitudes (on Venus) where the pressure is high and the atmosphere optically thick, and as radiation loss in the upper atmosphere which is optically thin. Radiation loss by CO_2, probably in the 15 μ band, is the most probable cause of the relatively low thermospheric temperatures found on Mars and Venus.

There is another component which is of some significance in the cooling of the lower thermosphere, and this is atomic oxygen. The principal transitions are the forbidden electronic transitions $^1S \rightarrow {}^1D$ (5577 Å) and $^1D \rightarrow {}^3P$ (6300, 6364 Å). The latter is, as a matter of fact, of some importance in the theory of energy deposition in the F_2 layer but neither transition is of much significance insofar as atmospheric heating is concerned. However, the $^3P_0$ and $^3P_1$ levels lie slightly above the $^3P_2$ ground level (0.028 and 0.020 eV, respectively) and very slow radiative transitions from $^3P_0$ or $^3P_1$ to the ground state are of importance. Because of the low transition probability, excited levels will be populated according to a Boltzmann distribution. Since thermal excitation is sufficient to populate the $^3P_0$ and $^3P_1$ states, atomic oxygen can be a fairly efficient agent of cooling by radiation because the atmosphere is optically thin to this radiation above ~120 km. This mechanism was first suggested by D. R. Bates.[3]

3.3 Mechanical and Chemical Equilibrium in the Earth's Upper Atmosphere

Although an upper atmosphere is never in a true state of equilibrium, either thermal, mechanical, or chemical, it does approach this condition closely enough in various altitude regions to make the concept a useful one.

There are actually three types of equilibrium among the constituents that we must consider, depending upon the component in question as well as upon the altitude range. In the thermosphere one finds a condition of gravitational separation, sometimes called diffusive equilibrium, above about 110 to 120 km for all atmospheric constituents. At these levels the molecular mean free paths are quite long (although still short compared to the scale height) and atoms and molecules move about as independent particles. Each component obeys a hydrostatic law characterized by its own molecular mass with the result that each also has its own scale height:

$$H_i = \frac{kT}{m_i g} \tag{3.31}$$

and the hydrostatic equations (3.6 and 3.8) hold for each component separately. The partial pressures of relatively light gases like helium will decrease slowly with altitude whereas the pressure of N_2, for example, will decrease more rapidly. Thus, we have a condition of gravitational (or diffusive) separation with the lighter gases on top. The mean scale height is the weighted average of all those corresponding to the various atmospheric constituents and varies with altitude because of this factor as well as changing temperature.

At an altitude between 110 and 120 km (turbopause) molecular diffusion is dominated by atmospheric turbulence which leads to a mixing of the various components so that their number density ratios do not vary with altitude. The theory of turbulence and eddy diffusion in the upper atmosphere is still in a rather unsatisfactory state, but the little knowledge that we do have will be briefly discussed in Chapter 5.

The last type of equilibrium to be mentioned is the photochemical equilibrium regime in which chemical production and loss processes are much faster than molecular and eddy diffusion. Because the reactions which remove a constituent are more rapid well below the turbopause, one expects this type of equilibrium to be a relatively low altitude phenomenon, and indeed it is. Atomic and molecular oxygen are in photochemical equilibrium near the mesopause, and ozone, which is a mesospheric and stratospheric component, always tends toward chemical equilibrium. Since the entire area of aeronomic chemistry will be discussed at length in the following chapter, we defer further development to that point.

3.4 Molecular Diffusion

Gaseous diffusion occurs by virtue of the random motion of the molecules and atoms in the gas and hence must be approached from the standpoint of the kinetic theory of gases which was briefly discussed in Section 2.3. We shall begin our study with the diffusion or continuity equation itself. Assuming vertical stratification only, this relation can be written

$$\frac{\partial N_i}{\partial t} + \frac{\partial}{\partial z}(N_i W_i) = J_i - L_i \tag{3.32}$$

where N_i is the number density of the ith constituent, W_i is the mean vertical velocity of the ith constituent, J_i is the rate of formation of the ith component, and L_i is its loss rate. According to Nicolet,[4] the diffusion velocity can be approximated by

$$W_i = -D\left[\frac{1}{N_i}\frac{\partial N_i}{\partial z} + \frac{1}{H_i} + (1 + \alpha_T)\frac{1}{T}\frac{\partial T}{\partial z}\right] \tag{3.33}$$

where D is the diffusivity discussed in Chapter 2, and α_T is the thermal diffusion factor which for practical purposes can be considered constant. In fact, it need be included only for the light gases (hydrogen and helium). The second term in brackets on the right hand side of Eq. (3.33) represents mass motion due to a temperature gradient. Under conditions of diffusive equilibrium the time derivative of N_i vanishes and we are left with the steady state equation

$$\frac{\partial}{\partial z}(N_i W_i) = J_i - L_i \tag{3.34}$$

In a real atmosphere the source and sink terms J_i and L_i, as well as the diffusion velocity W_i, are dependent upon the densities of other constituents. Furthermore, the loss terms are usually nonlinear so that we have a set of coupled nonlinear differential equations to deal with. Finally, winds, convection, and mixing (at altitudes below the turbopause) must be included. Needless to say, attempts to embrace the entire upper atmosphere in a scheme of this type are futile and appropriate simplifying approximations must be introduced in order to attain some degree of success. One of the most useful which does correspond to actual conditions in the upper atmosphere is to consider all components but the one in question to be fixed, known quantities at every altitude; this is the case for a minor constituent. Other simplifying assumptions are the absence of nonlinear loss mechanisms, turbulent mixing, horizontal winds, and convection.

It is instructive to consider the simple but artificial case characterized by:

(a) No loss mechanisms
(b) $D = D_0 e^{z/H}$ where H is the scale height of the principal species
(c) $J = J_0 e^{-z/H}$
(d) An isothermal atmosphere.

The diffusion equation then takes the simple form:

$$\frac{d^2N}{dz^2} + \frac{(c+1)}{H}\frac{dN}{dz} + \frac{cN}{H^2} = -\frac{J_0}{D_0}e^{-2z/H} \tag{3.35}$$

where c is the ratio of the molecular mass of the minor to the mass of the major species. The complementary function is easily found to be $N = C_1e^{-z/H} + C_2e^{-cz/H}$; the particular integral is also quite easy to obtain by one of the standard methods, e.g., variation of parameters, and the complete solution is

$$N = C_1e^{-z/H} + C_2\,e^{-cz/H} - \frac{J_0H^2}{D_0(2-c)}e^{-2z/H} \tag{3.36}$$

The constants of integration C_1 and C_2 are set by the upper and lower boundary conditions. The problem becomes somewhat more complicated but still remains tractable if one includes a linear loss term, e.g., $Ke^{-2z/H}N$.

It is interesting and instructive to consider the physical significance of the first two terms in Eq. (3.36) which determine the behavior of N at very high altitudes. If $C_1 \neq 0$, the number density of the minor species has scale height H in part of the profile, and, as we shall see from Eq. (3.37) below, is a consequence of a flux of minor species molecules into or out of the atmosphere. If $C_2 \neq 0$, such a flux exists. A more realistic but still soluble problem results if we include a linear loss term of the type mentioned above; this is left for the problems at the end of the chapter.

Returning to the equation for the vertical transport velocity W_i [Eq. (3.33)], one can gain considerable insight into the sign of W_i by writing the diffusion velocity in an isothermal atmosphere

$$W_i = -D_i\left(\frac{1}{N_i}\frac{\partial N_i}{\partial z} + \frac{1}{H_{ie}}\right) = D\left(\frac{1}{H_i} - \frac{1}{H_{ie}}\right) \tag{3.37}$$

where H_{ie} is the scale height at equilibrium. Obviously, if $H_i = H_{ie}$, the diffusion velocity vanishes; we expect this to be the case for equilibrium. If H_i should be greater than H_{ie}, i.e., the number density N_i decreases with increasing altitude more slowly than would occur at equilibrium, then W_i is negative. The gas diffuses downward to attain equilibrium. On the other hand, if $H_i < H_{ie}$, W_i is positive and constituent i diffuses upward.

The problem is somewhat more complicated if temperature gradients and nonlinear loss terms are involved. Nevertheless it is still amenable to solution on a computer. Mange[5] solved the problem some years ago for the case of O_2 initially in photochemical equilibrium in an atmosphere of O and N_2 with the assumptions that $H = 8$ km at 110 km and $dH/dz = 0.2$. It is evident from Figure 3.3 that the approach to equilibrium is quite rapid, particularly

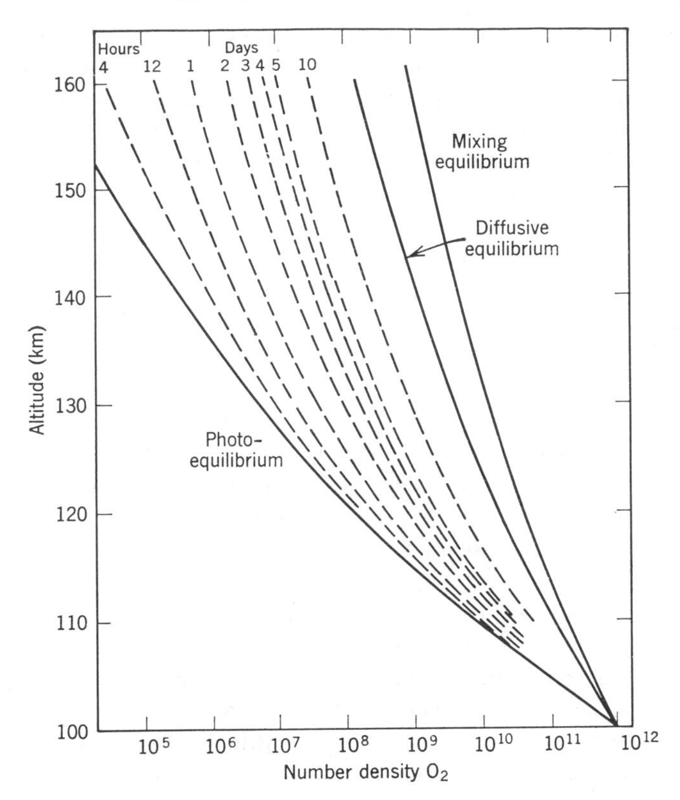

Figure 3.3 Time required for molecular oxygen, initially in photochemical equilibrium, to assume indicated vertical distributions as a result of diffusion, even though dissociations continue to occur. Oxygen is assumed to be a minor constituent in a main mixed atmosphere of nitrogen and atomic oxygen. (After Mange, *Ann. Géophys.* **11,** 153 (1955); reprinted by permission of Service des Publications du C.N.R.S., Paris.)

in the early stages. Recently, the effects of winds and convection have received some attention, but there is still considerable doubt as to their influence on the vertical distribution of atmospheric constituents.

3.5 Variations in the Earth's Atmosphere

Variations in upper air structure, e.g., density, pressure, and temperature have their origin in the rotation of the earth (diurnal variations), revolution of the earth about the sun (seasonal variations), and the sunspot cycle (solar-cycle variations). In addition, these physical properties also depend upon geographic and probably magnetic latitudes.

The diurnal variation is the simplest to analyze and will be considered first. It is, of course, caused entirely or almost entirely by variation of the heating rate due to changes in the flux of solar ultraviolet radiation. Although most of this radiation is absorbed at altitudes below 150 km the amplitude of the density and temperature variations is small at these heights because of the large heat capacity of the atmosphere. Higher in the thermosphere the heat capacity is smaller and daily fluctuations are greater. It is possible by solving the time-dependent heat equation (3.9) to actually compute the variations in temperature and density; this has been done, for example, by Harris and Priester whose model is discussed further in Section 3.7. The results of some day and night measurements of atmospheric density will be presented in Section 3.6.

Seasonal variations, which are in many ways similar to latitude variations, are more difficult to analyze. Certainly, there are variations arising from changes in the noon zenith angle as well as in the duration of daylight. One aspect of the latter was in fact mentioned earlier in connection with absorption and radiation by ozone. Such effects can easily be incorporated in the model of Harris and Priester but are sometimes overshadowed by other effects which are poorly understood or have a complicated origin. For example, the cause of the so-called sudden polar warming which has been observed in the winter stratosphere and mesosphere is unknown. Another seasonal phenomenon which would be extremely difficult to incorporate in a model is the chemical heating of the polar mesopause in winter. The production of heat by the chemical reactions (recombination of atomic oxygen) is well understood, but the seasonal changes in the downward transport of O atoms to the recombination region is not.

It is well known that solar activity has a very significant influence on upper atmospheric properties. Most of the effect is undoubtedly due to variations in the intensity of ultraviolet solar radiation incident on the earth. Satellite and rocket observations of the EUV solar spectrum carried out mainly since about 1960 support this assertion, although the stronger EUV lines, such as the Lα line of atomic hydrogen and the He II Lα line, appear to vary only slightly over a solar cycle. This work has been summarized through 1963 by Hinteregger.[6] Fluctuations in solar corpuscular radiation, specifically the solar wind, also apparently cause important fluctuations in atmospheric structure.

It has been found empirically that there is a strong correlation between solar radio flux† at wavelengths of 8 to 10 cm, the Zurich sunspot number,

† The intensity S of the solar radio flux is usually given in units of 10^{-22} watts $(\mathrm{m^2Hz})^{-1}$ (sometimes called a Jansky).

and upper air properties. This relationship is not really surprising since the radio flux is believed on theoretical grounds to be emitted from the same region as the ultraviolet radiation. Actually the correlation of the 27-day (solar rotational period) average of radio flux S_{27} with the 27-day average sunspot number (sometimes called the Wolf sunspot number R_{27}) is more significant than the correlation of the instantaneous values. It is customary to employ the radio flux at 8 and 10.7 cm since these seem to correlate extremely well with solar activity; however, it should be pointed out that the ratio between these fluxes does vary (approximately linearly) somewhat over a solar cycle. Figure 3.4 shows a plot of the 10.7 cm solar flux S_{27} as a function of the Wolf sunspot number for the years 1951–1962. It should be noted that the ratio of the 27-day average values of R and S almost always lies within 10 per cent of its mean value.

Because of the dependence of thermospheric temperature on the solar heating rate, one expects a correlation between the former and the 10.7 cm

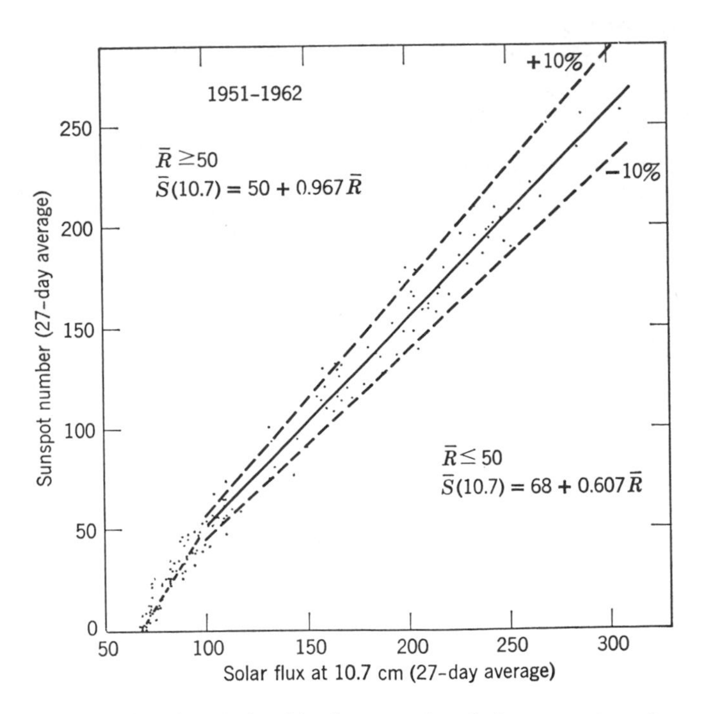

Figure 3.4 The relationships between the relative sunspot number and the radio flux at 10.7 cm (in Janskys). A linear relationship cannot be used for the entire solar cycle. (After Nicolet, *J. Geophys. Res.* **68,** 6121 (1963); reprinted by permission of the American Geophysical Union.)

flux. A linear relation was suggested by Jacchia[7] several years ago and has since been generally adopted:

$$T = T_0 + a\bar{S}_{27} + b\Delta S \tag{3.38}$$

where ΔS represents the oscillations of S during the 27-day period. For average nighttime conditions $T_0 = 280°\text{K}$, $a = 4.6$ m^2 Hz watt^{-1}, and $b = 2.5$ m^2 Hz watt^{-1}. For daytime conditions, Jacchia suggested

$$T_{\text{day}} = 1.3T_{\text{night}} \tag{3.39}$$

Average daytime thermopause temperature (27-day mean values) and the 27-day maximum temperature range for the years 1952 to 1962 are shown in Figure 3.5. This figure is a striking portrayal of the relationship between solar activity and upper atmospheric properties.

In addition to atmospheric variations due to ultraviolet radiation during the solar cycle, it is well known that such variations are also associated with geomagnetic activity. Since the latter also is correlated with the 8- and 10.7-cm solar flux, the effect on thermospheric temperatures is already included

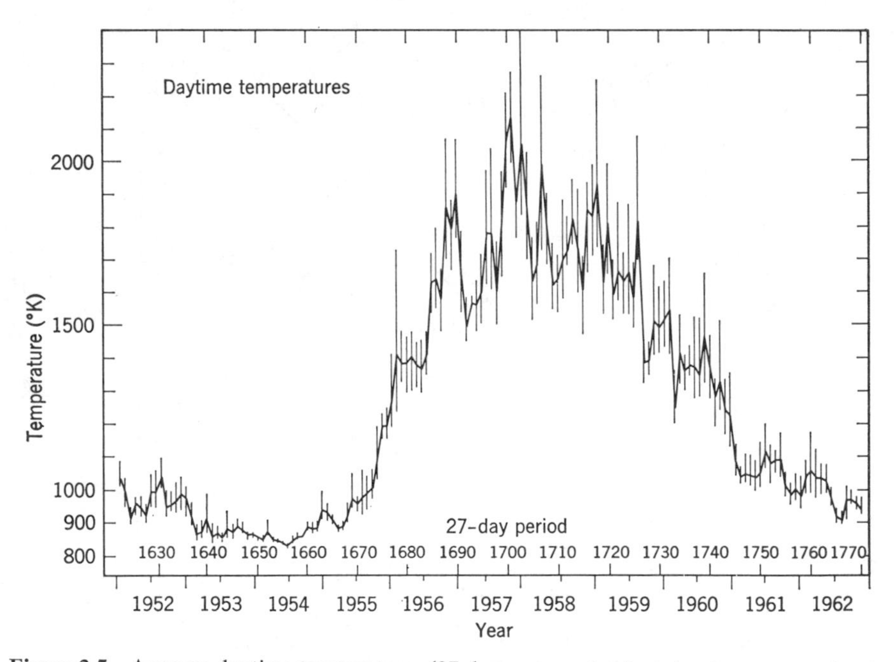

Figure 3.5 Average daytime temperatures (27-day mean values) at the thermopause level and the 27-day maximum temperature range from 1952 to 1962. Magnetic storm effects are not included. (After Nicolet, *J. Geophys. Res.* **68**, 6121 (1963); reprinted by permission of the American Geophysical Union.)

in the (empirical) Eqs. (3.38) and (3.39). For the case of magnetic storms, however, the correlation is not so good and these two equations cannot be expected to be very reliable. The heating mechanism that occurs during magnetic storms is not completely understood. Hydromagnetic waves as well as an influx of low energy charged particles have been suggested, but it is not yet known which of these, if either, is correct.

3.6 Measurements of Atmospheric Properties

Although numerous methods of measuring the various properties of the earth's upper atmosphere, such as pressure, density, and temperature, have been employed in recent years, we shall confine our discussion to a few which have been most popular and to the very interesting measurements on the Martian atmosphere by Mariner 4. For a more complete discussion of methods and results regarding the earth's atmosphere the interested reader is referred to Craig (see Sections 3.3 and 6.4).

3.6.1 Density

At low altitudes it is a relatively simple matter to measure pressure, density, and temperature directly using balloon-borne instruments. However, the use of balloons is limited to about 35 km, and more sophisticated techniques are necessary at higher altitudes. Many of these depend upon rockets to carry the instruments aloft although some ground-based methods of measuring density have been tried.

One of the older rocket experiments involves the use of a sphere which is ejected from the vehicle at some altitude above 60 km. The subsequent acceleration of the sphere as it first ascends and then descends through the atmosphere is measured by three orthogonal internal accelerometers and the data telemetered to ground by a small radio transmitter. The total acceleration which is, of course, the vector sum of the three observed components, is related to atmospheric density by the equality

$$\rho = \frac{2m}{v^2 C_D}(g - a) \tag{3.40}$$

neglecting the buoyant force which is small. Here m and C_D are, respectively, the mass and ballistic coefficient of the sphere, v is its speed, g is the acceleration of a body in free fall, and a is the magnitude of the measured acceleration. Since the drag force is velocity-dependent, the measured acceleration must be integrated once in order to obtain the air density. This is probably the largest source of error in the experiment. Very reliable results have been obtained with this technique up to altitudes of 135 km.[8] The older experiments employed metal spheres and were limited to ~85 km; but more recently, inflatable spheres of low mass and large ballistic coefficient have permitted extension of the method to the higher altitudes. The results of two of these experiments are shown in Figure 3.6.

A somewhat similar method which is limited to altitudes above 150 km depends upon observations of changes in the orbital periods of satellites due to atmospheric drag. This technique yields useful results up to as much as 1100 km where the density becomes too small to produce significant changes in the orbit. The simplest problem of this type to treat analytically is the case of a circular orbit and a spherically symmetric atmosphere. In practice, of course, a circular orbit is very difficult to attain and the atmosphere does not exhibit such symmetry. However, for illustrative purposes we shall derive the relation of orbital period to atmospheric density before mentioning the more practical case of an elliptical orbit and exponential atmosphere.

We begin the analysis by writing down an equation for the total energy E of the satellite (kinetic plus potential)

$$E = \tfrac{1}{2}mv^2 - \frac{GmM}{r} \tag{3.41}$$

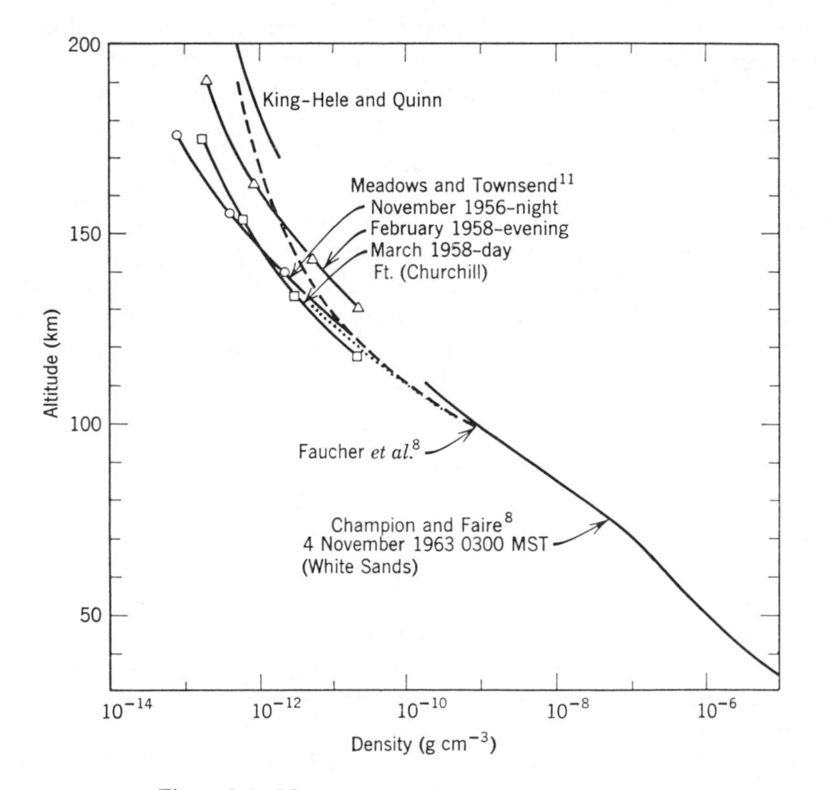

Figure 3.6 Measurements of mass density, 30 to 200 km.

where m and M are the mass of the satellite and the earth, v is the speed of the satellite, G is the gravitational constant, and r is the distance from the satellite to the center of the earth (we neglect asymmetries and inhomogeneities in the latter). E can be expressed entirely in terms of r by invoking the fact that the centripetal force and the gravitational force are equal:

$$\frac{mv^2}{r} = \frac{GmM}{r^2} \tag{3.42}$$

thus permitting one to write

$$E = -\frac{GmM}{2r} \tag{3.43}$$

According to the law of conservation of energy, the work done on the atmosphere by the satellite is equal to the energy lost. Since we are really dealing here with time rates of change, the power lost by the satellite dE/dt is equal to the power transferred to the atmosphere Fv (which ultimately appears as heat energy):

$$Fv = \frac{dE}{dt} \tag{3.44}$$

where F is the drag force which is related to atmospheric density by

$$F = \tfrac{1}{2}\rho v^2 C_D \tag{3.45}$$

Upon substituting the above forms of F and E into Eq. (3.44), we obtain

$$\frac{dr}{dt} = -\frac{\rho v C_D r}{m} \tag{3.46}$$

which is still not quite in usable form. In order to put Eq. (3.46) in terms of the orbital period T, we need Kepler's third law

$$T^2 = \left(\frac{4\pi^2}{GM}\right) r^3 \tag{3.47}$$

Differentiating and substituting into Eq. (3.46) yields

$$\frac{dT}{dt} = -\left[\frac{3\pi C_D r}{m}\right] \rho \tag{3.48}$$

The corresponding equation for an elliptical orbit in an exponential atmosphere has been developed by King-Hele and others[9] and can be expressed as

$$\rho_\pi \approx -\frac{dT}{dt}\frac{m}{3C_D}\left(\frac{2e}{\pi a H_\rho}\right)^{1/2}\left[1 - 2e - \frac{H_\rho}{8ae}\right] \qquad (0.02 \le e \le 0.2) \tag{3.49}$$

where ρ_π is the atmospheric density at perigee, e is the orbital eccentricity, a is the semimajor axis of the orbit, and H_ρ is the density scale height. The last quantity is identical to the pressure scale height H discussed in Section 3.1 if the latter is constant. On the other hand, if it is described by Eq. (3.5), then

$$H = (1 + \alpha)H_\rho \tag{3.50}$$

Calculation of accurate values of ρ_π from Eq. (3.50) requires accurate values of the scale height. With the large number of objects which are presently orbiting the earth, scale heights are routinely found by combining data from several satellites at different heights.

Several density profiles computed by King-Hele[10] from satellite data obtained during the years 1958–1964 are shown in Figure 3.7. The diurnal and solar-cycle (and scale height) variations are the outstanding features with much larger densities at high altitude occurring near solar maximum.

At this point we should perhaps say a few words about an apparent paradox associated with the orbital changes due to atmospheric drag: as the perigee altitude steadily decreases, the satellite velocity increases. At first glance we should expect the opposite to be true because drag opposes the motion. The explanation is as follows: The drag force is small enough initially that the satellite orbit is not perturbed very much. Hence Eq. (3.42) still holds approximately, and thus as potential energy is converted to kinetic energy, r decreases and v increases. Of course, when entry into the lower atmosphere occurs, the velocity does decrease due to the drag forces.

Before going on to discuss temperature measurements, we shall briefly mention two techniques for directly measuring number densities of atmospheric constituents. One of these is based on optical measurements of solar ultraviolet radiation at various wavelengths and various altitudes by a rocket-borne spectrograph.[12] Equations (3.12) and (3.13) could be used to compute the number densities from such measurements if the spectral lines were sharp. Actually they have finite widths and Eq. (3.12) must be integrated over the appropriate wavelength range. Hence the number densities are rather difficult to "unfold" from the intensity data. Some results obtained by Hinteregger and coworkers[13] are shown in Figure 3.8.

The other technique involves the use of a mass spectrometer to measure number density. Atmospheric atoms and molecules are ionized before entry into the instrument and the ions are mass analyzed, usually by a radio frequency quadrupole spectrometer. Numerous results such as those by Nier *et al.*[14] and Schaeffer and Nichols[15] shown in Figure 3.8 have been obtained, but one must be cautious in using atomic oxygen profiles; O tends to be under-estimated because of recombination on the walls of the device, although observers do indeed try to correct for this effect.

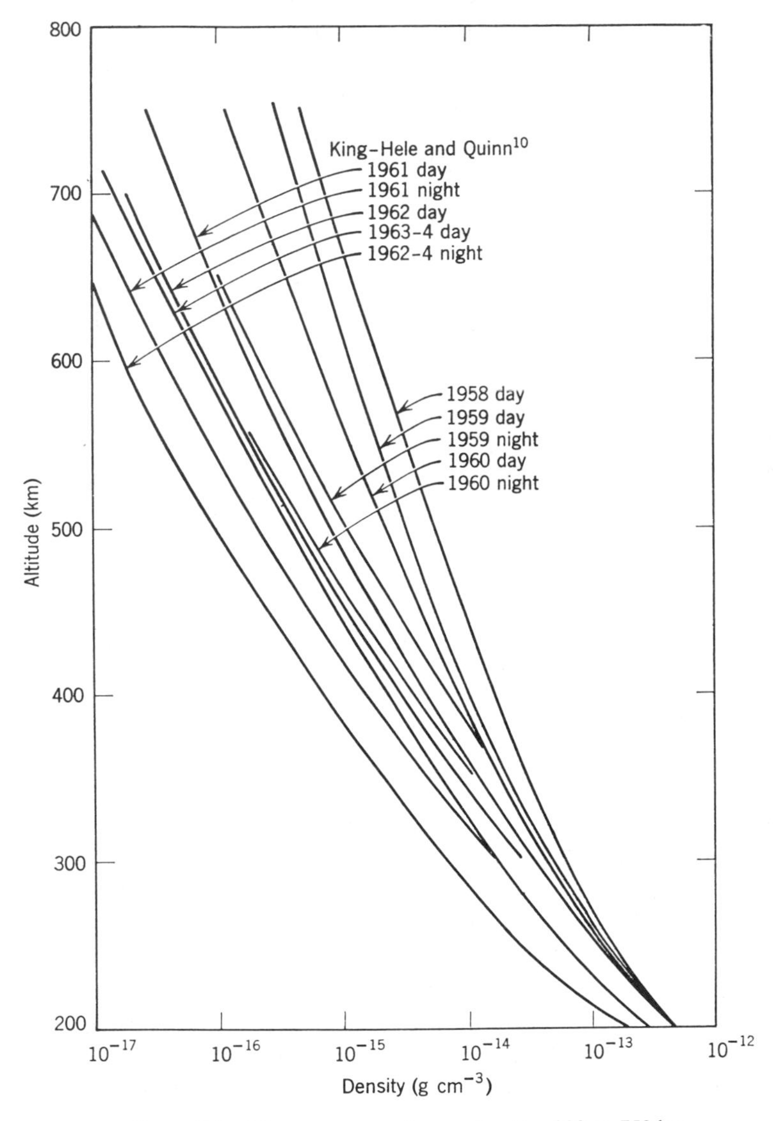

Figure 3.7 Measurements of mass density, 200 to 750 km.

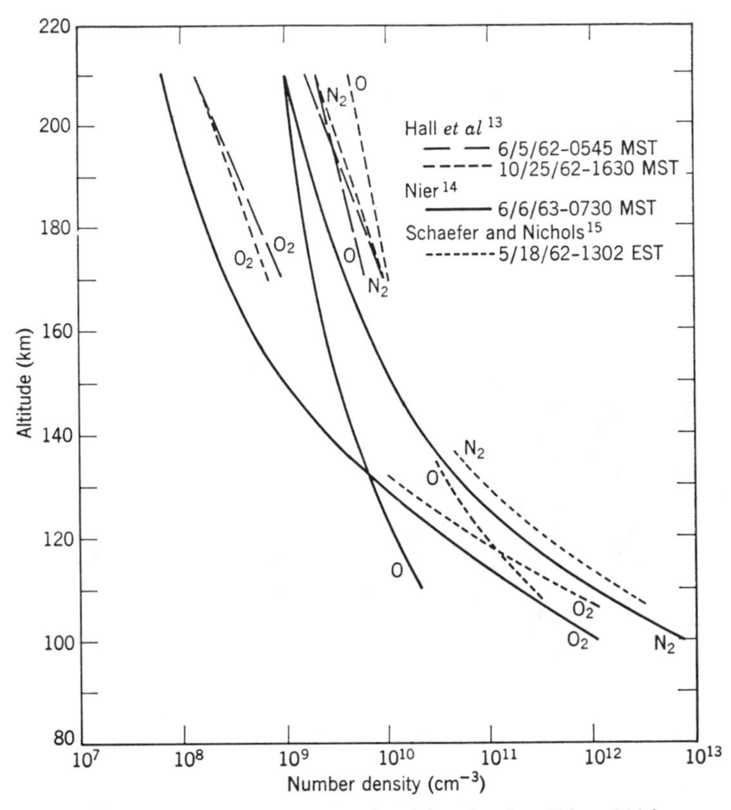

Figure 3.8 Measurements of number density, 100 to 200 km.

3.6.2 Temperature

The data available on upper air temperatures are rather sparse and the reliability of those which have been obtained is comparatively poor. However, several techniques have been used with some success and deserve mention here. The rocket grenade method which has been in use for many years is based on the relationship of the velocity of sound to absolute temperature. As the rocket ascends from perhaps 60 to 90 km, grenades are ejected and exploded every few kilometers and the sounds from the explosion are detected by an array of microphones on the ground. The direction from which the waves are received as well as the measured time intervals between the arrival of successive detonation waves are then employed to compute upper air temperatures and wind velocities. The results of Stroud[16] shown in Figure 3.9 are particularly interesting because they exhibit very prominently the winter warming of the high latitude mesosphere. The interested reader is referred to Craig for a more detailed discussion of the rocket grenade method.

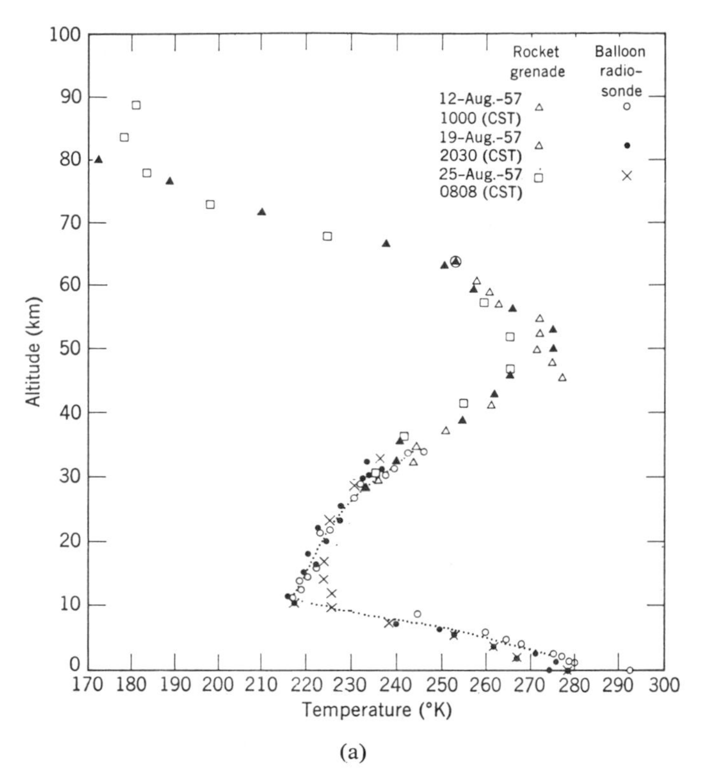

(a)

Figure 3.9 Temperature as a function of altitude at Churchill, determined from the rocket grenade experiment on the indicated days in (a) August 1957, (b) December 1957. (After Stroud *et al. J. Geophys. Res.* **65,** 2307 (1961); reprinted by permission of the American Geophysical Union.)

Spectroscopic methods have been employed for measurements at higher altitudes, but their reliability is open to some question. One of the earliest and best of these measurements was carried out by Blamont[17] who released a cloud of sodium vapor at an altitude of about 370 km. Observation of the Doppler width of the *D*-lines emitted from the cloud yielded a temperature of $1450 \pm 75°K$. Other techniques employ line width measurements of airglow line spectra as well as rotational band spectra of OH lines from the airglow. In the latter case it is assumed that the rotational levels of the OH molecules are populated according to a Boltzmann distribution and that the rotational temperature is equal to the gas kinetic temperature. The major error here is height determination. There is good reason to believe that the population of the rotational states of OH is in thermal equilibrium with the

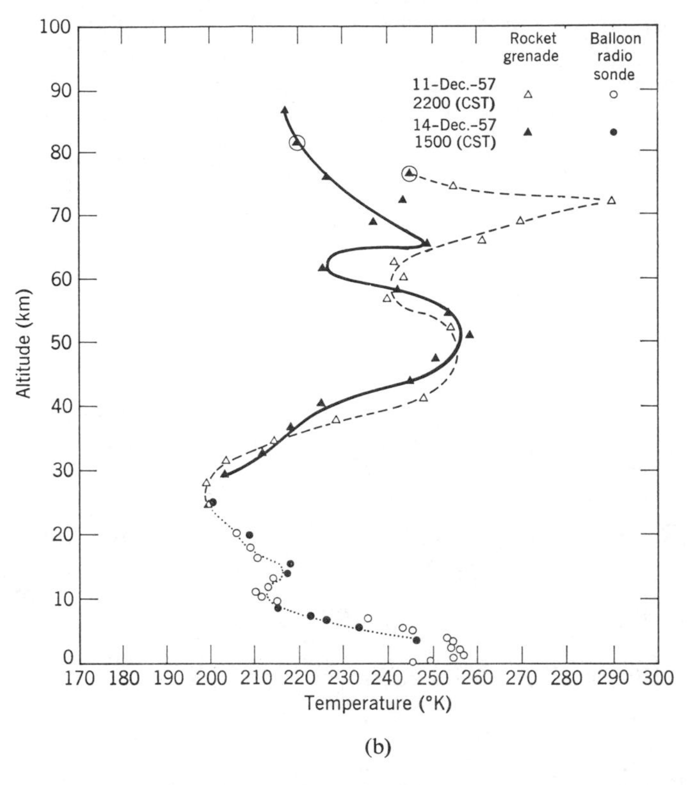

(b)

Fiurge 3.9 (Contd.)

atmosphere so that the rotational temperature is indeed representative of the true gas kinetic temperature.

Reasonably reliable estimates of upper thermospheric temperatures can be made from density measurements, provided that the composition is reasonably well known. In the altitude range 300 to 500 km, the principal constituent is atomic oxygen, and neglect of others such as O_2, N_2, and He is usually justified. By means of this approach which involves Eq. (2.21), we estimate the temperature at 450 km during the daytime in 1961 to be 920°K (see Figure 3.7).

3.6.3 Diffusion

Most of the work on the pertinent diffusion parameters has been carried out in the laboratory or has been done by computational methods. However, there have also been some very useful *in situ* measurements of the effective molecular diffusion coefficients at high altitudes by observing the spread of a cloud of sodium or lithium vapor released from a rocket. At altitudes above ~110 km this motion proceeds in a very uniform manner and is undoubtedly due to molecular diffusion. By suitable photographic techniques employing triangulation for accurate determination any part of the cloud can be tracked with a high degree of precision. The time rate of development of the cloud leads directly to the diffusion coefficient. Some of the more recent results which were summarized in a paper by Golomb and McLeod[18] are shown in Figure 3.10. These authors also discuss the basic theory of the cloud diffusion process (which is merely an extension of our development in Section 3.4) and the interested reader is referred to their paper for details.

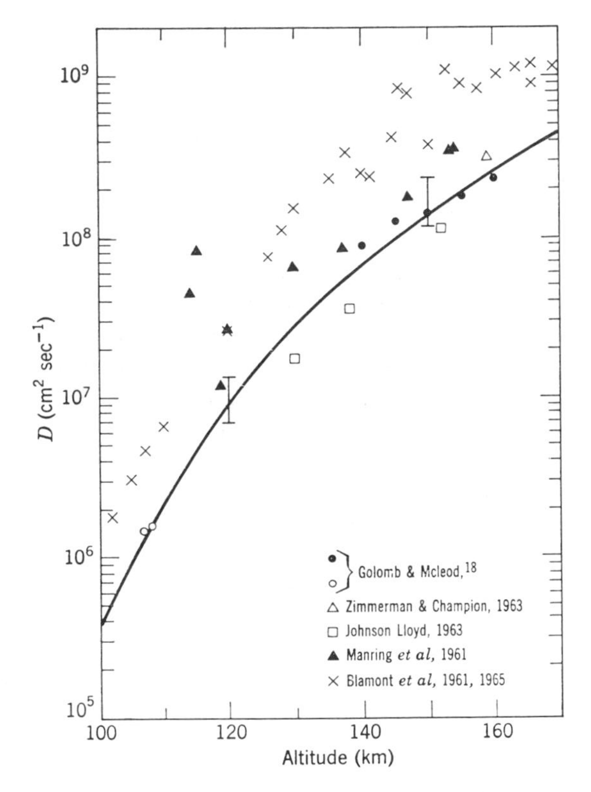

Figure 3.10 Molecular diffusion coefficients as measured in the altitude range 100-170 km by workers using various diffusing species. The bars represent theoretical extremes of variation with season and solar cycle. *See* Golomb and MacLeod for a list of the references. (After Golomb and MacLeod, *J. Geophys. Res.* **71,** 2299 (1966); reprinted by permission of the American Geophysical Union.)

3.6.4 Observation of Planetary Atmospheres by Spacecraft

With the aid of interplanetary spacecraft, investigators have been able to probe the atmospheres of Mars and Venus in much greater detail than was ever possible previously. The first successful observations of Mars were made by the Mariner 4 spacecraft in 1965. More recently (1967) the atmosphere of Venus was probed[20,21] by Mariner 5 and by the Russian "Venera 4." The principle employed by the Mariner craft for observations of atmospheric density was the refraction of radio signals (illustrated in Figure 3.11) when they passed through the atmosphere and ionosphere on their way from Earth to the spacecraft (see Section 11.7.8).

The results from Mariner 4, which are presented in Figure 3.12, were rather surprising. Prior to the experiment most investigators had estimated the pressure at the Martian surface to be about 0.1 earth atmosphere and to decrease more slowly with altitude than does the earth's atmosphere. The reason for assuming a larger scale height for Mars is logical enough—the gravitational acceleration is smaller on Mars while the temperature and mean molecular mass was thought to be approximately that for Earth. The "occultation" experiment, however, yielded a scale height which did not differ greatly from the earth's. One can account for this result by assuming a sufficiently low temperature or sufficiently large mean molecular mass. The observed surface pressure was also smaller than previously thought, only about 0.01 earth atmosphere. The abundance of CO_2 in the Martian atmosphere was measured by spectroscopic techniques some years before Mariner 4. As it turned out, the particle concentration obtained from the spacecraft was quite close to the result of the CO_2 experiment. Hence, one can assume that the lower atmosphere is entirely CO_2 and obtain results in quite good agreement with all experiments. The surface temperature was then estimated from the observed scale height to be $\sim$180°K. As we shall see, carbon dioxide is not appreciably dissociated in the upper atmosphere of Venus; by implication it should not be dissociated in the upper atmosphere of Mars either.

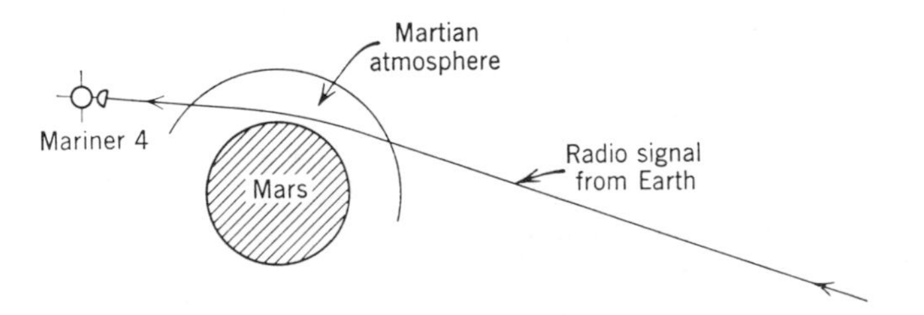

Figure 3.11 Schematic of the Mariner 4 occultation experiment. The radio signal from the spacecraft to earth is refracted by the Martian atmosphere as the spacecraft is occulted.

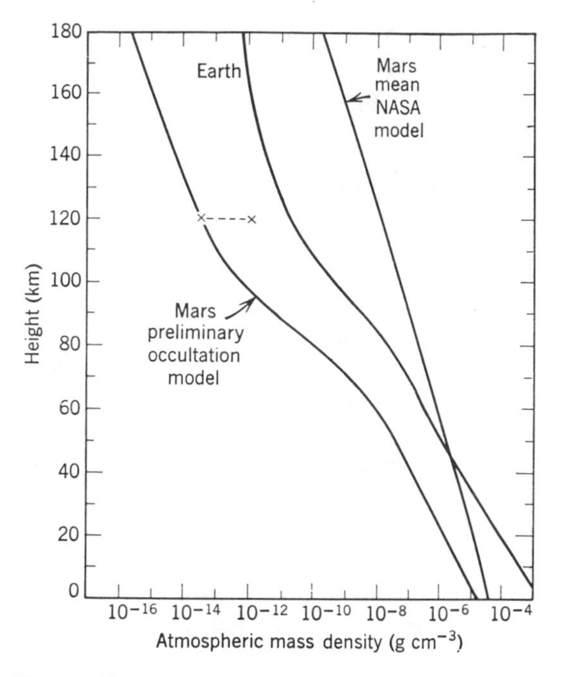

Figure 3.12 Atmospheric mass density versus altitude above Electris at the time of immersion into occultation. The NASA engineering model shown here was used for planning purposes prior to the Mariner 4 mission. (After Fjeldbo, Fjeldbo, and Eshleman, *J. Geophys. Res.* **71**, 2307 (1966); reprinted by permission of the American Geophysical Union.)

The atmosphere of Venus is considerably denser near the planetary surface than is that of Mars. This leads to a rather curious phenomenon in the transmission of signals from Mariner 5; when the signal rays passed through the level at which the pressure was about 8 earth atmospheres, their radius of curvature due to refraction was equal to the planetary radius. Hence, they were trapped in the atmosphere and could not escape. In fact, no signals were obtained when rays passed below the 5 or 8 atmospheres level. It is thus quite certain that the cut-off was due to critical refraction. Atmospheric temperatures were inferred from the measured scale height. They extended from a high of 450°K at the 5-atmosphere level to 240°K at an altitude 30 km higher. The vehicle was also instrumented to detect atomic oxygen by looking for resonance scattering of sunlight of appropriate frequency by the oxygen atoms. However, none was observed to within the sensitivity of the detectors. This was a surprising result because it was expected that solar UV radiation would dissociate CO_2 rather strongly at high altitudes.

The findings of Mariner 5 were confirmed dramatically by the USSR craft (Venera 4) which made direct measurements of atmospheric properties at low altitudes; apparently it stopped transmission at about 25 km above the surface. The most important of these was composition: CO_2—90 to 95 percent; $N_2 < 7$ percent; $O_2 \sim 0.4$ percent; $H_2O + O < 1.6$ percent; $O < 3 \times 10^3\ cm^{-3}$ at 300 km altitude. Profiles of temperature and pressure obtained by Mariner 5 and Venera 4 are shown in Figure 3.13.

3.7 Model Atmospheres

Numerous models of the earth's atmosphere which present temperature, pressure, mass density, and other properties as a function of altitude have been constructed within recent years. The earlier ones (i.e., before 1957) underestimated the density at high altitudes; the use of satellite orbital observations beginning with Sputnik I have greatly improved this situation. The mass density profiles shown in Figure 3.7 can, in fact, be used as a basis for the construction of atmospheric models. This is generally the approach of the empirical models such as the COSPAR† International Reference Atmosphere and the U.S. Standard Atmosphere to be briefly discussed later. Before doing this, however, it is instructive to study in some detail the

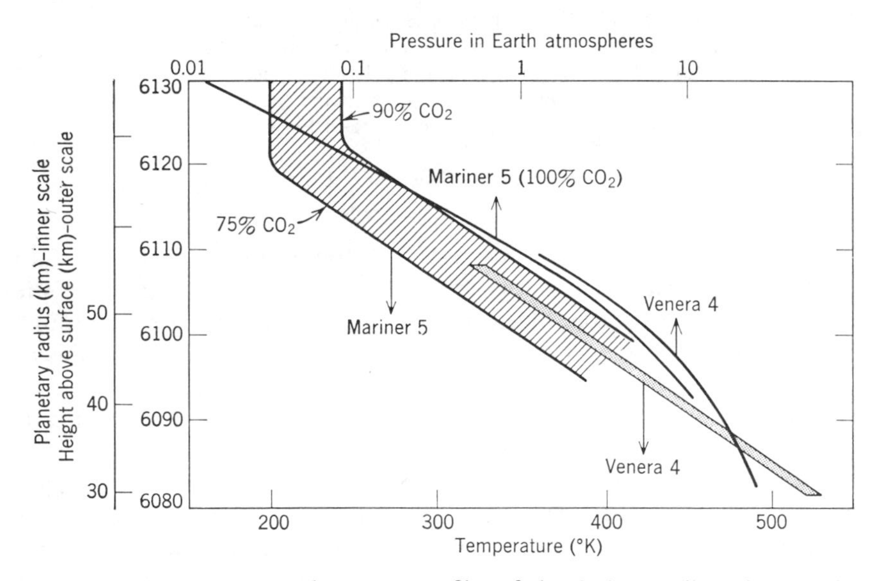

Figure 3.13 Temperature and pressure profiles of the Cytherean (Venus) atmosphere derived from observations by Mariner 5 and Venera 4.

† COSPAR stands for Committee on Space Research.

semiempirical model of Harris and Priester in order to elucidate the physical principles involved in model construction. We use the adjective "semi-empirical" because it is based in part on empirically determined boundary conditions.

Model atmospheres for Venus and Mars have also been constructed from time to time over a number of years. The most recent ones will be discussed briefly in Section 3.7.3.

3.7.1 The Model of Harris and Priester[23]

These two investigators have constructed a very comprehensive model extending from 120 to 2050 km altitude for various conditions of solar activity and times of the day. It is essentially a perturbation solution of the thermal balance problem of the thermosphere and thus must start with an equation describing the temperature dependence of a flowing ideal gas in an external gravity field.

$$\frac{\partial T}{\partial t} + \frac{\partial T}{\partial z} w + \frac{RT}{C_v}\frac{\partial w}{\partial z} = \frac{q}{C_v} \tag{3.51}$$

This is a variant of Eq. (2.31) with the assumption that the expansion of the rising air is adiabatic. Hence the change in entropy is entirely due to heat input. The actual derivation from thermodynamic and hydrodynamic principles is left as an exercise.

Here w is the vertical gas velocity, C_v is the specific heat capacity at constant volume, R is the universal gas constant (R = Boltzmann constant × Avogadro's Number), and q is the net gain of heat due to thermal sources, conduction and radiation. The factor R/C_v should actually be replaced by $C_p/C_v - 1$ (C_p is the specific heat capacity at constant pressure), if the exact expression is desired, but the form R/C_v is good enough for the purpose at hand. In order to be tractable with standard mathematical methods, Eq. (3.41) must be "linearized," i.e., $T(z, t) = T_0(z) + T_1(z, t)$ where $T_1 \ll T_0$

$$\frac{\partial T_1}{\partial t} + \frac{\partial T_0}{\partial z} w + \frac{RT_0}{C_v}\frac{\partial w}{\partial z} = \frac{q}{C_v} \tag{3.51a}$$

In order to solve this equation, it is necessary to obtain an expression for w. Harris and Priester attacked this aspect of the problem by considering a gas cell which moves vertically, all the while remaining at constant pressure

$$p(z + w\,\Delta t, t + \Delta t) = p(z, t) \tag{3.52}$$

If the motion is slow enough, we have (to first order)

$$p_0 \exp\left[-\int_0^{z+w\Delta t} \frac{gm}{kT(t + \Delta t)}\,dz'\right] = p_0 \exp\left[-\int_0^z \frac{gm}{kT(t)}\,dz'\right] \tag{3.53}$$

Expanding $[T(t+\Delta t)]^{-1}$ as

$$\frac{1}{T(t+\Delta t)} \approx \frac{1}{T(t)}\left[1 - \frac{1}{T(t)}\frac{\partial T}{\partial t}\Delta t\right]$$

and substituting into Eq. (3.53) yields the following value for the vertical velocity gradient:

$$\frac{\partial w}{\partial z} = \frac{\partial T_0}{\partial z}\int_0^z \frac{1}{T_0^2}\frac{\partial T_1}{\partial t}\,dz' + \frac{1}{T_0}\frac{\partial T_1}{\partial t} \tag{3.54}$$

Substituting this form of $\partial w/\partial z$ into Eq. (3.51a) yields

$$\frac{\partial T_1}{\partial t} + \frac{\partial T_0}{\partial z}T_0\int_0^z \frac{1}{T_0^2}\frac{\partial T_1}{\partial t}\,dz' = \frac{q}{C_p} \tag{3.51b}$$

which is the basic equation to be integrated numerically.

Before doing this, one must obtain an expression for q

$$q = R_{\text{cond}} + R_{\text{rad}} + R_{\text{EUV}} + R' \tag{3.55}$$

where R_{cond} is the thermal conduction term discussed in Section 3.2.

$$R_{\text{cond}} = \frac{\partial}{\partial z}k(T)\frac{\partial T}{\partial z} \tag{3.56}$$

$$k = \frac{\sum_i A_i n_i T^{1/2}(z)}{\sum_i n_i} \tag{3.57}$$

The symbols k, n_i, and A_i refer, respectively, to the thermal conductivity, the number density of the ith constituent, and a constant whose value varies from one constituent to another. The term denoted by R_{rad} is the radiation term of Bates which was discussed in Section 3.3 and is written as

$$R_{\text{rad}} = -n(\text{O})E_1A_{12}\left[\frac{W_1e^{-E_1/kT}}{W_2 + W_1e^{-E_1/kT} + W_0e^{-E_0/kT}}\right] \tag{3.58}$$

where $n(\text{O})$ is the number density of atomic oxygen, E_0 and E_1 are, respectively, the energy differences between the $^3P_0$ and $^3P_2$ levels and between the $^3P_1$ and $^3P_2$ levels of atomic oxygen, W is the statistical weight of the level indicated by the subscript, and A_{12} is the Einstein coefficient for the transition $^3P_1 - {}^3P_2$ (see Section 2.4.1 for a definition of Einstein coefficients). The R_{EUV} term is the heat input due to solar EUV radiation obtained in a manner similar to Eq. (3.14) except that we assume a multi-component atmosphere and a continuous EUV spectrum,

$$R_{\text{EUV}} = \sum_i \epsilon_i n_i(z)\int_0^\infty d\lambda\, F_\lambda Q_i(\lambda)e^{-\tau_i(\lambda,z,t)} \tag{3.59}$$

where τ is the optical depth

$$\tau_i(\lambda, z, t) = Q_i(\lambda) \int_z^\infty \frac{n_i(z')\, dz'}{\cos \chi(t)} \tag{3.60}$$

(See Section 3.1 for definition of the symbols.)

Finally, R' is the heat source term representing other heating mechanisms such as corpuscular radiation. The time dependence of this term was adjusted until agreement with observation was obtained (peak thermosphere temperature occurs at 2 p.m.). This necessitated the assumption of an R' with a maximum occurring at 9 a.m. and was postulated to be due to corpuscular radiation. Actually, there is considerable doubt concerning this interpretation. A better one is that horizontal pressure gradients produced by horizontal temperature gradients in the upper atmosphere cause mass motions (advection). It is these winds which provide the required heat transport.

In the integration procedure Harris and Priester started with an arbitrary temperature profile as an initial condition, the upper boundary condition $\partial T/\partial z = 0$ at 1000 km and a set of lower boundary conditions based on empirical data. In order to obtain convergence of the temperature profile, the heat equation had to be integrated over a time span of 4 to 5 days.

Aside from the temperature profile, the most important quantities given by the model are the number density profiles $n_i(z)$ for each species. With the assumption that the atmosphere is in diffusive equilibrium above 120 km, one can easily write an equation for $n_i(z)$ using the ideal gas law (3.3) and the barometric equation (3.4)

$$n_i(z, t) = n_i(0) \frac{T(0)}{T(z, t)} \exp \left[-\int_0^z \frac{m_i g(z')\, dz'}{kT(z', t)} \right] \tag{3.61}$$

Other properties such as mean molecular mass, scale height, pressure, mass density, etc., are then easily obtained from the hydrostatic theory outlined in Section 3.1. Some predictions from this model are shown in Table 3.1. The interested reader is referred to the original works of Harris and Priester for the complete model atmosphere.[23,24]

3.7.2 *Empirical Models*

The COSPAR International Reference Atmosphere[25] and the U.S. Standard Atmosphere[26] were prepared, respectively, under the aegis of COSPAR and various agencies of the U.S. Government. These models are essentially the work of committees using in their judgment the best data available from rockets and satellite measurements. Hence one cannot describe the origin of the models in quantitative terms. We shall content ourselves with a presentation of some of the principal features of the COSPAR

TABLE 3.1. PROPERTIES OF THE UPPER ATMOSPHERE FOR 14:00 h LOCAL TIME AS A FUNCTION OF ALTITUDE FROM 120 TO 2050 km. [After Harris and Priester, *J. Atmos. Sci.* **19,** 286 (1962); reprinted by permission of the American Meteorological Society.]

Alt km	Temp K	Density gm cm^{-3}	Pressure dyne cm^{-2}	Scale ht km	Mean mol wt	$N(N_2)$ cm^{-3}	$N(O_2)$ cm^{-3}	N(O) cm^{-3}	N(He) cm^{-3}	N(H) cm^{-3}
120	355	3.536E–11	3.802E–02	11.4	27.46	5.800E 11	1.200E 11	7.600E 10	2.500E 07	4.356E 04
130	509	1.180E–11	1.849E–02	16.6	26.98	1.928E 11	3.591E 10	3.470E 10	1.568E 07	2.960E 04
140	652	5.365E–12	1.095E–02	21.7	26.55	8.695E 10	1.498E 10	1.979E 10	1.131E 07	2.265E 04
150	789	2.889E–12	7.241E–03	26.8	26.17	4.630E 10	7.495E 09	1.272E 10	8.777E 06	1.843E 04
160	916	1.742E–12	5.138E–03	31.6	25.81	2.755E 10	4.231E 09	8.866E 09	7.168E 06	1.566E 04
170	1030	1.137E–12	3.824E–03	36.1	25.47	1.774E 10	2.602E 09	6.554E 09	6.085E 06	1.376E 04
180	1130	7.876E–13	2.943E–03	40.3	25.15	1.210E 10	1.703E 09	5.059E 09	5.138E 06	1.241E 04
190	1218	5.697E–13	2.322E–03	44.1	24.84	8.610E 09	1.167E 09	4.034E 09	4.751E 06	1.141E 04
200	1293	4.261E–13	1.867E–03	47.5	24.54	6.329E 09	8.285E 08	3.297E 09	4.318E 06	1.065E 04
220	1413	2.562E–13	1.257E–03	53.5	23.96	3.667E 09	4.499E 08	2.323E 09	3.700E 06	9.585E 03
240	1502	1.648E–13	8.799E–04	58.6	23.39	2.263E 09	2.616E 08	1.718E 09	3.277E 06	8.886E 03
260	1567	1.111E–13	6.334E–04	63.0	22.84	1.457E 09	1.591E 08	1.311E 09	2.967E 06	8.397E 03
280	1615	7.736E–14	4.655E–04	66.9	22.31	9.646E 08	9.977E 07	1.022E 09	2.725E 06	8.037E 03
300	1650	5.525E–14	3.478E–04	70.4	21.79	6.518E 08	6.397E 07	8.093E 08	2.528E 06	7.760E 03
320	1677	4.025E–14	2.635E–04	73.6	21.29	4.470E 08	4.168E 07	6.479E 08	2.363E 06	7.539E 03
340	1696	2.980E–14	2.019E–04	76.7	20.81	3.100E 08	2.749E 07	5.228E 08	2.219E 06	7.357E 03
360	1711	2.237E–14	1.563E–04	79.6	20.36	2.169E 08	1.830E 07	4.246E 08	2.092E 06	7.201E 03
380	1723	1.699E–14	1.221E–04	82.3	19.93	1.528E 08	1.228E 07	3.465E 08	1.978E 06	7.065E 03
400	1732	1.304E–14	9.613E–05	84.9	19.52	1.082E 08	8.289E 06	2.838E 08	1.874E 06	6.944E 03
420	1739	1.010E–14	7.623E–05	87.5	19.15	7.702E 07	5.624E 06	2.333E 08	1.779E 06	6.833E 03

TABLE 3.1. *(continued)*

Alt km	Temp K	Density gm cm^{-3}	Pressure dyne cm^{-2}	Scale ht km	Mean mol wt	$N(N_2)$ cm^{-3}	$N(O_2)$ cm^{-3}	N(O) cm^{-3}	N(He) cm^{-3}	N(H) cm^{-3}
440	1744	7.887E–15	6.084E–05	89.9	18.80	5.504E 07	3.833E 06	1.922E 08	1.690E 06	6.731E 03
460	1748	6.208E–15	4.885E–05	92.2	18.47	3.947E 07	2.623E 06	1.587E 08	1.608E 06	6.635E 03
480	1752	4.920E–15	3.943E–05	94.5	18.18	2.839E 07	1.801E 06	1.314E 08	1.531E 06	6.545E 03
500	1755	3.925E–15	3.198E–05	96.7	17.90	2.048E 07	1.241E 06	1.089E 08	1.459E 06	6.459E 03
520	1757	3.149E–15	2.606E–05	98.7	17.65	1.481E 07	8.572E 05	9.043E 07	1.391E 06	6.376E 03
540	1759	2.540E–15	2.133E–05	100.7	17.42	1.074E 07	5.939E 05	7.521E 07	1.327E 06	6.297E 03
560	1760	2.059E–15	1.752E–05	102.7	17.20	7.806E 06	4.126E 05	6.264E 07	1.267E 06	6.220E 03
580	1761	1.677E–15	1.444E–05	104.5	17.00	5.686E 06	2.873E 05	5.223E 07	1.210E 06	6.145E 03
600	1763	1.372E–15	1.195E–05	106.4	16.82	4.151E 06	2.006E 05	4.362E 07	1.156E 06	6.073E 03
620	1763	1.126E–15	9.917E–06	108.1	16.65	3.036E 06	1.404E 05	3.646E 07	1.104E 06	6.002E 03
640	1764	9.276E–16	8.255E–06	109.9	16.48	2.225E 06	9.845E 04	3.052E 07	1.056E 06	5.933E 03
660	1765	7.667E–16	6.891E–06	111.6	16.33	1.634E 06	6.920E 04	2.558E 07	1.010E 06	5.865E 03
680	1765	6.357E–16	5.768E–06	113.3	16.17	1.202E 06	4.875E 04	2.146E 07	9.659E 05	5.799E 03
700	1766	5.285E–16	4.842E–06	115.1	16.03	8.862E 05	3.441E 04	1.802E 07	9.243E 05	5.735E 03
720	1766	4.406E–16	4.075E–06	116.8	15.88	6.545E 05	2.434E 04	1.515E 07	8.848E 05	5.671E 03
740	1767	3.682E–16	3.438E–06	118.6	15.73	4.842E 05	1.726E 04	1.275E 07	8.471E 05	5.609E 03
760	1767	3.084E–16	2.908E–06	120.5	15.58	3.588E 05	1.226E 04	1.074E 07	8.113E 05	5.548E 03
780	1767	2.590E–16	2.467E–06	122.4	15.43	2.664E 05	8.723E 03	9.057E 06	7.772E 05	5.489E 03
800	1767	2.179E–16	2.097E–06	124.4	15.26	1.981E 05	6.221E 03	7.645E 06	7.448E 05	5.430E 03
820	1768	1.837E–16	1.788E–06	126.5	15.10	1.476E 05	4.445E 03	6.460E 06	7.139E 05	5.372E 03
840	1768	1.552E–16	1.529E–06	128.7	14.92	1.101E 05	3.182E 03	5.464E 06	6.844E 05	5.315E 03
860	1768	1.314E–16	1.311E–06	131.0	14.74	8.233E 04	2.282E 03	4.625E 06	6.564E 05	5.260E 03
880	1768	1.114E–16	1.127E–06	133.6	14.54	6.164E 04	1.640E 03	3.920E 06	6.296E 05	5.205E 03
900	1768	9.469E–17	9.714E–07	136.3	14.33	4.622E 04	1.180E 03	3.325E 06	6.041E 05	5.151E 03
920	1768	8.062E–17	8.401E–07	139.2	14.11	3.472E 04	8.514E 02	2.822E 06	5.797E 05	5.098E 03

TABLE 3.1. (*continued*)

Alt km	Temp K	Density gm cm^{-3}	Pressure dyne cm^{-2}	Scale ht km	Mean mol wt	$N(N_2)$ cm^{-3}	$N(O_2)$ cm^{-3}	N(O) cm^{-3}	N(He) cm^{-3}	N(H) cm^{-3}
940	1768	6.877E–17	7.288E–07	142.3	13.87	2.612E 04	6.151E 02	2.398E 06	5.565E 05	5.046E 03
960	1768	5.877E–17	6.342E–07	145.7	13.62	1.968E 04	4.452E 02	2.040E 06	5.343E 05	4.995E 03
980	1768	5.033E–17	5.538E–07	149.4	13.36	1.485E 04	3.228E 02	1.736E 06	5.131E 05	4.945E 03
1000	1768	4.319E–17	4.853E–07	153.4	13.08	1.123E 04	2.345E 02	1.479E 06	4.929E 05	4.895E 03
1050	1768	2.972E–17	3.542E–07	164.9	12.34	5.612E 03	1.062E 02	9.951E 05	4.462E 05	4.775E 03
1100	1768	2.074E–17	2.647E–07	179.0	11.52	2.832E 03	4.865E 01	6.728E 05	4.044E 05	4.659E 03
1150	1768	1.469E–17	2.026E–07	196.0	10.66	1.442E 03	2.251E 01	4.573E 05	3.670E 05	4.548E 03
1200	1768	1.058E–17	1.589E–07	216.3	9.79	7.407E 02	1.052E 01	3.124E 05	3.335E 05	4.440E 03
1250	1768	7.759E–18	1.276E–07	239.9	8.94	3.839E 02	4.967E 00	2.145E 05	3.035E 05	4.337E 03
1300	1768	5.801E–18	1.047E–07	266.8	8.15	2.007E 02	2.368E 00	1.480E 05	2.765E 05	4.237E 03
1350	1768	4.426E–18	8.762E–08	296.5	7.43	1.058E 02	1.140E 00	1.026E 05	2.522E 05	4.140E 03
1400	1768	3.449E–18	7.464E–08	328.3	6.79	5.622E 01	5.538E–01	7.149E 04	2.303E 05	4.047E 03
1450	1768	2.745E–18	6.456E–08	361.4	6.25	3.012E 01	2.716E–01	5.003E 04	2.106E 05	3.958E 03
1500	1768	2.230E–18	5.656E–08	394.7	5.80	1.627E 01	1.344E–01	3.517E 04	1.927E 05	3.871E 03
1550	1768	1.847E–18	5.008E–08	427.4	5.42	8.855E 00	6.710E 02	2.483E 04	1.766E 05	3.788E 03
1600	1768	1.557E–18	4.473E–08	458.6	5.12	4.857E 00	3.379E–02	1.761E 04	1.620E 05	3.707E 03
1650	1768	1.334E–18	4.025E–08	487.8	4.87	2.684E 00	1.717E–02	1.254E 04	1.488E 05	3.629E 03
1700	1768	1.158E–18	3.643E–08	514.7	4.67	1.494E 00	8.793E 03	8.973E 03	1.368E 05	3.553E 03
1750	1768	1.018E–18	3.313E–08	539.3	4.52	8.376E–01	4.542E–03	6.444E 03	1.258E 05	3.480E 03
1800	1768	9.037E–19	3.026E–08	561.6	4.39	4.730E–01	2.365E–03	4.647E 03	1.159E 05	3.409E 03
1850	1768	8.092E–19	2.772E–08	581.7	4.29	2.689E–01	1.241E–03	3.365E 03	1.069E 05	3.341E 03
1900	1768	7.296E–19	2.547E–08	600.0	4.21	1.540E–01	6.564E–04	2.446E 03	9.867E 04	3.275E 03
1950	1768	6.618E–19	2.346E–08	616.8	4.15	8.874E–02	3.449E–04	1.784E 03	9.116E 04	3.210E 03
2000	1768	6.032E–19	2.166E–08	632.2	4.09	5.149E–02	1.879E–04	1.307E 03	8.430E 04	3.148E 03
2050	1768	5.520E–19	2.003E–08	646.5	4.05	3.006E–02	1.016E–04	9.607E 02	7.803E 04	3.088E 03

model in Figure 3.14. The two models agree so closely in most respects that a separate figure for the U.S.S. model is not given here.

3.7.3 Atmospheric Models of Venus and Mars

Atmospheric models for Venus and Mars can also be constructed using the principles outlined earlier in this chapter. One must, of course, have information concerning composition and boundary conditions, e.g., surface pressure and thermospheric temperature. The most recent have been

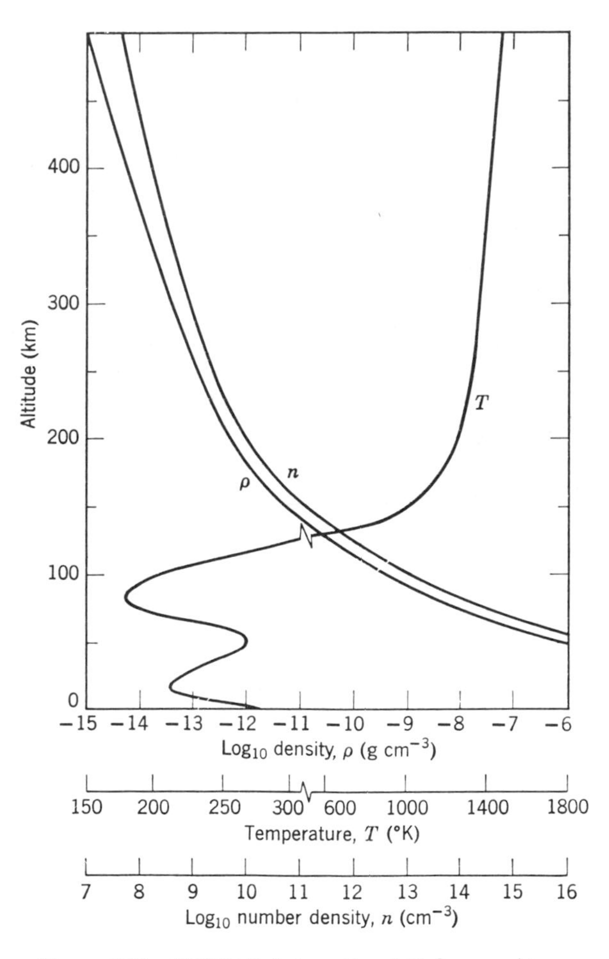

Figure 3.14 COSPAR International Reference Atmosphere.[24] Mean values of temperature, density, and number density.

developed by McElroy[27] who employed spectroscopic observations of CO_2 abundance together with results[19,20] from Mariner 4 and 5 as lower boundary conditions. In order to try to learn something about the composition, McElroy repeated his computations for varying amounts of CO_2, N_2, CO, O, and A. He obtained the best results in best agreement with the Mariner 5 observations by using a pure CO_2 model, which is presented in Figure 3.15a. Indeed the computed electron number density, assuming that the ionosphere was formed by solar EUV radiation, agreed extremely well with the observed ionosphere (see Section 8.5).

A similar program for Mars was also carried out with the aid of the Mariner 4 observations. Although the best agreement with observation was obtained with a pure CO_2 atmosphere shown in Figure 3.15b, the computed electron number density profile differed considerably from the actual one. Since models of the ionospheres of Mars and Venus are discussed in some detail in Chapters 8 and 9, we shall not consider them further in this chapter.

3.7.4 Atmospheric Models of Jupiter

Present knowledge of the structure of the Jovian atmosphere is based entirely upon ground-based spectroscopic observations. We know that the atmosphere contains H_2, CH_4, and NH_3, although there is uncertainty as to the abundances of each; we suspect that it contains helium also. Theoretical

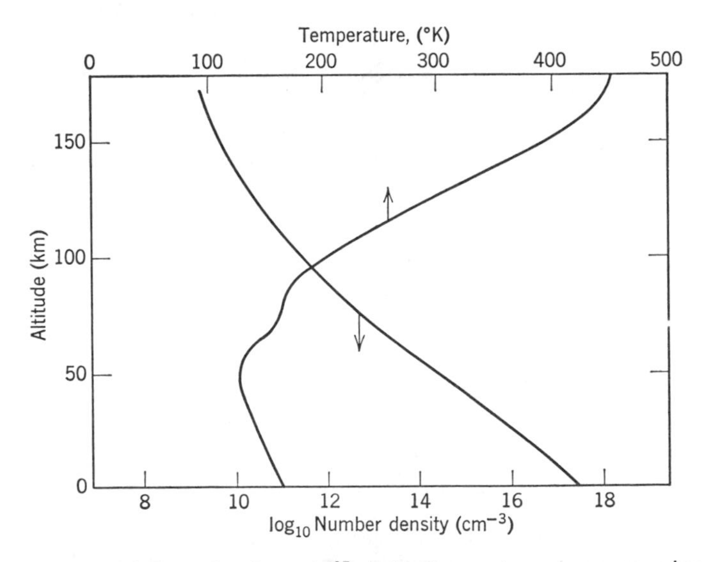

Figure 3.15a McElroy's model[27] of the Venus atmosphere, assuming that it is composed of pure CO_2.

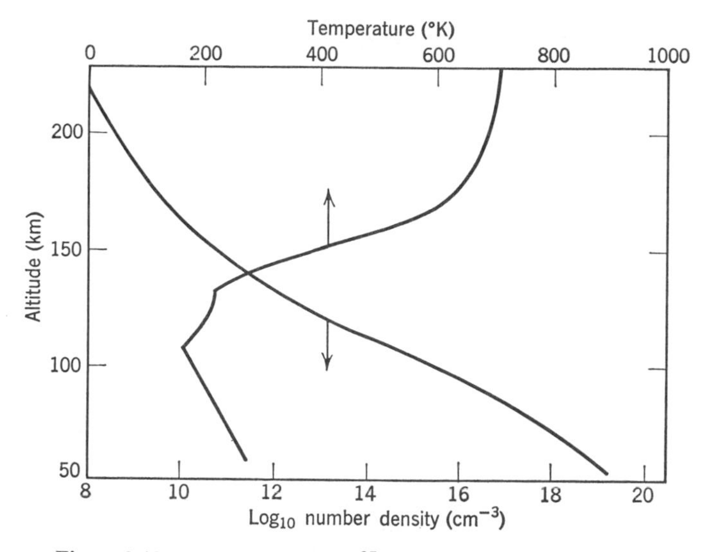

Figure 3.15b McElroy's model[27] of the Mars atmosphere, assuming that it is composed of pure CO_2.

considerations yield an He/H_2 ratio between 1/2 and 2. We also have a rough idea of the temperature from infrared measurements.

Because of the paucity of observational data, it is not yet possible to construct atmospheric models which have the same degree of reliability as for Mars and Venus. One such model[28] is presented in Figure 3.16. It is based in part on the temperature measurements and on a model of the lower atmosphere developed by Trafton.[29] Since the surface of Jupiter is cloud-covered and thus not visible, it was difficult to choose a reference altitude. Because of the very high thermal conductivity of H_2 it has been conjectured that the high atmosphere is isothermal, but below a certain altitude (the mesopause) the temperature increases with decreasing height; the mesopause is usually taken as the reference altitude. In the model shown it is assumed that turbulence keeps the atmosphere well mixed below the mesopause but that it is in diffusive equilibrium at higher altitudes.

Molecular hydrogen, like N_2 in the earth's atmosphere, is only weakly photodissociated. The little atomic hydrogen which does occur is produced mainly from H_2^+ by various ionic reactions; the H_2^+ is formed by ionization of H_2 by solar EUV radiation.

3.8 Planetary Exospheres

In a general qualitative way, the exosphere (sometimes called the planetary corona) can be described as the region where the mean free path of the neutral

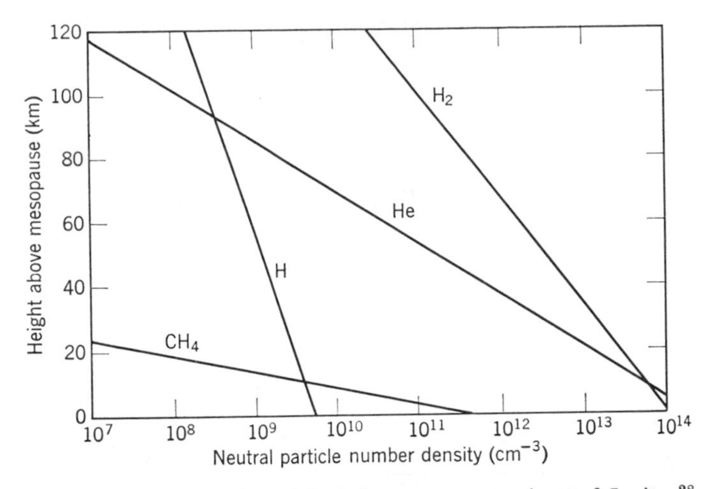

Figure 3.16 Proposed model of the upper atmosphere of Jupiter.[28] Diffusive equilibrium is assumed above the mesopause, mixing equilibrium below. The relative abundances of H_2 and He are 1 to 3. The thermospheric temperature is taken to be 96°K.

particles is so long that a substantial number are able to escape from the atmosphere. More specifically, we define it as the region in which the mean free path l exceeds the scale height H. Although this supposition of a critical level where $l = H$ is not fully justified, the theoretical predictions are surprisingly accurate. The rigorous result can only be obtained by solving the appropriate form of the Boltzmann equation. Because the simple theory developed originally by Jeans is adequate for most purposes and because it illustrates the physics of the problem extremely well, we shall present it in some detail.

3.8.1 Theory of the Exosphere

The demonstration that the mean free path is equal to the scale height at the critical level z_c is quite straightforward. We begin by defining z_c as the height at which a group of fast upward-moving particles will suffer no more collisions. To a good approximation the molecules can be treated as hard spheres of radius a; hence, their collision cross section is $Q_c = \pi(2a)^2 = 4\pi a^2$ and the probability of collision when crossing a region of height dz is

$$4\pi a^2 N(z)\, dz$$

where N is the particle number density. We have already stated that no collisions occur above z_c. Thus the probability for free passage to $z = \infty$ is unity, or

$$1 = \int_z^\infty 4\pi a^2 N(z)\, dz \tag{3.62}$$

Assuming the atmosphere to be isothermal so that

$$N(z) = N(z_c)e^{-(z-z_c)/H} \tag{3.63}$$

we obtain the result

$$N(z_c) = (4\pi a^2 H)^{-1} \tag{3.64}$$

which serves as a definition of the critical height. Since the mean free path (m.f.p.) between collisions is given by

$$l \approx \frac{1}{4\pi a^2 N} \tag{3.65}$$

we see that the scale height is approximately equal to the m.f.p. at the critical height. For typical temperatures ($\sim$1500°K) and densities in the earth's upper atmosphere, $z_c \approx 550$ km which may be taken as the lower boundary of the exosphere.

Exospheric particles are conveniently classified according to the character of their trajectories:

a. Ballistic particles coming up from the critical level on elliptic trajectories. They have less than escape velocities.
b. Trapped particles in bound orbits. These particles do not descend to the critical level. If there were no collisions above the critical level, this class would not be populated. Such collisions do in fact occur and trapped particles cannot be neglected in an accurate treatment of the problem.
c. Escaping particles which come up from the critical level.
d. Particles coming in from interplanetary space, whose trajectories pass through the critical level.
e. Particles coming in from interplanetary space, whose trajectories lie above the critical level.

Ordinarily, classes d and e are insignificant and can be neglected but class b may be of importance. Also, the velocity distribution of particles in the exosphere may depart significantly from Maxwellian. It is surprising then that the simple Jeans theory of atmospheric escape, which neglects collisions and assumes the gas to be Maxwellian, works as well as it does. Computation of the escape flux Φ is quite straightforward; merely compute the mean upward velocity of particles (see Section 2.3) in class c and multiply by the number density at the critical level

$$\Phi = 2\pi N(z_c)\left(\frac{m}{2\pi kT}\right)^{3/2} \int_{v_{\text{escape}}}^{\infty} \int_0^{\pi/2} v^3 \exp\left(-\frac{mv^2}{2kT}\right) dv \cos\theta \sin\theta \, d\theta \tag{3.66}$$

Here, θ is the angle between $\mathbf{v}$ and the vertical and the angular integration is carried out over the upper hemisphere only. Since the total energy of a particle which can just escape is zero, the escape velocity must be related to the gravitational constant G, the mass of the earth M, and the radius R_c of the critical level by

$$v_{\text{escape}} = \sqrt{\frac{2GM}{R_c}} \tag{3.67}$$

The escape flux is then

$$\Phi = \tfrac{1}{2}N(z_c)\sqrt{\frac{2kT}{\pi m}}\left[1 + \frac{GMm}{kTR_c}\right]\exp\left(-\frac{GMm}{kTR_c}\right) \tag{3.68}$$

for an isothermal atmosphere at temperature T.

We have already indicated that the Jeans approach is not rigorous. Collisions do occur in the exosphere, the critical level is not really well-defined, and the velocity distribution is not Maxwellian. Numerous attempts to improve the simple theory have been made with considerable success. However, their development is quite complex and lack of space precludes their further consideration. In any case the final results do not materially differ from those of the Jeans theory. The reader who wishes to pursue the topic is referred to the comprehensive paper by Chamberlain.[30]

3.8.2 The Earth's Exosphere (Geocorona)

One can gain a general idea of the composition of the exosphere from Figure 3.17. Although the exact values should not be taken literally because the Harris and Priester modeling technique is not valid in the exosphere, they are representative. The dominance of helium and atomic hydrogen at great altitudes is the outstanding feature. It is well-known that the former is a product of the radioactive decay of the heavier nuclei such as uranium, thorium, and radium that rises to the exosphere by upward transport. Hydrogen originates at the earth's surface in the form of water vapor and methane, and because these compounds are carried aloft by turbulent diffusion, they approximate a mixing distribution up to the levels where dissociation occurs. In the thermosphere hydrogen is dissociated and both H and He are in diffusive equilibrium. Because of their small masses, their scale heights are much larger than those of O and N_2, thus accounting for the greater relative abundance of H and He except perhaps near the bottom of the exosphere.

Of the two constituents, helium is in some ways the more interesting because it poses a problem which has not yet been solved. If the helium content of the atmosphere is indeed in equilibrium, then some nonthermal

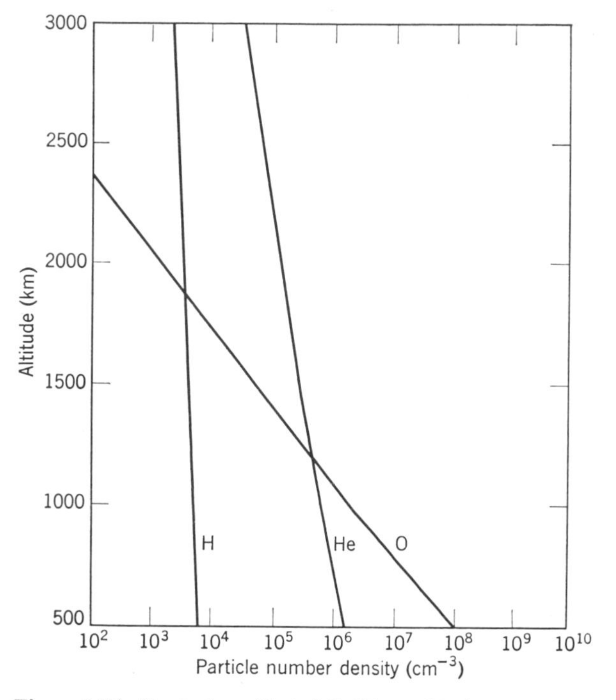

Figure 3.17 Typical profiles of O, He, and H in the earth's exosphere. Number densities were taken from Table 3.1 and extrapolated above 2000 km.

mechanism must dominate helium escape: The rate of escape predicted by the classical Jeans formula [Eq. (3.68)] is too low by at least an order of magnitude.[29] Since there is no good reason to doubt that equilibrium has existed for the last million years (the time required for the helium content to reach its present value if only thermal escape is effective), the only alternative is a nonthermal loss mechanism. The suggestion[29] that helium is lost in ionized form to interplanetary space via open magnetic field lines may be the answer. So far, it is too early to be sure.

Exospheric hydrogen is important because of its charge transfer reactions with atomic oxygen. As we shall show in Chapter 9, most of the ions present at exospheric altitudes are H^+.

3.8.3 The Exosphere of Venus

As of this writing, the only observation of the exosphere of another planet was made by Mariner 5 when it flew past Venus. The experiment was performed by measuring the intensity of solar Lyman α radiation which was

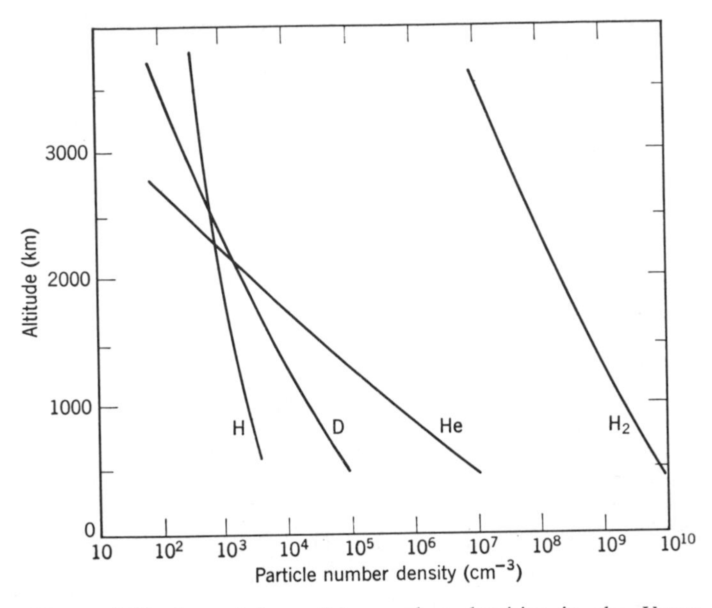

Figure 3.18 Suggested particle number densities in the Venus exosphere.

scattered by the gases there.[30] In order to explain the measurements, it is necessary to assume that two components, one with half the scale height of the other, are present. If the exosphere is composed mainly of hydrogen, as seems reasonable, it is necessary to assume (a) that atomic hydrogen at two different temperatures, one of which is twice the other, is present, (b) that atomic hydrogen and deuterium are present, or (c) that molecular, as well as atomic, hydrogen is present. It is difficult to see why two temperatures, one almost exactly twice the other, should exist; hence, we rule out (a). On the other hand, the concentration of H_2 which must be present in order to cause the observed intensity of scattered Lyman α is $\sim 10^{10}$ cm^{-3} at 6250 km radius. As we shall see in Chapter 9, this value would lead to too large an electron concentration above 6250 km. We are, thus, left with choice (b). The principal difficulty with choice (b) lies in the required abundance of D: it is about 20 times larger than H at the base of the exosphere. McElroy and Hunten[32] have argued that isotopic enrichment may have occurred through differential thermal escape. More measurements are urgently required. It is also probable that helium is present in the exosphere. In fact, as we shall show in Chapter 9, He seems to be necessary in order to account for the observed electron concentration above 6250 km. Some proposed number density profiles for H, D, H_2, and He are shown in Figure 3.18.

PROBLEMS

3.1. The scale height in a particular atmosphere is given (to second order) by $H = H_0(1 + \alpha z + \beta z^2)$. Derive an expression for the pressure and number density as a function of altitude z. Assume the gravitational acceleration to be constant. Repeat for the case in which the temperature lapse rate is adiabatic. *Hint:* Use Eq. (2.31).

3.2. Show that the distance through which a particle, whose energy is equal to the mean energy of an assembly of particles, will rise is one-half the scale height. Assume that the assembly is in thermal equilibrium.

3.3. Derive an expression for the optical depth for a given radiation in an atmosphere whose scale height varies with altitude as in Eq. (3.5).

3.4. The atmosphere is irradiated by a black body source. In the wavelength region near the peak of the spectrum and above some frequency threshold ν_0 the absorption cross section of the atmosphere varies as (frequency)$^{-3}$. Obtain an expression for the local energy deposition. This problem is important in atmospheric energy absorption during a high altitude nuclear explosion.

3.5. A single-component atmosphere is characterized by a scale height which varies according to Eq. (3.5). It is irradiated by monochromatic radiation for which the absorption cross section is Q. Obtain an expression for the altitude z_m at which maximum absorption occurs.

3.6. Consider an atmosphere consisting of oxygen. What is the effect on the concentration of atomic oxygen if each of the following quantities is doubled?. (a) the intensity of the dissociating radiation; (b) the absorption cross section; (c) the scale height.

3.7. An atmosphere with scale height H and a coefficient of thermal conductivity which is independent of temperature absorbs energy at the rate $qe^{-z/h}$ watts m^{-3}. If the upper boundary is $\lim_{z\to\infty} dT/dz = 0$ and the lower boundary condition is $T(0) = T_0$, derive an expression for the temperature profile. Assume quasiequilibrium.

3.8. An atmosphere is in diffusive and photochemical equilibrium, respectively, at very high and very low altitudes. Compute the density N as a function of altitude of a minor constituent whose scale height is H and which is characterized by a linear loss rate $\alpha e^{-2z/H}$ and production term $J_0 e^{-z/H}$. The boundary conditions are $\lim_{z\to\infty} N \sim e^{-z/H}$, $\lim_{z\to-\infty} N = 0$.

3.9. Compute the vertical diffusion velocity in Problem 3.8, assuming an isothermal atmosphere.

3.10. A minor component in an atmosphere in diffusive equilibrium is allowed to decay from some initial profile. Obtain an equation for the number density N at altitude z after time t, assuming the boundary conditions $\lim_{z\to-\infty} N = 0$ and $\lim_{z\to\infty} \sim e^{-z/H}$, and the loss rate given in Problem 3.8.

3.11. Assuming that atomic oxygen is the dominant species present during the night at an altitude of 400 km in 1963, compute the temperature. Use the appropriate profile in Figure 3.7 and assume diffusive equilibrium.

3.12. Suppose that diffusive separation occurs at about 110 km where the temperature is 300°K. Assume further that the number densities of O and O_2 are equal at that altitude and that the temperature increases at a rate of 3°/km up to 200 km. Compute the relative number densities of O and O_2 at 200 km.

3.13. Obtain Eq. (3.51a) from the basic principles of thermodynamics and hydrodynamics discussed in Chapter 2.

3.14. What is the escape energy in eV for hydrogen atoms? For helium atoms? To what kinetic temperatures do these values correspond?

3.15. Find the escape flux of atomic hydrogen from the exosphere of Venus according to classical theory. Take the exospheric temperature as 1000°K, the molecular diameter as 1.5×10^{-8} cm, and the surface gravitational acceleration as 880 cm $\sec^{-2}$.

3.16. Modify Eq. (3.68) to include the effect of planetary rotation (i.e., include centrifugal force).

3.17. By what fraction does the density of atomic hydrogen change between the altitudes 500 and 3000 km, assuming a thermospheric temperature of 2000°K?

REFERENCES

1. S. Chapman, *Proc. Phys. Soc.* **43,** 26, 484 (1931).
2. M. Nicolet, *Planet. Space Sci.* **5,** 1 (1961).
3. D. R. Bates, *Proc. Phys. Soc.* **B64,** 805 (1951).
4. M. Nicolet, The properties and composition of the upper atmosphere, in *Physics of the Upper Atmosphere*, Ed. by J. A. Ratcliffe, Academic Press, New York, 1960.
5. P. Mange, *Ann. Géophys.* **11,** 153 (1955).
6. H. E. Hinteregger, *Space Sci. Revs.* **4,** 461 (1965).
7. L. G. Jacchia, *Nature* **192,** 1147 (1961).
8. The interested reader is referred to the papers by L. M. Jones in *Space Research II*, North-Holland Publishing Co., Amsterdam, 1961; Faucher, Procunier, and Sherman, *J. Geophys. Res.* **68,** 3437 (1963); and Champion and Faire, unpublished report AFCRL 64–554, Air Force Cambridge Research Laboratories (July 1964).
9. D. G. King-Hele, *Planet. Space Sci.* **11,** 261 (1963); *Ann. Géophys.* **22,** 40 (1966); also see T. E. Sterne, *Astronomical Journal* **63,** 424 (1958).
10. D. G. King-Hele, *Nature* **203,** 959 (1964).
11. Meadows and Townsend, *Space Research I*, North-Holland Publishing Co., Amsterdam, 1960.
12. H. E. Hinteregger, *J. Atmos. Sci.* **19,** 351 (1962); a brief summary of the method is also contained in the authors' book *Physics of the Lower Ionosphere*, Prentice-Hall, Englewood Cliffs, N.J., 1965.

13. Hall, Schweizer, and Hinteregger, *J. Geophys. Res.* **68,** 6413 (1963).
14. Nier, Hoffman, Johnson, and Holmes, *J. Geophys. Res.* **69,** 979 (1964).
15. Schaeffer and Nichols, *J. Geophys. Res.* **69,** 4649 (1964).
16. Stroud, Nordberg, Bandeen, Bartman, and Titus, *J. Geophys. Res.* **65,** 2307 (1961).
17. Blamont, Lory, Schneider, and Courtes, in *Space Research II*, North-Holland Publishing Co., Amsterdam, 1961.
18. Golomb and MacLeod, *J. Geophys. Res.* **71,** 2299 (1966).
19. Fjeldbo, Fjeldbo, and Eshleman, *J. Geophys. Res.* **71,** 2307 (1966); Kliore, Cain, Levy, Eshlemen, Fjeldbo, and Drake, *Science* **149,** 1243 (1965).
20. Kliore, Levy, Cain, Fjeldbo, and Rasool, *Science* **158,** 1683 (1967); also see the article by "Mariner Stanford Group" on page 1678 of the same issue. Also in the same issue, the atomic oxygen results are reported by Barth, Pearce, Kelley, Wallace, and Fastie.
21. Vinogradov, Surkov, and Florensky, *J. Atmos. Sci.* **25,** 535 (1968); Kurt, Dostovalow, and Sheffer, *J. Atmos. Sci.* **25,** 668 (1968).
22. Chamberlain and McElroy, *Science* **152,** 21, (1966); T. M. Donahue, *Science* **152,** 763 (1966).
23. Harris and Priester, *J. Atmos. Sci.* **19,** 286 (1962).
24. Harris and Priester, *Theoretical Models for the Solar Cycle Variation of the Upper Atmosphere*, unpublished report, Goddard Space Flight Center (April 1962).
25. H. Kallman-Bijl *et al.*, *COSPAR International Reference Atmosphere*, North-Holland Publishing Co., Amsterdam, 1961.
26. *U.S. Standard Atmosphere*, U.S. Government Printing Office, Washington, D.C. 1962, and its 1966 Supplements; also see Faire and Champion, *Revs. Geophys.* **1,** 57 (1963).
27. M. B. McElroy, *J. Geophys. Res.* **74,** 29 (1969).
28. M. Shimizu, *Icarus* (to be published 1970).
29. L. M. Trafton, *Ap. J.* **147,** 765 (1967).
30. J. W. Chamberlain, *Planet. Space Sci.* **11,** 901 (1963).
31. *See* W. I. Axford, *J. Geophys. Res.* **73,** 6855 (1968) for a brief but thorough discussion.
32. C. A. Barth, *J. Atmos. Sci.* **25,** 564 (1968).
33. M. B. McElroy and D. M. Hunten, *J. Geophys. Res.* **74,** 1720 (1969).

GENERAL REFERENCES

R. A. Craig, *The Upper Atmosphere, Meteorology and Physics*, Academic Press, New York, 1965, is an excellent general treatise on upper atmospheric physics.

S. K. Mitra, *The Upper Atmosphere*, 2nd Edition, The Asiatic Society, Calcutta, 1952, is a classic in the field. Although most of the detail is now out of date, the general principles presented are not.

M. Nicolet, La Constitution et la Composition de l'Atmosphère Supérieure (in French), in *Geophysics, the Earth's Environment*, Ed. by DeWitt, Hieblot, and

Lebeau, Gordon and Breach, New York, 1963 is a general review article on physical and chemical aeronomy by an outstanding worker in the field.

J. C. Brandt and P. W. Hodge, *Solar System Astrophysics*, McGraw-Hill, New York, 1964. Chapter 4 contains a simple but lucid treatment of radiative transfer, including methods of solving the radiative transfer equation, as well as the development of the classical Jeans theory for the escape of planetary gases.

T. L. Swihart, *Astrophysics and Stellar Astronomy*, John Wiley & Sons, New York, 1968, Sections 19 and 20; the former discusses radiative transfer and the latter spectral line broadening.

L. H. Aller, *Astrophysics—The Atmospheres of the Sun and Stars*, 2nd Edition, Ronald Press Co., New York, 1963; Chapter 4; and

R. M. Goody, *Atmospheric Radiation I. Theoretical Basic*, Oxford University Press, Oxford, 1964, Chapters 2 to 5 are more advanced treatises on radiative transfer theory.

4

Chemical Aeronomy

It is somewhat artificial to classify some types of aeronomic processes as physical and other types as chemical, because both types are so tightly intertwined in the complex system of interactions that occur in an atmosphere. However, it is convenient to consider separately the processes that mainly affect the distribution of neutral species and those that mainly affect the distribution of ionic species and electrons. The former reactions will be considered in this chapter. The processes responsible for the distribution of neutral constituents are predominantly chemical in the sense that the breaking of chemical bonds is involved, (i.e., molecules are dissociated or atoms are rearranged to form new molecular species, neutral or ionic). A review of some basic concepts such as reaction rates, lifetimes, photodissociation, etc., is presented in Chapter 2.

One of the recurring themes in ionospheric discussions is the interaction of radiation and matter, e.g., solar photons, galactic or solar protons, and auroral electrons react with atmospheric constituents. This interaction is the essential ingredient of aeronomy; chemical processes proceed from that and continue until some kind of steady state is achieved. Solar radiation has a further role in the chemistry of atmospheric constituents; aside from the original interactions, the allowable processes and their rates are determined by the gas temperature, which is in turn a function of interaction with solar energy. In order to occur in the atmosphere, chemical reactions must proceed with "thermal energies," i.e., the interacting particles have kinetic energies determined by the temperature of the ambient gas. Only exothermic reactions are considered, i.e., energy must be released by the reaction. The released energy is found in several forms, viz., kinetic energy of the products, excitation energy of the products, or emitted photons.

It should be noted that there are two conditions that may appear to be exceptions to the rule of considering only those reactions that are exothermic and whose rates are determined by ambient temperatures. Reactions that are endothermic with ground state reactants may be exothermic if one or more of the reactants is in an excited state. The energy of excitation would be added to the thermal energy and if the total is sufficient the reaction could proceed; the rates would probably be different from the rates with ground-state reactants. The other "exception" occurs in the upper ionosphere (F-region) where the ionic, electronic, and neutral components of the atmosphere are not in thermal equilibrium. Hence, all the reactants do not have the same mean energies, and "thermal energy" reaction rates may not apply. Actually, the "exceptions" noted are not really exceptions to the rule but illustrations of the need to account for all energy sources and conditions before applying rules.

4.1 Dissociation and Recombination†

Although, in theory, any process that supplies a sufficient amount of energy to break the chemical bonds between atoms can cause dissociation, the principal mode of dissociation in the atmosphere is *photo*dissociation. Proton collisions and electron impact are relatively unimportant except, perhaps, in the polar night. Following dissociation, the products either recombine in the presence of a third body M or combine with other species.

For example, in a pure oxygen atmosphere, the following reactions occur.‡ The oxygen molecule is photodissociated:

$$O_2 + h\nu \rightarrow O + O, \text{ with a rate } \rho_a \tag{4.1a}$$

The product atoms can recombine in the presence of a third body:

$$O + O + M \rightarrow O_2 + M, \text{ with rate constant } = k_b \tag{4.1b}$$

Or the oxygen molecule can combine with an oxygen atom to produce ozone:

$$O_2 + O + M \rightarrow O_3 + M, \text{ with a rate constant } = k_c \tag{4.1c}$$

The ozone molecule can also be dissociated:

$$O_3 + h\nu \rightarrow O_2 + O, \text{ with a rate } \rho_d \tag{4.1d}$$

or it can combine with an oxygen atom:

$$O_3 + O \rightarrow O_2 + O_2, \text{ with a rate constant } = k_e \tag{4.1e}$$

From the above reactions, we see that atomic oxygen is produced by the photolysis of molecular oxygen (reaction 4.1a) and by the photolysis of ozone (reaction 4.1d). At the same time, atomic oxygen disappears via the three-body recombination reactions 4.1b and c and the two-body reaction

†The term "recombination" (or "combination") refers to the reuniting (or uniting) of neutral or ionized atoms or molecules or to the reuniting of ions and electrons. The term "reassociation" (or association) is often used when only neutral species are involved.

‡ These reactions and rates are considered in more detail in Sections 4.1.1 and 4.1.2.

4.1e. Hence, the following equation can be written to describe the rate of change of the concentration of oxygen atoms:

$$\frac{d[O]}{dt} = 2\rho_a[O_2] + \rho_d[O_3] - 2k_b[O]^2[M] - k_c[O][O_2][M] - k_e[O_3][O] \tag{4.2a}$$

where [ ] designates the concentration of the species; $\rho = \int_\lambda F_\lambda Q_\lambda d\lambda$; F_λ = the spectral irradiance (or the flux of photons within the wavelength band $\lambda + d\lambda$ that penetrates to the altitude for which the calculations are made; and Q_λ is the cross section for photolysis in the same band.

Similar reasoning leads to rate equations for oxygen and ozone molecules

$$\frac{d[O_2]}{dt} = \rho_d[O_3] + k_b[O]^2[M] + 2k_e[O_3][O] - \rho_a[O_2] - k_c[O_2][O][M] \tag{4.2b}$$

$$\frac{d[O_3]}{dt} = k_c[O_2][O][M] - \rho_d[O_3] - k_e[O_3][O] \tag{4.2c}$$

In addition, we know that the total number of oxygen atoms must remain constant through the series of reactions:

$$\frac{d[O]}{dt} + 3\frac{d[O_3]}{dt} + 2\frac{d[O_2]}{dt} = 0 \tag{4.2d}$$

Equations (4.2) can be solved with appropriate computer programs.

However, under carefully selected conditions, the rate equations can be solved more easily. For example, we can consider only steady state conditions, i.e., $d[\ \ ]/dt = 0$:

$$2\rho_a[O_2] + \rho_d[O_3] = 2k_b[O]^2[M] + k_c[O][O_2][M] + k_e[O_3][O] \tag{4.3a}$$

$$-\rho_a[O_2] + \rho_d[O_3] = -k_b[O]^2[M] + k_c[O_2][O][M] - 2k_e[O_3][O] \tag{4.3b}$$

$$-\rho_d[O_3] = -k_c[O_2][O][M] + k_e[O_3][O] \tag{4.3c}$$

$$[O] + 2[O_2] + 3[O_3] = \text{Const} \tag{4.3d}$$

In the upper terrestrial atmosphere $[O_3] \ll [O_2]$ or $[O]$; hence, if we restrict our interest to O and O_2, we can consider $[O_3] \approx 0$. From Section 4.1.2 we find that $k_b = 2.7 \times 10^{-33}\ \text{cm}^6\ \text{sec}^{-1}$ and $k_c = 8 \times 10^{-35} \exp(890/RT)\ \text{cm}^6\ \text{sec}^{-1} \approx 8 \times 10^{-35}\ \text{cm}^6\ \text{sec}^{-1}$; then we can see from Eq. (4.3a) that for $[O]/[O_2] \gg 0.1$, $2k_b[O]^2[M] \gg k_c[O][O_2][M]$ or the rate of the three-body recombination to form O_2 is much faster than the three-body reaction to form O_3 and is, therefore, the dominant loss process. The photochemical equilibrium equations then can be reduced to:

$$\rho_a[O_2] \approx k_b[O]^2[M] \tag{4.4}$$

In the lower terrestrial atmosphere $[O] \ll [O_2]$ and, therefore, the three-body recombination of O to form O_2 is insignificant; also both [O] and $[O_3]$ are small compared with $[O_2]$. Using these considerations, we can derive approximate formulas for [O] and $[O_3]$ in the lower atmosphere.

Rather than comparing reaction rates of competing candidate processes in order to select the important reactions, it is often instructive to consider species "lifetimes" (see Section 2.5.1). That is, how long will a particular atomic, molecular, ionic, or excited state exist in an atmosphere? To judge whether a candidate reaction is important, we can ask what would be the lifetime of the species of interest if the candidate reaction is the one responsible for the loss of the species; then by comparing the calculated lifetimes of a particular species for a number of possible loss processes, we can select the controlling reactions. For example, the loss rate of atomic oxygen through the reaction (4.1b) is

$$\frac{d[O]}{dt} = -k_b[O][O][M] \tag{4.5a}$$

The lifetime of atomic oxygen "against" reaction (4.1b) is

$$\tau_b(O) = \frac{[O]}{k_b[O][O][M]} = \frac{1}{k_b[O][M]} \tag{4.5b}$$

If we consider reaction (4.1c)

$$\tau_c(O) = \frac{[O]}{k_c[O][O_2][M]} = \frac{1}{k_c[O_2][M]} \tag{4.5c}$$

and for reaction (4.1e)

$$\tau_e(O) = \frac{[O]}{k_e[O][O_3]} = \frac{1}{k_e[O_3]} \tag{4.5d}$$

From earlier discussions we know that $k_b \approx 30\ k_c$; hence, we conclude that so far as atomic oxygen is concerned, reactions (4.1b) and (4.1c) are of equal importance when $[O_2] \approx 30[O]$. When [O] and $[O_2]$ are approximately equal, as is the case in the altitude region of 100 to 110 km in the terrestrial atmosphere, the lifetime of O against reaction (4.1b) is much shorter than it is against reaction (4.1c); hence, we conclude that reaction (4.1c) represents a process that is not important when $[O] > [O_2]$, which is the case at altitudes greater than 110 km in the terrestrial atmosphere.

When $[O_2] \gg [O]$, as is the case in the lower terrestrial atmosphere, τ_b becomes very long compared to τ_c and, therefore, reaction (4.1b) becomes unimportant. This means that any atomic oxygen produced in the lower atmosphere will disappear via reaction (4.1c) to produce ozone or it will disappear via reaction (4.1e) to destroy ozone; hence, high ozone concentrations are favored under conditions where τ_c is short and τ_e is long, or when $\tau_e \gg \tau_c$ or $k_c [O_2] [M] \gg k_e [O_3]$. Referring to Section 4.1.2, we see that k_e is very temperature-dependent, and, hence, we conclude that the concentration of ozone is very sensitive to temperature.

We see, then, that by considering species lifetimes some insight into the importance of atmospheric processes can be obtained even without solving a formidable series of equations and the effort in calculating approximate concentrations can be significantly lessened. The concept of species lifetimes is a very convenient one, and will be used in several sections of this book.

Some specific dissociation and recombination reactions will be considered in Sections 4.1.1 and 4.1.2. Application to various planetary atmospheres will be discussed in Section 4.3.

4.1.1 Photodissociation Reactions

The process of photodissociation was introduced in Section 2.5.2 in a general way. It was pointed out that the dissociation cross sections cannot be computed successfully from first principles. Hence, we will review some of the more important reactions by utilizing spectroscopic results and deductions based on spectroscopic observations.

Oxygen. The photodissociation of molecular oxygen is a critical process inasmuch as it is the principal source of oxygen atoms in the terrestrial ionosphere; in the terrestrial ozonosphere, the dissociation of ozone also becomes significant. The absorption characteristics of molecular oxygen are important in another way: The absorption of solar ultraviolet radiation by oxygen controls the radiation intensity throughout the upper atmosphere. It is mainly oxygen that determines the ultraviolet flux available for the dissociation and ionization of other important species.

A potential energy level diagram is shown as Figure 4.1. The diagram is not complete but it illustrates some of the photodissociative processes. For example, absorption in the Herzberg bands will cause a transition from the $X^3\Sigma_g^-$ ground state to the $A^3\Sigma_u^+$ state. Because of the greater internuclear distance and shallower potential well, this state can lead easily to the production of two ^3P ground state atoms. Absorption of the Schumann-Runge band on the other hand would allow two production processes; a transition would occur to the $B^3\Sigma_u^-$ state which, for the same reason given for the $A^3\Sigma_u^+$ state, will easily lead to the production of one ^3P ground state and one ^1D metastable excited state atom. It can also, however, lead to a radiationless transition to the $^3\Pi_u$ repulsive state which would yield two ^3P ground state atoms. This latter process is called predissociation. Figure 4.2

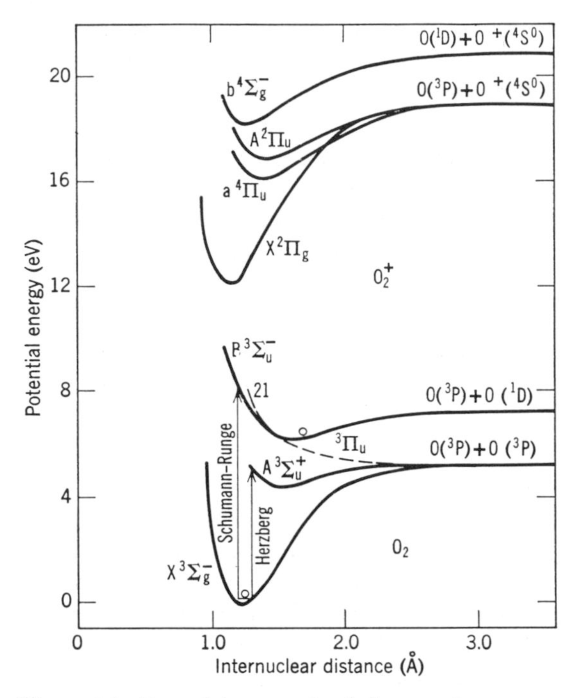

Figure 4.1 Potential energy level diagram for oxygen. (After Gilmore, *J. Quant. Spect. Rad. Trans.* **5**, 369 (1965); reprinted by permission of Pergamon Press.)

is a simplified energy level diagram that provides additional information regarding possible states and band emissions.

Absorption in the Schumann-Runge bands 1759–1950 Å leads to the production of $O(^3P)$, and absorption in the 1290–1750 Å Schumann-Runge continuum leads to $O(^1D)$ and $O(^3P)$; cross sections for these are shown in Figure 4.3a. Absorption in the Herzberg bands and continuum 2600 to 1850 Å produces $O(^3P)$; absorption cross sections are shown in Figure 4.3b. Below 1300 Å, absorption of radiation may lead to $O(^3P)$ and $O(^1S)$; and below 923 Å, to two $O(^1S)$ atoms.

Nitrogen. Nitrogen is a weak absorber of radiation in the wavelength range above 900 Å. Photodissociation cross sections are not known but it appears[1] that the important photo reaction for atomic nitrogen production is predissociation following absorption in the Lyman-Birge-Hopfield band. (See Figures 4.4a and 4.4b for energy level diagrams.) This band is weak. The maximum absorption cross section is approximately 10^{-21} cm^2 at 1226 Å, and the predissociation reaction that produces atomic nitrogen must occur

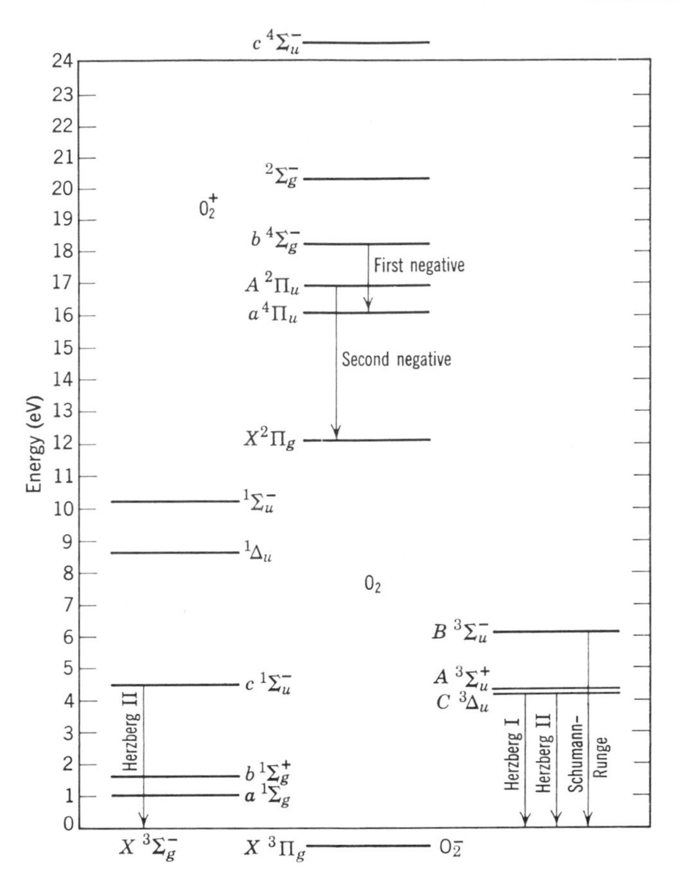

Figure 4.2 A simplified energy level diagram for O_2. (After Gilmore, *Rand Memo* RM-4034-PR (1964); reprinted by permission of the Rand Corp.)

with an efficiency less than 100 percent; Bates[1] estimates a photodissociation rate coefficient of 10^{-12} sec^{-1} at zero optical depth in the terrestrial atmosphere. Maximum cross sections in the Lyman-Birge-Hopfield bands, which begin at 1450 Å, are[2] about 7×10^{-21} cm^2, and between bands have upper limits of 3×10^{-22} cm^2. At the important solar emission lines Lα and Lβ, the cross sections are[3] 6×10^{-23} cm^2 and 1.1×10^{-20} to 4×10^{-21} cm^2, respectively. The absorption cross sections below 1000 Å wavelength are shown in Figure 4.5. Ion-neutral and ion-electron reactions such as 4.13 are much more important than photolytic reactions as sources of atomic nitrogen in the terrestrial atmosphere.

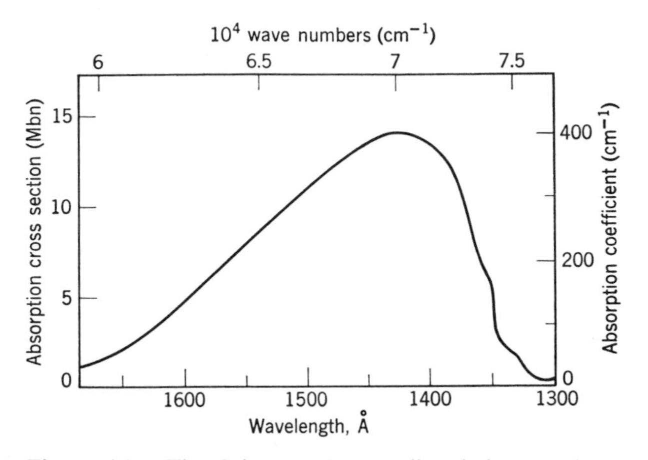

Figure 4.3a The Schumann-Runge dissociation continuum cross-section curve. (After Metzer and Cook, *J. Quant. Spect. Rad. Trans.* **4,** 107 (1964); reprinted by permission of Pergamon Press.)

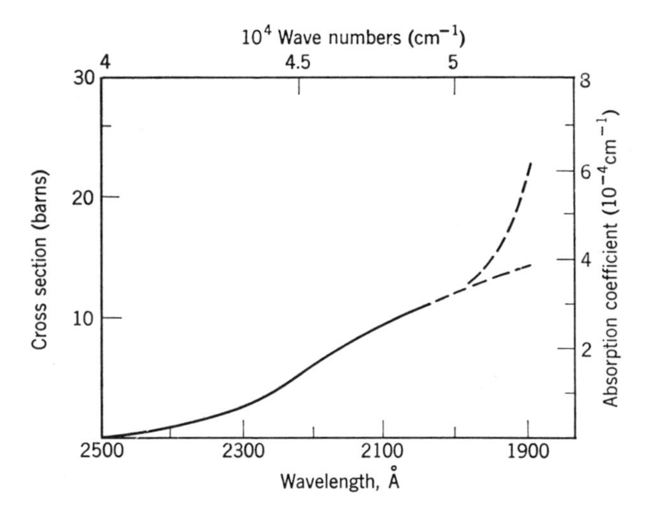

Figure 4.3b The Herzberg dissociation continuum cross-section curve. (After Ditchburn and Young *JATP* **24,** 127 (1967); reprinted by permission of Pergamon Press.)

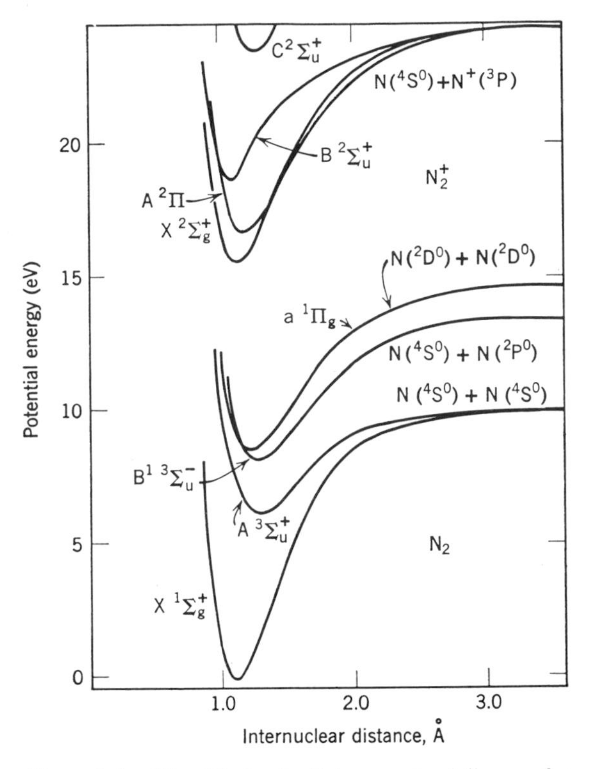

Figure 4.4a Simplified potential energy level diagram for nitrogen. (After Gilmore, *Rand Memo* RM-4034-PR (1964); reprinted by permission of the Rand Corp.)

Ozone. Ozone is produced by reaction (4.9b), which is controlled by the photolysis of molecular oxygen; ozone in turn, is a strong absorber of ultraviolet radiation. Absorption coefficients are presented in Figure 4.6. In the diffuse ultraviolet Hartley bands (2100–3200 Å), absorption of radiation leads to dissociation as follows:[4,5]

$$O_3(S_0) + h\nu \rightarrow O_3(S_1) \xrightarrow{\text{below 3100 Å}} O_2(^1\Delta_g) + O(^1D) \qquad (4.6a)$$

or

$$\xrightarrow{\text{below 2600 Å}} O_2(^1\Sigma_g^+) + O(^1D) \qquad (4.6b)$$

or

$$\xrightarrow{\text{below 1690 Å}} O_2(^1\Delta_g) + O(^1S) \qquad (4.6c)$$

or

$$\xrightarrow{\text{below 1790 Å}} O_2(^1\Sigma_g^+) + O(^1S) \qquad (4.6d)$$

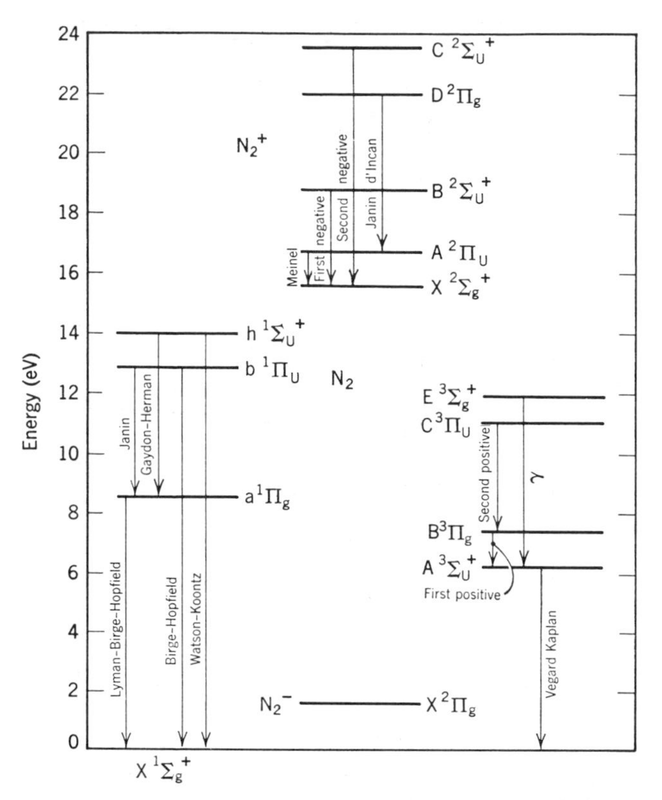

Figure 4.4b A simplified energy level diagram for N_2. (After Green, *The Middle Ultraviolet* (1966); reprinted by permission of John Wiley & Sons.)

Strong absorption by ozone also occurs in the vacuum ultraviolet, below 1600 Å. Weaker, but well-defined absorption occurs in the Huggins bands between 3200 and 3600 Å, and in the weak Chappius bands of the visible spectrum (4400–7400 Å). In the visible band,[5] the triplet state of ozone is probably excited leading to the formation of ground state products:

$$O_3 + h\nu \xrightarrow{\text{below } 11{,}400 \text{ Å}} O_3(T_1) \rightarrow O_3(^3\Sigma_g^-) + O(^3P) \qquad (4.6e)$$

The following reaction is energetically possible but forbidden by the Wigner spin rule (see Section 2.6.5).

$$O_3(S_1) + h\nu \xrightarrow{\text{below } 5900 \text{ Å}} O_3(T_1) \rightarrow O_2(^1\Delta_g) + O(^3P) \qquad (4.6f)$$

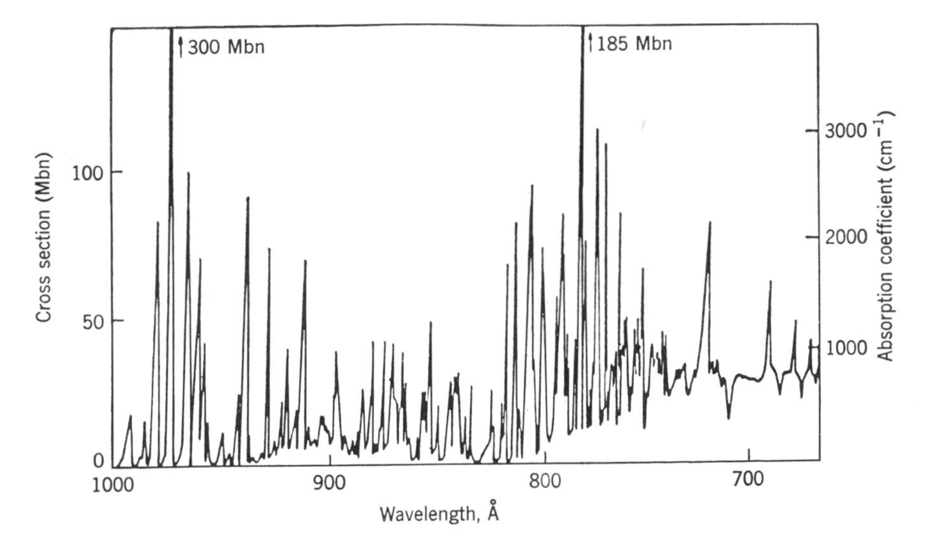

Figure 4.5 Molecular nitrogen absorption cross section. (After Marr, *Photoionization Processes in Gases* (1967); reprinted by permission of Academic Press.)

Nitric Oxide. Nitric oxide photodissociation is not important to ionospheric chemistry but for completeness and future reference, we include a potential energy diagram and the absorption cross sections in Figures 4.7a and 4.7b. It will be discussed again in Sections 4.2 and 4.3; rates are quoted from Nicolet[6]:

$$NO + h\nu \xrightarrow{\sim 1900\ \text{Å}} N + O, \quad \rho = 10^{-7}\ \text{sec}^{-1} \text{ at zero optical depth in the terrestrial atmosphere} \tag{4.6g}$$

$$NO + h\nu \xrightarrow{\text{below } 1340\ \text{Å}} NO^{+} + e \rightarrow N + O, \quad \rho = 5 \times 10^{-7}\ \text{sec}^{-1} \text{ at zero optical depth} \tag{4.6h}$$

$$NO_2 + h\nu \xrightarrow{\text{below } 3975\ \text{Å}} NO + O, \quad \rho = 5 \times 10^{-3} \text{ at zero optical depth} \tag{4.6i}$$

Carbon Dioxide. The photolysis of carbon dioxide is important to the chemistry of the Mars and Venus ionospheres. In some respects, carbon dioxide is the Mars and Venus counterpart of terrestrial oxygen. Photodissociation occurs in at least three regions of the ultraviolet spectrum.[5] This

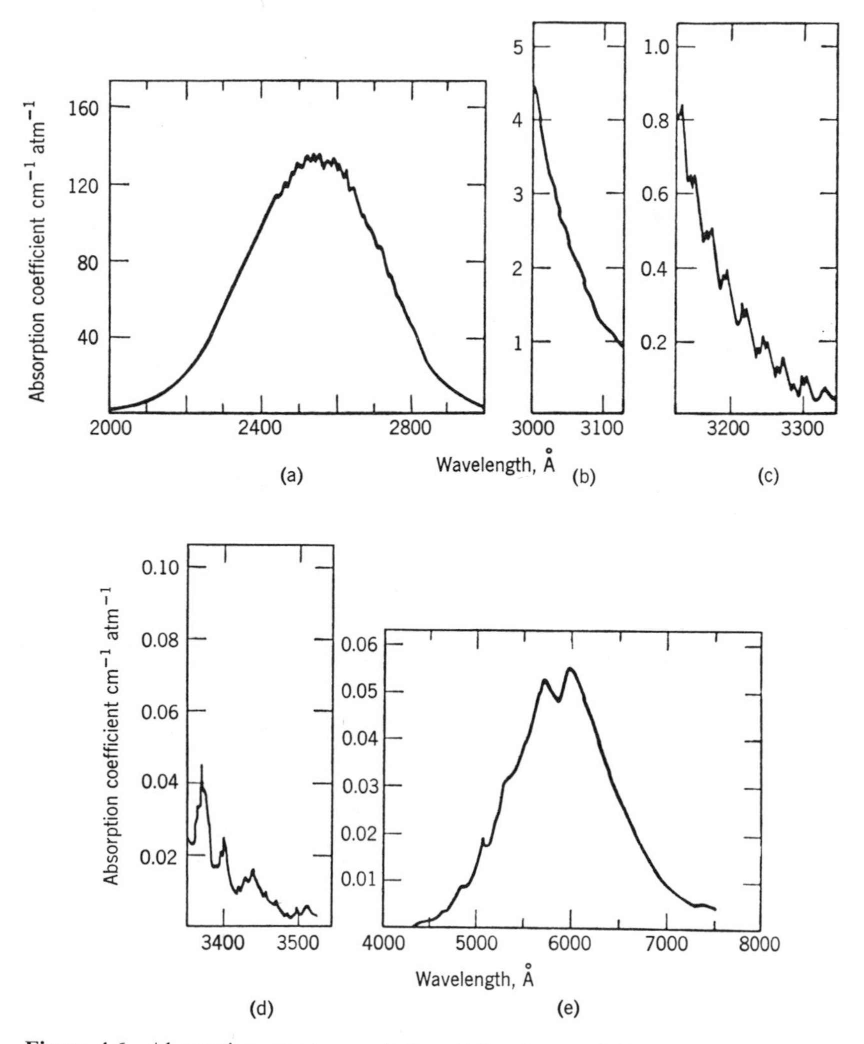

Figure 4.6 Absorption spectrum of O_3. (After Inn and Tanaka, *Advances in Chemistry*, *Series* **21,** 266 (1959); reprinted by permission of the American Chemical Society.)

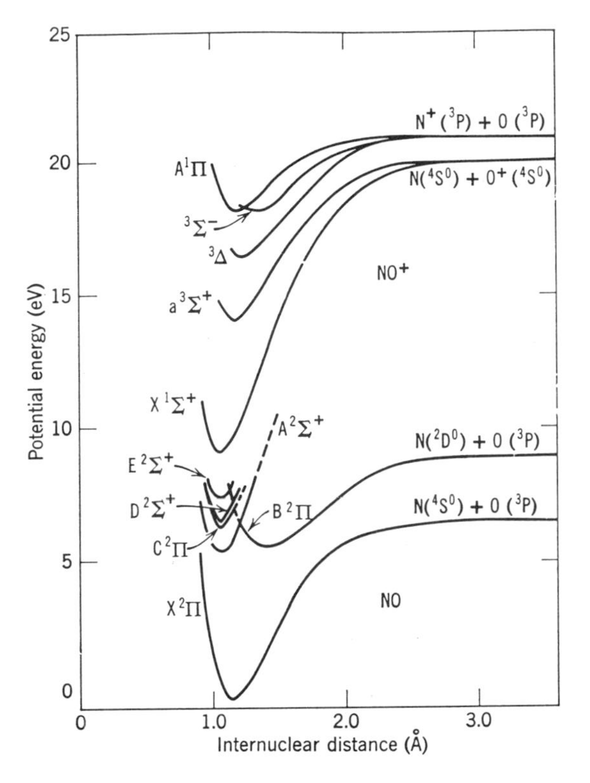

Figure 4.7a Simplified potential energy diagram for nitric oxide. (After Gilmore, *J. Quant. Spect. Rad. Trans.* **5**, 369 (1965); reprinted by permission of Pergamon Press.)

is illustrated in Figure 4.8a, b, c. At wavelengths below 1650 Å, the following dissociation reactions are possible.[5]

$$CO_2 + h\nu \xrightarrow{\text{below 1650 Å}} CO(X^1\Sigma^+) + O(^1D) \tag{4.7a}$$

$$\xrightarrow{\text{below 1373 Å}} CO(X^1\Sigma^+) + O(^1S) \tag{4.7b}$$

$$\xrightarrow{\text{below 1070 Å}} CO(a^3\Pi) + O(^3P) \tag{4.7c}$$

It should be noted that because of the low resolution of the absorption cross section measurements presented in Figures 4.8a, b, c, it was not possible to separate the absorption of line radiation (which could cause excitation and fluorescence) and the absorption of continuum radiation

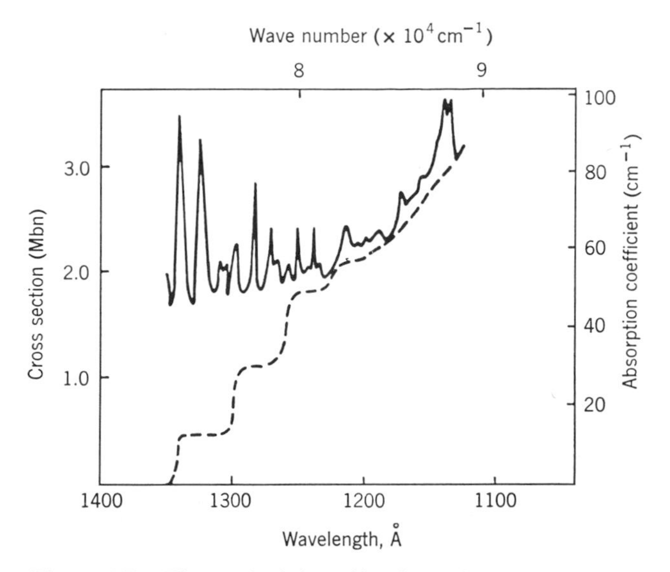

Figure 4.7b The total nitric oxide absorption cross-sectional curve. ---- represents the photoionization cross-sectional curve. (After Watanabe, *Adv. Geophys.* **5**, (1958); reprinted by permission of Academic Press.)

(which could cause dissociation). Hence, photodissociation cross sections based on published spectra may be seriously over-estimated.

Hydrogen. Hydrogen photochemistry is important in the upper fringes of the atmospheres of Earth, Mars, and Venus, and even more important in the atmospheres of Jupiter and Saturn. Absorption cross sections are presented in Figure 4.9a. Potential energy curves are presented in Figure 4.9b.

Water Vapor. Water vapor appears to be present in the atmosphere of Mars and Venus as well as in the terrestrial atmosphere. If so, it may play an important role in both neutral and ion chemistry and may be a source of the hydrogen found in the outer fringes of planetary atmospheres. The photodissociation of water vapor results in the following reactions[5]:

$$H_2O + h\nu \xrightarrow{\text{below 2420 Å}} H(^2S) + OH(^2\Pi) \tag{4.8a}$$

$$\xrightarrow{\text{below 1356 Å}} H(^2S) + OH(^2\Sigma^+) \tag{4.8b}$$

$$\xrightarrow{\text{observed at 1236 Å}} H_2 + O(^1D) \tag{4.8c}$$

$$H_2O_2 + h\nu \xrightarrow{\text{below 5650 Å}} OH + OH \tag{4.8d}$$

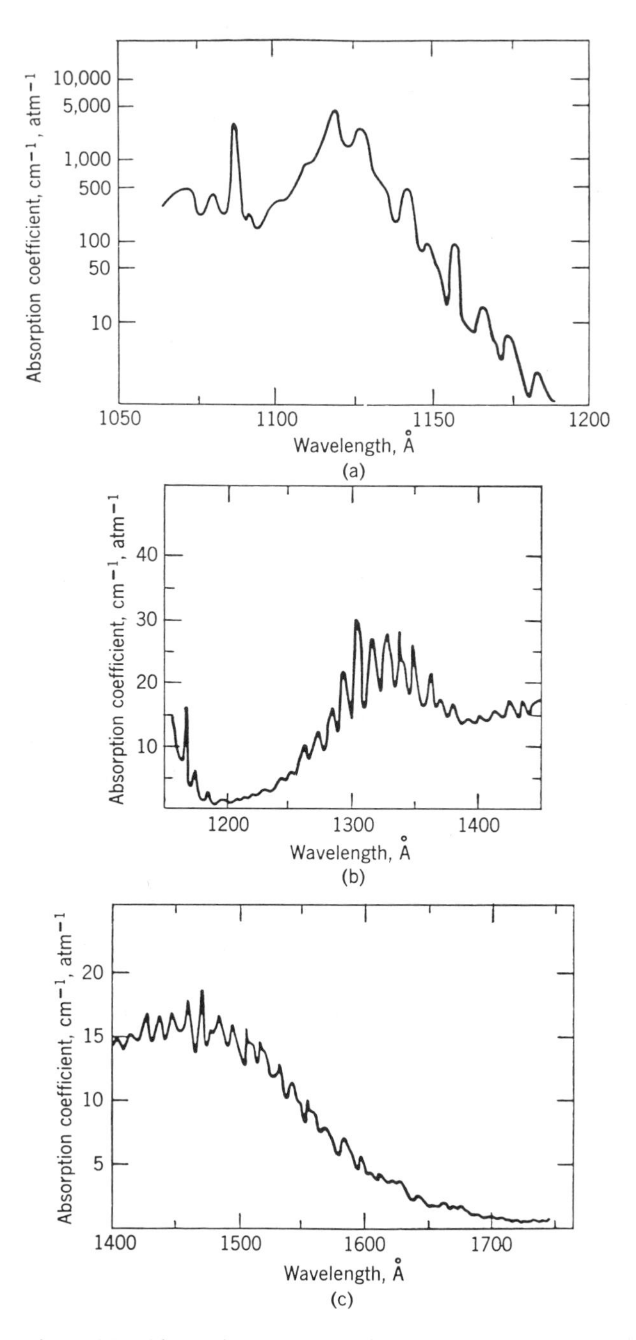

Figure 4.8 Absorption spectrum of carbon dioxide. (After Inn, Watanabe, and Zelikoff *J. Chem. Phys.* **21,** 1648 (1953); reprinted by permission of The American Institute of Physics.)

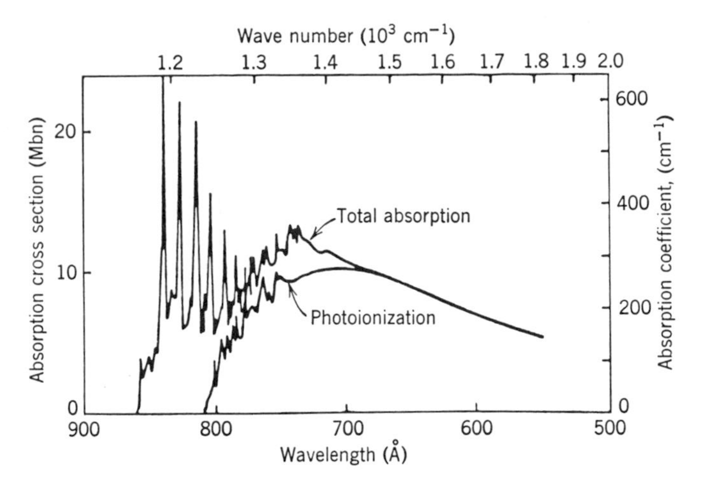

Figure 4.9a Absorption cross-sectional plots for molecular hydrogen. (After Cook and Metzger *J. Chem. Phys.* **41,** 321 (1964); reprinted by permission of The American Institute of Physics.)

Absorption coefficients are presented in Figure 4.10. The quantum yields of atomic hydrogen (reactions 4.8a and b) are near unity. The importance of reaction (4.8c) throughout the dissociation spectra is not known.

4.1.2 Recombination Reactions

As with the photodissociation reactions, only limited success has been achieved in determining allowable recombination reactions and rates from first principles. Hence, we shall review the important reactions as they have been determined by experiment. However, hypothetical reactions that have been seriously suggested will also be noted. It will be apparent from the lack of discussion that the chemistry of excited species, though possibly critical in some areas, is largely an area of ignorance.

For convenience, the reactions will be grouped according to the principal reactants. Where temperature dependence is given, T is in degrees Kelvin and R is the universal gas constant ($R = 1.986$ cal. (mole °K)$^{-1}$).

Oxygen, Ozone. Atomic oxygen recombines to form molecular oxygen and combines with molecular oxygen to form ozone. These are all three-body reactions:

$$O(^3P) + O(^3P) + M \rightarrow O_2 + M, \; k = 2.7 \times 10^{-33} \text{ cm}^6 \text{ sec}^{-1} \quad (4.9a)$$

$$O(^3P) + O_2 + M \rightarrow O_3 + M, \; k = 8 \times 10^{-35} \exp(890/RT) \text{ cm}^6 \text{ sec}^{-1} \quad (4.9b)$$

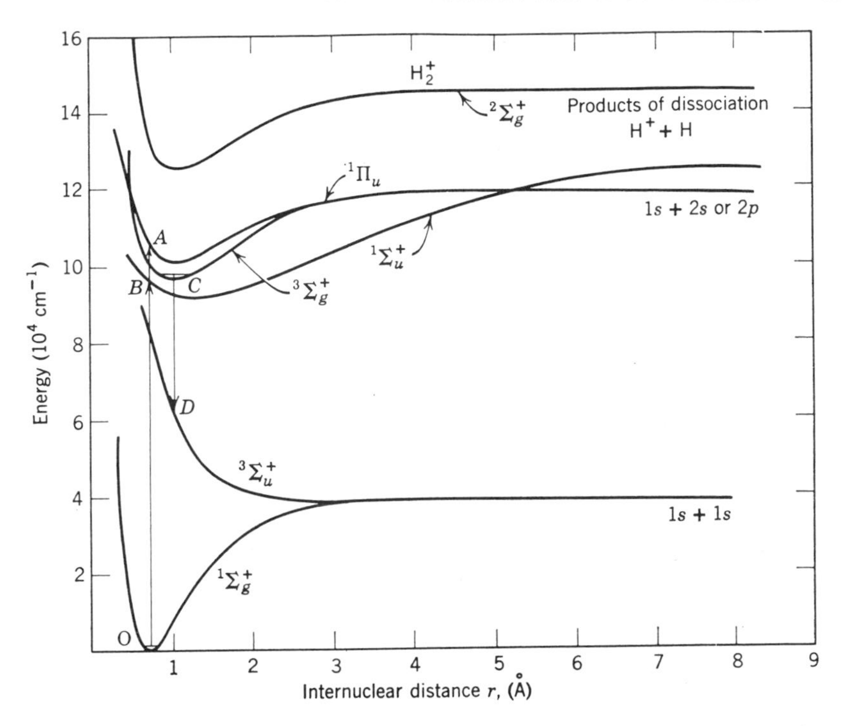

Figure 4.9b Potential energy curves for several electronic states of H_2 and H_2^+ of photochemical interest; transitions B ← 0 and A ← 0 correspond to important absorption bands at 1109 and 1002 Å, respectively. The transition C → D results in the continuum which is of great value as an ultraviolet source for absorption spectroscopy. (After Calvert and Pitts, *Photochemistry* (1966); reprinted by permission of John Wiley & Sons.)

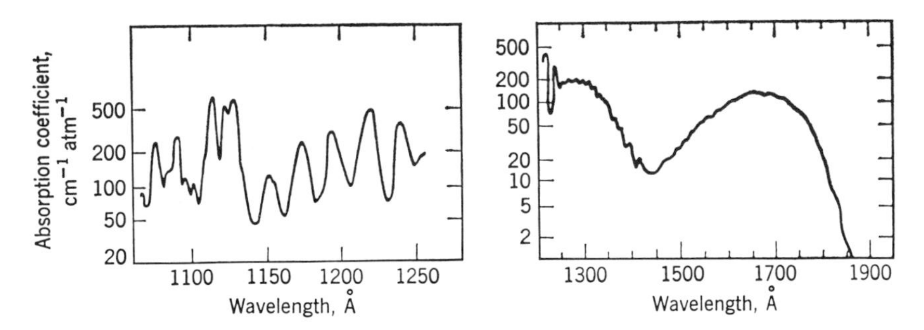

Figure 4.10 The absorption spectrum of water vapor. (After Watanabe and Zelikoff *J. Opt. Soc. Am.* **43**, 753 (1953); reprinted by permission of The American Institute of Physics.)

A special case of (4.9a) is thought to be responsible for the nightglow green (5577 Å) line at low altitudes and is known as the Chapman reaction. (See Section 6.2.3.)

$$O + O + O \rightarrow O_2 + O(^1S),\ k \text{ must be } >2 \times 10^{-35}\ \text{cm}^6\ \text{sec}^{-1} \text{ and is probably } \sim 10^{-34}\ \text{cm}^6\ \text{sec}^{-1} \quad (4.9c)$$

Ozone and oxygen can interact

$$O(^3P) + O_3 \rightarrow 2O_2,\ k = 5.6 \times 10^{-11} \exp(-5700/RT)\ \text{cm}^3\ \text{sec}^{-1} \quad (4.9d)$$

The existence of atomic oxygen in the metastable state $O(^1D)$ is an important consideration in planetary atmospheres because $O(^1D)$ may enter into reactions that the ground state atom cannot; and in cases where it enters the same reactions, it is thought to react at rates that are different from those measured for ground state reactions. To date, however, reaction rates involving the metastable state have not been measured because of experimental difficulties. The concentrations of $O(^1D)$ found in atmospheres depend not only on subsequent chemical reactions but also on de-excitation reactions; these are discussed briefly in Section 4.1.3.

Nitrogen, Oxygen. Atomic nitrogen can disappear by recombination:

$$N + N + M \rightarrow N_2 + M,\ k = 7.0 \times 10^{-33}\ \text{cm}^6\ \text{sec}^{-1} \quad (4.10a)$$

However, nitrogen plays a much more important role in the earth's atmosphere by reacting with oxygen to form nitric oxide, which, though present in minute amounts, has the potential of affecting the ionization balance significantly. We will present a few of the important reactions but the reader is referred to the many review papers by M. Nicolet[6]

$$N + O + M \rightarrow NO + M,\ k = 1 \times 10^{-32}\ \text{cm}^6\ \text{sec}^{-1} \quad (4.10b)$$

$$N + O \rightarrow NO + h\nu,\ k = 1 \times 10^{-17}\ \text{cm}^3\ \text{sec}^{-1} \quad (4.10c)$$

$$O + NO + M \rightarrow NO_2 + M,\ k = 3 \times 10^{-33} \exp(1000/T)\ \text{cm}^6\ \text{sec}^{-1} \quad (4.10d)$$

$$O + NO \rightarrow NO_2 + h\nu,\ k = 6 \times 10^{-17}\ \text{cm}^3\ \text{sec}^{-1} \quad (4.10e)$$

$$O + NO_2 \rightarrow NO + O_2,\ k = 1.5 \times 10^{-12}\, T^{1/2} \exp(-500/T)\ \text{cm}^3\ \text{sec}^{-1} \quad (4.10f)$$

$$O_3 + NO \rightarrow NO_2 + O_2,\ k = 5 \times 10^{-14}\, T^{1/2} \exp(-1200/T)\ \text{cm}^3\ \text{sec}^{-1} \quad (4.10g)$$

$$N + NO \rightarrow N_2 + O,\ k = 1.5 \times 10^{-12}\, T^{1/2}\ \text{cm}^3\ \text{sec}^{-1} \quad (4.10h)$$

$$N + O_2 \rightarrow NO + O,\ k = 2 \times 10^{-13}\, T^{1/2} \exp(-3000/T)\ \text{cm}^3\ \text{sec}^{-1} \quad (4.10i)$$

Hydrogen, Oxygen. The products of water vapor photolysis can react with oxygen and ozone. Thus, hydrogen-oxygen reactions must be considered in atmospheres containing water vapor. The importance of such processes to ozone equilibrium chemistry in the upper terrestrial atmosphere was demonstrated by Hunt,[7] who considered some seventeen possible reactions. We list only two as examples:

$$H + O_3 \rightarrow OH + O_2, \; k = 2.6 \times 10^{-11} \text{ cm}^3 \text{ sec}^{-1} \quad (4.11a)$$

$$OH + O(^3P) \rightarrow H + O_2, \; k = 5 \times 10^{-11} \text{ cm}^3 \text{ sec}^{-1} \quad (4.11b)$$

Carbon, Oxygen. Considerable uncertainties exist in knowledge of the recombination reactions involving the photolytic products of CO_2. This uncertainty involves both lack of knowledge of which reactions are possible and lack of measurements on reactions that are thought to be possible, particularly those involving $O(^1D)$. Rates are mostly unknown; the listed reactions are speculative if a rate is not given.

$$CO + O(^3P) + M \rightarrow CO_2 + M, \; k \leq 8 \times 10^{-36} \, T^{1/2} \text{ cm}^6 \text{ sec}^{-1} \quad (4.12a)$$

$$CO + O(^1D) \rightarrow CO_2 + h\nu \quad (4.12b)$$

$$CO + OH \rightarrow CO_2 + H \quad (4.12c)$$

$$CO + HO_2 \rightarrow CO_2 + OH \quad (4.12d)$$

$$CO_2 + O(^1D) \rightarrow CO_2 + O(^3P), \; k = 1 - 3 \times 10^{-10} \text{ cm}^3 \text{ sec}^{-1} \quad (4.12e)$$

$$CO_2 + O(^1D) \rightarrow CO_3 + h\nu \quad (4.12f)$$

$$CO_3 + CO \rightarrow CO_2 + CO_2 \quad (4.12g)$$

$$CO_3 + h\nu \rightarrow CO_2 + O(^1S) \quad (4.12h)$$

4.1.3 Deactivation of Excited Oxygen Atoms

In the above reactions we have noted that some reactions involve an excited metastable oxygen atom $O(^1D)$. Where possible, we have attempted to differentiate between reactions with $O(^1D)$ and those with the ground state atom $O(^3P)$ because the excited state is more energetic and more reactive chemically.

The metastable state is a product of both photolysis and dissociative molecular ion-electron recombination and is produced quite plentifully in planetary atmospheres: It is thought to be the predominant form of atomic oxygen produced in CO_2-rich atmospheres. The $O(^1D)$ state is long-lived (on the order of 100 seconds) so the probability of radiative de-excitation is low and the probability of entering into chemical reactions is great.

However, data regarding the rates of chemical reaction with $O(^1D)$ are qualitative, if they do indeed exist. Laboratory investigators have concluded that the reason for not being able to study $O(^1D)$ reactions is that the metastable state is rapidly quenched, i.e., it is de-excited to the ground state by a collision in which the excitation energy of the oxygen atom is transferred to a molecule. For example, the following is often suggested:

$$O(^1D) + O_2(^3\Sigma_g^-) \rightarrow O_2(^1\Delta_g \text{ or } ^1\Sigma_g^+) + O(^3P) \tag{4.9f}$$

Deactivation rates are inferred from reactions observed in the atmosphere in which the emission of the 6300 Å doublet of the $^3P \rightarrow {}^1D$ transition can be observed and correlated with other parameters. Laboratory investigations seem to be compatible with the atmospheric results. Present data indicate that $O(^1D)$ is deactivated very efficiently by O_2, N_2, and CO_2; the rates for the deactivation reaction are on the order of 10^{-11} cm^3 sec^{-1} for O_2 and N_2, and 10^{-10} for CO_2. Obviously this is very rapid and emphasizes the possible importance of such reactions in atmospheric chemistry.

Because of the still speculative nature of the subject, we shall not discuss it further here. The reader is referred to two excellent reviews of the problem as it is presently understood: McGrath and McGarvey[14]; and Hunten and McElroy.[15]

4.2 Ionization, Recombination, and Interchange

The ionization of atmospheric constituents and the subsequent recombination, charge transfer, and interchange reactions that produce ionospheres are introduced in Chapter 2. Although these processes can properly be classified as chemical. we shall restrict the discussion in this chapter to the chemistry of neutral species. Detailed discussions of ionic processes and how they affect the ion and electron distribution in the atmosphere is postponed to Chapters 8, 9, and 10; ionic processes relevant to optical emissions in the atmosphere are discussed in Chapter 6.

The possibility exists, however, that the distribution of neutral species, particularly minor constituents, in the atmosphere is affected and even controlled by ionic processes. In general, this possibility is discounted because photodissociation and reactions with major atmospheric constituents can be shown to dominate in most cases. Nonetheless, the influence of ionic processes on neutral species chemistry cannot be dismissed out of hand. As noted in Chapter 6, recombination of molecular ions is an important source of metastable atomic species. The products of most ionic recombination and interchange processes include neutral species; hence for completeness, they must be considered in detailed balance computations.

For example, Nicolet[8] has shown that some ionic reactions are required to account for observed neutral nitric oxide concentrations in the D and E regions and that the source of atomic nitrogen in the terrestrial atmosphere is more likely to be dissociative recombination of NO^+ or atomic oxygen ion interchange with molecular nitrogen. Thus, in the terrestrial atmosphere, the following ionic reactions must be considered:

$$O_2^+ + N_2 \rightarrow NO^+ + NO \tag{4.13a}$$

$$NO^+ + e \rightarrow N + O \tag{4.13b}$$

$$O^+ + N_2 \rightarrow NO^+ + N \tag{4.13c}$$

$$N_2^+ + O \rightarrow NO^+ + N \tag{4.13d}$$

4.3 Photochemical Equilibrium

If at a specific altitude, or within a narrow altitude range, the production and loss of an atmospheric constituent is controlled by photo and chemical reactions and conditions are such that no net change occurs in the concentration of that constituent, it is said to be in photochemical equilibrium. Under these narrowly defined conditions, pertinent reactions can be selected and, subsequently, calculations can be made of the concentration to be expected of that constituent. The simplest model that can be computed for the distribution of a constituent in the upper atmosphere is prepared by making such calculations for a range of altitudes; quite often it is the only model that can be calculated. The criterion for applicability of such a model is that the chemical processes involving the constituent of interest must be rapid compared to transport processes such as diffusion and mixing; in other words, the lifetimes of the species against chemical reactions must be short compared to lifetimes against transport processes. The lifetime against chemical processes τ_c has been discussed in Sections 4.1 and 2.5.1. The lifetime against transport processes is determined by the time required for the species to move vertically through a distance equal to its own scale height; thus, the diffusion (molecular or eddy) lifetime would be:

$$\tau_D = \frac{H}{W} = \frac{H^2}{D} \tag{4.14a}$$

where H is the scale height, W is the diffusion velocity and D is the diffusion coefficient (see Chapter 3), and the criterion for validity of a photochemical equilibrium model is

$$\tau_c < \tau_D \tag{4.14b}$$

Models accounting for minor atmospheric species have been computed and, although they are thought to be in error because of the neglect of diffusion, they are useful departure points for additional studies. Several such models will be discussed in the succeeding sections.

4.3.1 Terrestrial Atmosphere

Oxygen. Reference to Section 4.1, together with the assumption of equilibrium conditions, indicates that the atomic oxygen concentration can be calculated by

$$\rho[O_2] = k[O]^2[M] \tag{4.4}$$

i.e., rate of production of oxygen by photolysis is equal to the loss rate by recombination, where $\rho = \int_\lambda Q_{O_2} F_\lambda(z)\, d\lambda$; k is the three-body recombination rate; Q_{O_2} is the cross section for photolysis; and the brackets indicate concentrations of the species contained within them, or

$$[O] = \left(\frac{\rho[O_2]}{k[M]}\right)^{1/2} \tag{4.15}$$

The principal difficulty with this approach becomes apparent when we consider the photochemical equilibrium criterion discussed above. The rate k of the three-body recombination of atomic oxygen is 2.7×10^{-33} cm^3 sec^{-1} and the lifetime of the oxygen atom against chemical recombination would be

$$\tau_0 = \frac{1}{k[O][M]} = \frac{10^{33}}{2.7[O][M]} \text{ sec} \tag{4.16}$$

It is obvious that when the product of [O] and [M] is less than $\sim 10^{28}$, the lifetime of O atoms will exceed one day. Reference to Figure 3.8 indicates that this will occur at altitudes even below 100 km and, furthermore, the lifetimes will increase exponentially with altitude; according to the data in Figure 3.8, $\tau_0 \sim 10^8$ sec at 100 km. Clearly, O atoms would never recombine; they would disappear by other processes and the notion of chemical equilibrium is irrelevant. Considerations of this sort led Nicolet[9] to conclude that one can utilize the concept of Eq. (4.15a) up to altitudes where chemical lifetimes are short (when compared with transport processes); but above those altitudes, transport processes become important and control the concentrations of O and O_2. The observed peak atomic oxygen concentration altitudes could be explained by assuming that perfect mixing occurs above the altitude at which the recombination lifetime becomes too long. Mixing would provide a way for oxygen atoms to descend to higher pressure regimes where they could recombine; at the same time, the molecules ascend to provide a source of molecular oxygen at the higher altitudes. However, the result of perfect mixing down to a specified altitude would be a constant O/O_2 ratio above that altitude, and this is known not to be the case.

Thus, one is forced to include diffusive transport (see Section 3.4) of molecular oxygen upward into the higher altitudes to achieve the proper balance. At the higher altitudes, molecular oxygen is treated as a minor species diffusing upward into the major species, atomic oxygen, whereas at the lower altitudes, atomic oxygen is treated as the minor species diffusing downward; obviously, there should be intermediate altitudes where diffusion occurs in both directions. A complete treatment would require chemical, mixing, and diffusion theory. The problem is still not completely solved for the oxygen allotropes (O, O_2, and O_3) in the terrestrial atmosphere.

As can be seen from Figure 3.8 the atomic and molecular oxygen concentrations are observed to become equal at about 120 km; the peak atomic oxygen concentration probably occurs at about 100 km. Although the peak concentration altitude can be calculated fairly well from photoequilibrium assumptions, the cross-over point between atomic and molecular oxygen profiles cannot. To be complete, eddy diffusion and many other reactions involved in the oxygen chemistry must be also accounted for. The model in the following section neglects transport but includes many additional important reactions.

Ozone. Hunt[7] has shown that water vapor must be included in calculations of atomic oxygen and ozone profiles below 100 km. Although the neglect of eddy diffusion and turbulence may cause errors in the atomic oxygen and hydrogen profiles the photochemical model appears to account well for observed ozone profiles at low altitudes. Hunt's models include reactions of oxygen photolysis (Section 4.1.1) ozone photolysis (reactions 4.6a and e) water vapor and peroxide photolysis (reactions 4.8a, b, and d), molecular-oxygen and ozone formation (reactions 4.9a, b, and d), and hydrogen-oxygen combinations. The models attempt to account for the important metastable oxygen atom $O(^1D)$ even though knowledge of the chemistry of $O(^1D)$ is still incomplete. Computed profiles are shown in Figure 4.11 for night and day.

Is the apparent confirmation (by observations) of the ozone profiles predicted by Hunt's models sufficient evidence that the atomic oxygen profiles are correct? The criterion for applicability of photoequilibrium models (Eq. 4.14b) indicates that the profiles become less reliable at the higher altitudes where atomic oxygen lifetimes approach the order of 10^5 seconds. The agreement between the ozone observations and calculations for the lower altitudes does show, however, that the important chemical reactions were included in the model, that water vapor chemistry cannot be neglected, and that in the lower altitudes, where ozone is important, lifetimes against chemical reactions are indeed short compared to lifetimes against transport.

Nitric Oxide. Several attempts have been made to account theoretically for nitric oxide in the atmosphere. In fact, until very recently,[10] the existence

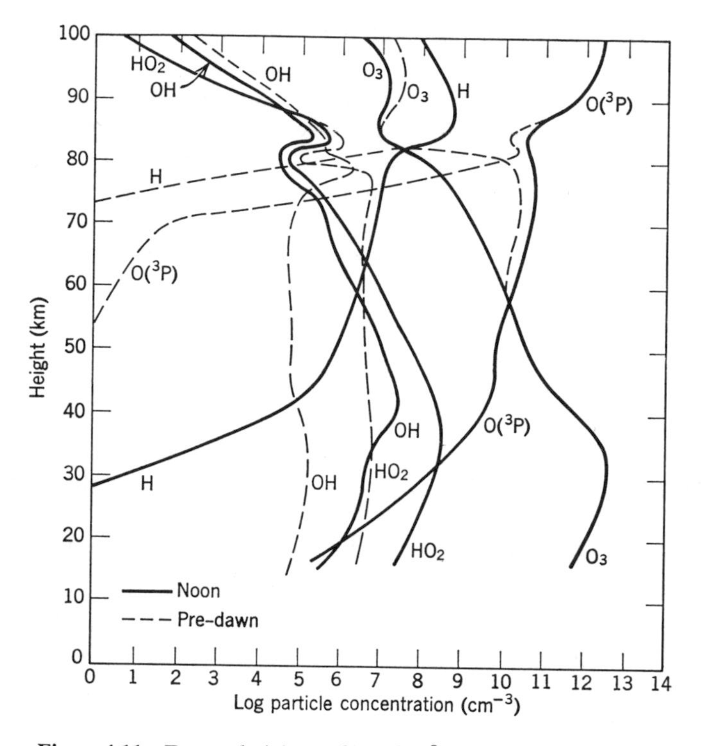

Figure 4.11 Day and night profiles of $O(^3P)$, O_3, H, OH, and HO_2. Dashed lines are from the CIRA atmosphere. (After Hunt, *JGR* **71,** 1385 (1966); reprinted by permission of the American Geophysical Union.)

of nitric oxide could be demonstrated only theoretically. The theory was first developed by Nicolet;[11] we will follow the later work of Barth,[12] however, in this summary. Time-dependent rate equations (such as those in Section 4.1) can be assembled to describe the loss of N, O, NO, O_2, O_3, and NO_2 after sunset using an upper atmosphere model such as Figure 4.12 for the initial concentrations of pertinent species and assuming that the following reactions occur: (4.9a), (4.9b), (4.9d), (4.10a), (4.10d), (4.10f), (4.10h), and (4.10i). The rates of decrease of the various constituents can be calculated for periods of several hours; when the results are analyzed it is found that the important production and loss reactions for NO are (4.10i) and (4.10h), respectively, and that the concentration of NO can be obtained approximately from the equilibrium conditions:

$$k_h[NO] \cong k_i[O_2] \tag{4.17}$$

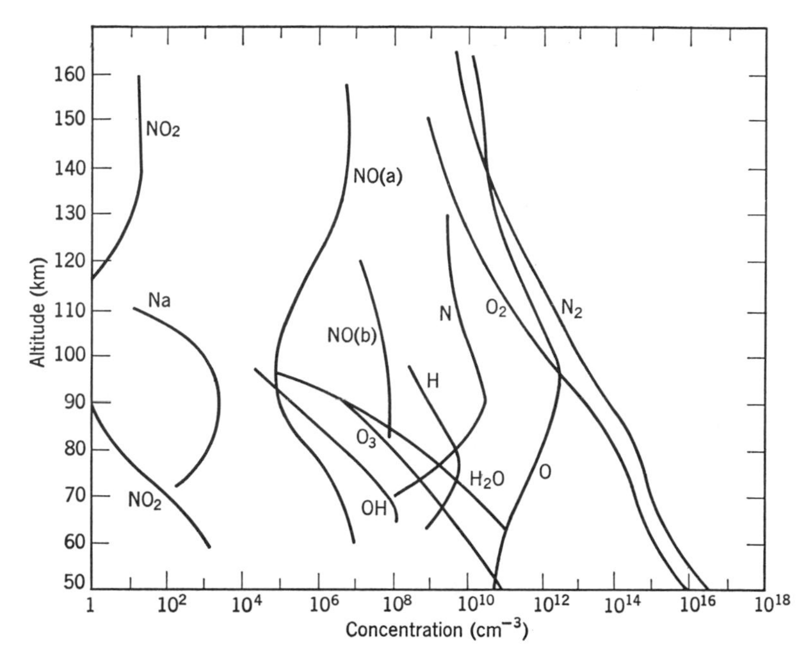

Figure 4.12a Particle number densities of various constituents in the upper atmosphere.

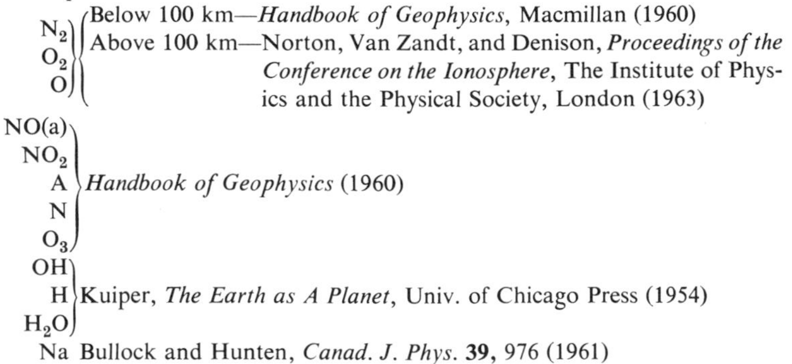

N_2, O_2, O: Below 100 km—*Handbook of Geophysics*, Macmillan (1960); Above 100 km—Norton, Van Zandt, and Denison, *Proceedings of the Conference on the Ionosphere*, The Institute of Physics and the Physical Society, London (1963)

NO(a), NO_2, A, N, O_3: *Handbook of Geophysics* (1960)

OH, H, H_2O: Kuiper, *The Earth as A Planet*, Univ. of Chicago Press (1954)

Na Bullock and Hunten, *Canad. J. Phys.* **39**, 976 (1961)

NO(b) Barth, *J. Geophys. Res.* **69**, 3301 (1964)

(After Whitten and Poppoff, *Physics of the Lower Ionosphere* (1965); reprinted by permission of Prentice-Hall, Inc.)

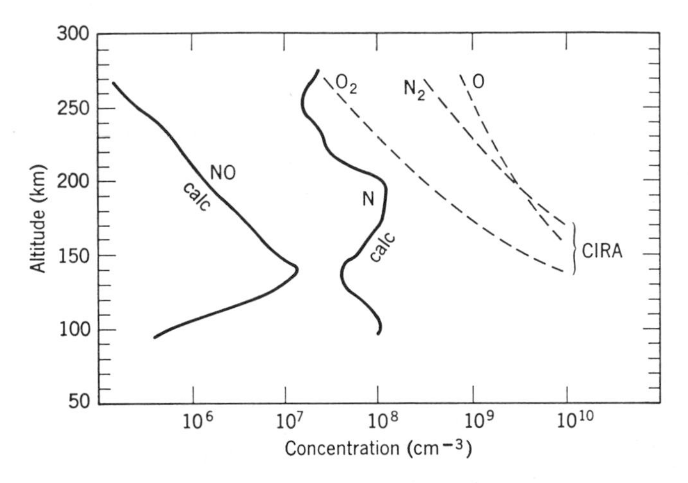

Figure 4.12b Computed Profiles of N and NO. (After Gosh, *JGR* **73,** 309 (1968); reprinted by permission of the American Geophysical Union.)

where k_h is the rate constant for Eq. (4.10h) and k_i is the rate constant for Eq. (4.10i). Atomic nitrogen is obviously necessary to allow reactions (4.10h and i), but as long as there is enough to maintain reaction (4.10h) (rather than transport processes) as the principal loss mechanism, the concentration of NO is independent of N. Hence, according to this analysis, the ratio of [NO] to $[O_2]$ is constant and determined by reaction rates k_h and k_i:

$$\frac{[NO]}{[O_2]} \simeq \frac{k_i}{k_h} \tag{4.18}$$

The next step in this analysis is to determine the range of altitudes over which atomic nitrogen densities are adequate to allow Eq. (4.17) to be valid. Consider the following sources of atomic nitrogen:

$$N_2 + h\nu \rightarrow N + N \tag{4.19a}$$

$$N_2^+ + e \rightarrow N + N \tag{4.19b}$$

$$O^+ + N_2 \rightarrow NO^+ + N \tag{4.13c}$$

$$NO^+ + e \rightarrow N + O \tag{4.19c}$$

$$N_2^+ + O \rightarrow NO^+ + N \tag{4.13d}$$

The atomic nitrogen production rates based on the preceding equations can be calculated to obtain a production rate profile. The loss rate of atomic nitrogen based on Eqs. (4.10a, b, h, and i) can also be calculated as a function of altitude. Again assuming equilibrium, the nitrogen atom density can be obtained. Unlike Figure 4.12a, this calculated profile of nitrogen atom concentrations shows atomic nitrogen concentrations to be a relatively uniform 10^5 cm^{-3} above 110 km (up to 190 km); it peaks at approximately 10^6 cm^{-3} at 90 km and falls rapidly to zero below 80 km. As a result, one is led to the conclusion that any nitric oxide that exists below 80-90 km must have been transported downwards from higher altitudes; the NO concentrations below 90 km, down to some undetermined mixing level, must, therefore, be computed by using transport theory and can be calculated with Eq. (4.18) for the regions above 90 km. This is the profile shown in Figure 4.12a, as NO(a).

The profile labeled NO(b) is the one measured by Barth[13] in an airglow experiment; the concentrations are obviously much greater than those in the theoretical curve. Several factors may account for the difference. First, the nitrogen production mechanisms used in the theoretical model were technically correct, but the yields were incorrectly apportioned; the important sources were more likely to be 4.13c, 4.19c, and 4.13d, whereas Barth emphasized 4.19a, b, and 4.13c. Reaction 4.19b is not likely to be noticeable inasmuch as the nitrogen ion transfers charge to the oxygen atom faster than the latter can recombine with electrons (see Chapter 8). Laboratory measurements of 4.13c and 4.19c made since Barth's work reveal faster reaction rates; hence, their greater importance.

Second, Nicolet has pointed out (Section 4.2) that the ionic rearrangement reactions are probably much more important than they have been thought to be, and that the measured NO concentrations can be reconciled if ionic reactions are considered. It has also been suggested that the rate of the production reaction 4.10i may be faster in the upper atmosphere than in the laboratory because either N or O_2 may be in an excited energy state.

Ghosh[28] has computed both N and NO concentrations using all available information for 31 ion, neutral, and ion-neutral reactions together with O, O_2, and N_2 profiles from the CIRA 1965 Standard Atmosphere. He finds that if he uses temperature-dependent rate equations and atmospheric temperature profiles in this computations, the results are compatible with the column density measurements integrated above 125 km by Barth[10]; they obviously do not match the observed profile below 125 km. The profiles for N, NO computed by Ghosh together with the CIRA profiles for O, O_2, and N_2 are presented in Figure 4.12b.

Sodium. The existence of nitric oxide discussed in the preceding section was theorized long before direct observational evidence was found. Sodium, on the other hand, has been observed in the atmosphere for many years but no definitive theory has yet been derived to explain the observations. The hypothesis by Donahue[25] of an aerosol layer source appears to have wide acceptance and will be summarized here; however, a meteor ablation sodium source theory is also widely accepted.

Resonance scattering by the sodium doublet has been observed in twilight glow and dayglow for over a decade (see Chapter 6). Rocket measurements[26,27] have confirmed the ground station observations and refined the previous derivations of concentration profiles; see Figure 4.13. Rocket measurements of positive ions further confirm the existence of atomic sodium (see Chapter 8).

The question is where does the sodium originate and what controls the observed concentrations of both neutral and ionized atoms? It is generally thought that atmospheric sodium originates in the interplanetary matter that penetrates the terrestrial atmosphere (some theories have proposed that the oceans are the source) and that a narrow layer (<10 km thick) of matter exists at an altitude of approximately 90 km. Sunlight presumably vaporizes sodium from this layer of aerosol particles; the sodium atoms subsequently disappear by chemical and ionic reactions. The details of both the production and loss are speculative, but Donahue[25,26] proposes three loss mechanisms (oxidation, ionization, and diffusion) and three production mechanisms (vaporization, reduction, and dissociative ion-electron recombination). The chemical and ionic reactions are the following (rates of chemical reactions are

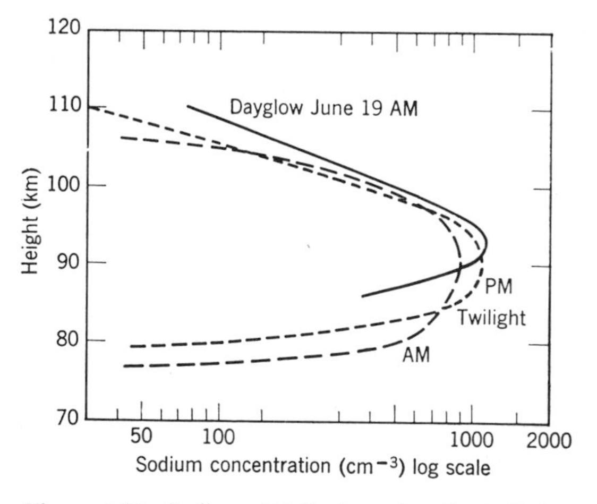

Figure 4.13 Sodium distributions for the twilights and dayglow of June 19, 1964. (After Hunten and Wallace *JGR* **72,** 69 (1967); reprinted by permission of the American Geophysical Union.)

estimated from analogous hydrogen reactions):

$$Na + O_3 \rightarrow NaO + O_2, \; k = 6.5 \times 10^{-12}\ cm^3\ sec^{-1} \quad (4.20a)$$

$$NaO + O \rightarrow Na + O_2, \; k = 4 \times 10^{-11}\ cm^3\ sec^{-1} \quad (4.20b)$$

$$Na^+ + O_3 \rightarrow NaO^+ + O_2 \quad (4.20c)$$

$$NaO^+ + e \rightarrow Na + O \quad (4.20d)$$

Hunten and Wallace[27] suggest that the topside of the sodium layer is controlled by evaporation and condensation of sodium and that the bottom side is controlled by chemical reactions.

4.3.2 The Atmospheres of Mars and Venus

Considering the general lack of success in computing theoretical models of the terrestrial atmosphere, one seems foolhardy in attempting models based on photochemical equilibrium for the atmospheres of other planets. However, such models, though tentative and probably in error, serve as guides in attempts to explain the fragmentary measurements that are made, and are expected to be made, of the atmospheres of Mars and Venus. They can also serve as starting points for more complex analyses.

It is now well established that carbon dioxide is a major component of the Mars atmosphere; however, estimates of the CO_2 composition vary from 100 percent to 20 percent of the total. The additional major component or components, if any, have not been identified by spectroscopic techniques; nitrogen and argon are usually suggested because they are nonreactive, cosmologically abundant, and cannot be identified with present observational techniques. Water vapor is present though it may vary seasonally[16] and a small fraction of oxygen has been tentatively identified.[17]

Spectroscopic observations,[18] radio-occultation measurements by the U.S. Mariner V planetary probe,[19] and the U.S.S.R. Venera IV interplanetary station[20] disagree on many details but there is now abundant evidence that the Venus atmosphere is almost pure CO_2. HCl and HF have been identified, whereas other minor constituents such as O, O_2, and H_2O have been reported but are not yet confirmed.

Thus, it appears that the photochemistry of both Mars and Venus is dominated by carbon dioxide. A reliable photochemical model calculation cannot be made for Mars or Venus until the photochemistry of carbon dioxide is better understood and the physical conditions (temperature and pressure profiles and vertical transport processes) are better delineated. However, let us consider some of the current problems and inconsistencies.

By utilizing the photolytic reactions (4.7a) to (4.7c) and the recombination reactions (4.12a) and (4.9a) photochemical equilibrium models can be computed for both the Mars and Venus upper atmosphere. Such a model is shown in Figure 4.14a for Mars[21] and in Figure 4.14b for Venus.[22] The first point to be considered is that the recombination reaction (4.9a) leading to the production of molecular oxygen is faster (by at least a factor of 25) than the recombination reaction (4.12a) for the production of carbon dioxide. Hence, we would expect the concentration of molecular oxygen to increase rapidly and absorb strongly in the CO_2 photodissociation spectral region (below 1700 Å); this would result in a "screening" of CO_2 by O_2 and a decrease in the yield of CO_2 photolytic products. Obviously O_2 will also be photolyzed and the concentration of O_2 will thereby be reduced; therefore, these reactions must also be taken into account for calculations of photochemical models.

Using the reasoning of Section 4.3.1. and the profiles of Figure 4.14a, we see that if the recombination of CO and O goes according to equation 4.12a, the lifetime of CO and O is in excess of 10^8 seconds above 105 km on Mars and in excess of 10^7 seconds above 80 km (altitude above cloud tops) on Venus. If we consider the three-body recombination of O, the lifetime is in

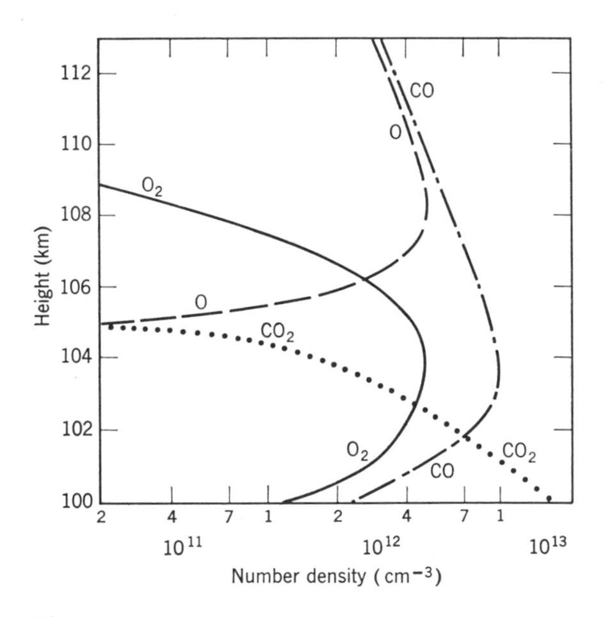

Figure 4.14a Photochemical equilibrium model of Mars upper atmosphere. (After Chamberlain and McElroy, *Science* **152,** 21 (1966); reprinted by permission of the American Association for the Advancement of Science; copyright 1966 by the AAAS.)

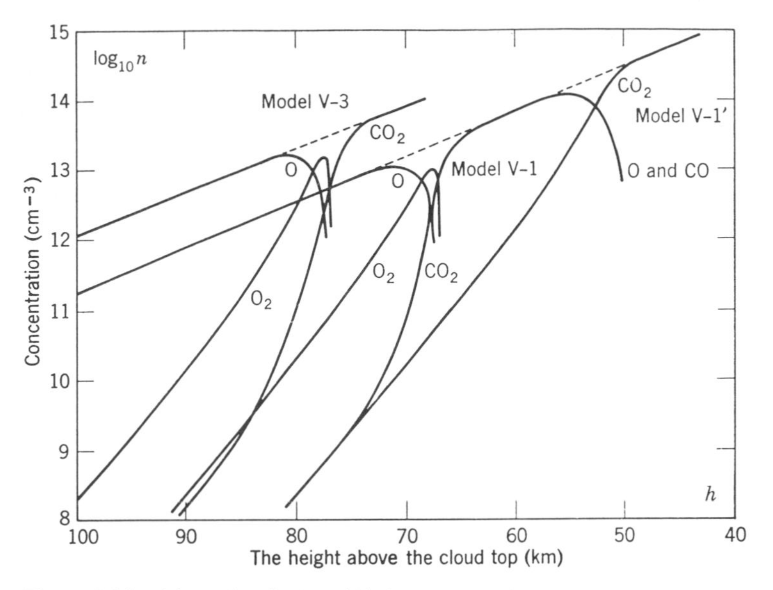

Figure 4.14b Photochemical equilibrium model of Venus upper atmosphere. (After Shimizu, *Planet. Space Sci.* **11,** 269 (1963); reprinted by permission of Pergamon Press.)

excess of 10^7 seconds above 105 km on Mars and in excess of 10^6 seconds above 80 km on Venus.

A similar situation exists with regard to the production of O and CO (or, in other words, the loss of CO_2) by photolysis. Chamberlain and McElroy[21] point out that in the upper Mars atmosphere, the lifetime of the CO_2 molecule against photodissociation is in excess of 3×10^6 seconds ($\sim$one month) and that if screening by O_2 is taken into account the lifetime may exceed 10^3 years.

Clearly, the concept of photochemical equilibrium does not appear to be applicable in the upper atmosphere; and, unlike the oxygen-rich atmosphere of the earth, there appears to be no important photolytic reactions at lower altitudes where equilibrium conditions might be justified.

From this heuristic argument it is tempting to conclude that transport processes control the chemistry of the Mars and Venus upper atmosphere. This is indeed the conclusion reached by Chamberlain and McElroy[21] in their analysis of the Mars atmosphere; they reason, therefore, that the gases in the Mars upper atmosphere are homogeneously mixed and that CO_2 is predominant.

On the other hand, our discussion in Section 4.1.2 suggests that other (and, possibly, faster) recombination reactions should be considered in CO_2

photochemistry and that the controlling loss reactions might have very high rates. If that were the case, we could conclude that chemical recombination rather than mass transport controls the upper Mars and Venus atmosphere. However, considering the low production rate, a rapid chemical loss rate would also lead to the same conclusion as the preceding paragraph.

Obviously, theoretical models of Mars and Venus upper atmospheres are still speculative. The fragmentary observational data regarding CO_2 photochemistry and transport processes that are available to date indicate only that atomic oxygen could not be detected in the Venus atmosphere by either U.S. or U.S.S.R. experiments;[20,24] these data may provide crude constraints on models but they are not definitive. On the other hand, atomic oxygen and carbon monoxide have definitely been observed[28] in the Mars upper atmosphere.

PROBLEMS

4.1. Calculate the kinetic energies ("thermal energies") of atmospheric particles in thermal equilibrium at temperatures of 140, 200, 300, 500, 1000, and 10,000°K. Express energies in units of eV.

4.2. The planet Glitch has an atmosphere of pure smellerium (PU). The radiation flux responsible for the photolysis of smellerium into phooium and urpium is found to be 10^{16} photon cm^{-2} sec^{-1} incident to the atmosphere. The surface temperature on Glitch is 400°K and the atmosphere is isothermal, the molecular weight is 83.1, the neutral particle density at the surface is 10^{20} cm^{-3}, the acceleration due to gravity is 400 cm sec^{-2}, and the following reactions are known to occur

$$\mathrm{PU} + h\nu \rightarrow \mathrm{P} + \mathrm{U},\ Q_a = 10^{-23}\ \mathrm{cm}^2 \quad \text{(a)}$$

$$\mathrm{P} + \mathrm{U} + \mathrm{PU} \rightarrow 2\,\mathrm{PU},\ k_b = 10^{-35}\ \mathrm{cm}^6\ \mathrm{sec}^{-1} \quad \text{(b)}$$

$$\mathrm{P} + \mathrm{P} + \mathrm{PU} \rightarrow \mathrm{P}_2 + \mathrm{PU},\ k_c = 10^{-29}\ \mathrm{cm}^6\ \mathrm{sec}^{-1} \quad \text{(c)}$$

$$\mathrm{U} + \mathrm{U} + \mathrm{PU} \rightarrow \mathrm{U}_2 + \mathrm{PU},\ k_d = 10^{-31}\ \mathrm{cm}^6\ \mathrm{sec}^{1} \quad \text{(d)}$$

$$\mathrm{P}_2 + h\nu \rightarrow \mathrm{P} + \mathrm{P},\ Q_e = 10^{-26}\ \mathrm{cm}^2 \quad \text{(e)}$$

$$\mathrm{U}_2 + h\nu \rightarrow \mathrm{U} + \mathrm{U},\ Q_f = 10^{-28}\ \mathrm{cm}^2 \quad \text{(f)}$$

a. Construct a neutral particle density profile. Assume constant molecular weight and neglect photochemistry.
b. If the absorption cross section is 10^{-23} cm^2, determine the unit optical depth (altitude at which the radiation intensity is reduced to $1/e$) in the smellerium atmosphere for dissociating radiation (assume that photolysis is 100 per cent efficient, i.e., all the radiation that is absorbed causes dissociation, and that the zenith angle is 0°). Determine altitudes at which radiation is reduced by $(1/e)^2$ $(1/e)^3$, etc., and construct an intensity profile.
c. Construct profiles of the loss rate of PU owing to photodissociation.
d. Write all the appropriate rate equations.

e. If $[P] = 10^{-6}$ [PU] and $[U] = 10^{-5}$ [PU], find the lifetimes of P and U against all the possible P and U loss reactions at 0, 10, 40, 80, 120 km.
f. If U_2 and P_2 are dissociated with 100 per cent efficiency by the same radiation that photolyzes PU, determine the lifetimes of PU, U_2, and P_2 against photolysis at the same altitudes as for part e.
g. According to the results of e and f, select the most important production and loss reactions for P and U, if the Glitchian day is 10^9 secs long.
h. Using the loss reactions selected in g, write the equilibrium equations for P and U, derive expressions for [P] and [U] and construct profiles of P and U concentrations from 0 to 120 km.

4.3. If the concentration of XY in an atmosphere is controlled by the following reactions,

$$X + Y_2 \xrightarrow{k_1} XY + Y$$

$$XY + X \xrightarrow{k_2} X_2 + Y$$

Derive the equation that relates the equilibrium concentrations of XY and Y_2.

4.4. Assume that the dissociating radiation flux at altitude z is $dI(z)/d\lambda$, the cross section for dissociation of X_2 is $Q_{X_2}(\lambda)$ and that X disappears by a three-body recombination reaction with a rate k. Find expressions for:
a. The lifetime of X_2 against photolysis.
b. The lifetime of X against recombination.
c. The equilibrium concentration of X.

4.5. List the five reactions involved in the production and loss of ozone in a dry atmosphere, the rate equations, and the equilibrium equations. Consider only the ground state reactions.

4.6. From the equilibrium equations listed in Problem 4.5, derive the following approximate expression for $[O_3]$ in terms of $[O_2]$

$$[O_3] = [O_2]\left[\rho_{O_2} k_{O,O_2}[M] \Big/ \left(\rho_{O_3} k_{O,O_3} + \frac{\rho_{O_3}^2 k_{O,O}}{k_{O,O_2}[O_2]}\right)\right]^{1/2}$$

4.7. Using the atomic oxygen concentration profiles computed by Hunt (Figure 4.11) and the molecular oxygen and nitrogen concentration profiles in Figure 4.12a, calculate for 50, 70, 80, 90, and 100 km the lifetime of the O atom against three-body recombination and against the oxidation reaction with O_3. Compute the transport velocity that must be exceeded for diffusion to become dominant.

4.8. If all the atomic oxygen in an atmosphere were produced in the metastable ^1D state and removed by the ground-state ^3P three-body recombination process, show that the equilibrium concentration of $O(^1D)$ is proportional to the square of the $O(^3P)$ concentration. Assume that the de-excitation collision is with the same species that acts as the third body for the recombination reaction.

4.9. Using the molecular oxygen profile from Figure 4.12a, plot the nitric oxide concentration between 70 and 120 km according to Barth's model.

4.10. From Table 8.1 find the rate constants for reactions (4.19b, d), and (4.13c). At an imaginary altitude where $[N_2] = 10^{15}$ cm^{-3} and $[N_2^+] = [O^+] = [NO^+] = 10^4$ cm^{-3}, find the production rates of atomic nitrogen by each of the three processes. Find the rate for $N_2^+ + O_2 \rightarrow O_2^+ + N_2$ at the same altitude (assume $[N_2]/[O_2] = 5$) and compare lifetimes of N_2^+ against this reaction and against reaction 4.19b. Which of the three are the most important sources of N? What is the important N_2^+ loss process?

Note: $[e] = [N_2^+] + [O^+] + [NO^+]$

4.11. Assume that Mars has an isothermal atmosphere of 180°K, a surface neutral particle concentration of 2×10^{18} cm^{-3}, and is composed of CO_2 and A in equal parts by volume. Assume also that the total flux of incident ultraviolet radiation in the 1350 to 1750 Å photodissociation region is 10^{12} photon cm^{-2} sec^{-1}. The acceleration due to gravity is 380 cm sec^{-2}.

a. Construct a CO_2 molecule concentration profile from 0 to 70 km, if *no* photolysis occurs.

b. Find the volume rate of photolysis over the above altitude range, if the absorption cross section in the 1350–1750 Å range is 10^{-16} cm^2 and all the absorption is caused by CO_2 and results in photolysis.

c. Assume that only the following reactions occur:

$$CO_2 + h\nu \xrightarrow{Q} CO + O$$

$$CO + O + M \xrightarrow{k} CO_2 + M$$

Write the set of differential equations that relate the concentrations of CO, CO_2, and O.

d. Write the set of equations that describes equilibrium conditions.

e. If transport processes are insignificant, what are the equilibrium concentrations of CO and O for the altitudes range specified in Part a?

f. What values must the vertical transport velocity exceed to be dominant at those altitudes?

g. If the atomic oxygen produced in photolysis is in the $O(^1D)$ state and is deactivated by collisions with CO_2 with a rate constant of 10^{-12} cm^3 sec^{-1}, derive an expression relating the equilibrium concentrations of $O(^1D)$ and $O(^3P)$.

h. If the following reaction were added to the list of reactions in part c

$$CO + O \rightarrow CO_2 + h\nu, \; k = 10^{-12} \text{ cm}^{-3} \text{ sec}^{-1}$$

what would the principal atomic oxygen loss process be?

i. Assuming that the process listed in Part h is the only loss process for atomic oxygen, calculate the equilibrium concentrations of CO and O at the altitudes specified in Part a.

REFERENCES

1. D. R. Bates, The physics of the upper atmosphere, in *The Earth as a Planet*, G. P. Kuiper, ed., The University of Chicago Press, Chicago, 1954.
2. K. Watanabe, Ultraviolet absorption in the upper atmosphere, *Adv. in Geophysics* **5,** 153, Academic Press, New York and London, 1958.
3. Geoffrey V. Marr, *Photoionization Processes in Gases*, Academic Press, New York and London, 1967.
4. A. E. S. Green, *The Middle Ultraviolet: Its Science and Technology*, John Wiley & Sons, New York, 1966.
5. J. G. Calvert, and J. N. Pitts, Jr., *Photochemistry*, John Wiley & Sons, New York, 1966.
6. M. Nicolet, *J. Geophys. Res.* **70,** 679 (1965).
7. B. G. Hunt, *J. Geophys. Res.* **71,** 1385 (1966).
8. M. Nicolet, *J. Geophys. Res.* **70,** 691 (1965).
9. M. Nicolet, The properties and constitution of the upper atmosphere, in *Physics of the Upper Atmosphere*, Ratcliffe, ed., Academic Press, New York, 1960.
10. C. A. Barth, *Ann. Geophys.* **22,** 198 (1966).
11. M. Nicolet, Nitrogen oxides and the airglow, in *The Threshold of Space*, M. Zelikoff, ed., Pergamon Press, New York, 1956.
12. C. A. Barth, Nitrogen and oxygen atomic reactions in the chemosphere, in *Chemical Reactions in the Lower and Upper Atmosphere*, Interscience, New York, 1961.
13. C. A. Barth, *J. Geophys. Res.* **69,** 3301 (1964).
14. W. D. McGrath and J. J. McGarvey, *Planet. Space Sci.* **15,** 427 (1967).
15. D. M. Hunten and M. B. McElroy, *Rev. Geophys.* **4,** 303 (1966).
16. R. A. Schorn, H. Spinrad, R. Moore, H. J. Smith, and L. P. Giver, *Astrophys. J.* **147,** 743 (1967).
17. M. J. S. Belton, and D. M. Hunten, *Astrophys. J.* **153,** 963 (1968).
18. D. M. Hunten, *J. Geophys. Res.* **73,** 1093 (1968).
19. A. Kliore, G. S. Levy, D. L. Cain, G. Fjeldo, S. I. Rasool, *Science* **158,** 1683 (1967).
20. Venus 4: An automatic Interplanetary Station (Translation of TASS report), *Trans. Amer. Geophys. Union* **48,** 931 (1967).
21. J. W. Chamberlain, and M. B. McElroy, *Science* **152,** 21 (1966).
22. M. Shimizu, *Planet. Space Sci.* **11,** 269 (1963).
23. S. N. Ghosh, *J. Geophys. Res.* **73,** 309 (1968).
24. C. A. Barth, J. B. Pearce, K. K. Kelly, L. Wallace, and W. G. Fastie, *Science* **158,** 1675 (1967).
25. T. M. Donahue, *J. Geophys. Res.* **71,** 2237 (1966).
26. T. M. Donahue, and R. R. Meier, *J. Geophys. Res.* **72,** 2803 (1967).
27. D. M. Hunten, and L. Wallace, *J. Geophys. Res.* **72,** 69 (1967).
28. C. A. Barth, W. G. Fastie, C. W. Hord, J. B. Pearce, K. K. Kelly, A. I. Stewart, G. E. Thomas, G. P. Anderson, and O. F. Raper, *Science* **165,** 1004 (1969).

GENERAL REFERENCES

J. A. Ratcliffe, ed., *Physics of the Upper Atmosphere*, Academic Press, New York, 1960. Chapter 2: The properties and constitution of the upper atmosphere, by M. Nicolet.

J. G. Calvert, and J. N. Pitts, Jr., *Photochemistry*, John Wiley & Sons, New York, 1966. Chapters 1 and 3 are especially useful, but many sections of this book are useful in understanding the photochemistry of atmospheric species.

Geoffrey V. Marr, *Photoionization Processes in Gases*, Academic Press, New York, 1967. Chapters 1, 2, and 8.

R. C. Whitten and I. G. Poppoff, *Physics of the Lower Ionosphere*, Prentice-Hall, Englewood Cliffs, N.J., 1965. Chapter 2 is a useful survey of solar radiation spectra in the photodissociation wavelengths.

G. P. Kuiper, *The Earth as a Planet*, University of Chicago Press, Chicago, Illinois, 1954. Chapter 9: The absorption spectrum of the atmosphere, by L. Goldberg. Chapter 12: The physics of the upper atmosphere, by D. R. Bates. Chapter 13: Dynamic effects in the high atmosphere, by M. Nicolet. This is an especially good treatment of the role of transport processes in aeronomy.

A. E. S. Green, *The Middle Ultraviolet: Its Science and Technology*, John Wiley & Sons, New York, 1966. Chapter 2: Atomic and molecular species, by R. T. Brinkman, *et al.* Chapter 4: Atmospheric ozone, by M. Griggs. Chapter 10: The ultraviolet spectroscopy of planets, by C. A. Barth. Chapter 11: Astronomical ultraviolet radiation, by K. L. Hallam.

5

Fluid Aeronomy

In the preceding two chapters planetary upper atmospheres were considered from the standpoint of static phenomena such as compression due to gravitational forces acting on the air, or of molecular scale processes such as diffusion, solar energy absorption and chemical reactions. Phenomena of a collective nature such as winds and oscillations are of comparable significance but, owing to their different character, are best treated as a separate topic. They are the subject of this chapter.

The basic laws which underlie classical fluid dynamics are, of course, Newton's laws of motion, the laws of thermodynamics, and the conservation laws. Unfortunately, the full investigation of the application to fluids of these principles would be excessively lengthy. We shall restrict the discussion to a simple treatment of general theory and those aspects of interest to the earth's upper atmosphere. In particular, these are acoustic-gravity waves, atmospheric tides, winds and circulation, and turbulent transport.

Although separately discussed, all of these phenomena are closely related. For example, thermospheric winds are the result of thermally created tidal motions at those altitudes; gravity waves are probably closely associated with irregular wind components in the upper mesosphere and lower thermosphere; and wind shears may be the source of turbulence in the same altitude regime. Further examples of such interrelationships will be introduced as we proceed.

Undoubtedly the dynamic aspects of the atmospheres of other planets are just as complex as those of the earth. Models of lower atmospheres have indeed been constructed by a number of investigators, but they are still entirely conjectural and of little aid in studying atmospheric dynamics at high altitudes; when such studies do become fruitful, the general principles of fluid motion developed in this chapter will apply.

5.1 Fundamentals of Fluid Dynamics

Before launching into a discussion of fluid phenomena in the atmosphere, we shall develop briefly the pertinent aspects of the theory of fluid motions.

Unfortunately, there is no really satisfactory general theory of fluid dynamics because of the nonlinear character of the equations. This fact should be kept in mind by the reader when perusing the rest of this chapter.

5.1.1 The Equations of Motion

We commence our introduction to the subject by writing a statement of Newton's second law of motion and then identifying the forces involved. According to this principle the motion of an air parcel of specific volume α (i.e., volume per unit mass) is governed by the equation

$$\frac{D^2\mathbf{r}}{Dt^2} = \frac{D\mathbf{v}}{Dt} = -\alpha\nabla p + \mathbf{g} + \mathbf{f} \tag{5.1}$$

where use of the capital "D" denotes the total derivative with respect to an inertial reference frame, $\mathbf{g}$ is the gravitational acceleration, $\mathbf{f}$ is the acceleration resulting from frictional (i.e., viscous) forces, and p is the gas pressure. In the atmospheres of rotating planets such as the earth, it is more convenient to investigate the motion relative to the planetary surface rather than to inertial reference frames. Hence, we express $D\mathbf{v}/Dt$ in terms of *relative* acceleration. It is well known from basic mechanics that the absolute velocity $D\mathbf{r}/Dt$ of an air parcel is related to the relative velocity $(d\mathbf{r}/dt)_{\text{rel.}}$ by

$$\frac{D\mathbf{r}}{Dt} = \left(\frac{d\mathbf{r}}{dt}\right)_{\text{rel.}} + \mathbf{\Omega} \times \mathbf{r} \tag{5.2}$$

where $\mathbf{\Omega}$ is the angular velocity of the planet. Therefore, the absolute acceleration must be

$$\frac{D\mathbf{v}}{Dt} = \left(\frac{d\mathbf{v}}{dt}\right)_{\text{rel.}} + 2\mathbf{\Omega} \times \mathbf{v} + \mathbf{\Omega}(\mathbf{\Omega} \cdot \mathbf{r}) - \mathbf{r}\Omega^2 \tag{5.3}$$

Substitution into Eq. (5.1) then yields an equation for the relative acceleration

$$\left(\frac{d\mathbf{v}}{dt}\right)_{\text{rel.}} = -2\mathbf{\Omega} \times \mathbf{v} + \mathbf{g} - \mathbf{\Omega}(\mathbf{\Omega} \cdot \mathbf{r}) + \mathbf{r}\Omega^2 + \mathbf{f} - \alpha\nabla p \tag{5.4}$$

which, in terms of the "gravitational" potential

$$g\chi = \mathbf{g} \cdot \mathbf{r} + \tfrac{1}{2}\Omega^2 R^2 \tag{5.5}$$

becomes

$$\left(\frac{d\mathbf{v}}{dt}\right)_{\text{rel.}} + 2\mathbf{\Omega} \times \mathbf{v} + \alpha\nabla p + g\nabla\chi = \mathbf{f} \tag{5.6}$$

Here $\mathbf{R}$ is the axial position vector illustrated in Figure 5.1. Equation (5.6), frequently called Euler's equation, is the first member of the set which we need.

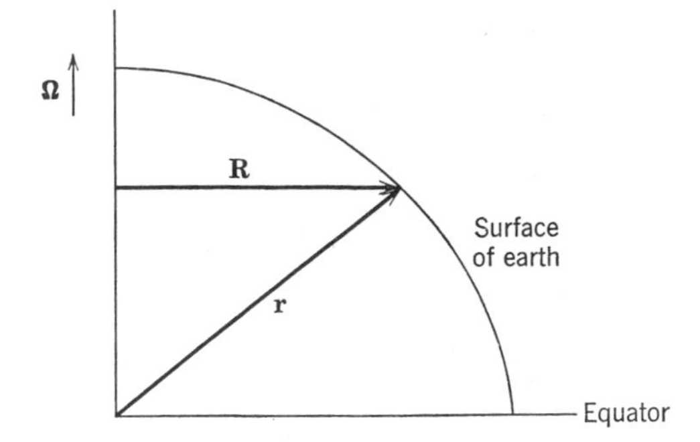

Figure 5.1 Relationship of the central position vector **r** to the axial position vector **R**.

The two others are statements of the conservation of mass (continuity equation)

$$\frac{D\alpha}{Dt} - \alpha\nabla \cdot \mathbf{v} = 0 \tag{5.7}$$

and the conservation of energy (first law of thermodynamics)

$$\frac{DS}{Dt} = \frac{q_T}{T} \tag{5.8}$$

in which q is the rate of energy deposition in the region, S is entropy, and T is the absolute air temperature. Equations (5.5), (5.7), and (5.8) can be combined into a single equation which represents the conservation of energy. First, we take the scalar product of Eq. (5.5) with $\mathbf{v}$, obtaining

$$\frac{D(\frac{1}{2}v^2)}{Dt} + \alpha\mathbf{v} \cdot \nabla p + g\mathbf{v} \cdot \nabla\chi = \mathbf{f} \cdot \mathbf{v} \tag{5.9}$$

The internal thermal energy ϵ is included by recasting Eq. (5.8) as

$$T\frac{DS}{Dt} = \frac{D\epsilon}{Dt} - p\frac{D\alpha}{Dt} \tag{5.10}$$

[see Eq. 2.31)] or, invoking the continuity equation,

$$T\frac{DS}{Dt} = \frac{D\epsilon}{Dt} + \alpha p\nabla \cdot \mathbf{v} \tag{5.11}$$

Since $\partial\chi/\partial t = 0$†, Eq. (5.9) and (5.11) can be added to yield [$\rho = 1/\alpha$ is the mass density]

$$\frac{DW}{Dt} + \frac{1}{\rho}\nabla\cdot(p\mathbf{v}) = q_T + \mathbf{f}\cdot\mathbf{v} \tag{5.12}$$

$$W = \tfrac{1}{2}v^2 + \epsilon + g\chi$$

which is the complete expression for the conservation of energy, the energy density (i.e., per unit mass) being represented by W. The vector quantity $p\mathbf{v}$ has the significance of "energy flux" density and the terms q_T and $\mathbf{f}\cdot\mathbf{v}$ on the right hand side represent an energy source and an energy transformation, respectively.

5.1.2 Perturbation Theory

In order to obtain solutions to the fundamental equations of fluid dynamics, it is necessary to evaluate the unknowns (i.e., velocity, density, and entropy) at particular points in space, thus requiring the replacement of the substantive derivatives D/Dt by $\partial/\partial t + \mathbf{v}\cdot\nabla$. It is evident that the resulting equations are nonlinear and thus, with one exception to be discussed below, not amenable to general analytic solution. This is most unfortunate because the properties which arise from the nonlinearity are perhaps the most interesting of all. Nevertheless, the mathematics of nonlinear differential equations has not been sufficiently developed to permit this, and we must resort to approximation methods which in unfavorable conditions obscure the nonlinear character. Specifically, we separate the dependent variables $\mathbf{v}$, α, and S into static and time-varying parts, treating the latter as small perturbations,

$$\mathbf{v} = \mathbf{v}_0 + \mathbf{v}_1;\ \mathbf{v}_0 = 0$$

assuming that all winds are due to the perturbation, and that

$$\alpha = \alpha_0 + \alpha_1$$

$$S = S_0 + S_1$$

$$p = p_0 + p_1$$

The zeroth order equation is the statement of hydrostatic equilibrium

$$\alpha_0\nabla p_0 = -g\nabla\chi = -g\hat{\boldsymbol{\zeta}} \tag{5.13}$$

† $\partial\phi/\partial t$, the time derivative of a function ϕ at a fixed point in space must be distinguished from the substantive derivative $d\phi/dt$ which applies to a parcel which moves along with the fluid. The two are related by

$$\frac{d\phi}{dt} = \frac{\partial\phi}{\partial t} + \mathbf{v}\cdot\nabla\phi$$

(ζ is a vertical unit vector) while those of first order are expressed as

$$\frac{\partial \mathbf{v}_1}{\partial t} + \alpha_0 \nabla p_1 + \alpha_1 \nabla p_0 + 2\Omega \times \mathbf{v}_1 = \mathbf{f}_1 \tag{5.14a}$$

$$\frac{\partial \alpha_1}{\partial t} + \mathbf{v}_1 \cdot \nabla \alpha_0 + \alpha_0 \nabla \cdot \mathbf{v}_1 = 0 \tag{5.14b}$$

$$\frac{\partial S_1}{\partial t} + \mathbf{v}_1 \cdot \nabla S_0 = q_T/T_0 \tag{5.14c}$$

It is apparent that we have too many variables (4) for the number of equations (3). In order to resolve this impasse, we introduce the thermodynamic relation

$$\delta p = \left(\frac{\partial p}{\partial \alpha}\right)_s \delta\alpha + \left(\frac{\partial p}{\partial s}\right)_\alpha \delta S = -\frac{p\gamma}{\alpha}\,\delta\alpha + \frac{p}{C_v}\,\delta S \tag{5.15}$$

(see Problem 2.7). In terms of the perturbed and unperturbed variables, Eq. (5.15) can be expressed as

$$p_1 = -\frac{p_0\gamma}{\alpha_0}\,\alpha_1 + \frac{p_0}{C_v}\,S_1 \tag{5.15a}$$

where γ is the ratio C_p/C_v of the specific heat capacities at constant pressure and volume. Note that all second order terms have been neglected in Eqs. (5.14) because they are "small."

We now eliminate α_1 and p_0 in the set of equations (5.14) with the aid of Eq. (5.15a), and the hydrostatic equation (5.13), obtaining

$$\frac{\partial \mathbf{v}_1}{\partial t} + \alpha_0\left(\nabla p_1 + \frac{g}{p_0\alpha_0\gamma}\,\hat{\zeta} p_1\right) - N^2 S\hat{\zeta} + 2\boldsymbol{\Omega} \times \mathbf{v}_1 = \mathbf{f}_1 \tag{5.16a}$$

$$\frac{\partial p_1}{\partial t} + \gamma p_0 \nabla \cdot \mathbf{v}_1 - \frac{g}{\alpha_0}\,\mathbf{v}_1 \cdot \hat{\zeta} = (\gamma - 1)\,\frac{q_T}{\alpha_0} \tag{5.16b}$$

$$\frac{\partial S}{\partial t} + \mathbf{v}_1 \cdot \hat{\zeta} = q_T/T_0 S_0' \tag{5.16c}$$

$$S = S_1/S_0'$$

where

$$N^2 = \frac{g}{C_p}\,S_0' \tag{5.17}$$

is the so-called "Brunt-Väisälä" frequency which is discussed in detail below; the derivative dS_0/dz is denoted by S_0'.

Although the force $\mathbf{f}_1$ has not yet been specified, it turns out that the only one in which we shall be interested later on is the viscous interaction

$$\mathbf{f}_1 = \nu \nabla^2 \mathbf{v} \tag{5.18}$$

for an incompressible fluid. The factor ν is the kinematic viscosity which is determined from the principles of kinetic theory.

5.1.3 The Significance of N

We have already noted that N has the dimensions of a frequency. In order to learn its nature it is convenient to consider a parcel of fluid initially at altitude z' to be displaced adiabatically to a height $z' + \xi$ where ξ is "small". The pressure differential between the two heights is given by the equation

$$\delta p = -g\rho\xi \tag{5.19}$$

where the density ρ is related to the specific volume α by $\rho = 1/\alpha$. Since the expansion of the gas is adiabatic, the density inside the parcel is

$$(\delta\rho)_{\text{in}} = \delta p/\gamma R T_0 = -(g/\gamma R T_0)\rho\xi \tag{5.20}$$

whereas that outside is, to the first order term of a Taylor series,

$$(\delta\rho)_{\text{out}} = \xi \frac{d\rho}{dz} \tag{5.21}$$

The net buoyant force per unit volume, f_B, acting on the parcel is obtained from the difference between Eqs. (5.21) and (5.20)

$$f_B = g[(\delta\rho)_{\text{in}} - (\delta\rho)_{\text{out}}] = g\xi\left(\frac{d\rho}{dz} + \frac{\rho g}{c_0^2}\right) \tag{5.22}$$

which must be equal to the inertial force on the parcel $f_I = \rho d^2\xi/dt^2$ or

$$\frac{d^2\xi}{dt^2} - g\left(\frac{1}{\rho}\frac{d\rho}{dz} + \frac{g}{\gamma R T}\right)\xi = 0 \tag{5.23}$$

Obviously the quantity

$$-g\left(\frac{1}{\rho}\frac{d\rho}{dz} + \frac{g}{\gamma R T_0}\right)$$

has the significance of the square of an oscillation frequency ω_b *if* the latter is real.

It remains to find the relationship between this frequency and the Brunt-Väisälä frequency N. It is left as an exercise to prove that

$$S_0' = -\frac{\gamma R}{\gamma - 1}\left(\frac{1}{\rho_0}\rho_0' + \frac{g}{\gamma R T_0}\right) \tag{5.24}$$

Comparison of this result with Eq. (5.22) shows that

$$\omega_b^2 = \frac{g}{C_p} S_0' = N^2 \tag{5.25}$$

or the frequency of oscillation of the parcel is just the Brunt-Väisälä frequency introduced previously. It is apparent that if S_0' is positive, N is real and the motion of the parcel is stable. However, if S_0' is negative, N is imaginary, the motion is unstable and the parcel will not return to its initial position. In an adiabatic atmosphere $S_0' = 0$ and thus $N = 0$. No forces act on the parcel if it is moved and it will remain at rest at whatever position it is placed. The Brunt-Väisälä frequency therefore has the significance of a stability parameter which represents the small scale dynamical properties. The profile of N shown in Figure 5.2 is typical of the earth's atmosphere although it does not conform to any particular model.

It is evident from Figure 5.2 that air in the upper atmosphere is usually stable (i.e., $N^2 > 0$) and thus resists vertical convective motions. In fact the stratosphere and thermosphere are stable by definition. However, it is possible for the mesosphere to be unstable, particularly at high latitudes in summer. The consequences of such instability for vertical motions are as yet unclear but may be related to the appearance of noctilucent clouds just below the mesopause.

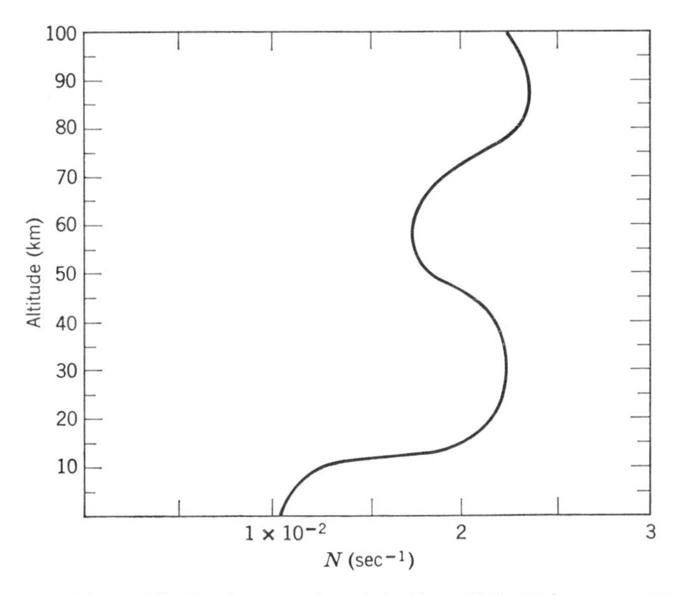

Figure 5.2 Typical profiles of the Brunt-Väisälä frequency N.

5.1.4 The Geostrophic Approximation and the Thermal Wind

It is frequently a surprisingly good approximation to consider atmospheric motion to be steady (i.e., $D\mathbf{v}_1/Dt = 0$, which is usually called the geostrophic approximation). If frictional forces are negligible ($\mathbf{f} = 0$), one obtains by taking the vector product of $\hat{\boldsymbol{\zeta}}$ with the right hand side of Eq. (5.16a)

$$2\boldsymbol{\Omega}(\mathbf{v}_g \cdot \hat{\boldsymbol{\zeta}}) - 2\mathbf{v}_g(\boldsymbol{\Omega} \cdot \hat{\boldsymbol{\zeta}}) = -\alpha\hat{\boldsymbol{\zeta}} \times \nabla p \tag{5.26}$$

where we have omitted the subscript from p in order to simplify the notation. If we further make the very well justified assumption that the wind is horizontal $(\mathbf{v}_g \cdot \hat{\boldsymbol{\zeta}}) = 0$, we have an equation for the geostrophic wind velocity

$$\mathbf{v}_g = \frac{\alpha_0}{2\boldsymbol{\Omega} \cdot \hat{\boldsymbol{\zeta}}} \hat{\boldsymbol{\zeta}} \times \nabla p \tag{5.27}$$

To what extent the geostrophic approximation in the upper atmosphere is valid is presently not known because of the lack of adequate synoptic data. Undoubtedly, it is not *everywhere* applicable—certainly it is not in the lower atmosphere—but probably is reasonably well justified in many cases, particularly on a planetary scale.

As we shall discover in Section 5.4, one of the outstanding features of the mesosphere and lower thermosphere is the presence of large wind shears. The *thermal wind* which is the vertical gradient of wind velocity

$$\mathbf{v}_T = \frac{\partial \mathbf{v}}{\partial z} \tag{5.28}$$

is a very useful concept in relating the wind shear to atmospheric properties. In the case where the geostrophic approximation is valid the thermal wind is of the form

$$\mathbf{v}_T = \frac{\partial}{\partial z}\left(\frac{\alpha_0}{2\boldsymbol{\Omega} \cdot \hat{\boldsymbol{\zeta}}} \hat{\boldsymbol{\zeta}} \times \nabla p\right) \tag{5.29}$$

or

$$\mathbf{v}_T = \frac{1}{2\boldsymbol{\Omega} \cdot \hat{\boldsymbol{\zeta}}} \hat{\boldsymbol{\zeta}} \times \frac{\partial}{\partial z}(\alpha_0 \nabla p) \tag{5.30}$$

Insofar as the geostrophic approximation is valid, one could employ Eq. (5.30) to compute the winds resulting from the atmospheric tides introduced in Section 5.3. For a number of reasons this is probably not a profitable venture and so we leave the subject at this point.

5.1.5 Vorticity and Cyclogenesis

Because of planetary rotation and some other effects to be introduced later, the atmosphere frequently exhibits rotational characteristics which are represented by the *vorticity* $\tilde{\boldsymbol{\omega}}$ which is defined as the curl of the velocity field.

$$\tilde{\boldsymbol{\omega}} = \text{curl } \mathbf{v} \tag{5.31}$$

We are usually interested in the vertical component of $\tilde{\boldsymbol{\omega}}$ which we denote by $\tilde{\varpi}$

$$\tilde{\varpi} = \bar{\boldsymbol{\zeta}} \cdot \text{curl } \mathbf{v} = \frac{\partial v_y}{\partial x} - \frac{\partial v_x}{\partial y} \tag{5.32}$$

If we choose the horizontal coordinate axes in such a way that

$$\boldsymbol{\Omega} = \Omega_H \hat{j} + \Omega_V \hat{k} \tag{5.33}$$

Euler's equation (5.16a) then becomes, (in component form)

$$\begin{aligned}
&\frac{\partial v_x}{\partial t} - 2\Omega_V v_y + 2\Omega_H v_z = -\alpha_0 \frac{\partial p}{\partial x} \\
&\frac{\partial v_y}{\partial t} + 2\Omega_V v_x = -\alpha_0 \frac{\partial p}{\partial y} \\
&\frac{\partial v_z}{\partial t} - 2\Omega_H v_x = -\alpha_0 \left(\frac{\partial p}{\partial z} + \frac{g}{p_0 \gamma} p\right) - N^2 \frac{S_1}{dS_0'}
\end{aligned} \tag{5.34}$$

Taking the curl of $\mathbf{v}$, we obtain the following expression for the time rate of change of $\tilde{\varpi}$ to first order

$$\frac{\partial \tilde{\varpi}}{\partial t} + 2\Omega_V \left(\frac{\partial v_x}{\partial x} + \frac{\partial v_y}{\partial y}\right) - \Omega_H \frac{\partial v_z}{\partial y} = 0 \tag{5.35}$$

or, using Eq. (5.16b) to remove the factor $(\partial v_x/\partial x + \partial v_y/\partial y)$,

$$\frac{\partial \tilde{\varpi}}{\partial t} = 2\Omega_V \left(\frac{1}{\gamma p_0} \frac{\partial p}{\partial t} + \frac{\partial v_z}{\partial z} - \frac{q_T}{C_p T_0}\right) \tag{5.36}$$

We have neglected terms containing v_z and $\partial v_z/\partial y$ because they are expected to be much smaller than the others. Equation (5.36) is a mathematical expression which relates "first order" cyclogenesis (the formation and growth of atmospheric vortices) to planetary rotation (Ω_V), local temporal pressure changes ($\partial p/\partial t$), vertical velocity gradients or wind shears ($\partial v_z/\partial z$), and heating (q_T). Evidently planetary rotation and at least one of the last three phenomena must be present for the generation of vorticity. By taking the second derivative of $\tilde{\omega}$ with respect to time and substituting Eqs. (5.34) into (5.35), an equation describing *vorticity oscillations* results:

$$\frac{\partial^2 \tilde{\omega}}{\partial t^2} + 4\Omega_V^2 \tilde{\omega} = 2\Omega_V \alpha_0 \left(\frac{\partial^2 p}{\partial x^2} + \frac{\partial^2 p}{\partial y^2} \right) \tag{5.37}$$

where we have neglected the terms containing v_z because they are "small" compared to the others. Equations (5.36) and (5.37) directly relate horizontal pressure variations to cyclogenesis.

The principal vortex motion observed in the upper atmosphere occurs in the lower stratosphere[1] and is evidently subject to considerable oscillation. We shall have more to say about this effect in Section 5.4.

5.1.6 *Atmospheric Oscillations*

Under suitable conditions (e.g., real N) a perturbation of the atmosphere will result in an oscillatory motion. If so, the field variables have periodic time dependence of the form $e^{-i\omega t}$ or superpositions of such terms. The equations of motion (5.16a,b) and (5.14c) become

$$-i\omega \mathbf{v} + \alpha_0 \left(\nabla + \frac{g}{\gamma p_0 \alpha_0} \hat{\zeta} \right) p - N^2 S \hat{\zeta} + 2\mathbf{\Omega} \times \mathbf{v} = \mathbf{f} \tag{5.38a}$$

$$-i\omega p + \gamma p_0 \left(\nabla - \frac{g}{\gamma \alpha_0 p_0} \hat{\zeta} \right) \cdot \mathbf{v} = (\gamma - 1) q_T / \alpha_0 \tag{5.38b}$$

$$-i\omega S + \mathbf{v} \cdot \hat{\zeta} = q_T / T_0 S_0' \tag{5.38c}$$

The form of the spatial functions used to represent $\mathbf{v}$, p, and S are, of course, dependent upon the shape of the boundaries. For example, if the wavelengths associated with the oscillations are very small compared to the radius of the earth, as in the case of internal gravity waves, a Fourier expansion in the functions $e^{i\mathbf{k}\cdot\mathbf{r}}$ is quite suitable. On the other hand, the shape of the earth is significant in the case of tidal oscillations and spherical functions are appropriate for the expansion. In a medium like the atmosphere which is bounded on one surface only, the appropriate boundary condition is

$$\hat{\zeta} \cdot \mathbf{v} = 0 \tag{5.39}$$

at the surface of the planet. The following two sections deal in some detail with the two principal types of oscillation.

5.2 Internal Gravity Waves

Internal gravity waves are disturbances which are allowed to propagate as a consequence of buoyant forces present in the atmosphere. They typically have periods of the order of several minutes to perhaps three hours. Since the periods of oscillation are in any case substantially less than the earth's rotational period, it is a good approximation to neglect the latter. Furthermore, the wavelengths are much less than the planetary radius, allowing us to express the variables as

$$A \exp\left[-i\omega t + i\mathbf{k}\cdot\mathbf{r}\right] \tag{5.40}$$

Under such conditions the free equations of motion are simple algebraic expressions

$$-i\omega v_x + i\alpha_0 k_x p = 0 \tag{5.41a}$$

$$-i\omega v_y + i\alpha_0 k_y p = 0 \tag{5.41b}$$

$$-i\omega v_z + \alpha_0\left(ik_z + \frac{g}{\gamma p_0 \alpha_0}\right)p - N^2 S = 0 \tag{5.41c}$$

$$-\omega p + \gamma p_0(ik_x v_x + ik_y v_y) + \gamma p_0\left(ik_z - \frac{g}{\gamma p_0 \alpha_0}\right)v_z = 0 \tag{5.41d}$$

$$-i\omega S + v_z = 0 \tag{5.41e}$$

for which the only nontrivial solutions are the roots of the determinantal or secular equation

$$\begin{vmatrix} \omega & 0 & 0 & -\alpha_0 k_x & 0 \\ 0 & \omega & 0 & -\alpha_0 k_y & 0 \\ 0 & 0 & \omega & -\alpha_0\left(k_z - \dfrac{ig}{\gamma p_0 \alpha_0}\right) & -N^2 \\ -\gamma p_0 k_x & -\gamma p_0 k_y & -\gamma p_0\left(k_z + \dfrac{ig}{\gamma p_0 \alpha_0}\right) & \omega & 0 \\ 0 & 0 & i & 0 & \omega \end{vmatrix} = 0 \tag{5.42}$$

These give rise to a *dispersion relation* which relates the phase velocity $= \omega/k$ to the wave number k

$$c^2 = \frac{c_0^2}{2}\left\{\left(\frac{N^2}{k^2 c_0^2} + 1 + \frac{\Gamma^2}{k^2}\right) \pm \left[\left(\frac{N^2}{k^2 c_0^2} + 1 + \frac{\Gamma^2}{k^2}\right)^2 - \frac{4N^2}{c_0^2 k^2}\left(\frac{k_x^2 + k_y^2}{k^2}\right)\right]^{1/2}\right\} \tag{5.43}$$

where

$$c_0 = \sqrt{\gamma p_0 \alpha_0} = \sqrt{\gamma R T_0} \tag{5.44}$$

is the speed of sound in a gas of temperature T_0,

$$\Gamma = \frac{g}{c_0^2} \tag{5.45}$$

and R is the universal gas constant divided by the *molecular weight of the gas*. Furthermore, we have replaced k_z by $k_z - i\Gamma$. The two possible modes of oscillation correspond to the sign of the second term on the right hand side of Eq. (5.43). The plus sign is characteristic of the *acoustic mode*, so-called because these waves become sonic waves in the high frequency limit, while the negative sign corresponds to *internal gravity waves*. The former has a cutoff frequency at

$$N_1 = \sqrt{N^2 + c_0^2\Gamma^2} \tag{5.46a}$$

When $\omega < N_1$, the wave number K is imaginary and propagation is forbidden. It is easily shown that at high frequencies the phase velocity c is asymptotic to the sonic velocity c_0.

For the gravity mode the phase velocity is given approximately by

$$c^2 = \frac{N^2}{k^4}(k_x^2 + k_y^2) \tag{5.46b}$$

at low frequencies. This equation shows, incidentally, that wave propagation is isotropic about a vertical axis but not about a horizontal axis. This is to be expected because of the crucial role of the (vertical) buoyant forces. On the other hand, at relatively high frequencies the wave frequency approaches N as a maximum allowed value. These effects are illustrated in Figure 5.3 which shows the "allowed" modes as shaded regions, and the "forbidden" region in which waves cannot propagate, as an unshaded diagonal band. The direction and rate of energy flow is, of course, specified by the group velocity $v_g = \partial\omega/\partial k$; the investigation of which is left as an exercise.

In general, the amplitude of an unattenuated atmospheric disturbance such as a gravity wave increases with height in such a way that the product of the amplitude and the square root of the density is always a constant. If attenuation is present, this product will decrease slowly with increasing height because the wavelengths are large (of the order of tens to hundreds of km). At heights above 85 to 90 km, the variations in pressure and density become comparable to the ambient values of these quantities and the perturbation methods fail. In the vernacular of mathematical physics, the "perturbation expansion" does not converge. Some investigators have indeed attempted to employ this method at thermospheric heights, but their results are probably unreliable.

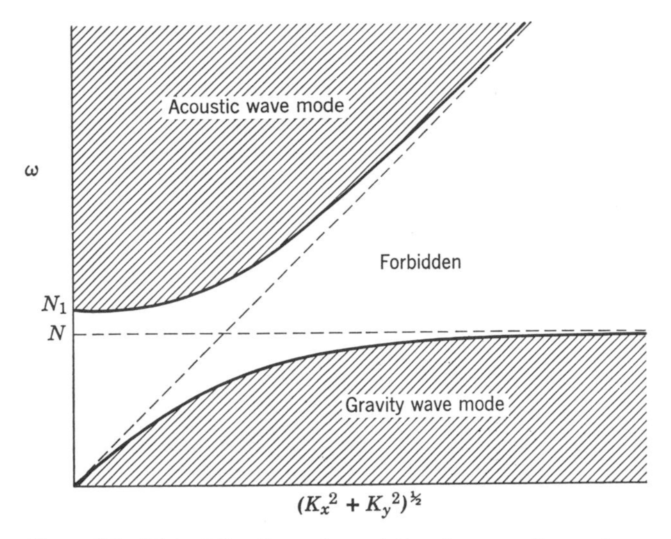

Figure 5.3 Plot of the dispersion relation for acoustic-gravity waves.

The mechanisms which generate internal gravity waves are poorly understood. Possibilities are tidal motions in the upper atmosphere or disturbances in the troposphere, but such suggestions are speculative. So far the only phenomena which are known to have generated such oscillations are large nuclear explosions in the atmosphere such as the 60-megaton device detonated by the Russians in 1961. Gravity waves which originate in the lower atmosphere can penetrate to the upper atmosphere if they are not first refracted back to Earth. Although we have not done so in this text, it is possible to show that positive wind shears (i.e., velocity increases with height) can cause "trapping." Since these conditions exist in the earth's atmosphere, one would expect "trapping" of the gravity waves at relatively low altitudes. However, Gossard[2] has shown that a fairly broad window can exist at wave periods of about 10 minutes to 2 hours. A Brunt-Väisälä frequency which *decreases* with increasing height can also act as a trapping mechanism. For this reason, one would expect internal gravity waves to be ducted in the region near the mesopause (lower thermosphere and upper mesophere). Hines[3] has suggested that the irregular wind component at these heights is a consequence of internal gravity waves.

5.3 Tidal Oscillations

The occurrence of tidal motions in the Earth's atmosphere has been known for many years. The earliest evidence for such an effect was the observation of small diurnal variations in atmospheric pressure near sea level at rather low latitudes. In regions far from the equator, weather changes are sufficient to mask the tidal effect. The theory of ocean and atmospheric tides was originated in 1799 by Laplace. He correctly attributed the atmospheric tides to the sun because the variations followed the solar rather than the lunar day. Because the gravitational forces exerted by the sun are much weaker than those of the moon, he concluded that the oscillation is of thermal rather than gravitational origin. A harmonic analysis of the data reveals that the dominant mode has a period of one-half a solar day rather than an entire day which one would ordinarily expect from the nature of the driving force. In 1882 Lord Kelvin suggested that while the variations are indeed due to thermal effects, the large amplitude of the semidiurnal mode is due to resonance. The atmosphere was assumed to have a natural period of oscillation very close to half a day.

The resonance theory was in vogue for many years and has only recently died because upper air investigations have proved that the atmosphere simply does not have the requisite characteristics. Recent developments in tidal theory have fairly definitely established that the motions are of thermal origin and that the diurnal mode is *suppressed* by the atmosphere.

Tidal motions in the high atmosphere are also responsible for the flow of electric currents in the ionosphere. This relationship between tides and geomagnetic variations which was first advanced by Balfour Stewart nearly a century ago is the subject of Chapter 7.

The frequencies of the tidal modes are of the same order as the rotational frequency of the earth, and therefore the Coriolis term $2\mathbf{\Omega} \times \mathbf{V}$ cannot be omitted from Eq. (5.38a). Because the boundary shape (i.e., the planetary surface) is spherical, it is convenient to express the equations of motion in the coordinates shown in Figure 5.4;

$$-i\omega v_\theta + \frac{\alpha_0}{r}\frac{\partial p}{\partial \theta} - 2\Omega\cos\theta\, v_\phi = f_\theta \tag{5.47a}$$

$$-i\omega v_\phi + \frac{\alpha_0}{r\sin\theta}\frac{\partial p}{\partial \phi} - 2\Omega\cos\theta\, v_\theta + 2\Omega\sin\theta\, v_r = f_\phi \tag{5.47b}$$

$$-i\omega v_r + \alpha_0\frac{\partial p}{\partial r} + \alpha_0\Gamma p - N^2 S - 2\Omega\sin\theta\, v_\phi = f_r \tag{5.47c}$$

$$-i\omega p + \frac{p_0\gamma}{r\sin\theta}\left[\frac{\partial}{\partial\theta}(v_\theta\sin\theta) + \frac{\partial v_\phi}{\partial\phi}\right] + p_0\gamma\frac{1}{r^2}\frac{\partial}{\partial r}(r^2 v_r) - \gamma p_0\Gamma v_r = \frac{(\gamma-1)}{\alpha_0}q_T \tag{5.47d}$$

$$-i\omega S + v_r = \frac{q_T}{T_0 S_0'} \tag{5.47e}$$

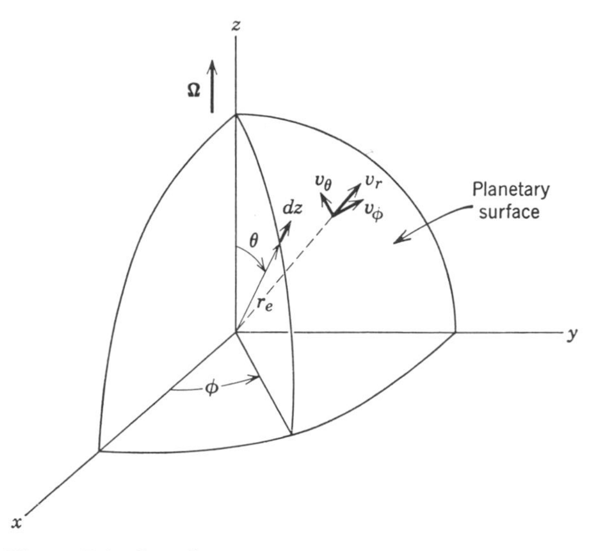

Figure 5.4 Coordinate system used in the equations of motion for tidal oscillations; θ is the colatitude and ϕ is the longitude. The spherical components of the velocity are illustrated on the surface of the octant.

We now introduce two approximations. The first, that $r = r_e$ (radius of the earth) is very good because of the small vertical extent of the atmosphere. The second, that $2\Omega \sin \theta\, v_\phi$ is negligible, is much poorer but does greatly simplify the equations. Fortunately, for our purposes this simplification which was habitually made by all investigators until relatively recently does not yield misleading results. With these provisos and the replacement of $\partial/\partial r$ by $\partial/\partial z$, the vertical component of the gradient, the equations of motion become

$$-i\omega v_\theta - 2\Omega \cos \theta v_\phi + \frac{\alpha_0}{r_e}\frac{\partial p}{\partial \theta} = f_\theta = -\frac{1}{r_e}\frac{\partial \Phi}{\partial \theta} \tag{5.48a}$$

$$-2\Omega \cos \theta v_\theta - i\omega v_\phi + \frac{im\alpha_0}{r_e \sin \theta}\, p = f_\phi = -\frac{im}{r_e \sin \theta}\Phi \tag{5.48b}$$

$$-i\omega v_r + \alpha_0\left(\frac{\partial p}{\partial z} + p\Gamma\right) - N^2 S = -f_r = -\frac{\partial \Phi}{\partial z} \tag{5.58c}$$

$$-i\omega p + \frac{\gamma p_0}{r_e \sin \theta}\left[\frac{\partial}{\partial \theta}(\sin \theta v_\theta) + imv_\phi\right] + \gamma p_0\left(\frac{\partial v_r}{\partial z} - \frac{\Gamma}{c_0}v_r\right) = \frac{(\gamma - 1)q_T}{\alpha_0} \tag{5.48d}$$

$$-i\omega S + v_r = q_T/T_0 S_0' \tag{5.48e}$$

It has been assumed here that $p \propto e^{im\phi}$ where m is an integer and that Φ is the periodic gravitational potential which serves as a driving mechanism. The heating terms containing q_1 are, of course, also periodic.

Although the problem of tidal oscillations can be approached from the standpoint of pressure and velocity perturbations, it is simplified by using the velocity divergence

$$\chi = \frac{1}{r_e \sin\theta}\left[\frac{\partial}{\partial\theta}(\sin\theta v_\theta) + \frac{\partial v_\phi}{\partial\phi}\right] + \frac{\partial v_r}{\partial z} \tag{5.49}$$

It is left as an exercise to show that in an adiabatic atmosphere the equation satisfied by χ is

$$H_0 \frac{\partial^2\chi}{\partial z^2} + \left(\frac{dH_0}{dz} - 1\right)\frac{\partial\chi}{\partial z} - \frac{g}{4r_e^2\Omega^2} F\left[\left(\frac{\gamma - 1}{\gamma} + \frac{dH_0}{dz}\right)\chi\right] = 0 \tag{5.50}$$

where

$$H_0 = \frac{c_0^2}{\gamma g} \tag{5.51}$$

is the scale height and F is the operator

$$F = \frac{1}{\sin\theta}\frac{\partial}{\partial\theta}\left(\frac{\sin\theta}{f^2 - \cos^2\theta}\frac{\partial}{\partial\theta}\right) - \frac{m}{f^2 - \cos^2\theta}\left(\frac{m}{\sin^2\theta} + \frac{1}{f}\frac{f^2 + \cos^2\theta}{f^2 - \cos^2\theta}\right) \tag{5.52}$$

where $f = \omega/2\Omega$.

Since Eq. (5.50) is linear, it can be solved by the method of separation of variables in which we set

$$\chi = Z(z)\Theta(\theta)e^{im\phi} \tag{5.53}$$

The two ordinary differential equations which result are

$$F\Theta(\theta) + \frac{4r_e^2\Omega^2}{gh}\Theta(\theta) = 0 \tag{5.54}$$

(Laplace's tidal equation), and the radial equation

$$H_0\frac{d^2Z}{dz^2} + \left(\frac{dH_0}{dz} - 1\right)\frac{dZ}{dz} + \left(\frac{\gamma - 1}{\gamma} + \frac{dH_0}{dz}\right)\frac{Z}{h} = 0 \tag{5.55}$$

where h is the separation constant; h has the dimensions of length and for tides in a continuous uniformly deep ocean, it corresponds to the depth. Hence, it is called the *equivalent* depth. These two equations describe the modes of oscillation of the "free" atmosphere, i.e., no external forces are acting on it. We shall return shortly to the case where forcing terms are included.

The separation constants h specify the modes of oscillation of the system and are termed the *eigenvalues*[4] of the angular equation (5.54). The functions

$\Theta(\theta)$ are the *eigenfunctions*, each of which is associated with or "belongs to" a particular eigenvalue. The functions $\Theta(\theta)e^{im\phi}$, called Hough functions after their discoverer, can be shown to be linear combinations of spherical harmonics

$$\Theta_n^m(\theta)e^{im\phi} = \sum_{l=m}^{\infty} a_{ln}^m Y_m^l(\theta, \phi) \tag{5.56}$$

The separation constants h which we should label $h_{m,n}$ have a discrete set of values due to the boundary conditions which must be imposed upon $\Theta(\theta)$. These are $\Theta_n^m(0) = 0$ (i.e., it vanishes at the pole), and either $\Theta_n^m(\pi/2) = 0$ or $d\Theta_n^m/d\theta|_{\theta=0} = 0$ (Θ or its first derivative vanishes at the equator). The first is necessary to insure finite wind velocities at the pole and the second that Θ be finite and single-valued over the entire globe. The requirement of single valuedness also lies behind the specification that m be an integer. The two possibilities for Θ at the equator give rise to two different classes of modes, the so-called symmetric and anti-symmetric cases. Owing to space limitations we cannot pursue the topic of Hough functions in this text but refer the reader to one of the references on tidal theory and to the recent work of Kato[5] who carried out a very thorough investigation of these functions.

It has been known for some time that the dominant driving mechanism is a thermal one. Hence, for a first approximation to a "correct" theory of tidal motions it is adequate to retain the heating term $[(\gamma - 1)/\alpha_0]q_T$ and neglect the gravitational terms in Eqs. (5.48). In order to cast this set of equations into a tractable form, they are again analyzed in terms of the velocity divergence χ and the heating function q_T, which are assumed to be expandable as

$$\begin{aligned} q_T(z, \theta) &= \sum_{m,n} q_n^m(z)\Theta_n^m(\theta) \\ \chi(z, \theta) &= \sum_{m,n} Z_n^m(z)\Theta_n^m(\theta)e^{im\phi} \end{aligned} \tag{5.57}$$

Strictly speaking, this expansion is not legitimate because the Θ_n^m do *not* constitute a mathematically complete set of functions. In order to facilitate computation, however, the incompleteness is usually ignored with no apparent deleterious effects.

The radial equation proves to be quite complicated when written in terms of Z and z. However, by changing both the indedendent and dependent variables according to

$$\xi = \int_0^z \frac{dz'}{H_0(z')} \tag{5.58a}$$

and

$$\eta \exp(\xi/2) = Z - \frac{(\gamma - 1)q_T}{\gamma g H_0} \tag{5.58b}$$

the form of the radial equation is greatly simplified:

$$\frac{d^2\eta}{d\xi^2} - \frac{1}{4}\left[1 - \frac{4}{h}\left(\frac{(\gamma - 1)}{\gamma} H_0 + \frac{dH_0}{d\xi}\right)\right]\eta = \frac{(\gamma - 1)}{\gamma hg} q e^{-\xi/2} \qquad (5.59)$$

One such inhomogeneous equation results for each of the q_n^m and $h_{m,n}$. It is also possible to include the gravitational potential in the radial equation, but since it has only a minor influence on atmospheric tides, we ignore it in the remainder of our discussion.

Lindzen has solved Eq. (5.59) using heat inputs corresponding to absorption of solar radiation by water vapor and ozone. Each q_T was analyzed in terms of the Hough functions $\Theta_n^m(\theta)$, and the equation solved numerically subject to the boundary condition that $v_r(0) = 0$ and that (a) energy propagates upward for an oscillating mode, and (b) the solution decays exponentially with ξ for damped modes. His solutions showed why the semidiurnal tides are so much stronger than the diurnal motions: Most of the diurnal modes have associated with them small positive or large negative equivalent depths (the $h_{m,n}$). The former correspond to small oscillations while the latter correspond to strong exponential damping (see Problem 5.13). On the other hand, the semidiurnal modes are characterized by relatively large positive equivalent depths which result in large tidal amplitudes. Evidently radiation absorption by ozone is most closely associated with the diurnal modes, while absorption by water vapor is more closely related to the semidiurnal mode.

Although many details of tidal theory remain to be worked out, it appears that the problem which eluded so many investigators for so long is essentially solved. The principal reason for the long delay was lack of knowledge of important atmospheric parameters which have been observed relatively recently.

5.4 Winds and Circulation

The winds and circulation of the upper air are essentially of tidal (i.e., thermal) origin, but those in the thermosphere differ considerably from those at mesospheric heights. Hence we shall discuss the two regions separately.

5.4.1 Mesosphere

Circulation in the mesosphere and stratosphere has been found rather recently to be part of the same general pattern.[7] The principal features are East winds in summer and West winds in winter, with the summer winds being somewhat stronger than the winter winds. In contrast to the situation in the mesosphere, stratospheric zonal winds tend to be stronger in the winter. Typical wind speeds are of the order of 60 m sec^{-1}, with maxima near 200 m sec^{-1} at mid-latitudes. At the equinoxes the zonal wind which is westerly and much weaker, is of the order of 25 m sec^{-1} with the maximum velocity occurring near 60 km altitude. Despite rather regular behavior of the mean winds, the winds observed over short time spans are quite variable, the variability increasing with height. There are, incidentally, meridional as well as zonal components but the former are much weaker than the latter. The circulation patterns of the mesosphere-stratosphere system are illustrated in Figures 5.5 and 5.6.

The winds are apparently caused by the occurrence of meridional temperature gradients resulting from the absorption of solar radiation by ozone. Since this is one of the mechanisms which is responsible for tides, the mesospheric winds must be tidal in nature. One of the effects of the thermal gradients, which are quite large at high latitudes, is the production of a polar vortex[1] [see Eq. (5.36)]. This vortex, probably because of baroclinic instability, (i.e., owing to wind shears) is subject to large oscillations in its movements, particularly in winter. The simple equation for vortex oscillations presented in Section 5.1.5 probably does not apply here because of the influence of perturbations such as planetary tidal motions. These were not included in the first-order theory.

Wind shears are quite pronounced in the mesosphere as is evidenced by the profiles shown in Figure 5.7. One can relate the shears to the wind velocities by means of Eq. (5.28) which is, in integral form,

$$\mathbf{v} = \int \mathbf{v}_T \, dz \tag{5.60}$$

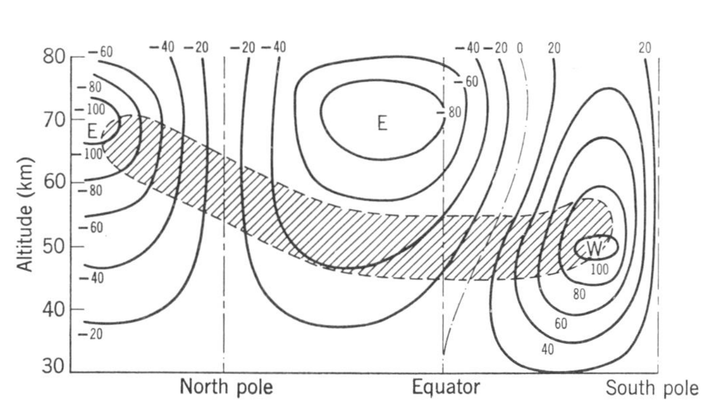

Figure 5.5 Meridional cross section of the stratospheric circulation system through the subsolar point at summer solstice in the northern hemisphere. Contours are zonal wind speeds (positive from the west) and the shaded area is the tidal circulation region projected on the meridional plane [after W. Webb, *Revs. Geophys.* **4**, 363 (1966); reprinted by permission of the American Geophysical Union.]

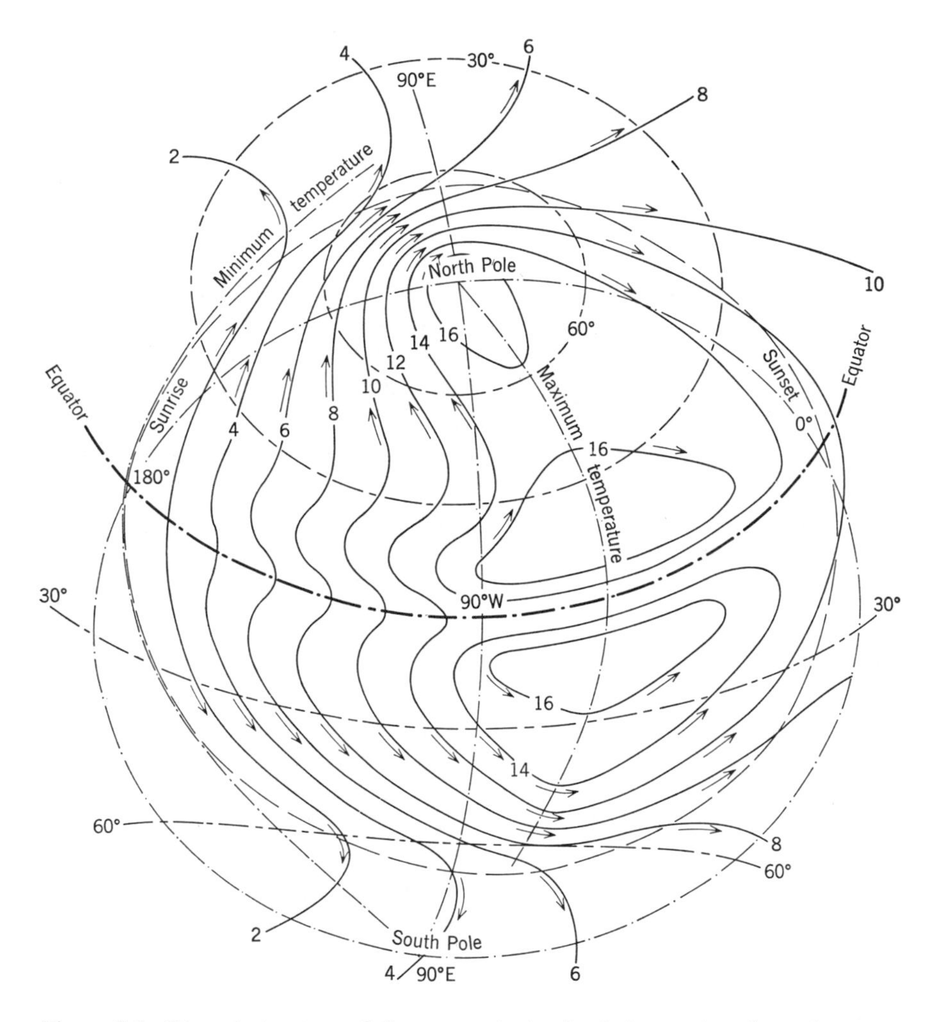

Figure 5.6 Diurnal structure of the stratospheric circulation projected on the stratopause plane as an equidistance projection centered at 42°N latitude and 103°W longitude. Temperature contours are in degrees C. [After W. Webb, *Revs. Geophys.* **4**, 363 (1966); reprinted by permission of the American Geophysical Union.]

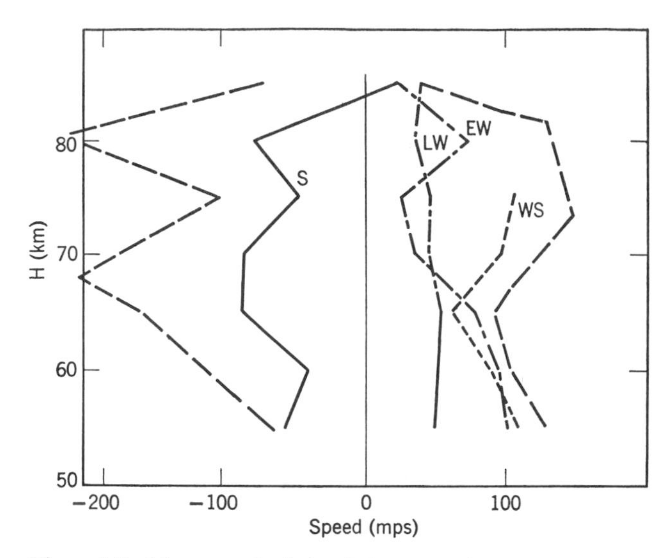

Figure 5.7 Mean zonal winds of the mesosphere over Wallops Island, Va., for each season of the stratospheric circulation. The light dashed lines indicate the extreme values thus far observed. Curves are for summer (S), early winter (EW), winter storm (WS), and late winter (LW) seasons. (After W. Webb, *Structure of the Stratosphere and Mesosphere;* reprinted by permission of Academic Press.)

Equation (5.60) was employed by Batten and Murgatroyd[8] to obtain wind speeds from the observed shear values; their results were in quite good agreement with the observed speeds. In recent years wind shears have come to play a central role in the theory of the mid-latitude "sporadic-E" layers. We shall discuss the mechanism in a later chapter and merely state here that the thin sheets of ionization result from interaction of the wind shears with the geomagnetic field and the ions which are normally present.

5.4.2 Thermosphere

The main components are again of tidal origin with the semidiurnal component the largest. This is hardly surprising because the main tidal component is semidiurnal. The diurnal component is much smaller and is more irregular. There is also present in the lower thermosphere an irregular component which was discovered in meteor trail observations, mainly above Jodrell Bank in the United Kingdom. Its characteristics which include a mean vertical scale of about 6 or 7 km has led Hines[3] to suggest gravity waves as the source.

Until recently little was known regarding thermospheric winds at middle heights (i.e., in the vicinity of the F_2 layer). Measurements have still not been carried out, but Geisler[9] has developed a theoretical model based on Eq. (5.16a) with the vertical components omitted

$$\frac{\partial \mathbf{v}}{\partial t} + 2\mathbf{\Omega} \times \mathbf{v} + \alpha_0 \nabla p = \mathbf{f} \tag{5.61}$$

in which the right hand term represents frictional drag as specified by Eq. (5.18). Hence, we can express f in the form

$$\mathbf{f} = \nu \nabla^2 \mathbf{v} - \lambda \mathbf{v} \tag{5.62}$$

of which the first represents the viscous drag of the atmosphere and the second the ionospheric drag. In order to compute the wind field $\mathbf{v}$ Geisler adopted a model of thermospheric pressure due to Jacchia and Slowey. He then integrated Eq. (5.61) subject to appropriate boundary conditions. The magnitude of the maximum value of $\mathbf{v}$ was found to be slightly greater than 200 m sec^{-1}.

As one may suspect, such air motions have ionospheric effects because of the viscous interaction between atmosphere and ionosphere. In particular they may raise or depress the altitude of the F_2 peak. They also have a pronounced effect on number densities of neutral components and on heat flow in the thermosphere.

5.4.3 *Wind Observations*

We shall not go into great detail on this topic but confine our remarks to several of the more widely used methods. For wind observations in the mesosphere (see Figure 5.7) the rocket grenade experiment is the only one used on a wide scale A rocket ejects a series of small grenades at specified altitudes and the difference in the times of arrival of the sound waves together with their incident directions are observed by a microphone array. From these data one can deduce both the temperature and the mean wind speed in the atmospheric slab between two successive bursts. At higher altitudes, i.e., in the lower thermosphere, observations of the motions of meteor trails and alkali metal clouds ejected from rockets lead directly to wind estimates. The meteor trails are observed by means of radar while the alkali metal clouds are tracked photographically. The rockets are usually launched during twilight when the thermosphere is still illuminated by the sun. The solar radiation which is resonance-scattered by the alkali metal atoms can be easily observed on the ground at that time. Lastly we mention the deduction of the tidal winds in the lower E-region from observations of diurnal magnetic observations. Because this method is discussed in chapter 7, we shall not develop it further here.

5.5 Turbulence in the Upper Atmosphere

So far in this chapter we have discussed only organized fluid motion in the upper atmosphere—acoustic and gravity waves, tidal oscillations, and winds. To be sure, such phenomena include most of the important aspects of fluid aeronomy, but large scale random fluctuations in the velocity and density (turbulence) are also significant because they tend to *mix* the atmosphere. This opposes the effect of gravity which, as we saw in Chapter 3, tends to separate the atmospheric constituents vertically according to their molecular masses. Mixing caused by random fluctuations in the fluid velocity field is called *turbulent* or *eddy* diffusion. At altitudes below about 105 km eddy diffusion and photochemical processes compete with each other. That is, photodissociation and various chemical reactions tend to cause changes with altitude in the relative concentrations of the various constituents, whereas mixing leads to uniform composition (no change with altitude). At altitudes above 105 km, molecular diffusion and photochemical processes are competitive.

5.5.1 Theory of Turbulence

In order to describe turbulent processes one must have a way of defining the fluctuations which are characteristic of turbulence. This is ordinarily done by writing each quantity as the sum of an average and a fluctuation. For example, wind velocity is written

$$\mathbf{v} = \bar{\mathbf{v}} + \mathbf{v}' \tag{5.63}$$

where T is the mean and $\mathbf{v}'$ the fluctuation. Each component $\bar{\mathbf{v}}_i$ of the former is defined as follows

$$\bar{v}_i = \frac{1}{T}\int_{-T/2}^{T/2} v_i(t)\,dt \tag{5.64}$$

where T is taken very large compared to the lifetime of a fluctuation. Since a repetition of the averaging operation has no effect on $\bar{\mathbf{v}}_i$ we must have $\bar{\bar{\mathbf{v}}} = \bar{\mathbf{v}}$ and thus $\overline{\mathbf{v}'} = 0$.

We now wish to show how fluctuations can develop in a fluid and for this purpose return to the equation of motion for a fluid, Eq. (5.4) which (neglecting the Coriolis force but writing out the advective term) is

$$\frac{\partial \mathbf{v}}{\partial t} + \mathbf{v} \cdot \nabla \mathbf{v} = -\frac{1}{\rho}\nabla p + \mathbf{g} + \mathbf{f} \tag{5.65}$$

where $\rho = \alpha^{-1}$ is the mass density. Although we shall not do so here, it can be shown that the friction or *viscous* term $\mathbf{f}$ due to transfer of momentum between adjacent layers of fluid can be expressed as

$$\mathbf{f} = \nu \nabla^2 \mathbf{v} \tag{5.66}$$

ν is called the *kinematic viscosity* of the fluid and is a measure of the rate at which momentum is transferred from layer to layer by random molecular motion. Substitution of Eqs. (5.63), (5.64), and

$$p = \bar{p} + p' \tag{5.67}$$

into Eq. (5.65) followed by the averaging operation yields

$$\frac{\partial \bar{\mathbf{v}}}{\partial t} + \bar{\mathbf{v}} \cdot \nabla \bar{\mathbf{v}} = -\frac{1}{\rho} \nabla \bar{p} + \mathbf{g} - \overline{\mathbf{v}' \cdot \nabla \mathbf{v}'} + \nu \nabla^2 \bar{\mathbf{v}} \tag{5.68a}$$

where we have assumed that $\rho'/\bar{\rho} \ll p'/\bar{p}$. We now assume that the fluid is incompressible (this was already implicit in the form of **f** used here) and that it is in hydrostatic equilibrium. These two conditions are equivalent to the statements that

$$\nabla \cdot \bar{\mathbf{v}} = \nabla \cdot \mathbf{v}' = 0 \tag{5.69}$$

and

$$\nabla \bar{p} = \mathbf{g} \rho \tag{5.70}$$

Hence Eq. (5.68a) can be recast as

$$\frac{\partial \bar{\mathbf{v}}}{\partial t} + \bar{\mathbf{v}} \cdot \nabla \bar{\mathbf{v}} = \nabla \cdot (\nu \nabla \mathbf{v} - \overline{\mathbf{v}'\mathbf{v}'}) \tag{5.68b}$$

The quantities $\tau_{ij} = \bar{\rho} v_i v_j$ are called the *Reynolds stresses* or *eddy stresses.* They represent the transport of momentum due to fluctuations in the velocity field.

The theory of turbulence was originated many years ago by Schmidt and Prandtl who introduced, respectively, the concepts of *exchange coefficients* and *mixing lengths* in analogy to molecular viscosity and mean free path in a gas. The exchange coefficients A_i are related to the Reynolds stresses by

$$A_i \frac{\partial \bar{v}_i}{\partial z} = -\overline{\rho v_i' v_z'} = \tau_{i'z} \tag{5.71}$$

and are thus measures of the vertical transport of the x and y components of momentum. In suggesting the idea of a mixing length, Prandtl assumed that an eddy originating at some level z carries the average momentum of that level to the new level $z + l$ where it is abruptly absorbed. Thus, just prior to absorption the fluctuation v_i' of the velocity component is

$$v_i' = \bar{v}_i(z + l) - \bar{v}_i(z) \tag{5.72}$$

or, expanding $v_i(z + l)$ in a Taylor series about l

$$v_i' = l\,\frac{\partial \bar{v}_i}{\partial z}, \tag{5.73}$$

Substitution into the equation for the Reynolds stresses yields

$$\tau_{ij} = -\rho\overline{v_i'v_j'} = -\rho\overline{v_j'}l\,\frac{\partial \bar{v}_i}{\partial z} \tag{5.74}$$

or, since all three eddy velocity components are comparable,

$$\tau_{ij} = -\rho l^2\left(\frac{\partial \bar{v}_i}{\partial z}\right)^2 \tag{5.75}$$

Here l is the mean mixing length for the vertical transport of momentum and

$$K_x = \frac{1}{\rho}A_x = l^2\,\frac{\partial \bar{v}_1}{\partial z} \tag{5.76}$$

is defined as the *kinematic eddy viscosity*. Thus, considering vertical transport only, Eq. (5.68b) becomes

$$\frac{\partial \bar{\mathbf{v}}}{\partial t} + \bar{\mathbf{v}}\cdot\nabla\bar{\mathbf{v}} = \frac{\partial}{\partial z}\left(\nu\,\frac{\partial \bar{\mathbf{v}}}{\partial z} + K\,\frac{\partial \bar{\mathbf{v}}}{\partial z}\right) \tag{5.77}$$

The very close analogy to molecular viscosity implicit in Eq. (5.77) is quite deceptive. First of all, the eddy does not acquire and lose momentum instantaneously. Since gain and loss are gradual processes, the concept of a mixing length is ill-defined. More importantly the distribution of eddy sizes which is the analogue of the molecular energy distribution function is itself subject to large fluctuations. As a result the averages which define the Reynolds stresses, for example, are not well-defined quantities. Nevertheless, it is necessary to assume that the averages are well defined if we are to make any headway in discussing turbulent transport.

The distribution of eddy sizes is best considered with the aid of the structure tensor

$$D_{ij}(\mathbf{r}) = \overline{[v_i(\mathbf{x})v_j(\mathbf{x}) - v_i(\mathbf{x})v_j(\mathbf{x} + \mathbf{r})]^2} \tag{5.78}$$

In the isotropic case we are interested only in the scalar which can be formed from $D_{ij}(\mathbf{r})$, namely its trace

$$D(\mathbf{r}) = \sum_i D_{ii}(\mathbf{r}) \tag{5.79}$$

With the aid of the definition of the *energy spectrum* of the turbulence $E(k)$ given by

$$\overline{\mathbf{v}(\mathbf{x}) \cdot \mathbf{v}(\mathbf{x} + \boldsymbol{\xi})} = 2\int_{-1}^{1}\int_{0}^{\infty} e^{ik\xi\mu}E(k)\, dk\, d\mu \tag{5.80}$$

we can express $D(r)$ as

$$D(r) = 4\int_{0}^{\infty} E(k)\left(1 - \frac{\sin kr}{kr}\right) dk \tag{5.81}$$

The parameter k is the wave number which is inversely proportional to eddy size. It is left as an exercise to find expressions for $D(r)$ when r is very small and when it is very large. By observing the expansion of chemical clouds released from rockets, Justus[10] has evaluated $D_{ii}(r)$ in three mutually perpendicular directions, one of which was vertical. Although the results were not very precise, he was able to conclude with some degree of confidence that the turbulence was reasonably close to isotropic.

It is instructive to inquire into the energetics of turbulence, for energy conservation considerations will give us insight into the production and dissipation mechanisms. Starting with the equation of motion, Eq. (5.65), it is easy to show that the equation of motion for the fluctuations is

$$\frac{\partial \mathbf{v}'}{\partial t} + \mathbf{v}' \cdot \nabla\bar{\mathbf{v}} + \bar{\mathbf{v}} \cdot \nabla\mathbf{v}' + \mathbf{v}' \cdot \nabla\mathbf{v}' - \overline{\mathbf{v}' \cdot \nabla\mathbf{v}'} = -\frac{1}{\rho_0}\nabla p' + \nu\nabla^2\mathbf{v}' + \frac{\mathbf{g}}{T_0}\theta \tag{5.82}$$

where θ is the potential temperature defined as

$$\theta = \text{constant} \times T/(p)^{1-1/\gamma} \tag{5.83}$$

Here T is the absolute temperature and γ is the ratio of specific heats (see Section 2.2). Now take the scalar product of both sides of Eq. (5.82) with $\mathbf{v}'$, carry out the averaging operation, and employ the condition for incompressibility

$$\nabla \cdot \mathbf{v} = \nabla \cdot \bar{\mathbf{v}} = \nabla \cdot \mathbf{v}' = 0 \tag{5.84}$$

in order to appropriately transform some of the terms. In the case of steady flow ($\partial \mathbf{v}'^2/\partial t = 0$), the energy conservation equation takes the simple form

$$\epsilon_s = \epsilon_g + \epsilon_d \tag{5.85}$$

where

$$\epsilon_s = -\sum_{i,j} v_i v_j \frac{\partial \bar{v}_i}{\partial x_j} \tag{5.86}$$

represents the energy supplied by wind shears, and

$$\epsilon_g = -\frac{g}{T_0}\overline{\theta v_3} \tag{5.87}$$

and

$$\epsilon_d = \nu \sum_{i,j} \overline{\left(\frac{\partial v_i}{\partial x_j}\right)^2} \tag{5.88}$$

are the rates at which energy is dissipated by buoyancy and by viscous forces, respectively. By a more detailed consideration of the energy conservation equation one can show that in the case of dissipation by viscous forces, the energy is successively transferred from larger to smaller eddies until the smallest ones transform it directly into the energy of random molecular motion.

A question which may have occurred to the reader is "why is turbulence sometimes present and sometimes not?" This point has never been satisfactorily settled. For some types of flow such as in a pipe, the criterion is whether or not the so-called Reynolds number

$$\mathrm{R_e} = \frac{\text{inertial forces}}{\text{viscous forces}} \sim \frac{\mathrm{VL}}{\nu} \tag{5.89}$$

exceeds a certain value, usually of the order 10^3; L is a characteristic length such as pipe diameter and V is the flow velocity. The pertinence of this ratio becomes evident when one realizes that the numerator is characteristic of independent particle (i.e., uncorrelated) motion while the denominator is characteristic of collective (i.e., correlated) behavior of the fluid. However, we are immediately faced with the difficulty of finding a characteristic length. The scale height has been suggested but its relation to onset of turbulence, if any, is far from apparent.

A second criterion which is often applied in the lower atmosphere is based on the Richardson number

$$\mathrm{R_i} = \frac{\epsilon_g}{\epsilon_s} \tag{5.90}$$

If $\mathrm{R_i}$ is of order unity or greater, buoyancy forces remove energy as fast as it is produced by wind shears and turbulence cannot develop. There have been strong objections raised against application of the Richardson number in the upper atmosphere. Nevertheless, wind shear is generally considered to be the principal mechanism for the production of turbulence. It is known from the theory of hydrodynamic stability (see the treatise by Chandrasekhar for a complete discussion of this topic) that fluid flow with velocity gradients is unstable and leads to turbulent motion. Unfortunately, theory is completely inadequate for a clear understanding of the development and maintenance of turbulence.

Numerous experiments with chemical cloud releases in the upper air have demonstrated that turbulence does cease very abruptly at altitudes of 105 to 110 km. Numerous attempts to interpret the *turbopause* in terms of Reynolds and Richardson numbers have been made, but they have not been satisfactory. Perhaps the best suggestion offered so far is that at some critical altitude molecular diffusion becomes dominant and turbulence is rapidly dissipated by molecular motion before it can develop.[11]

5.5.2 Mass and Heat Transport

The principal role of turbulence in the upper atmosphere is mass and heat transport in the region below the turbopause (~110 km). In particular, the turbulent mechanism tends to *mix* the atmosphere as opposed to molecular diffusion which tends to separate it. Although it is admittedly unsatisfactory in many respects, Prandtl's mixing length theory can be employed to give us at least a qualitative picture of the turbulent diffusion process.

We begin by considering the continuity equation in which fluctuations in number density n' and velocity $\mathbf{v}'$ are included

$$\frac{\partial \bar{n}}{\partial t} + \frac{\partial \bar{n}'}{\partial t} + \nabla \cdot (\bar{n} + n')(\bar{\mathbf{v}} + \mathbf{v}') = q - L \tag{5.91}$$

If we now carry out the averages defined by Eq. (5.64), we obtain

$$\frac{\partial \bar{n}}{\partial t} + \nabla \cdot (\bar{n}\bar{\mathbf{v}} + \overline{n'\mathbf{v}'}) = q - L \tag{5.91a}$$

Invoking the mixing length l defined in the analogue of Eq. (5.73)

$$n' = l\frac{\partial \bar{n}}{\partial z} \tag{5.92}$$

and substituting the above form for n' into Eq. (5.91a) yields

$$\frac{\partial \bar{n}}{\partial t} + \nabla \cdot \left(\bar{n}\bar{\mathbf{v}} + \overline{l\mathbf{v}'}\frac{\partial \bar{n}}{\partial z}\right) = q - L \tag{5.93}$$

In the upper atmosphere one may, at least in the vertical, set $\bar{v} = 0$ because there is no steady flow in that direction. Thus, variations in $\bar{n}$ are due to turbulent fluctuations only. Assuming that these are negligible except in the vertical, Eq. (5.93) becomes

$$\frac{\partial \bar{n}}{\partial t} = -\frac{\partial \Phi_e}{\partial z} + q - L \tag{5.94}$$

where Φ_e is the vertical flux due to eddy transport

$$\Phi_e = -K_m\left(\frac{\partial \bar{n}}{\partial z} + \frac{\bar{n}}{H_{\text{ave}}}\right) \tag{5.95}$$

H_{ave} is the mean scale height, and

$$K_m = -\overline{lv'} \tag{5.96}$$

is the mass exchange coefficient. Equation (5.95) is *not* rigorous, nor is the theory of turbulence yet capable of producing a rigorous expression. The equation for the flux Φ_e is based purely on analogy with the corresponding expression in molecular diffusion, and, therefore, cannot be expected to yield accurate results. The parameter K_m, which is usually considered to be independent of altitude, can be determined only by empirical methods (direct observation) and is probably subject to large and unpredictable temporal fluctuations. With these limitations in mind, we can write Eq. (5.94) as

$$\frac{\partial \bar{n}}{\partial t} = K_m\left(\frac{\partial^2 \bar{n}}{\partial z^2} + \frac{1}{H_{\text{ave}}}\frac{\partial \bar{n}}{\partial z}\right) + q - L \tag{5.94a}$$

As an application of Eq. (5.94a), consider an oxygen atmosphere dominated by turbulent transport (as opposed to molecular diffusion). Let us assume that O_2 is photodissociated at the specific rate $Q(z_0)$ at altitude z_0 and is removed via the process

$$O + O_2 + O_2 \rightarrow O_2 + O_3 \tag{5.97}$$

at the specific rate $\beta(z) = \beta(z_0)e^{-(z-z_0)/H_{\text{ave}}}$; Eq. (5.94a) takes the form

$$\frac{\partial \bar{n}}{\partial t} = K_m\left(\frac{\partial^2 \bar{n}}{\partial z^2} + \frac{1}{H_{\text{ave}}}\frac{\partial \bar{n}}{\partial z}\right) + Q(z_0)e^{-(z-z_0)/H_{\text{ave}}} - \beta(z_0)e^{-(z-z_0)/H_{\text{ave}}}\,\bar{n} \tag{5.94b}$$

where $\bar{n}$ is the number density of atomic oxygen. It is left as an exercise to find the steady state solution of Eq. (5.94b).

The equation for heat transfer by convection is derived in a similar manner. In terms of the potential temperature θ [see Eq. (5.83)]

$$\frac{\partial \bar{\theta}}{\partial t} = \frac{1}{\rho}\frac{\partial}{\partial z}\left(\rho K_h \frac{\partial \bar{\theta}}{\partial z}\right) \tag{5.98}$$

where ρ is the mass density. In a gas $K_m/K_h \approx 1$ so that mass and heat transport by turbulence may be expected to occur at comparable rates. However, we have tacitly assumed that the gas does not lose any heat energy. If it does (e.g., by radiation from CO_2 in the Martian and Cytheran atmospheres), heat transport by convection may be much less than the corresponding mass transport.

Colegrove, Hanson, and Johnson[12] have computed profiles of atomic and molecular oxygen from 120 km downward, assuming certain values of the $[O]/[O_2]$ ratio at that height. They included photodissociation and recombination processes, and employed values of K_m near 10^6 cm^2 sec^{-1}. For example, a ratio of $[O]/[O_2] = 1$ at 120 km employed with the value $K_m = 4 \times 10^6$ cm^2 sec^{-1} agreed quite well with observations. In fact, they found that any value of K_m given by

$$K_m([O]/[O_2])_{120\text{ km}} = 4 \times 10^6 \text{ cm}^2 \text{ sec}^{-1}$$

is in very good accord with experiment.

PROBLEMS

5.1. Derive the relations

$$\frac{D\mathbf{r}}{Dt} = \left(\frac{d\mathbf{r}}{dt}\right)_{\text{rel}} + \boldsymbol{\Omega} \times \mathbf{r}$$

$$\frac{D\mathbf{v}}{Dt} = \left(\frac{d\mathbf{v}}{dt}\right)_{\text{rel}} + 2\boldsymbol{\Omega} \times \mathbf{v} + \boldsymbol{\Omega}(\boldsymbol{\Omega} \cdot \mathbf{r}) - \mathbf{r}\Omega^2$$

where D/Dt represents the time derivative in an inertial reference frame and $(d/dt)_{\text{rel}}$ represents the time derivative relative to a rotating reference frame.

5.2. Obtain the second order perturbation equations of motion for the atmosphere.

5.3. What is the relationship of the parameter Γ defined in Eq. (5.45) to the scale height?

5.4. Starting with the basic laws of thermodynamics as applied to the upper atmosphere, derive Eq. (5.24). (*Hint:* Use the "*TdS*" equations of Section 2.2.)

5.5. Compute and plot a profile of N for an isothermal atmosphere in which $\gamma = \frac{7}{5}$ and $T = 300°$K. Discuss the possible ducting of internal gravity waves in such an atmosphere.

5.6. Compute and plot a profile of N for the U.S. Standard Atmosphere (1962) or COSPAR International reference atmosphere (see Chapter 3 for references), from 20 to 100 km altitude.

5.7. What is the condition for resonance in an oscillating vortex produced by pressure gradients?

5.8. Obtain expressions for the components of the group velocity of acoustic-gravity waves and discuss their behavior in the low and high frequency limits.

5.9. Discuss qualitatively the manner in which wind shears affect the propagation of acoustic-gravity waves. How does the group velocity depend upon wind shears when the velocity gradient is small compared to the wave frequency? What effect would a random distribution of wind shears have on the wave if the mean vertical wind gradient is much larger than the wave frequency?

5.10. Prove that for unattenuated waves the pressure amplitude must vary with height as $\exp\left[-\frac{1}{2}\int(dz/H)\right]$ where H is the scale height.

5.11. With the aid of Eqs. (5.48a–e), (5.49), (5.51), and (5.52) derive Eq. (5.50) under the assumption that $\Phi = q_T = 0$.

5.12. Use the transformations given by Eqs. (5.58a) and (5.58b) to obtain the tidal radial equation (5.59) with q_T included.

5.13. Show that if the "h" which appears in Eq. (5.59) is negative, the solution corresponds to a damped mode.

5.14. Find approximate expressions for the structure function $D(r)$ in the limiting cases of small r and large r.

5.15. Obtain an expression for the vertical distribution of atomic oxygen in a mixing isothermal atmosphere under the following assumptions: (a) O is formed by the photodissociation of O_2, (b) O is destroyed by the reaction (5.97), (c) the eddy diffusivity K_m is independent of altitude, (d) there is no influx of O molecules from above the atmosphere, and (e) the number density of O vanishes at an infinitely low altitude. Neglect atmospheric attenuation of the radiation which dissociates O_2. [*Hint:* See Eq. (5.96)].

REFERENCES

1. B. W. Boville, Planetary waves in the stratosphere and their upward propagation," in *Space Research VII*, North-Holland Publishing Co., Amsterdam, 1967.
2. E. E. Gossard, *J. Geophys. Res.* **67,** 745 (1962).
3. C. O. Hines, *Canad. J. Phys.* **38,** 1441 (1960).
4. P. M. Morse and H. Feshbach, *Methods of Theoretical Physics*, Vol. 1, McGraw-Hill, New York 1953, especially Chapter 6.
5. S. Kato, *Report of Ionosph. and Space Research in Japan* **20,** 448 (1966).
6. R. S. Lindzen, *Quar. J. Roy. Meteorol. Soc.* **93,** 18 (1967).
   ———, *Proc. Roy. Soc.* **A303,** 299, (1968).
7. W. Webb, *Revs. Geophys.* **4,** 353 (1966).
8. E. S. Batten, *J. Meteorol.* **18,** 283 (1961).
   R. J. Murgatroyd, *Quar. J. Roy. Met. Soc.* **83,** 417 (1957).
9. J. E. Geisler, *J. Atmos. Terr. Phys.* **28,** 703 (1966); **29,** 1469 (1967).
10. C. G. Justus, *J. Geophys. Res.* **71,** 3767 (1966); **72,** 1035 (1967).
11. G. J. McGrattan, *Planet. Space Sci.* **15,** 811 (1967).
12. Colegrove, Hanson, and Johnson, *J. Geophys. Res.* **70,** 4931 (1965).

GENERAL REFERENCES

General Fluid Dynamics

Carl Eckart, *Hydrodynamics of Oceans and Atmospheres*, Pergamon Press, New York and London, 1960, is a standard work in field. Unfortunately, the use of mercator rather than spherical coordinates obscures some of the developments.

Sir Horace Lamb, *Hydrodynamics*, 6th Ed., Dover Publications, New York, 1945, is a classic work on general fluid dynamics by one of the pioneers in the field.

G. J. F. MacDonald, Motions in the High Atmosphere, in *Geophysics. The Earth's Environment*, Ed. by Dewitt, Hieblot, and LeBeau; Gordon and Breach, New York and London, 1963, is a short but very readable account of the salient features of fluid aeronomy.

Journal of Atmospheric and Terrestrial Physics **30** (5), (May, 1968): Symposium on Upper Atmospheric Winds, Waves, and Ionospheric Drifts; Contains review papers by the leading investigators in each field.

Internal Gravity Waves

C. O. Hines, Internal atmospheric gravity waves at ionospheric heights, *Canadian Journal of Physics* **38,** 1441 (1960); M. L. V. Pitteway and C. O. Hines, The viscous damping of atmospheric gravity waves, *Canadian Jour. Phys.* **41,** 1935 (1963), and ——————, The reflection and ducting of atmospheric acoustic gravity waves, *Canadian Journal of Physics* **43,** 2222 (1965), are undoubtedly the most useful references in this area.

Tidal Oscillations

R. A. Craig, *The Upper Atmosphere. Meteorology and Physics*, Academic Press, New York and London, 1965, Chapter 8 contains a brief but very clear discussion of tidal theory.

Manfred Siebert, Atmospheric tides, *Advances in Geophysics* **7,** 105, Academic Press, New York and London, 1961, gives a more detailed account.

R. A. Lindzen and S. Chapman, Atmospheric tides, *Space Science Reviews* **10,** 3 (1969) is an excellent up-to-date review.

Winds and Circulation

R. A. Craig, *ibid.*, mainly chapters 3 and 8.

Willis Webb, *Structure of the Stratosphere and Mesosphere*, Academic Press, New York and London, 1966.

Willis Webb, Stratospheric tidal circulations, *Reviews of Geophys.* **4,** 363 (1966), are recent works upon which we have based Section 5.4.

Turbulence

G. K. Batchelor, *The Theory of Homogeneous Turbulence*, Cambridge University Press, Cambridge, 1952.

J. L. Lumley and H. A. Panofsky, *The Structure of Atmospheric Turbulence*, John Wiley (Interscience), New York, 1964, are the standard works on this topic.

S. Chandrasekhar, *Hydrodynamic and Hydromagnetic Stability*, Oxford University Press, London, 1961 is a standard work.

6

Optical Phenomena

Two types of optical phenomena occur in the terrestrial upper atmosphere: auroras and airglows. They are both characterized by the excitation of atmospheric species and subsequent radiation of photons, and have much in common on a microscopic scale. On a macroscopic scale, however, they appear to be quite distinct in terms of principal excitation mechanisms, temporal and spatial characteristics, intensity, and predominant emissions. The study of both auroras and airglows has significantly increased our knowledge of upper atmospheric species and processes.

Auroras were known in antiquity and have figured prominently in mythology and literature. They occur predominantly in an oval-shaped zone approximated by the circle of geomagnetic latitude at 23° and are observed in both the northern and southern latitudes but occasionally spread to mid-latitude areas. They are correlated in a general way with geomagnetic and solar activity in that the frequency of aurora increases with solar activity and the mechanisms that cause auroras also cause geomagnetic disturbances; unfortunately, the links between solar activity, geomagnetic perturbations, and auroras are not yet well established. On the other hand, a considerable amount of information does exist regarding the excitation of atmospheric species and the morphology of auroras.

Airglows were discovered much more recently. They are widespread and occur continuously, but are not visible to the naked eye. As with auroras, a considerable body of knowledge exists regarding morphology and characteristic emissions.

This chapter is intended as an introduction to the subject and a survey of the more salient aspects. Additional information such as energy level diagrams regarding excitation and ionization mechanisms can be found in Chapters 4, 8, and 10. Basic processes and the notation for atomic and molecular energy states are discussed in Chapter 2.

6.1 Emission Mechanisms

6.1.1 Resonance and Fluorescence

Resonance scattering is the absorption and emission of photons without a change in the photon energy, i.e., radiation is absorbed and re-emitted without a change in the wavelength. Fluorescence is the absorption of photons of one energy and the emission of photons of lesser energy; radiation of one wavelength is absorbed and radiation of a longer wavelength is emitted. In Figure 6.1 resonance scattering is represented by $h\nu_{21}$; an atom in the E_1 state absorbs a photon of energy $h\nu_{21}$, is excited to state E_2, then re-emits a photon of energy $h\nu_{21}$ and returns to state E_1. On the other hand, if a photon of energy $h\nu_{31}$ is absorbed, raising the atom to state E_3 and then de-excitation occurs by the emission of photons $h\nu_{32}$ and $h\nu_{21}$, the process is called fluorescence. In fluorescence, the emitted photons may not represent all of the emitted energy; for example, the absorbed photon may cause ionization and both an electron and a photon may be emitted. Resonance scattering and fluorescence may occur with ionic and molecular species as well. Resonance scattering is very efficient and provides a very sensitive means of detecting certain atmospheric species.

6.1.2 Chemiluminescence

The excess energy resulting from exothermic chemical reactions may be found as kinetic energy of the product particles, as vibrational, rotational, or electronic excitation of one or more of the products, or may be divided amongst all these possibilities. The energy that is stored temporarily in excited states will, usually, be emitted as electromagnetic radiation (collisional de-excitation is a possible exception, see Section 6.14). Chemical

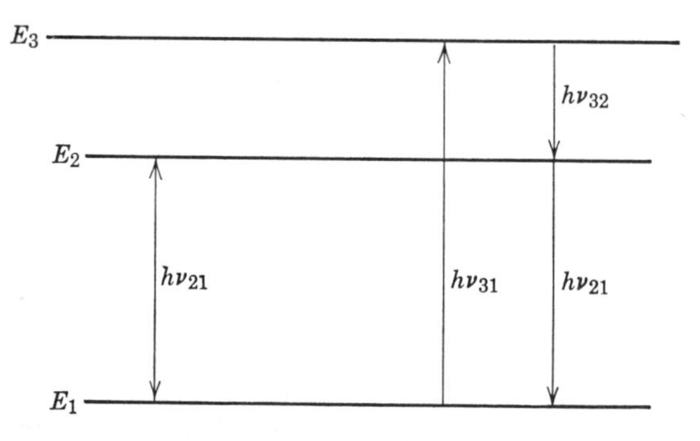

Figure 6.1 Schematic representation of resonance and fluorescence.

reactions that result in the emission of photons are said to produce chemiluminescence. For example,

$$X + Y + Z \rightarrow XY^* + Z \rightarrow XY + Z + h\nu \quad (6.1a)$$

$$X + Y + Z \rightarrow XY + Z^* \rightarrow XY + Z + h\nu \quad (6.1b)$$

$$XY + Z \rightarrow XZ^* + Y \rightarrow XZ + Y + h\nu \quad (6.1c)$$

6.1.3 Excitation by Charged Particles

Charged particles can cause excitation of atmospheric species via several modes.

Inelastic Collisions. Fast charged particles can excite atoms or molecules by giving up a small amount of energy on impact and exciting the atom or molecule to a higher energy state or by dissociating molecules and leaving one or more products excited, or by simultaneously ionizing and exciting molecules, or by simultaneously dissociating, ionizing, and exciting, or by charge exchange:

$$X + e \rightarrow X^* + e \quad (6.2a)$$

$$X_2 + e \rightarrow X + X^* + e \quad (6.2b)$$

$$X_2 + e \rightarrow X_2^{+*} + 2e \quad (6.2c)$$

$$X_2 + e \rightarrow X + X^{+*} + 2e \quad (6.2d)$$

$$X + Y^+ \rightarrow X^+ + Y^* \quad (6.2e)$$

Recombination. The recombination of thermal ions and thermal electrons or positive and negative thermal ions results in excess energy which may be dissipated in kinetic energy or in the excitation of one or more of the products:

$$X^+ + e \rightarrow X^* \quad (6.3a)$$

$$X_2^+ + e \rightarrow X^* + X^{**} \quad (6.3b)$$

$$X_2^+ + Y_2^- \rightarrow X_2^* + Y_2^* \quad (6.3c)$$

6.1.4 Energy Transfer

Under the proper conditions, excitation energy can be transferred from one particle to another during a collision. This is called collisional de-excitation or collisional deactivation or, more commonly, quenching. Of course, these terms all refer to the particle that loses energy; as far as the second particle is concerned, the process is one of collisional excitation. One example is the impact by a low energy electron that happens to have just exactly the energy (or a bit more) required to excite an atom to a specific electronic state; the atom is excited and the electron is thermalized (i.e., loses all but the kinetic energy determined by the ambient temperature). Another example would be an atom that is excited to a state that is X eV above its ground level and collides with an unexcited molecule that can be excited to a vibrational state that is also X eV above ground level; energy can be transferred from the atom to the molecule, de-exciting (or quenching) the atom and exciting the molecule; no radiation occurs during the process, although the molecule may emit radiation later.

Lifetimes for states that can radiate by means of allowed (i.e., electric dipole) transitions are typically microseconds or less, whereas metastable states (which must radiate by means of forbidden transitions such as magnetic dipole or electric quadrupole) can typically have lifetimes as long as seconds or more. Energy transfer reactions are important for particles that have been excited to metastable states because the time between collisions is much less than the radiative lifetimes of the states throughout much of the upper atmosphere. If deactivation rates, possible chemical reaction rates, and metastable lifetimes are known, one can determine whether an atom can be detected by radiation or be quenched (or disappear through chemical reaction) before it can radiate.

Because the radiative lifetimes of metastable states are long and the probability of quenching collisions are therefore likely to be high, radiation from metastable states is difficult to detect and study in the laboratory. Hence, the upper atmosphere (at altitudes where quenching collisions become insignificant) is usually the only place where metastable states can be studied experimentally.

6.1.5 Emission Rate

The basic unit of energy emission from auroras and airglows is the rayleigh, named after the fourth Lord Rayleigh (R. J. Strutt) who was a pioneer in airglow studies. Airglows and auroras have an apparent surface brightness, but, in fact, the emission is from a column of excited gas along the line of sight of the observing instrument. Assuming no absorption or energy transfer complications within the emitting column, the output is determined by

$$4\pi I = \int_0^\infty V(r)\, dr \tag{6.4a}$$

where $V(r)$, usually given in the units photons cm^{-3} sec^{-1}, is the volumetric emission rate at the distance r from the observer. If I is in units of 10^6 photons cm^{-2} (column) sec^{-1} ster^{-1}, a rayleigh (R) is defined as $4\pi I$ or the apparent emission rate of 10^6 photons cm^{-2} (column) sec^{-1}; a kilorayleigh, (kR) is 10^3 R.

Auroras are also classed according to an International Brightness Coefficient (IBC) scale. In this scale IBC I = 1 kR; IBC II = 10 kR; IBC III = 100 kR; and IBC IV = 1,000 kR.

6.2 Airglow

Although airglow is always present, detailed characteristics during day, twilight, and night are somewhat different. Accordingly we shall discuss this phenomenon in three sections.

6.2.1 Dayglow

The dayglow is the result of excitation of atmospheric constituents by solar radiation and, to a smaller extent, by low energy electrons. This means that all the mechanisms noted in Sections 6.1.2 to 6.1.5 may be taking place. Observations, however, are difficult because of the intense background of bright sky and solar Fraunhofer spectra. In order to observe dayglow from the ground, very high resolution and careful analysis is necessary. As a result, successful observations were not made until 1961 when Blamont and Donahue[1] observed resonance scattering by sodium. Since then rocket-borne and surface-based instruments have recorded emissions from several additional atmospheric sources; see Table 6.1.

Sodium. Resonance scattering from sodium produces a strong emission of the doublet at 5890 and 5896 Å arising from the allowed transitions $^2S_{1/2} - {}^2P_{3/2}$ and $^2S_{1/2} - {}^2P_{1/2}$, respectively; each of the lines has a hyperfine structure that should be accounted for in careful analyses.

The scattering comes from a narrow layer near 90 km. The emission rate is fairly constant between 10 and 40 kR. Although no systematic seasonal fluctuations are observed, there are day-to-day and diurnal variations. According to Hunten and Wallace,[2] the airglow emission from atomic sodium can be expressed (in Rayleighs) by

$$4\pi I = AN_c \tag{6.4b}$$

where N_c is the abundance of sodium atoms in a column of unit cross section (atoms cm^{-2}); $A = 1 \times 10^{-6}$ rayleighs cm^2 for dayglow and 7.65×10^{-7} rayleigh cm^2 for twilight glow (see Section 6.2.2). This is valid for a zenith observation. See Figure 4.13 for sodium distributions determined from airglow measurements.

Atomic Oxygen. The 1304 Å triplet is produced by the permitted resonance transition, $^3P - {}^3S$. Dayglow excitation is by resonance scattering and, possibly, electron collisions. A weaker UV line at 1355 Å is also seen from the forbidden (intercombinational) transition, $^3P - {}^5S$. Because of strong

TABLE 6.1. DAYGLOW OBSERVATIONS

Emission	Wavelength	Altitude	Zenith Intensity
Lyman α	1216 Å	>100 km	5–12 kR
OI($^3P-^3S$	1304 Å	>100 km	2–6 kR
OI($^3P-^5S$	1355 Å	100–300 km	0.4 kR
NO(γ)	2000–3000 Å	80–140 km	1 kR
N_2(2PG)	3000–4000 Å	obs > 170 km	0.4 kR
N_2^+(1 Neg)	3914 Å	130–300 km	2–7 kR
NI($^4S-^2D$)	5200 Å	>100 km	0.1 kR
OI($^1D-^1S$)	5577 Å	80–250 km	0.4–3 kR
Na(D)	5893 Å	85–95 km	2–40 kR
OI($^3P-^1D$)	6300 Å	>125 km possibly to >300 km	3–60 kR
$O_2(^1\Sigma_g^+ - {}^3\Sigma_g^-)$	7600 Å	peak at	~300 kR
	8640 Å	~50 km	~20 kR
$O_2(^1\Delta_g - {}^3\Sigma_g^-)$	1.27 μ	not measured, probably 40–80 km	~30 mR
OH	2.8–4.0 μ	not measured, probably 50–90 km	~5 mR

UV absorption by O_2, the 1304 and 1355 Å radiation can only be measured with rocket-borne instrumentation; maximum zenith intensity of both lines is in the 100 to 200 km region. The importance of the 1304 Å resonance line lies in the possibility (not yet realized because of complications from radiative-transport effects) of deducing atomic oxygen concentrations.

The 5577 Å oxygen green line is produced by the $^1D - {}^1S$ electric quadrupole transition. It is found to have a double maximum in altitude distribution with a minimum around 130 km. Above 130 km, both dissociative recombination of O_2 and excitation of atomic O by photoelectrons have been considered as mechanisms. Either one appears to be capable of explaining the observations. Below 130 km, the source of the ^1S state is undoubtedly the Chapman reaction (see Section 6.2.3).

The 6300 Å oxygen red line, arising from the intercombinational magnetic dipole transition $^3P - {}^1D$, and possibly the ultraviolet photolysis of O_2, is one of the brightest dayglow sources. As can be seen from the accompanying energy level diagram (Figure 6.2) the transition produces a triplet, of which

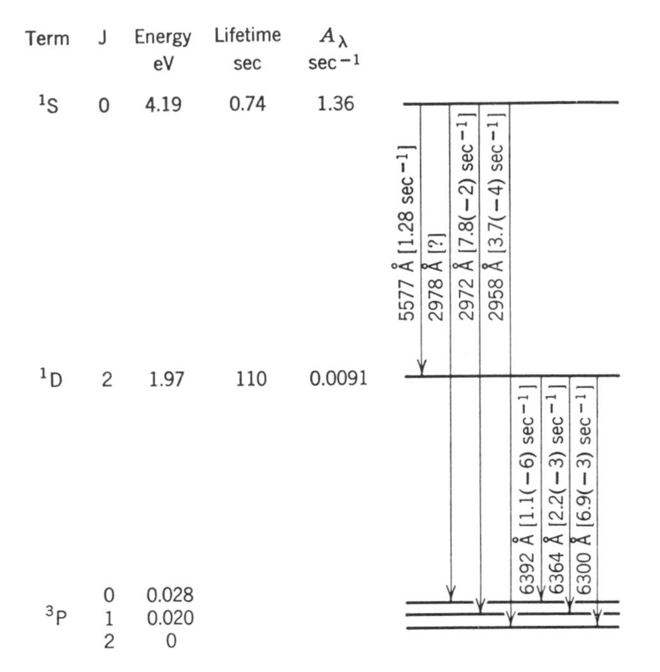

Figure 6.2 The low-energy portion of the atomic oxygen energy level diagram. The designation, total angular momentum quantum number, energy above ground level, and Einstein transition coefficient are indicated for each level. The wavelength (in Å) and the Einstein coefficient A_λ (in sec^{-1}) are also indicated for each transition. For the A_λ the exponent of 10 is in parentheses; thus, 7.8(−2) means 7.8×10^{-2}. (After Peterson, Van Zandt, and Norton, *JGR* **71**, 2255 (1966); reprinted by permission of the American Geophysical Union.)

the 6392 Å line is too weak to be observed; the 6364 and 6300 Å lines are observed. Their importance lies in the possibility of measuring the production of the ^{1}D state and thereby learning more about the importance of molecular-ion recombination in the *F* region. A serious complication, particularly at lower altitudes, is the fact that the ^{1}D state can be deactivated by collisions with other species. The sources of the ^{1}D state are thought to be recombination of NO^+ and O_2^+ with electrons, the photodissociation of O_2 by absorption of solar radiation in the Schumann-Runge bands, and excitation by photoelectrons.

Because of the importance of the atomic oxygen lines in the airglow, we will discuss the ^{1}D and ^{1}S airglow production processes in greater detail.

First, let us consider recombination:

$$O_2^+ + e \xrightarrow{\alpha_1} O + O + 6.96 \text{ eV} \tag{6.5a}$$

$$NO^+ + e \xrightarrow{\alpha_2} N + O + 2.76 \text{ eV} \tag{6.5b}$$

The molecular oxygen ion recombination can produce atomic oxygen in the ^3P, ^1S, or ^1D states; the nitric oxide ion recombination can only produce ^3P or ^1D oxygen states and production of the ^1D state violates conservation of spin rules† (see Section 2.6.5). Oxygen in the ^1D state is also produced by emission from the ^1S state. Photodissociation produces ^1D as follows:

$$O_2 + h\nu \xrightarrow{Q_1} O(^3P) + O(^1D) \tag{6.5c}$$

Photodetachment can produce ^1D by

$$O^- + h\nu \rightarrow O(^1D) + e \tag{6.5d}$$

but is probably unimportant as a dayglow source. The excitation of oxygen to the ^1D state by photoelectrons is theoretically possible because of the high cross section for such processes near 2 eV.

Collisional deactivation of the ^1D state is undoubtedly important but the exact process and rate have not been identified or measured unambiguously. Either molecular nitrogen or oxygen (or both) may be involved. The process (see Sections 6.1.4 and 4.1.3) is thought to be a near-resonance transfer of energy to a vibrational state; however, the formation of intermediate excited complexes has also been suggested. The reaction rates appear to lie in the range 10^{-11} to 10^{-10} cm^3 sec^{-1}.

Deactivation rates by photon emission are given by the Einstein coefficients (see Chapter 2) A_λ for the 6300 and 6364 Å lines as 6.9×10^{-3} sec^{-1} and 2.2×10^{-3} sec^{-1}, respectively.

The population N_i of a particular energy level at a given altitude z is given by a continuity equation

$$\frac{\partial N_i}{\partial t} = P_i(z) - L_i(z) - \text{div}\,(N_i) \tag{6.6a}$$

Following the treatment by Peterson, Van Zandt, and Norton[3] we neglect both time-dependent and divergent terms.

† The production of $O(^1D)$ by the interchange $O^+ + N_2 \rightarrow NO^+ + N$ followed by the recombination $NO^+ + e \rightarrow N^* + O^*$ would violate spin conservation. The inclusion of this source is therefore speculative. It has also been noted, however, that the excited nitrogen atom $N(^2D)$ that would be formed with a ^3P oxygen atom could, under optimum conditions, activate the $O(^1D)$ state by collisional excitation.

The production rate of O(^{1}D) at high altitudes (F-region) is

$$P_D = K_1\alpha_1[O_2^+]N_e + K_2\alpha_2[NO^+]N_e + A_{5577}[O(^1S)] \tag{6.7a}$$

and at low altitudes is

$$P_D = Q_1 I_{sr}[O_2] \tag{6.7b}$$

The production rate of O(^{1}S) at higher altitudes is

$$P_S = K_3\alpha_1[O_2^+]N_e \tag{6.7c}$$

while the production rate at lower altitudes is

$$P_S = k_c[O]^3 \tag{6.7d}$$

where $K_{1,2,3}$ are the numbers of excitations per recombination to the ^{1}D, ^{1}D, and ^{1}S states, respectively, $\alpha_{1,2}$ are the recombination coefficients for O_2^+ and NO^+, respectively, N_e is the electron concentration, [ ] denotes concentrations, Q_1 is the cross section for photodissociation, I_{sr} is the photon flux in the Shumann-Runge bands, and k_c is the reaction rate for the Chapman reaction.

The loss rates for ^{1}D and ^{1}S states are

$$L_D = (A_D + D_D)[O(^1D)] \tag{6.8a}$$
$$L_S = (A_S + D_S)[O(^1S)] \tag{6.8b}$$

where $A_{D,S}$ are the Einstein coefficients for the $^3P - {}^1D$ and $^1D - {}^1S$ transitions, respectively, and $D_{D,S}$ and D_S are the collisionless deactivation rates for O(^{1}D) and O(^{1}S), respectively. D_D is equal to $(k_{O_2} + k_{N_2}[N_2]/[O_2])$ where k_{O_2,N_2} are the deactivation reaction rates for O(^{1}D) with O_2 and N_2, respectively; and D_S is equal to k_S, where k_s is the deactivation reaction rate for O(^{1}S) with O_2.

The emission rate at a specific wavelength λ is given by

$$\epsilon_\lambda = A_\lambda[N_i] \tag{6.9a}$$

and the integrated emission rate in the vertical direction (in Rayleighs) is

$$I = 10^{-6}A_\lambda \int_0^\infty [N_i(z)]\, dz \tag{6.9b}$$

Nitric Oxide. The nitric oxide γ bands are excited by resonance radiation and can be observed above the ozone layer. Important information on the concentration of nitric oxide above 80 km was obtained (see Section 4.3.1) by measuring the resonance scattering in this band with rocket-borne instrumentation.

Hydrogen. The hydrogen Lyman-α line at 1216 Å is very strong in the solar radiation spectrum. Hence, the resonance scattering of this line is a sensitive indicator for atomic hydrogen. Because of the thick layer of atomic hydrogen, multiple scattering problems complicate the data analysis.

Hydroxyl. The strong emission of radiation in the hydroxyl bands 2.8 to 4μ is due to the reaction

$$O_3 + H \rightarrow OH^* + O_2 \tag{6.10a}$$

or possibly,

$$O_2^* + H \rightarrow OH^* + O \tag{6.10b}$$

followed by

$$OH^* \rightarrow OH + h\nu \tag{6.10c}$$

Interpretations of hydroxyl airglow are potentially valuable in that they should reveal a great deal about the distribution and roles of water vapor, atomic hydrogen, and ozone in the ionosphere: however, the possible reactions are numerous and complex.

Molecular Oxygen. Atmospheric oxygen band (7600, 8640 Å) emissions from the $b^1\Sigma_g^+$ state have been measured from several rocket flights. The band emissions are attributed to resonant scattering at lower altitudes. Above 90 km, the intensities appear too large to be explained by scattering; the excess has been attributed to collisional transfer of energy to molecular oxygen from the metastable oxygen atom ^1D (which is produced by photodissociation of molecular oxygen).

Strong infrared radiation at 1.27 μ has been measured from aircraft, balloons, and from rockets; it is identified as the infrared atmospheric band emission from the $^1\Delta_g$ state (see Figures 4.1 and 4.2). The source is thought to be the photodissociation of ozone by the 2000–3000 Å Hartley bands of solar radiation. Other suggestions have included fluorescence of O_2 and the photochemical reaction $O + O_3 \rightarrow 2O_2^*$. Interpretations would have to include transport processes if ozone is involved; this has not yet been done.

Nitrogen. The second positive bands of N_2 (3000–4000 Å, see Figures 4.4a and 4.4b) can apparently be accounted for by photoelectron excitation.

The atomic nitrogen doublet at 5200 Å arises from the intercombinational $^4S - {}^2D$ transition and produces a weak dayglow source above 100 km. The excitation mechanism is not clear. Either dissociative recombination or ion-interchange ($N_2^+ + O \rightarrow NO^+ + N^*$) is suspected. Apparently the excited nitrogen atom is strongly quenched by other atmospheric species (probably molecular oxygen). A quenching reaction rate of 10^{-12} cm^3 sec^{-1} is consistent with observations, but the evidence is scanty.

The molecular nitrogen-ion first negative band at 3914 Å is a prominent feature of the dayglow. It is observed mainly above 150 km with a maximum around 200 km. The emission at higher altitudes is almost, but not entirely, consistent with the theory that it is caused by resonance scattering of existing N_2^+ ions. At lower altitudes it is thought to be due to fluorescence scattering by N_2 which is both ionized and excited by UV solar radiation.

6.2.2 *Twilight Glow*

The twilight glow can be considered as a transition period between the dayglow in which resonance radiations are dominant and the nightglow in which chemiluminescence becomes important. It is also a transition in observation conditions because the sky brightness diminishes and weaker radiations become easier to distinguish from background. At the same time, parts of the atmosphere are still illuminated by the sun; resonance scattering and other photoprocesses continue, though in diminishing intensity as the solar depression angle increases.

Twilight conditions provide an opportunity to study ionospheric processes that are directly stimulated by sunlight as functions of altitude and time. Unraveling the observations is not a simple procedure; as a result, the opportunity may not have been fully exploited.

Consider Figure 6.3 which portrays a simplified version of twilight illumination.

$$R = (Z + R) \cos \alpha, \quad \text{or} \quad Z = R\,(\sec \alpha - 1) \tag{6.11a}$$

where R is the earth's radius and α is the solar depression angle, Z is the shadow height directly over an observer. Processes occurring as an immediate result of solar radiation must be taking place above altitude Z. If the observer is not looking toward the zenith but at an angle θ away from the zenith (and along the great circle that passes through both the observatory and the subsolar point), we use the van Rhijn formula which is derived from the law of sines:

$$\sin \phi = \frac{R}{R + Z} \sin \theta = \cos \alpha \sin \theta \tag{6.11b}$$

also

$$\beta = \theta - \phi \tag{6.11c}$$

if the non-zenith observation is of a point in the sky that is not along the great circle that passes through the observer and the subsolar point, but at an angle γ from the great circle, an additional relationship must be derived from spherical geometry:

$$\sin \alpha = \cos \beta \sin (\alpha + \beta) - \sin \beta \cos (\alpha + \beta) \cos \gamma \tag{6.11d}$$

In practice, corrections must be made for a number of additional considerations. An important consideration is screening by the atmosphere. If

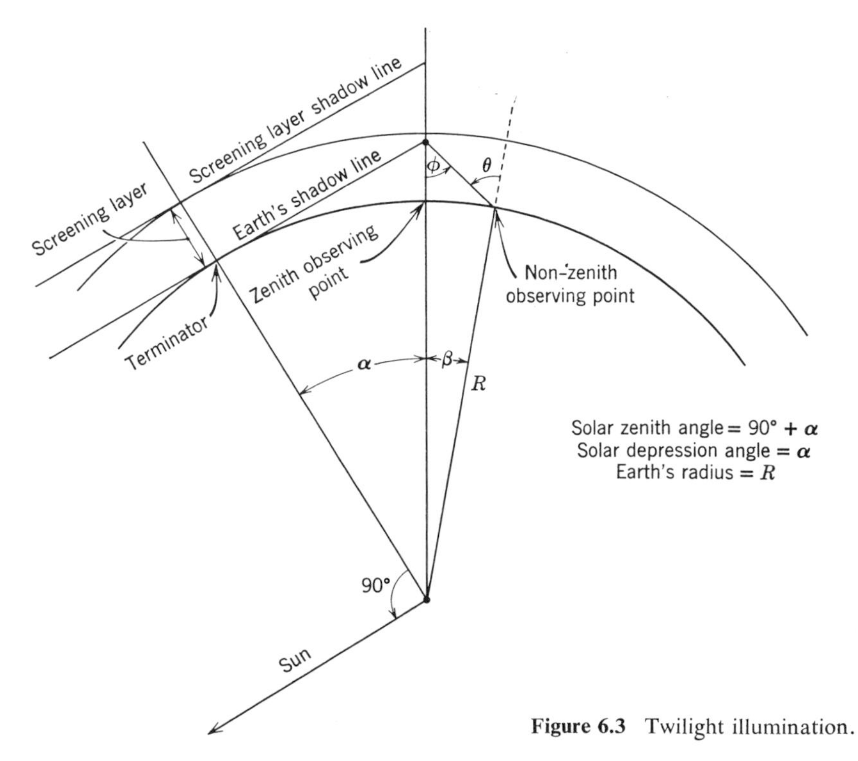

Figure 6.3 Twilight illumination.

the atmosphere totally absorbs a radiation, the opaque planetary disk is essentially increased in radius for that radiation and the shadow line is higher. This is especially important for processes that involve UV wavelengths inasmuch as they are easily absorbed in the atmosphere. The effective size of the opaque disk obviously varies with the absorption coefficient of the radiation and the distribution of absorbing atmospheric species. Corrections must also be made for refraction in the real atmosphere. For visible wavelengths, a ray will be bent approximately 0.5° on the way in through the atmosphere and 0.5° on the way out; hence, a ray grazing the ground will be bent approximately 1°.

In practice we must recognize that the radiation illuminating the upper atmosphere over the observer during twilight comes from below the illuminated volume and hence is affected by clouds and dust in the atmosphere and by scattering from the surface of the earth, as well as Rayleigh scattering by the atmosphere. Resonance scattering determinations should also account for the self-absorption of the resonance wavelengths if the radiation passes through long oblique paths. In addition to the movement of the shadow line, another difficulty that hinders interpretation of twilight data is that there is also a concurrent redistribution of atmospheric constituents: with respect to height because of the effect of changing temperatures on scale heights and vertical motion, and with respect to lateral redistribution because of strong zonal winds generated by movement of the terminator.

Some of the more important twilight emissions are described below.

Alkali Metals. Strong resonance scattering is detected from sodium, as would be expected from the preceding discussion of dayglow. Resonance scattered radiation is also detected from potassium and lithium. Emission has been detected from ionized calcium but not from neutral atomic calcium.

Atomic and Molecular Oxygen. As with the dayglow, emission from the $O(^1D)$ state is prominent. The sources are apparently the same; photodissociation by Schumann-Runge radiation in the lower ionosphere and dissociative recombination in the upper ionosphere. Similarly the $O(^1S)$ state emission is related to dissociative ion-electron recombinations in the upper atmosphere and the Chapman reaction in the lower atmosphere. The atmospheric infrared bands are observed from molecular oxygen excited to the $a^1\Delta_g$ level by the photodissociation of ozone.

Molecular Nitrogen Ions. The first negative bands of N_2^+ are observed in the twilight glow. As in the dayglow, this emission is due mainly to resonance scattering from existing nitrogen ions with a much smaller contribution from ionization processes.

Helium. One of the more interesting aspects of the twilight airglow spectrum is the emission of resonance scattered sunlight from the metastable 2^3S state of helium. The observed 10,830 Å line is produced by the $2^3S - 2^3P$ transition. Two possible sources of the 2^3S state have been suggested; both involve the absorption of extreme ultraviolet radiation (EUV) by the atmosphere. The first mechanism is the excitation of ground state 1^1S He atoms by photoelectrons exceeding 19.8 eV:

$$\mathrm{He}\ (1^1S) + e \rightarrow \mathrm{He}\ (2^3S) + e \tag{6.12}$$

The source of photoelectrons is the photoionization of atmospheric species by EUV, particularly by the He^+ solar line at 304 Å and shorter wavelength radiation.

6.2.3 Nightglow

The absence of sunlight obviously precludes direct solar influence on night airglow. Resonance scattering that predominated in day and twilight emissions is relegated to a minor role; chemical and ion-electron recombination reactions become dominant. Nonetheless, it should be remembered that the nightglow phenomena are mainly produced by sunlight; they originate during the daylight and persist throughout the night. Solar radiation is, in essence, stored in dissociated, ionized, and excited species during the daytime and released at night by various relaxation processes, i.e., neutral and ionic recombination, quantum emission, etc.

By measuring the altitude dependence of nightglow emissions, we hope to obtain additional knowledge of upper atmosphere chemical reactions, as well as some knowledge of upper atmosphere species. Three methods are used to determine emission altitude: rocket measurements, triangulation, and the van Rhijn method. Rocket measurements are the obvious choice for height determination; rockets have an added advantage in the ability to detect IR and UV emissions not observable from ground observatories. On the other hand, ground observatories provide opportunities to study weak emissions by using long integration times, to study the spatial distribution of emissions, and to study both short-term and long-term temporal variations. Hence, even with the availability of rockets, attempts to derive altitude information from ground observations are worthwhile. Triangulation methods, as the name implies, involve the use of two or more stations observing the same features. From the base path and observing angles, altitudes should be determinable; the principal limitation is the difficulty of uniquely identifying a nightglow feature from widely spaced observatories. The van Rhijn technique assumes a uniform thin, emitting layer; the opposite of the requirement for triangulation. Consider the non-zenith observing point in Figure 6.3. If the assumption of uniformity were correct, a scan from horizon-to-horizon should yield a variation of intensity as a function of the zenith angle θ inasmuch as the intensity of emission from the emitting layer varies as secant ϕ.

$$I(\theta) = I_{\text{zenith}} F(\theta, Z) \tag{6.13a}$$

$$F(\theta, Z) = \sec \phi = (1 - [R/(R + Z)]^2 \sin^2 \theta)^{-1/2} \tag{6.13b}$$

remembering that

$$\sin \phi = \left(\frac{R}{R + Z}\right) \sin \theta \tag{6.13c}$$

$F(\theta, Z)$ is called the van Rhijn function. Limitations and complications arise in the application of the van Rhijn method because the emitting layer may be thick or patchy and because of the need to correct for absorption within the emitting layer and between the layer and the observatory; correcting for the latter involves such particulars as absorption in the ozone layer and scattering by molecules and dust in the lower atmosphere.

The measurement of absolute intensities is limited by the need to correct for light from astronomical sources such as starlight or zodiacal light in order to account properly for the continuum.

Temperatures can also be derived from nightglow emissions either by analysis of rotational spectra or by measurement of the Doppler width of individual lines. These studies require very high resolution spectroscopic observations.

Some of the more important nightglow emissions are discussed below.

Atomic Oxygen. Both the 6300 Å and 5577 Å lines are prominent in the nightglow. The mechanisms for the production of $O(^1D)$ and $O(^1S)$ atoms by dissociative ion-electron recombination in the F-region were discussed under dayglow. The close relationship between F_2 ionization levels and the oxygen red line (6300 Å) is expressed in the semiempirical formula known as "Barbier's equation,"

$$I = C + D(f_0F_2)^2 \exp\left(\frac{h'F - 200}{H}\right) \tag{6.14}$$

where I is the emission rate of 6300 Å radiation, f_0F_2 is the F-region critical frequency in megahertz, $h'F$ is the virtual height in km (see Chapter 11) and H is the scale height at 200 km. The constant D apparently involves both the ion-recombination reactions that produce $O(^1D)$ and the $O^+ + N_2$ and $O^+ + O_2$ interchange reactions that control the electron density (see Chapter 9); the significance of C is not understood.

The principal source of the 5577 Å green line at low altitudes is the Chapman reaction

$$O + O + O \rightarrow O_2 + O(^1S) \tag{6.15}$$

which is a three-body oxygen recombination in which the third body is atomic oxygen and is excited by part of the energy released by the recombination. The $O(^1S)$ atoms either emit 5577 Å radiation through the $^1D - {}^1S$ transition or are deactivated by collisions with oxygen molecules and do not emit radiation; molecular nitrogen apparently does not deactivate $O(^1S)$ through collisions. As expected, the 5577 Å radiation is strongest near the 100 km level where atomic oxygen concentrations are highest during the daytime.

One of the more interesting features of the 6300 Å nightglow is the existence of the so called "red arcs" at high altitudes and mid-latitudes. Although they occur at latitudes that are well below the auroral zone, there is considerable evidence that the existence of these arcs is associated with aurora; hence, they will be discussed under aurora (Section 6.3.4).

Sodium. The existence of a sodium nightglow must be explained by photochemical reactions. However, the validity of proposed reactions has yet to be established. The following have been suggested:

$$NaO + O \rightarrow Na(^2P) + O_2 \tag{6.16a}$$

$$NaH + O \rightarrow Na(^2P) + OH \tag{6.16b}$$

$$NaH + H \rightarrow Na(^2P) + H_2 \tag{6.16c}$$

$$Na + O_2 \text{ (vibrationally excited)} \rightarrow Na(^2P) + O_2 \tag{6.16d}$$

Molecular Oxygen. Oxygen molecules can be formed in excited states (as well as in the ground state) by three-body recombination:

$$O(^3P) + O(^3P) + M \rightarrow O_2^* + M \tag{6.17}$$

This would be expected to give rise to the Herzberg ($A^3\Sigma_u^+ \rightarrow X^3\Sigma_g^-$) bands and the atmospheric systems ($b^1\Sigma_g^+ \rightarrow X^3\Sigma_g^-$ and $a^1\Delta_g \rightarrow X^3\Sigma_g^-$); emissions in the Herzberg bands and atmospheric systems are observed.

Hydrogen. Weak hydrogen Balmer Hα (6562.8 Å) emission is observed in the night sky from ground based observatories. Hydrogen Lyman α (1216 Å) emission is measured with rocket and satellite instruments. The origin of these radiations is sunlight scattered by hydrogen in the earth's outer atmosphere. Resonance scattering from the hydrogen geocorona is indeed observed from outside the atmosphere. The notion of scattering from the earth's outer atmosphere is also consistent with the observation that the hydrogen emissions are minimum in the direction away from the sun.

Hydroxyl. The hydroxyl emissions are prominent in the night sky. Production is attributed to the following cycle:

$$H + O_3 \rightarrow OH^* + O_2 \tag{6.10a}$$

Hydrogen may be produced again by

$$OH + O \rightarrow H + O_2 \tag{6.18a}$$

and ozone by

$$O_2 + O + M \rightarrow O_3 + M \tag{6.18b}$$

which, however, may be depleted by

$$O_3 + O \rightarrow O_2 + O_2 \tag{6.18c}$$

It should be noted that the nighttime supply of O atoms in the 60–90 km altitude region of OH emission is probably quite small (see Section 4.3.1).

6.2.4 Venus

Many reports have been made of a faint illumination of the dark side of Venus by an "ashen light." A seemingly analogous effect on the moon is due to earthshine; however, the probability of a similar phenomenon on Venus seems remote. Other suggestions have included aurora and airglow. The absence of a strong magnetic field on Venus would seem to eliminate the possibility of a strong aurora, but there is no *a priori* reason to doubt the possibility of an airglow.

In order to be observed by terrestrial observers, the intensity of a Venus airglow would have to be much stronger than that of the terrestrial nightglow. Venus spectra have been reported that show the 3914 Å emissions characteristic of N_2^+ and the definite absence of the 5577 Å atomic oxygen line; but reconsideration of the observations in recent years has lessened the credibility of these reports. More recent attempts to detect spectral emission from the dark side of Venus have been unsuccessful.

The flyby U.S. Mariner 5 probe and the U.S.S.R. Venera 4 automatic station have agreed that no atomic oxygen resonance radiation can be observed on Venus. If present, $O(^3P)$ must be less than 10^{-5} as abundant on Venus as on earth according to these measurements. On the other hand, a very definite hydrogen corona was observed; it was as intense as the terrestrial geocorona but had a smaller scale height. Unidentified airglow radiation in the 1350 to 1700 Å region was also measured. It can also be speculated that airglow might be produced by the dissociative recombination of CO_2^+, i.e.,

$$CO_2^+ + e \rightarrow CO(A^1\Pi) + O(^3P) \tag{6.19a}$$

which should cause the emission of the fourth positive system, $A^1\Pi \rightarrow X^1\Sigma^+$, of CO, or

$$CO_2^+ + e \rightarrow CO + O(^1D) \tag{6.19b}$$

which would lead to the emission of radiation at 6300 Å.

6.2.5 Mars

Airglow was observed[8] on Mars by spectrometers on the Mariner 6 spacecraft. Emissions identified as resonance and fluorescence scattering or electron impact excitation of the following species were observed: ionized carbon dioxide (feature at 2890 Å and Fox-Duffendack-Barker bands); carbon monoxide (Cameron bands and fourth positive bands); atomic oxygen (1304 Å, 1356 Å, and 2972 Å); and atomic hydrogen (1216 Å). No emissions from nitrogen or nitrogen compounds were reported. The atomic oxygen 2972 Å line emission is produced by the transition from the ^1S metastable state directly to the ^3P ground state. From Figure 6.2, we see that the transition $^1D - {}^1S$ is approximately 15 times more probable; hence, the 5577 Å line should be observable as should the 6300 Å line produced by the $^3P - {}^1D$ transition.

Why the ^3P oxygen atom can be seen on Mars, but not on Venus has not been explained. Inasmuch as the atmospheres of both Mars and Venus are presumably composed mainly of CO_2, the photochemistry of their upper atmospheres should be similar and, therefore, the airglow emissions should also be similar.

6.3 Aurora

A distinction between auroras and airglow seems easy. Everyone "knows" the difference, but if we attempt to specify a set of criteria to distinguish between the two phenomena, we find that it is much easier to relate the two phenomena than separate them. Basically, the same atmospheric species are involved and the same excitation mechanisms; geomagnetic effects are involved in both phenomena, and the geographic distributions overlap. The distinctions must be made on a relative basis. For example, it is clear that auroras result primarily from charged particle ionization or excitation processes, but resonance scattering and chemiluminescence are also important. Similarly, airglow results primarily from resonance scattering and chemiluminescence, but charged particle excitation is also important. Certainly, auroras are usually brighter than airglows, but at what intensity level do we distinguish between weak auroras and bright airglows? In some cases, such as the red arc mentioned briefly in the nightglow discussion it is difficult to decide whether it is aurora or airglow by any set of criteria. In the following, we will emphasize those facets of aurora that most obviously characterize that phenomenon, but it should be clear that auroras and airglows are not mutually exlusive phenomena.

6.3.1 Morphology

Whereas the principal source of airglow energy is the absorption of electromagnetic radiation, the principal source for auroral reactions is the energy dissipated by charged particles that bombard the atmosphere. There is a definite correlation with the frequency of geomagnetic activity which in turn is related to solar activity; hence, auroras are much more frequent during periods of high solar activity than during quiet solar periods. The ultimate source of the energy is undoubtedly the sun, but we are still looking for several of the links in the chain of events that causes the particles to arrive in the terrestrial auroral zone with the energies, frequencies, fluxes, etc., that are observed.

Auroras occur most frequently in a zone centered on the geomagnetic latitude of 67°. The frequencies of occurrence decrease to minima over the magnetic poles. Although the frequencies of occurrence decrease sharply toward mid-latitudes, aurora are not unknown at low latitudes and have been observed as far south as San Francisco, California, and Spetsoi, Greece. Because auroras are also frequent between 45° and 60° geomagnetic latitudes, this zone is called the subauroral belt. In fact, however, the auroral zone is not circular but oval, the maximum occurrence is around 67° on the nightside and near 78° on the dayside. The oval band tends to be thicker on the nightside. The thicker nightside portion pulsates by expanding explosively toward the pole and then breaking up into smaller patches. The pulsations have a period on the order of hours and are connected, in some unexplained way, with magnetic disturbances.

There appear to be definite temporal variations shorter than the 11-year solar cycle. Using radio absorption observations at College, Alaska, as a guide, it is seen that auroral ionization is relatively constant throughout the day during summer; but during winter and equinox periods, there is a pronounced maximum around 0800 local time and a minimum around 1900. A difference of approximately two in absorption is noted between summer and winter months, which cannot be explained simply by photodetachment effects inasmuch as absorption is greater in winter than summer. There is only a loose correlation between radio absorption and optical emissions.

The altitude of visible auroras is centered around 100 km; some are observed as low as 70–80 km and some at much higher altitudes (several hundred kilometers). The luminosity of an aurora at a particular altitude depends on the rate of excitation and the deactivation mechanisms. The more energetic the particles, the lower the aurora; but at the same time, collisional deactivation mechanisms increase rapidly with decreasing altitude and one would, therefore, expect that there would be an effective lower limit for visual observations.

Auroras are classified in terms of shape, structure, and luminosity. The shape classification is almost self-explanatory: A = arc, B = band, R = ray, D = drapery, G = glow, C = corona, and S = surface. The structure is classified as: H = homogeneous, R = rayed, D = diffuse, P = pulsating, and F = glowing. The letters are combined to describe a specific auroral form, e.g., RB = rayed bands, DS = diffuse surfaces, HA = homogeneous arc, PA = pulsating arc, etc. Some of the shapes and structures are illustrated in Figure 6.4.

Three special cases of red auroras, called Type A, Type B, and M-arcs are discussed in Section 6.3.4.

6.3.2 Excitation Mechanisms

Auroras are excited, basically, by charged particles: both protons and electrons. Earlier it was thought that only protons were the exciting particles because of observations of high altitude hydrogen emissions that were obviously caused by rapidly moving hydrogen atoms; the principal evidence was

Figure 6.4 Auroral types.
a. Multiple rayed and homogeneous bands. This was a quiet form, characteristic of the less active part of the sunspot cycle. There was essentially no movement of the form during the 40-second exposure.
b. Corona. Auroral rays are always directed parallel to the earth's magnetic field lines. Thus when a mass of auroral rays appears at the magnetic zenith, they are observed end on and perspective makes

them appear to converge to a point. The magnetic zenith is located 10° to the southwestward at College, Alaska, where the aurora was photographed.
c. Multiple homogeneous auroral bands. This aurora was photographed at a polar cap site, far north of the auroral zone.
d. Long rays and diffuse aurora. Auroral rays are always parallel to the earth's magnetic field lines.

(Reprinted by permission of Professor Victor P. Hessler.)

the Doppler shift on the order of 10 Å of the emitted lines. However, rocket measurements,[5] corroborated by balloon measurements of bremsstrahlung, have shown conclusively that electrons are the primary particles that cause most auroras. The energy spectra of incident electrons and their ionization characteristics are discussed in Chapter 10.

As charged particles are slowed by collisions with atmospheric constituents, the atmospheric constituents are ionized and excited. For each ion-electron pair produced in a collision, an average of 35 eV is dissipated by the primary particle and imparted to the ion-electron pair. Considering the ionization potential of atmospheric species (e.g., from 9.5 eV for NO to 15.5 eV for N_2), it is easily visualized that some secondary electrons are capable of producing another ion-pair. It has been calculated that perhaps half of the secondary particles are capable of additional ionization; some calculations show that for each ion-pair produced by primary particles, an average of two additional ion-pairs are produced by secondary electrons. As an electron slows to thermal velocities, excited species are produced by collisions with atmospheric constituents; during an aurora, the average electron energy is somewhat

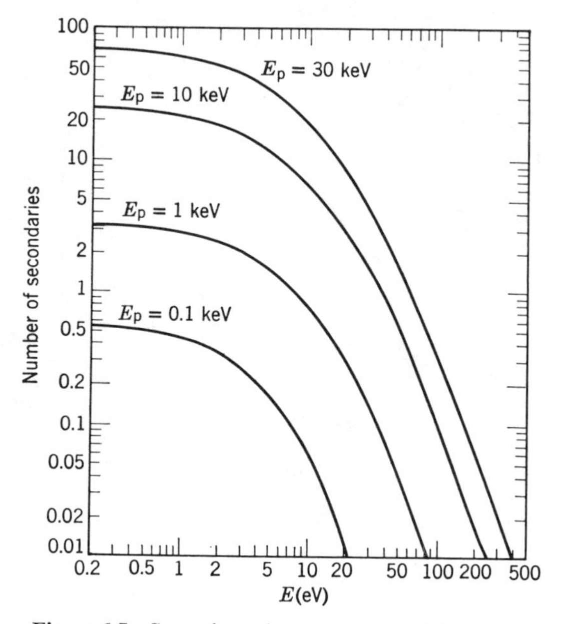

Figure 6.5 Secondary electron spectra. The number of secondaries per unit energy interval versus energy are shown for four incident energies. (After Stolarski and Green, *JGR* **72**, 3967 (1967); reprinted by permission of the American Geophysical Union.)

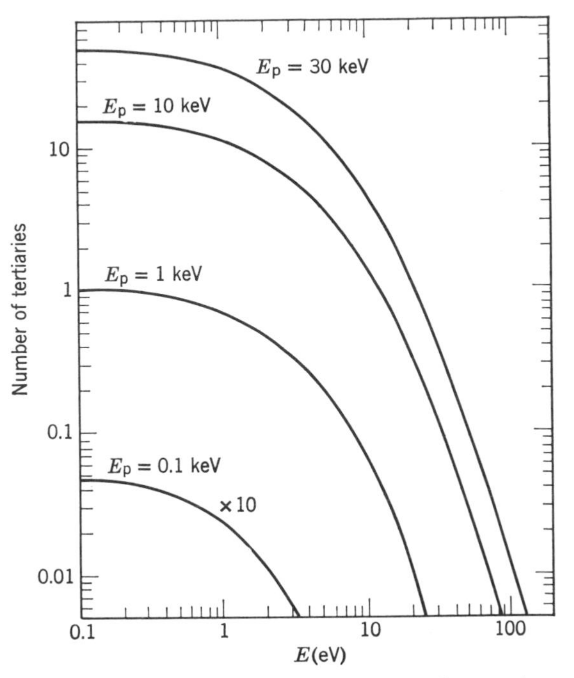

Figure 6.6 Tertiary electron spectra. The number of tertiaries per unit energy interval versus energy are shown for four incident energies. (After Stolarski and Green, *JGR* **72,** 3967 (1967); reprinted by permission of the American Geophysical Union.)

higher than thermal. Then, the thermalized electrons recombine with ions and thus produce additional excited species. Additional modes of ionization and excitation have been suggested such as photoionization of NO by the hydrogen Lyman α radiation emitted by bright proton auroras and by electrons heated by hydromagnetic waves. Figures 6.5 and 6.6 illustrate the production of secondary and tertiary electron production in a gas mixture representative of the 105 km altitude region.

6.3.3 Auroral Spectra

As might be expected, auroral spectra are produced mainly by excited nitrogen and oxygen, neutral and ionized, molecular and atomic. The spectra have been most thoroughly studied in the visible, near UV, and near infrared regions from ground stations; however, rocket and satellite measurements are being made of the far UV and infrared. The more prominent emissions are summarized in Table 6.2. Characteristics and production mechanisms of some of these emissions are reviewed below.

TABLE 6.2. PROMINENT AURORAL EMISSIONS

Species	Transition	Bands or Line
N_2	$B^3\Pi_g \rightarrow A^3\Sigma_u^+$	first positive (red-infrared)
	$C^3\Pi_u \rightarrow B^3\Pi_g$	second positive (violet-ultraviolet)
	$A^3\Sigma_u^+ \rightarrow X^1\Sigma_g^+$	Vegard-Kaplan (ultraviolet)
N_2^+	$B^2\Sigma_u^+ \rightarrow X^2\Sigma_g^+$	first negative (blue-ultraviolet)
	$A^2\Pi_u \rightarrow X^2\Sigma_g^+$	Meinel (infrared)
O_2	$b^1\Sigma_g^+ \rightarrow X^3\Sigma_g^-$	atmospheric (red)
	$a^1\Delta_g \rightarrow X^3\Sigma_g^-$	infrared atmospheric
O_2^+	$b^4\Sigma_g^- \rightarrow a^4\Pi_u$	first negative (green-red)
O	$^3P - {}^1D$	6300 and 6364 Å (red doublet)
	$^1D - {}^1S$	5577 Å (green line)
	$^5S - {}^5P$	7774 Å multiplets
	$^3S - {}^3P$	8446 Å multiplets
N	$^2D - {}^2P$	10,400 Å (infrared)
	$^4S - {}^2P$	3466 Å (violet)
H	$2^2P - 3^2D$	6563 Å (red) Balmer Hα
	$2^2P - 4^2D$	4861 Å (blue) Balmer Hβ
Na	$^2S - {}^2P$	5890 and 5896 Å (yellow doublet)

Molecular Nitrogen. The N_2 band systems are excited by electron collisions; most likely by the lower energy secondary electrons which can undergo electron-exchange reactions, inasmuch as electron spin reversal is involved. It is possible also for excitation to higher levels to occur in steps, i.e., excitation to a lower level by one collision followed by excitation to a higher level by a second collision.

Excited levels leading to the emission of Vegard-Kaplan, first positive, and second positive systems can be produced by (see Section 2.4.3 for an introduction to the nomenclature of molecular structure):

$$e + N_2(X^1\Sigma_g^+) \rightarrow e + N_2(A^3\Sigma_u^+) \quad (6.20a)$$

$$\rightarrow e + N_2(B^3\Pi_g) \quad (6.20b)$$

$$\rightarrow e + N_2(C^3\Pi_u) \quad (6.20c)$$

The excitation of N_2 bands may be a major factor in the slowing down of electrons in the 10–15 eV region, and hence deplete the population of electrons below 15 eV. This effect may, in turn, influence the population of other species, such as the $O(^1D)$, that are thought to be produced by collisions with slow electrons.

The N_2^+ excited states are probably produced simultaneously with ionization, i.e., from the neutral N_2 molecule rather than by excitation of the ground state N_2^+ ion. This view arises from the fact that if the N_2^+ observed in excited states were produced from N_2^+ ions in ground states, the parent population of nitrogen ions required would be unrealistically large. The states that emit the first negative and Meinel bands are produced as follows.

$$e + N_2(X^1\Sigma_g^+) \rightarrow N_2^+(B^2\Sigma_u^+) + 2e \tag{6.21a}$$

$$e + N_2(X^1\Sigma_g^+) \rightarrow N_2^+(A^2\Pi_u) + 2e \tag{6.21b}$$

Emission of the prominent 3914 Å bands of the first negative system occurs at the rate of one photon per 50 nitrogen ions formed.

As with airglow, sunlit auroras emit the 3914 Å band of the first negative system as resonance-scattered radiation from the N_2^+ ions.

Excitation functions (cross section vs. energy) for molecular nitrogen are shown in Figure 6.7.

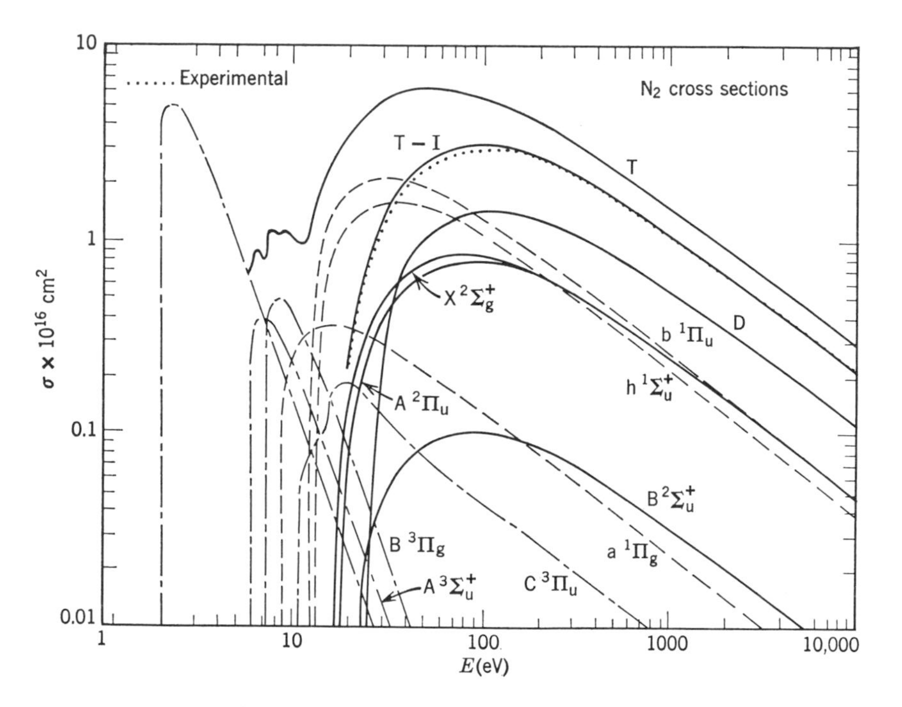

Figure 6.7 Excitation functions for molecular nitrogen. T = total of excitation cross sections (for neutral and ionized states). T − I = total of excitation cross sections for neutral states (I = total for ionization cross sections). D is cross section for excitation to the dissociative state. (After Green and Barth, *JGR* **70**, 1083 (1965); reprinted by permission of the American Geophysical Union.)

Molecular Oxygen. Metastable states of molecular oxygen are produced by secondary electrons as follows:

$$e + O_2(X^3\Sigma_g^-) \rightarrow e + O_2(b^1\Sigma_g^+) \tag{6.22a}$$

$$e + O_2(X^3\Sigma_g^-) \rightarrow e + O_2(a^1\Delta_g) \tag{6.22b}$$

The $b^1\Sigma_g^+$ state is also excited by thermal collisions with metastable oxygen atoms:

$$O(^1D) + O_2(X^3\Sigma_g^-) \rightarrow O(^3P) + O_2(b^1\Sigma_g^+) \tag{6.23}$$

As in airglows, the 8645 Å band of the atmospheric system emitted by the $b^1\Sigma_g^+$ state is observed; the 7619 Å band is absorbed by atmospheric oxygen. Similarly, 1.58 μ emission is observed for the Infrared Atmospheric System emitted by the $a^1\Delta_g$ state; the 1.27 μ band is absorbed. Excitation functions are presented in Figure 6.8.

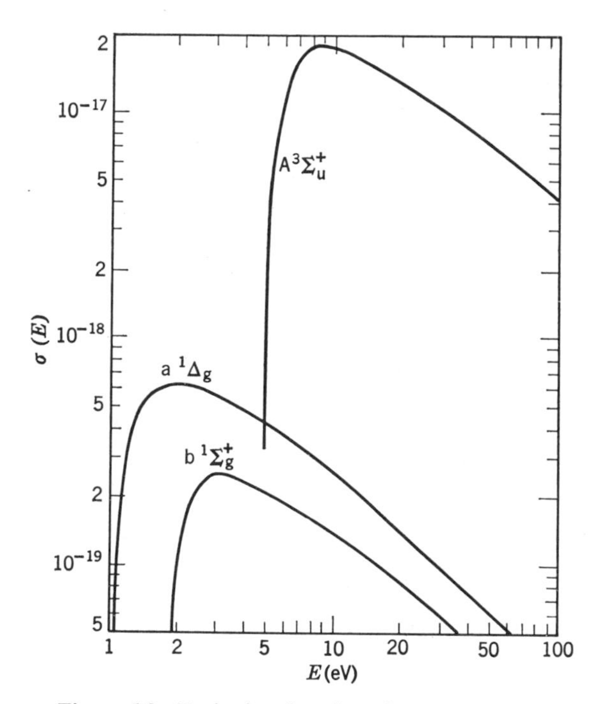

Figure 6.8 Excitation functions for molecular oxygen; units of σ are cm^2. (After Watson, Dulock, Stolarski, and Green, *JGR* **72,** 3961 (1967); reprinted by permission of the American Geophysical Union.)

The excited O_2^+ ions in auroras are produced in the same way as the excited N_2^+ ions, i.e., simultaneous ionization-excitation:

$$e + O_2(X^3\Sigma_g^-) \rightarrow O_2^+(b^4\Sigma_g^-) + 2e \tag{6.24}$$

Only the first negative system is observed in auroras. These bands are enhanced in the type B red auroras (see Section 6.3.4).

Atomic Nitrogen and Oxygen. Excited states of both neutral (NI) and ionized (NII) nitrogen atoms are produced by electron impact.

$$e + N_2 \rightarrow N + N^* + e \tag{6.25a}$$

$$e + N_2 \rightarrow N + N^{+*} + 2e \tag{6.25b}$$

NI is also produced by dissociative recombination

$$N_2^+ + e \rightarrow N + N^* \tag{6.26}$$

Similar reactions lead to the production of excited oxygen atoms except that electron impact on the atomic species is also possible. The metastable states can be populated as the result of impact with slow electrons with energies at or above the energies of the excited levels. Excitation functions for atomic oxygen and nitrogen are presented in Figure 6.9.

The metastable states of atomic oxygen that are formed by both electron impact and dissociative recombination can also be depopulated without the emission of radiation by collisions with nitrogen and/or oxygen molecules; this is one mechanism for exciting the Atmospheric system of molecular oxygen. The depopulation of metastable levels by collision is called "quenching" and is an important mechanism in accounting for the distribution of luminosity in certain auroral types. In the type A red aurora, the red atomic oxygen doublet ($^3P - {}^1D$) appears to be greatly enhanced at high altitudes. Rather than an enhancement at high altitudes, this is interpreted to be a great suppression of the $O(^1D)$ radiation at low altitudes by quenching. Inasmuch as the molecular nitrogen and oxygen concentrations increase with decreasing altitude, this effect is quite plausible and to be expected.

In type B red auroras, the red color is attributed to a suppression of $O(^1S)$ radiation at low altitudes (thus allowing the red N_2 first positive and O_2^+ first negative bands to dominate); this is due to quenching of metastable atomic oxygen.

Atomic Hydrogen. Observations of Doppler-shifted hydrogen lines emitted at high altitudes in the auroral zone provided the first direct evidence

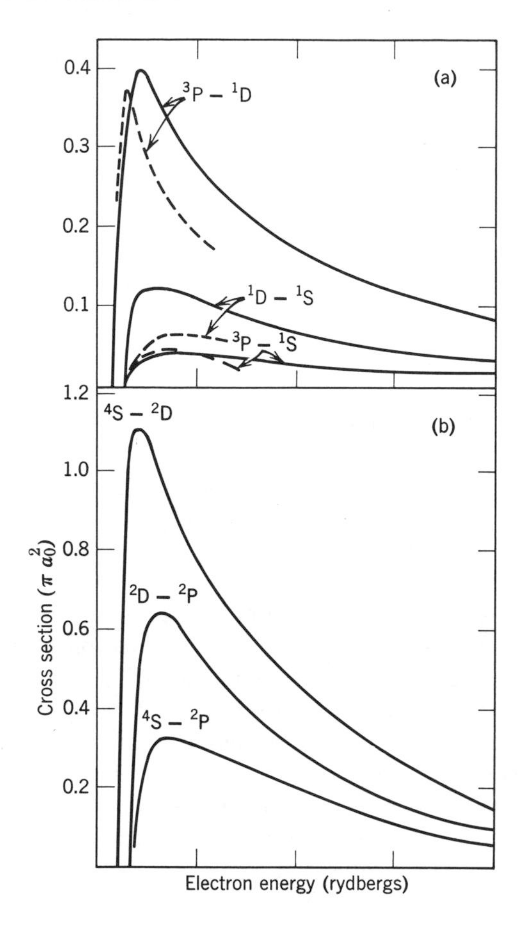

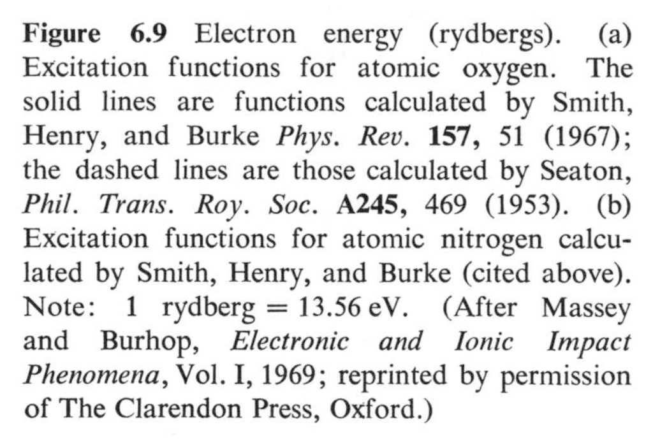

Figure 6.9 Electron energy (rydbergs). (a) Excitation functions for atomic oxygen. The solid lines are functions calculated by Smith, Henry, and Burke *Phys. Rev.* **157,** 51 (1967); the dashed lines are those calculated by Seaton, *Phil. Trans. Roy. Soc.* **A245,** 469 (1953). (b) Excitation functions for atomic nitrogen calculated by Smith, Henry, and Burke (cited above). Note: 1 rydberg = 13.56 eV. (After Massey and Burhop, *Electronic and Ionic Impact Phenomena*, Vol. I, 1969; reprinted by permission of The Clarendon Press, Oxford.)

of particle bombardment in the auroral zone. In fact, these direct observations of incoming protons led many researchers astray; but careful photometry and analysis of airglow emissions, corroborated by direct measurements with rockets and indirect measurements with balloons have shown that electrons are the primary sources of auroral excitation.

The hydrogen lines arise from the collisional excitation of hydrogen atoms to states that lead to the emission of Balmer lines or from charge exchange processes that leave the hydrogen atom in an excited state. After charge exchange, the hydrogen atom retains sufficient velocity to be ionized or excited by collision with neutral particles, hence

$$H^+ + M \rightarrow H^* + M^+, \qquad H^* \rightarrow H + h\nu \tag{6.27a}$$

followed by

$$H + M \rightarrow H^* + M^* \tag{6.27b}$$

or

$$H + M \rightarrow H^+ + M^* \tag{6.27c}$$

These reactions can occur repeatedly until the H atom energy becomes too small to produce either collisional excitation or ionization. Calculations indicate that approximately 50 Hα photons and Hβ photons are emitted per incident proton with initial energy over 100 keV.

As indicated earlier, a Doppler shift of the observed lines is caused by the high velocity of the emitting atom relative to the observer. Two considerations must be noted in this regard. First, the cycle of charge-exchange/excitation/ionization/etc. is most efficient at low proton energies (below 100 keV); therefore, velocities calculated on the basis of Doppler shift are lower limits. Second, because the protons spiral into the atmosphere along magnetic lines, calculation of the velocities of the emitting particles relative to the observer must account for the magnetic field as well as the zenith angle.

Sodium. The sodium D lines are occasionally observed in auroras; there appears to be some connection between the sodium enhancement and type B red auroras. Airglow mechanisms cannot account for the intensities observed but an auroral excitation theory has not been fully developed either. A plausible explanation attributes the excitation to energy transfer from excited, metastable atoms or molecules; the most probable energy source is metastable vibrationally excited molecular nitrogen.

6.3.4 Red Auroras

Three types of red auroras are observed that warrant special discussion; these are called the type A and type B red auroras and the M-arcs. The type A and type B auroras occur at high latitudes in the auroral zone; the

M-arcs occur at high latitudes, but well below the auroral zones and could possibly be termed airglows.

The type A red auroras are large, diffuse, relatively weak glows or arcs or the upper parts of rays. They tend to occur at high altitudes (up to 600 km) and during large geomagnetic storms. The red color is caused by an apparent enhancement of the $O(^1D)$ red doublet at high altitudes, but as discussed earlier, this apparent high altitude enhancement is really the result of a low altitude suppression by quenching processes.

The type B red auroras are short lived with vivid motions and thus difficult to study comprehensively. This form is a bright rayed band or arc with a low altitude red border which is due to the enhancement of the red first positive nitrogen bands and the first negative bands of ionized oxygen. As discussed earlier, this apparent enhancement may be due to a suppression of the oxygen green line because of quenching by molecular species; or it may simply be due to the increased concentrations of oxygen molecules at low altitudes. Some measurements of type B aurora have been made as low as 65 km, whereas normal displays seldom drop below 85 km.

The high altitude (several hundred kilometers) red M-arcs are very rarely observed and almost never visible; they occur well below the auroral zone (41° to 60° magnetic latitude) but are connected with auroral activity. M-arcs exist for about a day, are oriented along magnetic parallels, and extend East-West for thousands of kilometers (perhaps around the globe). The emission is almost purely that of the $O(^1D)$ red doublet. During these phenomena, the $O(^1S)$ green line intensity rarely exceeds nightglow intensities; the red doublet is 10^3–10^4 times stronger. Thus, the $O(^1D)$ level must be excited selectively. Two mechanisms have been suggested, photochemical and collisional. The photochemical explanation involves the F region ion reactions

$$N_2 + O^+ \rightarrow NO^+ + N$$
$$NO^+ + e \rightarrow N + O(^1D)$$

There are several objections to this, however. The principal problem is spin conservation;[6] however, it is also argued[3] that the nitrogen atom can be excited without violating spin conservation and can then be quenched by $O(^3P)$, resulting in $O(^1D)$; this requires that the nitrogen atom diffusion length be reasonably small (<10 km). This explanation also requires that the enhanced O^+ production must occur at altitudes that are high enough to relegate charge exchange with O_2 to a very minor role. The collisional explanation requires an enhancement of the population of electrons with energies above 2 eV (but below 4 eV) that can excite ground state oxygen to the lowest metastable level; an electron temperature of 3000–4000°K required for this seems somewhat too high. An increased flux of low energy

(<400 eV) electrons could be effective, but excitation of the first negative system should also occur and this has not been observed.

6.3.5 Polar Glow Auroras

During Polar Cap Absorption (PCA) events, the polar regions are bombarded by large fluxes of solar cosmic rays which consist of protons and alpha particles. Accompanying these events is an extensive polar glow; the most important emissions are the 3914 Å first negative bands of N_2^+, the 5577 Å $O(^1S)$ green line (weak at low altitudes because of quenching) and the hydrogen Balmer line. If alpha particles comprise half of the solar cosmic ray flux, the 5876 Å He I line should be detected. Since the helium line is not detected, it has been concluded[7] that alpha particles comprise less than 10 percent of the proton flux or that the alpha particles are incident at very low energies (<10 keV) or at very high energies (>1 MeV).

6.3.6 Other Planets

Inasmuch as no strong magnetic fields have been detected on Venus and Mars, it seems unlikely that strong, well-defined auroras exist on those planets, although the possibility of an analog to the polar glow aurora should not be excluded. Secondary electrons produced by the absorption of solar flare protons in the Mars or Venus atmosphere would be expected to excite and ionize CO_2, CO, O_2, OH, and He; therefore, a diffuse aurora composed of radiations characteristic of those constituents might be expected. The strong magnetic field observed by radio astronomy techniques would seem to justify speculations that strong auroral displays would be seen on Jupiter.

PROBLEMS

6.1. In the atmosphere of planet Glitch, there exists a minor constituent, Bunkium, which is present in the constant proportion of 10^{-8} of the total atmosphere at any altitude. See Problem 4.2 for details relevant to the Glitchian atmosphere. There is a resonance line at 1000 Å for Bunkium, the cross section for the resonance absorption is 10^{-18} cm^2 and the photon flux at 1000 Å is 10^{11} cm^{-2} sec^{-1} incident on the Glitchian atmosphere. The absorption cross section for 1000 Å photons by the principal components is 10^{-23} cm^2. The Einstein coefficient for the excited state is 10 sec^{-1}. Neglect absorption of the emitted resonance radiation.

a. Write an equation describing the production rate of the excited state of Bunkium as a function of altitude, assuming perfect mixing at all altitudes. What is the production rate at 50, 60, 70, 80, 90, 100 km?

b. Write an equation for the loss rate of the excited state as a function of altitude under equilibrium conditions.

c. Find, analytically, the altitude of maximum emission of the resonance radiation. Assume that the volume emission rate at the altitude of maximum emission is a good approximation for a 1 km layer; what would be the intensity measured (in rayleighs), at the Glitchian surface of the airglow produced by resonance scattering from that 1 km layer of Bunkium?

6.2. Using the nighttime concentrations of O computed by Hunt (Chapter 4), calculate:

a. The production rate of $O(^1S)$ at 80, 90, 100 km, using 10^{-34} cm^6 sec^{-1} as the reaction rate for the Chapman reaction.

b. The intensity (in rayleighs) of the 5577 Å $O(^1S)$ emission from the layer 75-105 km, assuming equilibrium between production and loss, that radiation is the only loss process, and that the volume emission rate is uniform over an altitude interval ± 5 km centered on the above altitudes (i.e., 75–85, 85–95, 95–105 km); neglect self-absorption and quenching of $O(^1S)$ radiation.

6.3. The equilibrium concentrations of electrons for a particular *D*-region model are given in Figure 8.6. Assume that the ion concentrations are equal to the electron concentrations, that half of the ions are NO^+, and that the oxygen atom produced by the reaction $NO^+ + e \rightarrow N + O$ is in the $O(^1D)$ state.

a. Find the intensity of $O(^1D)$ 6300 Å emission produced at sunspot maximum by solar X-rays in a layer 10 km thick centered on 85 km altitude. Assume that the $O(^1D)$ concentration throughout the layer is constant and equal to the value at 85 km. Assume also that the only loss of $O(^1D)$ is by radiation and that $O(^1D)$ production and loss are in equilibrium.

b. Find the intensity with the same conditions except that collisional de-excitation by both oxygen and nitrogen molecules is a second loss process. Assume that the de-excitation reaction by both N_2 and O_2 has a rate constant of 10^{-12} cm^3 sec^{-1}, that the neutral particle $(N_2 + O_2)$ concentration is 2×10^{14} cm^{-3}, and that the Einstein coefficient for the $O(^1D)$ state is 9.1×10^{-3} sec^{-1}.

6.4. Assume a simple Chapman-type ionization process (Chapter 8). Derive an expression for the intensity of the optical radiation emitted by the excited species formed by dissociative recombination of the ionized species that produce the Chapman layer. The intensity is that observed from the ground in the zenith direction. Assume equilibrium.

6.5. Find the height of the earth's shadow directly overhead (i.e., height, above which the sky is illuminated by the sun) at 15 minutes, 30 minutes, one hour, two hours, and three hours past local sunset. Neglect refractive effects and assume that local sunset occurs when the terminator passes the observing station.

6.6. Plot the variation of intensity (relative to the intensity in the zenith direction) from horizon to horizon for a uniform thin emitting layer at 50 km, 100 km, and 250 km altitude. Neglect absorption within the layer or between the layer and the station. Assume the horizon is $\pm 90°$ from the zenith, and neglect atmospheric refraction. Using the above plots, show how the layer heights can be deduced from measurements of intensity variations.

6.7. Find the abundance of sodium atoms in a column of unit cross section if the dayglow intensity is measured at 6.75 kilorayleighs.

6.8. What energies are required to ionize O_2 and N_2; what energies are required to produce the Meinel and first negative N_2^+ bands and the first negative O_2^+ band from ground state neutral molecules?

6.9. a. What are the basic mechanisms that produce type A and type B red auroras. Does either one have anything in common with the red M-cars?
b. What are the principal differences between auroral zone auroras and polar glow auroras?
c. Which of the other planets might be expected to have airglows, auroras, polar glow auroras?

6.10. Derive an equation to describe a vertical profile of $O(^1D)$. Assume that formation occurs via $O_2 + h\nu \rightarrow O(^1D) + O(^3P)$, that the quenching reaction is $O(^1D) + N_2 \rightarrow O(^3P) + N_2^*$, that molecular diffusion is important, and that the radiation which causes the formation of $O(^1D)$ is not appreciably attenuated by the atmosphere.

REFERENCES

1. J. E. Blamont and T. M. Donahue, *J. Geophys. Res.* **66,** 1407 (1961).
2. D. M. Hunten and L. Wallace, *J. Geophys. Res.* **72,** 69 (1967).
3. V. L. Peterson, T. E. Van Zandt, and R. B. Norton, *J. Geophys. Res.* **71,** 2255 (1966).
4. M. B. McElroy, *J. Geophys. Res.* **73,** 1513 (1968).
5. E. E. McIlwain, *J. Geophys. Res.* **65,** 2727 (1960).
6. A. Dalgarno and J. C. G. Walker, *J. Atmospheric Sci.* **21,** 463 (1964).
7. R. H. Eather, *Rev. Geophys.* **5,** 207 (1967).
8. C. A. Barth, W. G. Fastie, C. W. Hord, J. B. Pearce, K. K. Kelly, A. I. Stewart, G. E. Thomas, G. P. Anderson, *Science* **165,** 1004 (1969).

GENERAL REFERENCES

Except for specific references specifically cited in the text the source of material for this chapter was as follows. They are all recommended for supplementary reading on Airglows and Auroras.

Joseph W. Chamberlain, *Physics of the Aurora and Airglow*, Academic Press, New York and London, 1961.

D. M. Hunten, and M. B. McElroy, Quenching of metastable states of atomic and molecular oxygen and nitrogen, *Reviews of Geophysics.* **4,** 303 (1966).

Billy M. McCormac, ed., *Aurora and Airglow*, Reinhold Publishing, New York, Amsterdam, London, 1967: pp. 29–40, The worldwide morphology of the atomic oxygen nightglows, by F. E. Roach and L. L. Smith; pp. 59–74, The spectrum and excitation mechanisms in Aurora, by A. Omholt; pp. 93–106, Ionospheric implications of aurora and airglow studies, by Lance Thomas; pp. 109–122, Twilight observations, by M. Gadsden; pp. 123–132, Dayglow observations, by J. F. Noxon; pp. 267–280, The auroral oval and the internal structure of the magnetosphere, by S. I. Akasofu; pp. 305–314, Interpretation of twilight emissions, by M. Gadsden; pp. 315–321, Interpretation of the dayglow, by J. F. Noxon; pp. 651–666, Conference summary by M. Walt.

J. A. Ratcliffe, ed. *Physics of the Upper Atmosphere*, Academic Press, New York and London, 1960; pp. 219–267, The airglow, by D. R. Bates; pp. 269–296, General character of auroras, by D. R. Bates; pp. 297–353, The auroral spectrum and its interpretation, by D. R. Bates.

J. F. Noxon, Day airglow, *Space Science Reviews* **8,** 135 (1968).

Donald M. Hunten, Spectroscopic studies of the twilight airglow, Space Science Reviews **6,** 493 (1967).

7

Electric Currents in the Upper Atmosphere

The regular diurnal variations in the geomagnetic field were discovered almost 250 years ago by G. Graham, a London clockmaker. However, the cause of the variations was not discovered until Balfour Stewart proposed in the latter part of the nineteenth century that they are due to electric currents flowing in the upper atmosphere. Because the current flow requires fairly large conductivities, Stewart suggested that the corresponding region was ionized by solar radiation. This was, in fact, the first serious suggestion of the existence of an ionosphere. He then went on to develop to the extent possible at the time the theory by which the currents are generated: air motion caused by atmospheric tides sweeps the ions and electrons across the earth's magnetic field lines. This induces electric fields which in turn produce the current flow.

After the early investigations of the ionosphere by Appleton, Breit and Tuve, and others, the stage was set for further progress which led to the dynamo theory developed principally by Chapman and Bartels.[1] According to the dynamo theory, tidal winds produce the requisite motion of the charged particles. Under quiet conditions these winds and their associated currents can be separated into two components, one due to solar (Sq, quiet day solar) and lunar (L, quiet day lunar) variations. With rather crude knowledge of the ionospheric conductivity and tidal variations it is possible to calculate the current systems and magnitudes. This was done by Chapman and Bartels nearly 30 years ago and their work remains the cornerstone of the theory of ionospheric currents. Some solar daily variations of the geomagnetic field at various latitudes are shown in Figure 7.1 for the year 1902, a time of minimum solar activity when magnetic disturbances† were very infrequent and of small

† Magnetic activity is usually expressed in terms of the so-called "K index" which is a measure of the local fluctuations on a logarithmic scale, running from 0 to 9. Generally a given K value signifies a higher activity at low latitudes than at high latitudes. The worldwide weighted average is called K_p. The planetary daily index A_p is defined as the three-hourly planetary index averaged over a 24-hour period.

intensity. There is, in fact, a very close relationship, shown in Figure 7.2, between geomagnetic and solar activity. The disturbance variations are believed to be due mainly to magnetospheric phenomena and for that reason are discussed only in a cursory manner here. The interested reader is referred to one of the recent review articles[2] on magnetic storms for more detail.

Because of the solar cycle variations and because of magnetosphere-ionosphere coupling, it is impossible to completely separate the disturbance and quiescent magnetic field variations. Nevertheless, the concept of the dynamo action is a very useful one and we shall go on to develop it at some length in the present chapter. The theory of atmospheric tides was explained in considerable detail in Chapter 5 and we do not wish to repeat the discussion here. Suffice it to say that the solar tides, unlike the ocean tides, are stronger than the lunar ones and are caused mainly by heating rather than by gravitational effects. Although the semidiurnal mode is dominant, the Sq variations seem to have a largely diurnal character.[3] Because the Sq current systems are dependent upon the electrical conductivity of the ionosphere

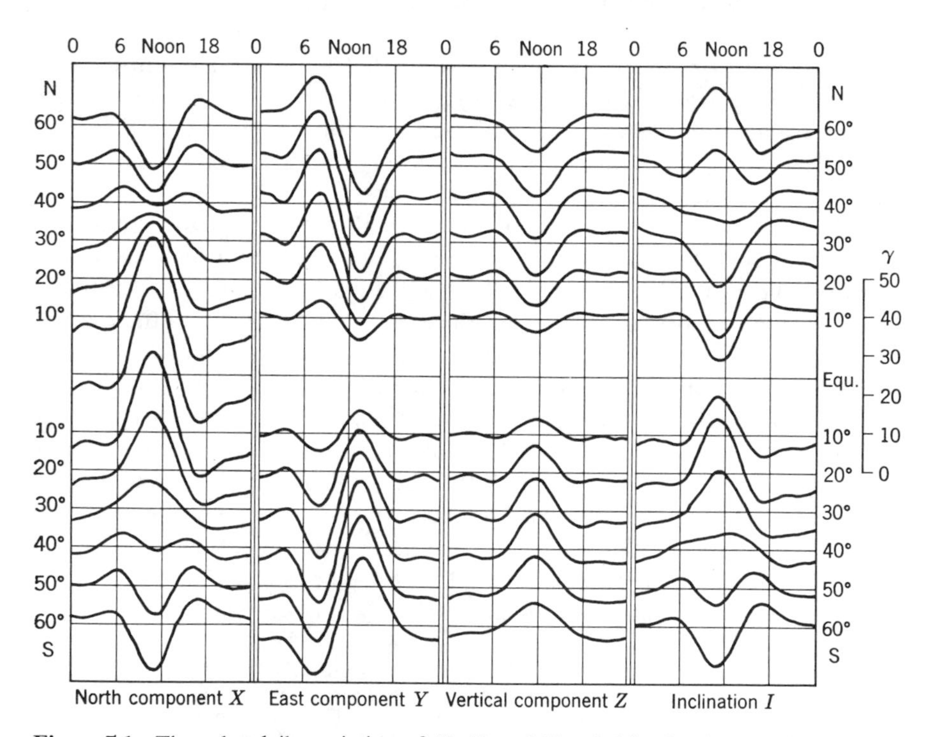

Figure 7.1 The solar daily variation of X, Y, and Z and I (inclination or dip angle) at latitudes 10° apart at the equinoxes, in the sunspot minimum year 1902. (After Chapman and Bartels, *Geomagnetism*; reprinted by permission of Oxford University Press.)

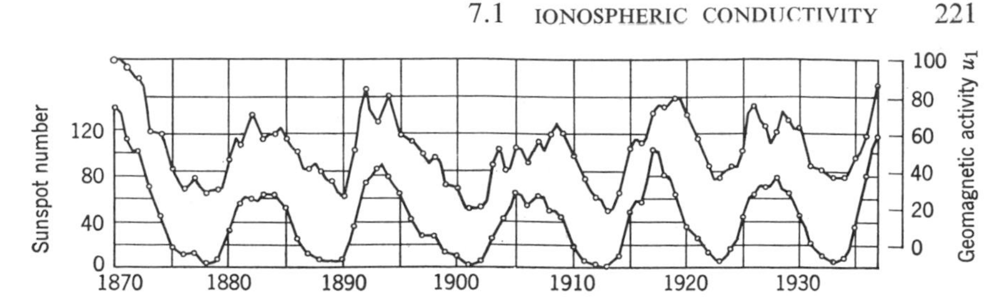

Figure 7.2 The magnetic activity u_1 (upper curve) and the sunspot number (lower curve); annual means 1870–1937. u_1 (not to be confused with K) is related to the monthly average diurnal variations U by $u_1 = U/\cos\lambda$ where λ is the angle between the magnetic axis and **H**. (After Chapman and Bartels, *Geomagnetism*; reprinted by permission of Oxford University Press.)

(which varies diurnally), as well as upon tidal winds, one should not expect the currents and tides to be necessarily characterized by the same modes. In any case, the tidal winds cause atmospheric motion relative to the geomagnetic field lines. Due to viscosity and the tidal motions of the air, the electrons and ions are dragged across the field lines; electric fields result due to dynamo action. In every case the motion obeys Newton's laws which provide us with ready means for attacking the problem. One can express the current density **j** as a function of the electric field **E** by use of a suitably generalized form of Ohm's law (see Section 2.1.1)

$$\mathbf{j} = \underset{\sim}{\boldsymbol{\sigma}} \cdot \mathbf{E} \tag{2.7}$$

where $\underset{\sim}{\boldsymbol{\sigma}}$ is the *tensor* conductivity to be discussed in the following section. It is a tensor because of the anisotropy introduced by the earth's magnetic field.

Application of the theory of ionospheric electric currents to other planets is mainly idle speculation at the present time. Mars and Venus apparently do not possess permanent magnetic fields, and Jupiter, which does have a strong magnetic field, probably experiences only weak tidal motions at most. However, it has been suggested that the ionospheres of the former two planets interact with the solar wind in such a way that electric currents flow in the boundary between the two media. We shall discuss this phenomenon in the final section.

7.1 Ionospheric Conductivity

Having introduced the fundamental equation (2.8) governing the flow of ionospheric currents, we now investigate the properties of the conductivity, i.e., its tensor form, the variation of each of the tensor elements with altitude, etc. This can be done by studying the motion of the ions and electrons with the aid of Newton's laws as justified by the appropriate Boltzmann transport equation (Eq. (2.62)). Knowledge of electron and ion density profiles will then permit us to calculate the various conductivity terms, setting the stage for a brief description of the dynamo theory.

7.1.1 The Motion of Charged Particles in a Magneto-Plasma

As we pointed out above, only the exact nature of the (electric) forces involved distinguishes this problem from any other in classical physics. These forces are, of course, those produced by electric fields and can be expressed as

$$\mathbf{F} = e(\mathbf{E}' + \mathbf{v} \times \mathbf{B}) \tag{7.1}$$

where $e(\mathbf{v} \times \mathbf{B})$ is the Lorentz force (see Section 2.1.1) arising from motion of the particle in an external magnetic field $\mathbf{B}$, and $\mathbf{E}'$ represents all the other electric fields which are present. The latter are, for the most part, derivable as the spatial gradient of a scalar potential field and are thus electrostatic in nature. In addition to the electric forces which tend to accelerate the particles, there are resistive forces—$\mathbf{F}_r$, arising mainly from charged-neutral particle collisions in the regions of interest. The complete equation of motion for a single particle is thus

$$\frac{d\mathbf{v}}{dt} = e(E' + \mathbf{v} \times \mathbf{B}) - \mathbf{F}_r \tag{7.2}$$

At the altitude range in which we are interested, the motion of the ions and electrons is steady because $\mathbf{F}_r$ just balances the electric forces. Hence

$$\mathbf{F}_r = e(\mathbf{E}' + \mathbf{v} \times \mathbf{B}) \tag{7.3}$$

The viscous force $\mathbf{F}_r$ is usually referred to as a stochastic force because it is statistical in nature; it is an average taken over the velocity distribution of the particles. As we saw in Chapter 2, one can express a stochastic force exerted on one type of particle by another as a function of the momentum transfer cross section Q_m for collisions between the two types

$$F_r = m_a \nu_m v \tag{7.4}$$

$$\nu_m = \overline{Q_m v} N_b \tag{2.149}$$

where v is the speed of the particles of type a, N_b is the number density of particles of type b, and m_a is the mass of a type a particle. In carrying out the averaging indicated in Eq. (2.149), the velocity distribution of the particles must be employed in the computation

$$N\overline{Q_m v} = \int f(\mathbf{v}) Q_m(v) v \, d\mathbf{v} \tag{7.5}$$

In general, the result ν will also be velocity-dependent.

The form of the distribution which should be employed in a situation of this type has been the subject of considerable investigation for many years. The procedure is to solve the Boltzmann transport equation (see Chapter 2) with appropriate simplifying assumptions. The form which is derived is dependent upon the collision integral used and the electric field strength. In any case, it will not be Maxwellian. If one then performs the computations indicated above, it is evident that the equation of motion of the type a particles (e.g., electrons or ions) can get rather complicated.

Fortunately, it isn't necessary to go this route since we only need to deal with the average speed $\mathbf{v}$ of an assembly of particles and the average rate of momentum loss (the stochastic force again). The latter can simply be written as $m\nu\mathbf{v}$ where m is the mass of a particle and ν is an average collision frequency for momentum transfer (see Chapter 2). With this simplification, Eq. (7.3) can be written as

$$m\nu\mathbf{v} = e(\mathbf{E}' + \mathbf{v} \times \mathbf{B}) \tag{7.6}$$

neglecting gravitational forces and gas pressure which are important only in the F-region. This, of course, accounts only for the average motion of a single charged particle. In order to obtain an expression which accounts for the motions of all the charged particles involved, it is necessary to sum Eq. (7.6) over the number densities of all the species

$$\sum_k m_k\nu_k N_k = \sum_k N_k e_k(\mathbf{E}' + \mathbf{v}_k \times \mathbf{B}) \tag{7.7}$$

where N_k is the number density and e_k is the charge of the kth species. Specializing to the case where the charged particles consist of electrons and one species of positive ion, we obtain

$$m_i\nu_i N_i\mathbf{v}_i + m_e\nu_e N_e\mathbf{v}_e = e(N_i - N_e)\mathbf{E} + e(N_i\mathbf{v}_i - N_e\mathbf{v}_e) \times \mathbf{B} \tag{7.8}$$

7.1.2 *The Conductivity Tensor for a Magneto-Plasma*

In order to obtain an expression for the ionospheric conductivity, it is necessary first to return to Eq. (7.6) and solve for the velocity of a single charged particle in terms of the parameters e, m, ν, $\mathbf{B}$, and $\mathbf{E}'$. This task is greatly facilitated by choosing $\mathbf{B}$ to point along one of the Cartesian coordinate axes (the z axis); this approach is perfectly general. Then Eq. (7.6) becomes in component form

$$\begin{aligned} m\nu v_x &= eE'_x + ev_y B_z \\ m\nu v_y &= eE'_y - ev_x B_z \\ m\nu v_z &= eE'_z \end{aligned} \tag{7.9}$$

or, solving for v_x, v_y, and v_z

$$\begin{aligned}
v_x &= \frac{(e/m)\nu}{\nu^2 + \omega_H^2} E'_x + \frac{(e/m)\omega_H}{\nu^2 + \omega_H^2} E'_y \\
v_y &= -\frac{(e/m)\omega_H}{\nu^2 + \omega_H^2} E'_x + \frac{(e/m)\nu}{\nu^2 + \omega_H^2} E'_y \\
v_z &= \frac{e}{m\nu} E'_z
\end{aligned} \tag{7.10}$$

where $\omega_H = eB/m$ is called the cyclotron or gyro frequency; it is the angular frequency at which the charged particle revolves about a magnetic line of force.

In order to arrive at an expression for the conductivity of the magneto-plasma, we must relate the velocity of the particles to the current density $\mathbf{j}$:

$$\mathbf{j} = \sum_k N_k e_k \mathbf{v}_k \tag{7.11}$$

Equation (7.10) is now employed to relate $\mathbf{j}$ to the electric field:

$$\begin{aligned}
j_x &= \sum_k \frac{\omega_k^2}{\nu_k^2 + \omega_{kH}^2} (\nu_k E'_x + \omega_{kH} E'_y) \\
j_y &= \sum_k \frac{\omega_k^2}{\nu_k^2 + \omega_{kH}^2} (-\omega_{kH} E'_x + \nu_k E'_y) \\
j_z &= \sum_k \frac{\omega_k^2}{\nu_k} E'_z
\end{aligned} \tag{7.12}$$

where $\omega_k = N_k e^2/m\epsilon_0$ is the plasma frequency (see Section 2.1.3 for further discussion) and ϵ_0 is the permittivity of free space. These three linear equations are of the form given by Eq. (2.7) if $\mathbf{j}$ and $\mathbf{E}$ are represented by column matrices and $\underset{\sim}{\boldsymbol{\sigma}}$ by a square 3×3 matrix:

$$\underset{\sim}{\boldsymbol{\sigma}} = \begin{pmatrix} \sigma_1 & -\sigma_2 & 0 \\ \sigma_2 & \sigma_1 & 0 \\ 0 & 0 & \sigma_0 \end{pmatrix} \tag{7.13}$$

where for a two-component plasma (electrons and one species of positive ion)

$$\sigma_1 = \frac{\epsilon_0 \omega_e^2 \nu_e}{\nu_e^2 + \omega_{eH}^2} + \frac{\epsilon_0 \omega_+^2 \nu_+}{\nu_+^2 + \omega_{+H}^2} \tag{7.14a}$$

is called the "Pedersen conductivity";

$$\sigma_2 = -\frac{\epsilon_0\omega_e^2\omega_{eH}}{\nu_e^2 + \omega_{eH}^2} + \frac{\epsilon_0\omega_+^2\omega_{+H}}{\nu_+^2 + \omega_{+H}^2} \tag{7.14b}$$

is called the "Hall conductivity"; and

$$\sigma_0 = \frac{\epsilon_0\omega_e^2}{\nu_e} + \frac{\epsilon_0\omega_+^2}{\nu_+} \tag{7.14c}$$

is the longitudinal conductivity. Because the magnetic field deflects positive ions and electrons oppositely, the two contributions to σ_2 tend to cancel. Hence the minus sign in the first term on the right hand side of Eq. (7.14b).

It is evident from the foregoing that the Cartesian coordinate system in some ways does not yield the best representation of the conductivity tensor because it is not "diagonal." The so-called spherical representation[4] which employs v_z and linear combinations of v_x and v_y does diagonalize $\underset{\sim}{\boldsymbol{\sigma}}$ and yields an interesting picture of current conduction in a magneto-plasma. Further investigation of this aspect is left as an exercise for the reader.

Before going on to discuss the application of the foregoing to the ionosphere it is instructive to investigate the nature of the Hall and Pedersen currents. Figure 7.3 shows a current **j** flowing in a magneto-plasma under

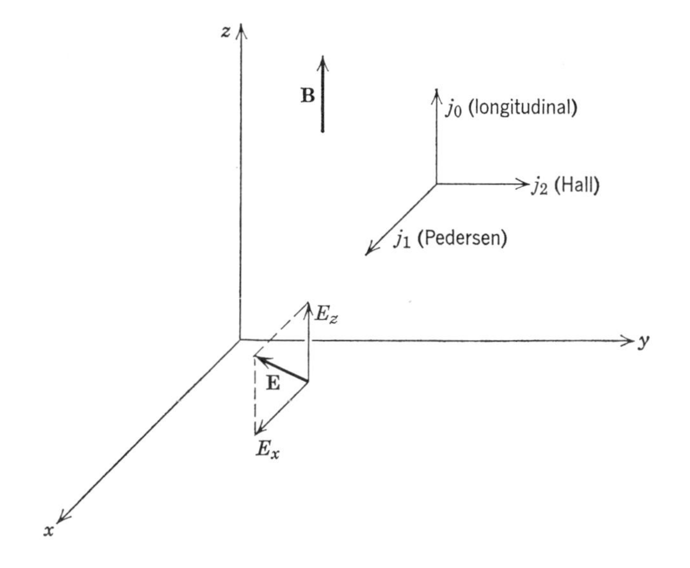

Figure 7.3 Electric fields and electric currents in the E-region. The magnetic field **B** is directed along the z-axis.

the influence of an electric field **E** which lies in the xz plane. The longitudinal component j_0 of the current density flows in the direction of the magnetic field **B** and is produced by the z-component of the electric field. It is the total current density which would flow if the magnetic field were removed. The Pedersen current j_1, which is in the x-direction in our diagram, is due to forced "diffusion" of the ions and electrons across the field lines in the direction of the x-component of **E**. If the magnetic field were removed, this component of the current would have the same form as σ_0; this statement should be obvious from Eq. (7.14a). The magnetic field, as well as the viscous force due to collisions, act as the damping mechanisms. The third component of **j**, the Hall current, is perpendicular to both **E** and **B** and thus is proportional to **E** × **B**.

The actual application of the foregoing to the ionosphere is straightforward with nothing new to be added. We must only employ the actual conductivities for the charged particle components and introduce the appropriate boundary conditions. This is done in the following section.

7.1.3 Electrical Conductivity of the Middle Ionosphere

It is convenient at this point to alter slightly the form of the conductivity tensor by rotating the x and z axes about the y axis which has already been chosen to point in the (magnetic) easterly direction. We do this such that the x axis assumes a (magnetic) southerly direction and the z axis is vertical. Then, in terms of the magnetic dip angle I, σ assumes the form

$$\underset{\sim}{\boldsymbol{\sigma}} = \begin{pmatrix} \sigma_1 \sin^2 I + \sigma_0 \cos^2 I & \sigma_2 \sin I & (\sigma_0 - \sigma_1) \sin I \cos I \\ -\sigma_2 \sin I & \sigma_1 & \sigma_2 \cos I \\ (\sigma_0 - \sigma_1) \sin I \cos I & -\sigma_2 \cos I & \sigma_1 \cos^2 I + \sigma_0 \sin^2 I \end{pmatrix} \tag{7.15}$$

We now introduce the boundary condition that under equilibrium conditions no current flows in the vertical direction ($j_z = 0$) because of the buildup of "polarization" charges at the top and bottom of the conducting layer. This modifies the electric field **E**′ to yield no vertical current component. This condition is completely justified in the E-region, but current can flow vertically at F-region heights.[3]

Using the elements in the bottom row of the conductivity tensor (7.15), the vertical electric field E_z can be related to E_x and E_y and eliminated from the equation for **j**.

$$E_z = \frac{(\sigma_1 - \sigma_0) \sin I \cos I E_x + \sigma_2 \cos I E_y}{\sigma_1 \cos^2 I + \sigma_0 \sin^2 I} \tag{7.16}$$

The conductivity matrix is thus reduced to the 2 × 2 form

$$\underset{\sim}{\boldsymbol{\sigma}} = \begin{pmatrix} \bar{\sigma}_{xx} & \bar{\sigma}_{xy} \\ \bar{\sigma}_{yx} & \bar{\sigma}_{yy} \end{pmatrix} \tag{7.15a}$$

where

$$\begin{aligned} \bar{\sigma}_{xx} &= \frac{\sigma_0\sigma_1}{\sigma_1 \cos^2 I + \sigma_0 \sin^2 I} \approx \frac{\sigma_1}{\sin^2 I} \\ \bar{\sigma}_{xy} &= -\bar{\sigma}_{yx} = \frac{\sigma_0\sigma_2 \sin I}{\sigma_1 \cos^2 I + \sigma_0 \sin^2 I} \approx \frac{\sigma_2}{\sin I} \\ \bar{\sigma}_{yy} &= \frac{\sigma_1\sigma_0 \sin^2 I + (\sigma_1^2 + \sigma_2^2) \cos^2 I}{\sigma_1 \cos^2 I + \sigma_0 \sin^2 I} \approx \sigma_1 \end{aligned} \tag{7.17}$$

except at very low magnetic latitudes. At the magnetic equator $I = 0$ and the elements of $\boldsymbol{\sigma}$ simplify to

$$\begin{aligned} \bar{\sigma}_{xx} &= \sigma_0 \\ \bar{\sigma}_{xy} &= 0 \\ \bar{\sigma}_{yy} &= \sigma_1 + \frac{\sigma_2^2}{\sigma_1} = \sigma_3 \end{aligned} \tag{7.17a}$$

of which the last, σ_3, is sometimes called the Cowling conductivity.

As we have already seen, the longitudinal, Pedersen, and Hall conductivities are rather complicated functions of the electron and ion number densities, and the collision frequencies, as well as the ion mass and magnetic field strength. Because of the variability of electron and ion number density particularly, and to a lesser extent because of the collision frequency, the conductivities vary substantially throughout the day and during the solar cycle. Figure 7.4 shows some representative electron and ion conductivities for an exponential atmosphere during daytime hours. Three features stand out clearly: the large value of the longitudinal conductivity σ_0 above 105 km, the dominance of the electron conductivity at low altitudes, and the extrema in σ_1 at $\nu_+ = \omega_{+H}$, $\nu_e = \omega_{eH}$, and $\nu_e\nu_+ = \omega_{eH}\omega_{+H}$. The demonstration of these relationships is left as an exercise for the reader. The large value of σ_0 above ~105 km has very important ramifications which will be briefly discussed in Section 7.1.4.

7.1.4 Ionospheric Conductivity at Very High Altitudes

It is apparent from Figure 7.3 that the only part of the conductivity tensor which is important at very high altitudes is the longitudinal conductivity σ_0.

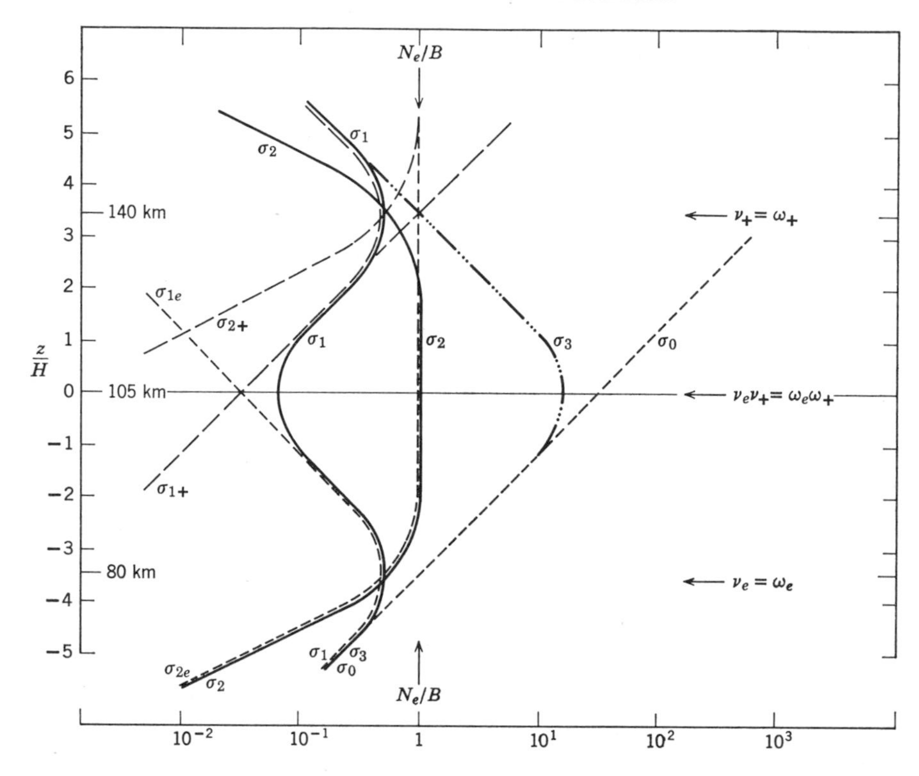

Figure 7.4 Conductivities per ion pair ($\sigma_0, \sigma_1, \sigma_2, \sigma_3$) plotted on a logarithmic scale relative to the value $N_e/B = \epsilon_0\omega_e^2/\omega_H$ as a function of reduced height z/H for an idealized isothermal model atmosphere. It is assumed that $\nu \propto e^{-z/H}$ and ω_H is independent of z for both positive and negative ions and electrons; also that $\nu_+\omega_e/N_e\omega_+ \approx 1000$(independent of z), this being a reasonable value for the actual ionosphere. The zero of z is the level where $\nu_+\nu_e = \omega_+\omega_e$ and the important levels where $\nu_+ = \omega_+$ and $\nu_e = \omega_e$ are therefore situated at $z = \pm\frac{1}{2}\ln(1000)$; also shown are the approximate altitudes at which they occur in the actual ionosphere. Subscripts + and e refer to ions and electrons. (After Rishbeth and Garriott, *Introduction to Ionospheric Physics*, reprinted by permission of Academic Press, Inc.)

Thus, charged particles are to a first approximation confined to individual lines of force along which they move freely because of the large conductivity. In fact, at sufficiently high altitudes in the absence of external electric fields the lines of force are lines of equipotential so that charge separation electric fields capable of redistributing the charges are nonexistent. The electrons and ions are in a very real sense "frozen" to the lines of force. The result is a strong coupling between the plasma and the magnetic field which may, in fact, support a mode of wave propagation, the hydromagnetic mode, which is to be discussed further in Chapter 11. Actually, one may encounter motions of charged particles across field lines if electrostatic fields are present. The resulting drifts are of considerable interest insofar as ionospheric structure is concerned and will be discussed at greater length later.

7.2 The Dynamo Theory

As we very briefly mentioned in the introduction to this chapter, it is the interaction of tidal winds with the geomagnetic field and the charged particles in the upper atmosphere which produces the current flows reflected in the diurnal magnetic variations. Because the mechanism which induces the current flow is very similar to the action of an electric generator, the name "dynamo theory" was given to the theory of diurnal geomagnetic fluctuations outlined here. A fairly accurate picture of the quiescent ionospheric current patterns was in fact deduced from magnetic variations many years ago.[1] A more recent determination[5] is portrayed in Figure 7.5 which shows the average Sq current system computed from magnetic observations taken during the IGY (solid line—counterclockwise flow; broken line—clockwise flow).

In order to compute the magnetic fluctuations produced by tidal motions, we need, as a function of altitude, estimates of the electric fields $\mathbf{E}'$ which produce the currents. In practice, such information is not available; only some mean value in the height range 90 to 130 km where the currents flow can be estimated. Therefore, we obtain the linear current density $\mathbf{J}$ which determines the magnetic field changes ΔH by integrating the conductivity σ over the appropriate altitude range

$$\Sigma_{ij} = \int \sigma_{ij} \, dz \tag{7.18}$$

and

$$\mathbf{J} = \underset{\sim}{\mathbf{\Sigma}} \cdot \mathbf{E} \tag{7.19}$$

where $\mathbf{E}$ is the mean electric field

$$\mathbf{E} = \mathbf{E}_s + \mathbf{v} \times \mathbf{B}; \;\; \mathbf{E}_s = -\nabla\phi \tag{7.20}$$

arising from a scalar potential ϕ and an induced field $\mathbf{v} \times \mathbf{B}$. The former is established by the polarization charges mentioned previously and constrains the current flow to a horizontal direction only. The magnetic variations $\Delta\mathbf{H}$ produced at ground level by ionospheric currents can easily be computed with the aid of Ampere's law

$$\text{curl } \mathbf{H} = \mathbf{j} \tag{7.21}$$

Direct integration yields

$$\begin{aligned} \Delta H_x &= \alpha J_y \\ \Delta H_y &= \alpha J_x \end{aligned} \tag{7.22}$$

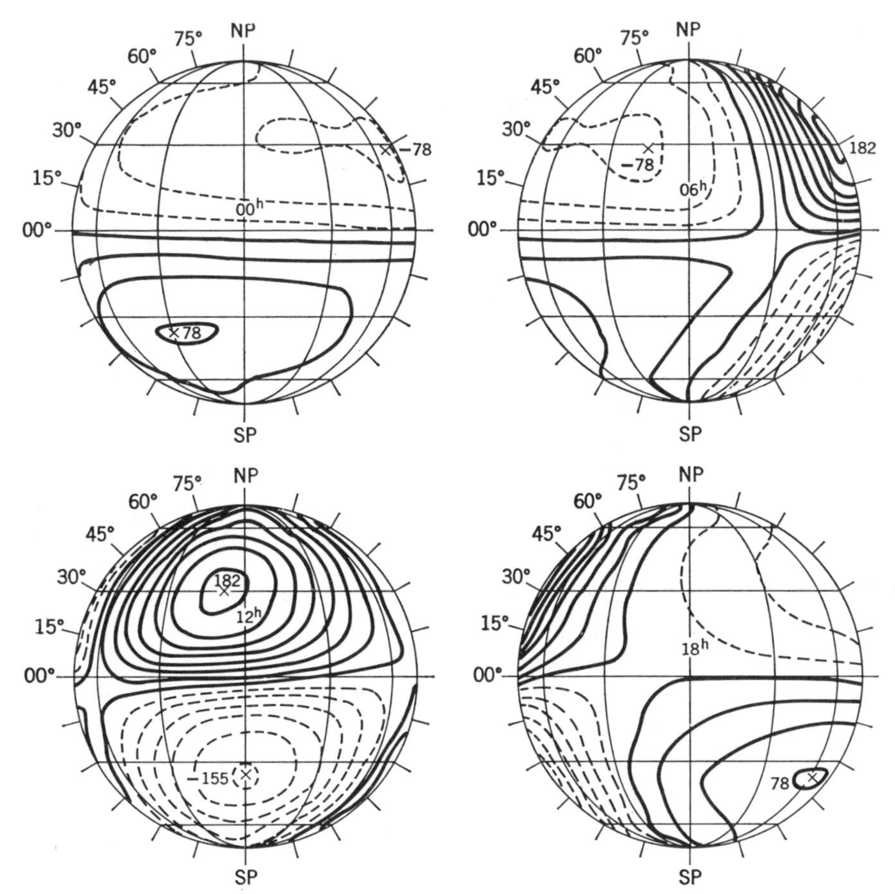

Figure 7.5 Average Sq current system during the IGY viewed from the magnetic equatorial plane at the 00, 06, 12, and 18^h meridians; numbers near the crosses indicate vortex currents in units of 10^3 amperes. [After S. Matsushita, *J. Geophys. Res.* **70**, 4395 (1965); reprinted by permission of the American Geophysical Union.]

where α is $\frac{1}{2}$ if all magnetic effects are due to current flow in the ionosphere and $\sim\frac{3}{4}$ when ground currents are included. If the ground were a perfect conductor, ground and ionospheric currents would contribute equally and α would be unity.

Profiles of the height-integrated conductivities Σ_{xx}, Σ_{yy}, and Σ_{xy} are shown in Figure 7.6 as functions of latitude for a typical midday ionosphere. It is evident that Σ_{xy} dominates over most of the earth but that Σ_{xx} and Σ_{yy} reach very high values at low magnetic latitudes. Σ_{xy}, on the other hand, is much smaller in this zone, vanishing (see Eq. (7.17)) at the equator.

Because of the large conductivity, we expect to find strong east-west current sheets just to the north and south of the magnetic equator. This "equatorial electrojet" has been a well-known phenomenon for many years and is partially explained, at least qualitatively, by the theory of ionspheric conductivity presented in this chapter.

It is a straightforward matter to compute the magnetic variations by means of Eqs. (7.19) and (7.20) together with the assumptions that the velocity field $\mathbf{v}$ and current $\mathbf{J}$ are derivable from potential functions

$$\mathbf{v} = -\boldsymbol{\nabla}\psi \tag{7.23a}$$

$$\mathbf{J} = -\boldsymbol{\nabla}\chi \tag{7.23b}$$

Ohm's law can thus be written in terms of ψ, χ, and ϕ as

$$\boldsymbol{\nabla}\chi = \boldsymbol{\Sigma} \cdot (\boldsymbol{\nabla}\phi + \boldsymbol{\nabla}\psi \times B) \tag{7.24}$$

In spherical coordinates where λ is the magnetic longitude, and θ the magnetic

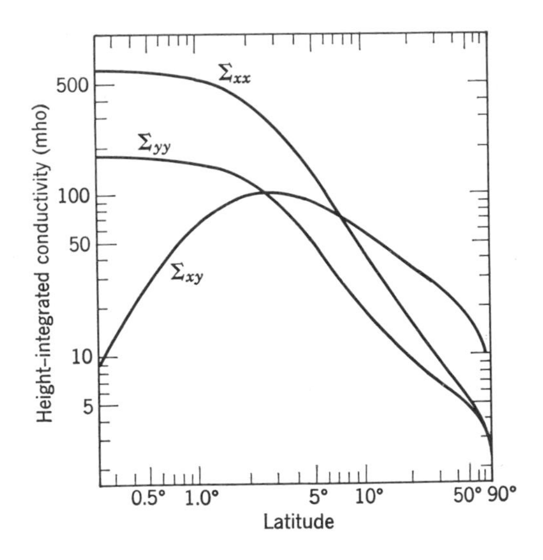

Figure 7.6 The height-integrated conductivities Σ_{xx}, Σ_{yy}, and Σ_{xy} as functions of magnetic latitude. [After J. A. Fejer, *Revs. Geophys.* **2**, 275 (1964); reprinted by permission of the American Geophysical Union.]

colatitude, Eq. (7.24) can be written explicitly as two coupled equations

$$\frac{1}{\sin\theta}\frac{\partial\chi}{\partial\lambda} = \Sigma_{xx}\left(-\frac{1}{\sin\theta}\frac{\partial\psi}{\partial\lambda}B_z - \frac{\partial\phi}{\partial\theta}\right) + \Sigma_{xy}\left(\frac{\partial\psi}{\partial\theta}B_z - \frac{1}{\sin\theta}\frac{\partial\psi}{\partial\lambda}\right)$$
$$-\frac{\partial\chi}{\partial\theta} = -\Sigma_{xy}\left(-\frac{1}{\sin\theta}\frac{\partial\psi}{\partial\lambda}B_z - \frac{\partial\phi}{\partial\theta}\right) + \Sigma_{yy}\left(\frac{\partial\psi}{\partial\theta}B_z - \frac{1}{\sin\theta}\frac{\partial\psi}{\partial\lambda}\right) \quad (7.25)$$

in which **B** is considered to have a north-south (i.e., "z") component only. In order to solve for χ, which determines the current, one eliminates ϕ and substitutes the appropriate functional form of B_z (usually taken as $\propto \cos\theta$). The result is an inhomogeneous differential equation for χ in terms of the velocity potential function ψ. Derivation of the equation with the simplifying assumption that Σ_{xx}, Σ_{xy}, and Σ_{yy} are constants is left as an exercise.

Actually the wind velocity is not irrotational. For realistic results, it must contain a rotational component which can be expressed in terms of a stream function π. The entire velocity field is then represented by

$$\mathbf{v} = -\nabla\psi - \mathbf{e}_r \times \nabla\pi \quad (7.26)$$

where $\mathbf{e}_r$ is a unit vector pointing radially outward from the earth. Calculations which include π have in fact been carried out by Maeda and Kato.[6] Some wind patterns for irrotational velocities and for velocities given by Eq. (7.26) are shown as a function of latitude in Figure 7.7. The principal component is the diurnal component which is characterized by a speed of about 30 m sec^{-1} and a weaker semidiurnal component of about 10 m sec^{-1}.

In the auroral zones and just to the north and south of the magnetic dip equator, especially strong currents called electrojets flow. The auroral zone electrojets are well-correlated with geomagnetic activity and are undoubtedly related to magnetospheric disturbances; for that reason we shall not discuss them further. The equatorial electrojet probably reflects enhanced E-layer conductivity in the regions where the dip angle is very small. That the conductivity in the east-west direction ($\bar{\sigma}_{yy}$) is increased can be seen from Eq. (7.17a), which states that $\bar{\sigma}_{yy}$ approaches the Cowling conductivity in magnitude as I approaches zero. Conductivity $\bar{\sigma}_{yy}$ is further enhanced by a factor of 2 to 5 because the magnetic lines of force are horizontal.[7] The equatorial electrojet is characterized by irregularities which are thought to result from a "two-stream plasma instability." Small wave-like disturbances can, for a certain range of wave parameters, grow to quite large amplitudes owing to the large relative velocity of ion and electron streams (see Section 11.6). According to Farley's theory,[8] the irregularities are observed as equatorial "sporadic-E" (see Section 10.4).

The dynamo theory as it exists today provides us with only a qualitative explanation of the Sq variations. The present inadequacies in the theory are

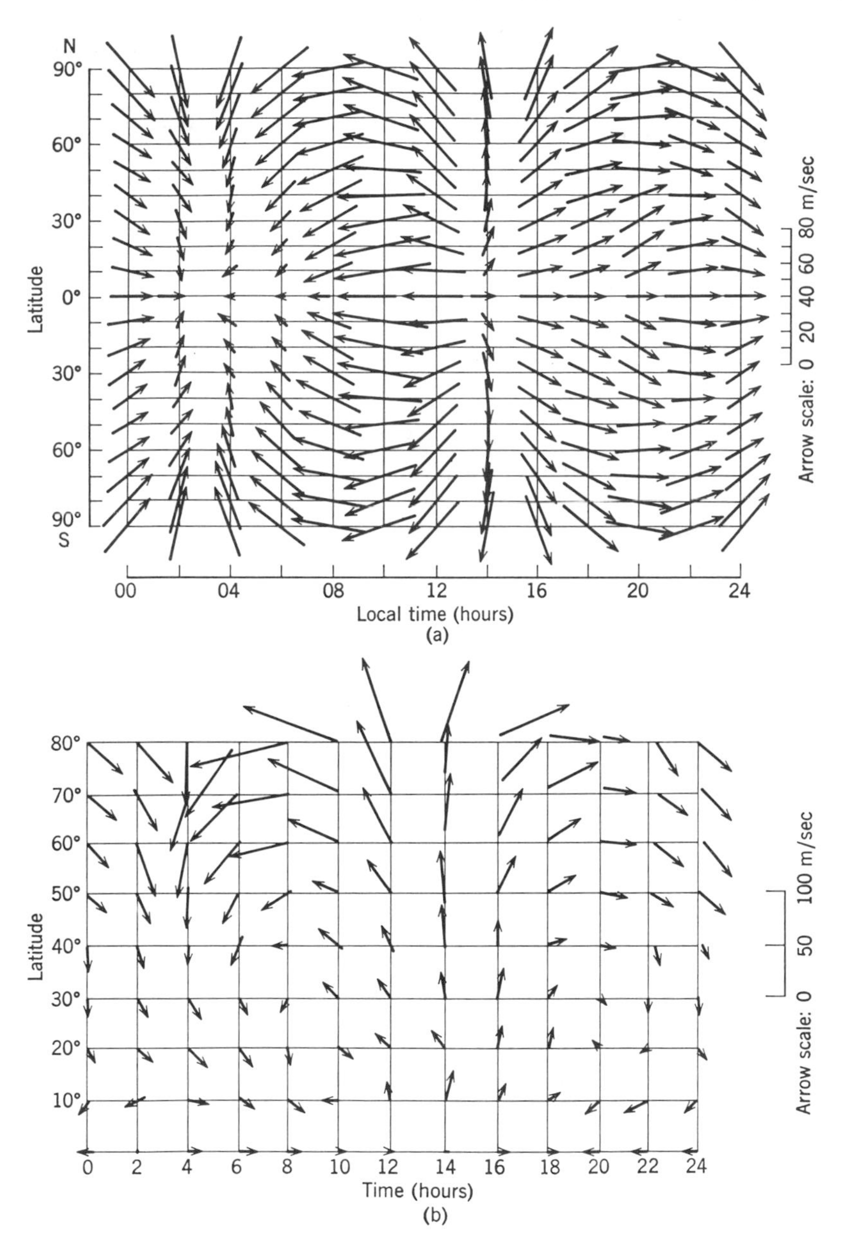

Figure 7.7 Wind systems deduced from the Sq variations (a) curl $\mathbf{V} = 0$, (b) both rotational and irrotational systems are considered. [After Maeda and Kato, *Space Science Revs.*, **5**, 57 (1966); reprinted by permission of Reidel Publishing Co., Amsterdam.]

probably a result of neglecting motions of the neutral atmosphere caused by the current systems, and distortion of the latter by interaction of the polarization electric field with ring currents in the magnetosphere. The latter are also closely associated with magnetic disturbances which are discussed briefly in the following section.

7.3 Magnetic Disturbances

It is not our purpose in this section to present a comprehensive treatment of magnetic disturbances. Because of the very significant role played by magnetospheric currents, such as the so-called "ring-current," this topic is not properly included in a study of the upper atmosphere. Magnetospheric phenomena are discussed in greater detail in a review article by Akasofu and Chapman[9] which the interested reader is encouraged to consult. Accordingly, we shall give here only a very brief summary of the characteristics of such disturbances.

The disturbance variation D is defined as the residue left after the S_q variations are averaged over five international magnetically quiet days and subtracted from the observed variations. These variations have two components of which one, the daily disturbance variation D_s, has a period of one solar day. The other component, the magnetic storm-time variation, D_{st}, is obtained from D by averaging it over 24 hours of local time for a given storm time (this removes the D_s component). One then computes D_s as a function of local time by subtracting D_{st} from D at a given storm time. It has been found that D_s predominates at high latitudes, especially near the auroral electrojet, while D_{st} is usually dominant at low latitudes. Investigations of the current systems responsible for the D_s variation has suggested that an enhanced current flow in the *E*-layer is undoubtedly associated with it, but that magnetospheric motions also play a very important role. The D_{st} variations are probably due entirely to magnetospheric effects such as the ring current mentioned earlier. However, the picture is more complicated than we have indicated here. In particular, the D_s component is also a function of storm time and is not always well-correlated with the auroral zone electrojet. The subject of magnetic storms is a very interesting one and an extensive literature dealing with it has developed in recent years. However, in accordance with the spirit of our earlier remarks, we shall not pursue it further.

7.4 *F*-Region Drifts

In Section 7.1.4 we mentioned that very large values of σ_0 compared to σ_1 and σ_2 signify a "freezing" of the charged particles to the lines of force if electrostatic fields are absent. If such fields are present, however, there may be very significant drifts of electrons and ions across the magnetic field. The charge redistribution brought about by such drifts may be very important in the determination of the structure of the upper ionosphere. It is true in general that if electrostatic fields exist in the E-region, they are also present at F_2 region heights because of the large conductivity along lines of force.

The appropriate starting point for our brief discussion of drifts is the continuity equations for the charged particles present in the ionosphere. If we call the number density of positive charges, for example, N_+ and the drift velocity of positive charges V_+, the continuity equation is written

$$\frac{\partial N_+}{\partial t} = \operatorname{div}(N_+\mathbf{V}_+) \tag{7.27}$$

if sources and sinks are not present.

The drift velocity $\mathbf{V}_+$ can be written as the sum of two terms, one ($\mathbf{W}_D$) due to ambipolar diffusion of the charges along the magnetic field lines, and the other, $\mathbf{U}_+$, due to convective effects of electric fields. We have already considered the diffusion of neutral particles in Chapter 3, and shall further discuss ambipolar diffusion in the F_2 region in Chapter 9. We merely remark here that the diffusion velocity is dependent upon the gravitational and pressure terms which were neglected in Eq. (7.6). We shall confine our remarks here to electrically induced drifts.

The total electric field $\mathbf{E}$ in the F_2 region is very nearly zero because of the very large longitudinal conductivity σ_0, and Eq. (7.20) becomes

$$\mathbf{E}_s \approx -\mathbf{V} \times \mathbf{B} \tag{7.28}$$

Since $\mathbf{V}$ does not include the diffusion velocity, we must have $\mathbf{V} = \mathbf{U}$ and $\mathbf{V} \perp \mathbf{B}$. Then Eq. (7.28) can easily be transformed to

$$U = \frac{\mathbf{E}_s \times \mathbf{B}}{B^2} \tag{7.29}$$

At the present time we know very little about the drift velocity $\mathbf{U}$ in the upper ionosphere, but there is much indirect evidence that it plays a very important role in the temporal development of the F_2 region. We shall consider it further in Chapter 9.

7.5 Currents in the Interface between the Solar Wind and Ionosphere for a Planet with no Magnetic Field

All attempts to detect a magnetic field on Mars and Venus have been unsuccessful; that is, any permanent field which those planets may have is less than about 10 gamma. Thus, induced electric currents at E-layer heights probably do not exist. The currents which one may expect are associated with the interaction between the solar wind and the upper ionospheres. The former carries magnetic field lines which are frozen into and move with the bulk velocity. When the field lines encounter the planetary plasma, they induce currents which flow down the solar wind-ionosphere interface, and produce a magnetic field which adds to the solar wind field outside the ionosphere and cancels it on the inside.[10] If the bulk velocity of the interplanetary plasma is large enough, the ionosphere is diamagnetic, and the solar wind magnetic field "piles" up around the upstream sector of the ionospheric cavity.

In order to see whether the solar wind and its associated magnetic field really are excluded from the ionosphere of Venus and Mars, let us write down Faraday's induction law

$$\nabla \times \mathbf{E} = -\frac{\partial \mathbf{B}}{\partial t}, \tag{2.3}$$

and invoke Ohm's law

$$\mathbf{E} = \mathbf{j}/\sigma \tag{2.8a}$$

and Ampere's law without the displacement current

$$\nabla \times \mathbf{H} = \mathbf{j} \tag{2.4a}$$

to eliminate the electric field $\mathbf{E}$:

$$\frac{\partial \mathbf{B}}{\partial t} = \frac{1}{\mu_0 \sigma} \nabla^2 \mathbf{B} \tag{7.30}$$

The resulting equation is obviously of the diffusion type. One may obtain an order of magnitude estimate of the "diffusion time constant," t_D, (time required for the magnetic field to penetrate the ionosphere) by using dimensional arguments to "solve" Eq. (7.30):

$$t_D \approx \mu_0 \sigma R^2, \tag{7.31}$$

where R is the planetary radius. The value of t_D so obtained is to be compared with the time required for the solar wind of velocity V to sweep past the planet

$$t \approx R/V \tag{7.32}$$

if $t_D \gg t$, then we can be reasonably sure that the solar wind and its magnetic field are excluded from the ionosphere.

Let us plug in some numbers for Venus:

$\sigma \approx 4 \times 10^{-7}$ mhos m.$^{-1}$ (see Problem 7.11)

$R \approx 6 \times 10^{6}$ meters

$V \approx 3 \times 10^{5}$ meters sec^{-1}

we immediately find $t_D \approx t$ and conclude that the interface current is large enough to screen the ionosphere from the solar wind. We shall discuss evidence for such screening in Chapter 9.

PROBLEMS

7.1. Show how to analyze the daily variation ΔH of the horizontal component of the geomagnetic field at a given point on the surface of the earth into diurnal, semidiurnal, etc. components. How can solar and lunar components be distinguished. Obtain a series expansion for the magnetic variations which relates ΔH at various points on the earth's surface and at various times.

7.2. Show that the maxima in the Pedersen conductivity σ_1 occur where $\nu_+ = \omega_{+H}$ and $\nu_e = \omega_{eH}$ and a minimum at $\nu_+\nu_e = \omega_{+H}\omega_{eH}$ for an isothermal atmosphere. Assume that the ion and electron number densities are constant and that $\nu_+/\nu_e = 1000\ (\omega_{+H}/\omega_{eH})$.

7.3. By means of a suitable linear transformation diagonalize the conductivity tensor in Eq. (7.15). Discuss the significance of the current density and electric field in this representation.

7.4. Apparent electric fields in the upper atmosphere can result from the Coriolis effect. That is, the motion of charged particles can be affected by the earth's rotation. Express this field as the curl of a vector field **w** and obtain an expression for **w**.

7.5. Derive Eq. (7.29) from Eq. (7.28).

7.6. Prove that if the conductivity of a plasma is infinite, magnetic lines of force are "frozen" into it.

7.7. Using data presented in Chapter 8 (e.g., electron number density and collision frequency profiles) construct profiles of σ_0, σ_1, σ_2, and σ_3. Compare with the idealized curves in Figure 7.4.

7.8. Discuss the effect on the local magnetic field of a sudden increase in the electron number density. Assume (a) a sharp pulse of ionization which decays by recombination; by attachment (see Chapter 8); (b) a gradual increase and decrease of ionization, e.g., in the case of a sudden ionospheric disturbance (see Chapter 10).

7.9. Discuss the effect on the current density **j** caused by varying the electric field intensity **E** if the electron collision frequencies are energy-dependent.

7.10. Use Eqs. (7.25) to show that

$$\frac{1}{\sin\theta}\frac{\partial}{\partial\theta}\left(\sin\theta\,\frac{\partial\chi}{\partial\theta}\right) + \frac{1}{\sin^2\theta}\,\frac{\partial^2\chi}{\partial\lambda^2}$$

$$= C\Sigma_0\left[\frac{1}{\sin\theta}\frac{\partial}{\partial\theta}\left(\sin\theta\cos\theta\,\frac{\partial\psi}{\partial\theta}\right) + \cot\theta\csc\theta\,\frac{\partial^2\psi}{\partial\lambda^2}\right]$$

where it is assumed that

$$\Sigma_{xx} = \Sigma_{yy} = \text{constant}$$
$$\Sigma_{xy} \text{ is constant}$$
$$\Sigma_0 = \Sigma_{xx} + \Sigma_{xy}^2/\Sigma_{yy}$$
$$B_z = C\cos\theta$$

7.11. Show that the conductivity of the upper ionosphere of Venus is $\sim 4 \times 10^{-6}$ mhos m^{-1} (*Hint:* See Chapter 9 for the electron and ion number densities in the upper ionosphere of Venus.)

REFERENCES

1. S. Chapman and J. Bartels, *Geomagnetism*, Oxford University Press, Oxford, 1940.
2. For example, S. Matsushita, Geomagnetic Storms and related phenomena, in *Research in Geophysics*, Vol. 1, Edited by H. Odishaw, MIT Press, Cambridge, Mass. (1964); S. Chapman, Solar plasma, Geomagnetism and Aurora, in Geophysics, *The Earth's Environment*, Ed. by DeWitt, Hieblot, and Lebeau, Gordon and Breach, New York (1963); J. H. Piddington, Geomagnetic storms, auroras, and related effects, *Space Sci. Revs.* **3,** 724 (1964).
3. K. Maeda and S. Kato, *Space Sci. Revs.* **5,** 57 (1966).
4. See, for example, any text on the quantum theory of angular momentum such as A. R. Edmonds', *Angular Momentum in Quantum Mechanics*, Princeton University Press, Princeton, N.J., 1957.
5. S. Matsushita, *J. Geophys. Res.* **70,** 4395 (1965).
6. K. Maeda, *J. Geomag. and Geoelec.* **9,** 86 (1957); S. Kato, *J. Geomag. and Geoelec.* **9,** 107 (1957).
7. A. Onwumechilli, Geomagnetic variations in the equatorial zone, in *Physics of Geomagnetic Phenomena*, Ed. by Matsushita and Campbell, Academic Press, New York, 1967.
8. D. T. Farley, *J. Geophys. Res.* **68,** 6083 (1963).
9. S-I. Akasofu and S. Chapman, Geomagnetic storms and auroras, in *Physics of Geomagnetic Phenomena*, Ed. by Matsushita and Campbell, Academic Press, New York, 1967.
10. A. J. Dessler, Ionizing plasma flux in the Martian upper atmosphere, in *Atmospheres of Venus and Mars*, Ed. by Brandt and McElroy, Gordon and Breach, New York, 1968.

GENERAL REFERENCES

K. Maeda and S. Kato, Electrodynamics of the ionosphere, *Space Science Reviews*, **5,** 57 (1966) is an excellent review of the current status of ionospheric electrodynamics.

J. A. Fejer, Atmospheric tides and associated magnetic effects, *Reviews of Geophysics* **2,** 275 (1965) also discusses ionospheric electrodynamics as well as the nature and theories of magnetic disturbances and atmospheric tides.

J. A. Fejer, Motions of ionization, in *Physics of the Earth's Upper Atmosphere*, Ed. by Hines, Paghis, Hartz, and Fejer, Prentice-Hall, Englewood Cliffs, N.J. (1965).

S. K. Mitra, *The Upper Atmosphere*, 2nd Ed., The Asiatic Society, Calcutta, 1952, Chapter 7 and

S. Chapman and J. Bartels, *Geomagnetism*, Oxford University Press, Oxford (1940) are old references which nevertheless contain a wealth of information which is still valid.

S. Matsushita and W. H. Campbell, (Editors), *Physics of Geomagnetic Phenomena*, Academic Press, New York (1967) is a recent monograph on geomagnetic phenomena by a number of outstanding investigators. Chapters III-1 and III-2, which discuss electric currents in the quiet ionosphere are particularly useful.

H. Rishbeth and O. K. Garriott, *Introduction to Ionospheric Physics*, Academic Press, New York, 1969; Chapter 3 is an excellent introduction to the subject of ionospheric electric currents.

8

Structure of the Lower and Middle Ionosphere

Ionizing radiations from the sun and from the galaxy penetrate planetary atmospheres and, as a consequence, produce ions and electrons. The nature of the ions that are produced depends upon the energy of the radiation and the composition of the atmosphere at the altitude at which ionization occurs. Immediately following the production of an ionic species, a number of processes occur that change the species of ions and decrease the numbers of electrons and ions that exist. The total population of ions and electrons depends on the competition between ionization processes, ion rearrangement processes, electron removal processes, ion and electron diffusion, geomagnetic effects, and mass motion. These processes superimposed upon the chemical and physical processes discussed in previous chapters determine the structure of the ionosphere. This sequence and combination of reactions is illustrated schematically for a simple ionosphere in Figure 8.1.

Specific processes and the various methods of treating them will be discussed in two chapters (8 and 9). The principal difference between the chapters lies in the importance of diffusion and geomagnetic effects in the F_2 region, and protonosphere which is the topic of Chapter 9. Because most of the available information concerns the terrestrial atmosphere, we shall generally use it for illustrative purposes; however, the principles that will be covered are fundamental to any application.

The discussion in this chapter will deal generally with quiescent low and middle latitude ionospheres. The effects of perturbations both in electromagnetic and corpuscular solar radiation cause significant ionospheric effects at all latitudes; these will be discussed in Chapter 10.

8.1 Ion-Electron Pair Production

The sources of ionization include both electromagnetic and corpuscular radiation. Because of magnetic shielding effects, corpuscular radiation is

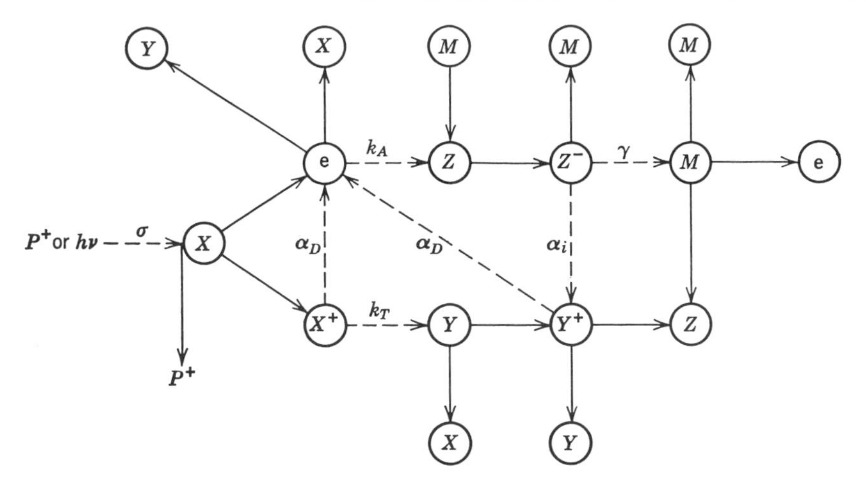

Figure 8.1 Schematic of ionospheric processes.

important only at high latitudes on the earth; on Mars or Venus, where no magnetic fields have been detected, corpuscular radiation should be important at all latitudes. An exception is the very energetic Galactic Cosmic Radiation (GCR) which penetrates the terrestrial magnetic shield to become a relatively important low altitude ionization source (especially at sunspot minima) and is likely to be important in other planetary atmospheres as well.

Electromagnetic radiation is generally confined to the sunlit hemisphere of a planet† and the latitude distribution depends only on zenith angle (which controls the slant length atmospheric attenuation within the atmosphere). The ionizing portion of the electromagnetic spectrum (generally wavelengths shorter than 1216 Å) represents a minute fraction of the total solar flux; for example, the 1216 Å Lα solar radiation flux is on the order of 10^{-2} times the flux emitted per angstrom in the middle of the visible spectrum, and the flux in the X-ray spectrum below 10 Å is on the order of 10^{-4} times the Lα flux. The energy emitted by the sun in the spectral region below 1500 Å is approximately 10^{-5} the energy emitted over the entire spectrum. Nonetheless, this seemingly insignificant energy has profound effects; it is responsible for the ionized state of the upper atmosphere and this provides the means for worldwide radio communication. On the other hand, an increase in flux of less than an order of magnitude in the solar X-ray spectrum can make communications impossible in many frequency bands.

† Important exceptions are UV radiations scattered into the dark hemisphere of a planet by atmospheric constituents such as H, He, and O.

8.1.1 *Photoionization*

Thus, except for very low altitudes and high latitudes, the ionization process in the lower and middle ionosphere is mainly photoionization. For a particular species, and monochromatic radiation, the absorption of solar radiation per atmospheric path increment dl is proportional to the radiation intensity I, species number density n, and the absorption cross section Q_a or

$$dI = Q_a \, nI \, dl = Q_a \, nI \, dz \sec \chi \tag{8.1}$$

where z is the altitude, and χ the zenith angle, or (see Section 3.2.1)

$$I(z) = I_\infty \exp \left(-Q_a \sec \chi \int_z^\infty n \, dz \right) \tag{8.2}$$

where I_∞ is the intensity of radiation incident to the atmosphere, and $I(z)$ is the intensity of radiation that penetrates to z.

If an isothermal exponential atmosphere with a scale height† $H = RT/Mg$ (see Chapter 3) is assumed, the number density has an exponential behavior

$$n = n_0 \exp(-z/H)$$

If we also assume that the absorption of radiation is by photoionization, $Q_a = Q_i$, and that one electron is produced per W eV of absorbed radiation, the ionization rate would be

$$q(z) = \frac{1}{W} \frac{dI}{dz} \cos \chi = \frac{Q_i n}{W} I(z) \tag{8.3a}$$

$$= \frac{Q_i n_0}{W} I_\infty \exp \left[-\frac{z}{H} - Q_i n_0 H \sec \chi \exp \left(-\frac{z}{H} \right) \right] \tag{8.3b}$$

The maximum rate of ionization occurs where $dq/dz = 0$, or

$$z_m = H \ln (H Q_i n_0 \sec \chi) \tag{8.4}$$

and the peak ionization rate is

$$q_m = \frac{I_\infty}{eWH} \cos \chi \tag{8.5}$$

where $e = \exp(1)$ or the base of natural logarithms. Note that q_m occurs at unit slant optical depth.

Equations (8.3), (8.4), and (8.5) describe a Chapman production function and the above treatment is called the Chapman Theory.[18] This treatment is a review of the derivation of a Chapman function which was given in Chapter

† This is equivalent to the form of H $(= kT/mg)$ given in Chapter 3 and often more convenient. R is the universal gas constant and M is the mean molecular weight.

3, but in a different context. The following inherent assumptions must be seriously considered when the Chapman Theory is applied. The equations assume a flat planetary surface which is adequate for small angles of incidence ($\chi < 75°$). They also assume isothermal single species atmospheres, monochromatic radiations, and that all absorbed radiation produces ions. Thus, the Chapman Theory describes a simple system, but can be applied to a real, complex system by describing the real system by a linear combination of simple systems.

In fact, the lower and middle ionosphere is best treated by using measurements of spectral irradiance F_λ ($dI/d\lambda$ integrated over solid angle) and species number densities, and computing q as follows.

$$q(z) = \frac{1}{W} \sum_j n_j(z) \int_\lambda Q_i^j(\lambda) F_{\lambda(\infty)} \exp\left[-\sum_j Q_a^j(\lambda) \int_z^\infty n^j(z)\, dz \sec \chi\right] d\lambda \quad (8.6a)$$

where n_j is the number density of species j and $F_{\lambda(\infty)}$ is the spectral irradiance above the atmosphere. The quantity in brackets is called the optical depth.

The reader is advised to consult the recent literature for values of spectral irradiance and species number densities, although the values used for illustrative purposes in this book (see, for example, Chapter 3) are believed to be reasonably valid.

Photoionization of atmospheric species is caused by solar radiation in the range of wavelengths between 1350 Å and 2 Å. Absorption cross sections in the ultraviolet wavelength region are found in the figures of Chapter 4, and ionization potentials and products are indicated in the energy level diagrams of Chapter 4. Cross sections for the ionization of air by soft X-rays and for the ionization of NO by ultraviolet radiation are shown in Figures 8.2a and 8.2b.

8.1.2 Corpuscular Ionization

With corpuscular radiation,† energy is deposited in the atmosphere as the particle interacts with atmospheric constituents. On the one hand, the problem is simpler than it is for electromagnetic radiation because the range of a particular particle is nearly independent of minor constituents of the gas mixture in which it is being absorbed; thus, the atmosphere can be treated as a single constituent which varies according to the hydrostatic equation and possesses a single set of absorption and ionization characteristics. On the other hand, the problem is much more complex because charged particles are

† Radiation is defined as the process of emitting energy in the form of waves or particles (see *Webster's Third New International Dictionary*). We use the term "corpuscular radiation" to designate energy radiated in the form of particles and "electromagnetic radiation" to designate energy radiated in the form of waves.

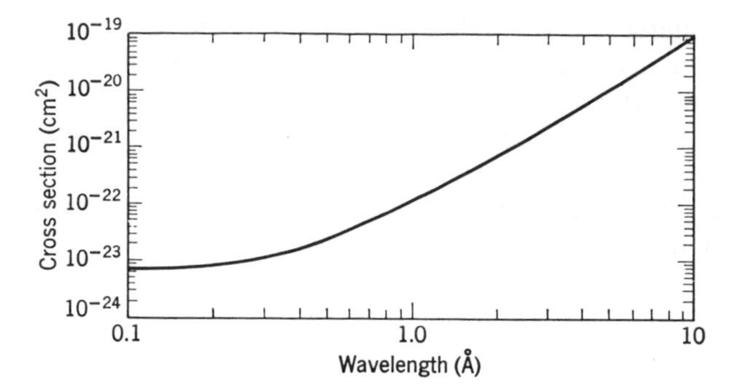

Figure 8.2a Absorption cross section for X-rays in air as a function of wavelength.

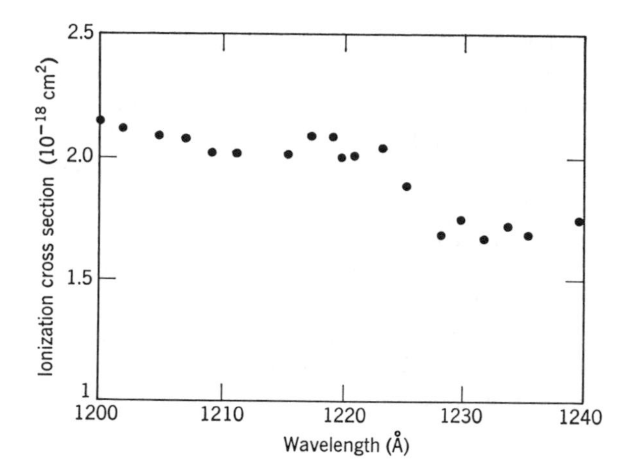

Figure 8.2b Ionization cross section of NO. (After Watanabe, *J. Chem. Phys.* **22,** 1564 (1954); reprinted by permission of The American Institute of Physics.)

influenced by magnetic fields and spiral toward the atmosphere guided by magnetic lines of force; they are incident on the atmosphere over a range of angles that depend on energy spectra and local magnetic fields as well as the original angular distribution. In addition, the energy spectra are dependent on the magnetic field and altitude inasmuch as particles with different energies are mirrored at different altitudes. Bremsstrahlung produced in the absorption of electrons may penetrate to lower altitudes than the electrons and constitute an additional source of photoionization. Thus, the problem becomes rather complex and defies solution via the neat analytical approach of the Chapman Theory.

The ionization rate for charged particles at altitude h is

$$q(z) = \rho(z) \iint_{E\Omega} \left(\frac{dE}{dx}\right) \frac{1}{W} \frac{dI}{dE}\, dE\, d\Omega \tag{8.6b}$$

where dI/dE is the particle differential flux in $cm^{-2}\, sec^{-1}\, eV^{-1}\, sterad^{-1}$, $\rho(z)$ is the mass density at altitude z, and (dE/dx) is the energy loss (stopping power) of the particle in $eV\, g^{-1}\, cm^2$. Of course, the differential flux must be known or calculated as a function of altitude. This should be much simpler for Mars and Venus, which are thought not to possess magnetic fields, than it is for Earth.

Ionization caused by Galactic Cosmic Radiation (GCR) is important at lower altitudes. GCR ionization rates are not usually calculated for terrestrial ionospheres but are measured directly by balloon-borne instrumentation and then extrapolated to the altitudes of interest.[1] Obviously this cannot be done for Mars or Venus.

8.2 Ion-Kinetics

The interaction between radiation and atmospheric species produces a source of ions; the ions rapidly undergo a series of reactions that eventually result in the disappearance of both electrons and ions. Intermediate steps often entail the conversion of one ionic species to another.

In some respects the most important of these reactions are the charge transfer or rearrangement reactions because they result in a change of composition of the basic pool of ions from which deionization processes proceed. For example, although molecular nitrogen is the most common species in the terrestrial atmosphere at the altitudes of interest, charge transfer processes quickly convert molecular nitrogen ions to atomic or molecular oxygen ions; rearrangement processes quickly convert both molecular nitrogen ions and atomic oxygen ions to nitric oxide ions. The charge tends to collect on the nitric oxide molecules because they have a lower ionization potential than nitrogen or oxygen molecules or oxygen atoms. These reactions are illustrated below. Their relative importances depend on species concentrations and reaction rates.

$$N_2^+ + O_2 \rightarrow O_2^+ + N_2 \tag{8.7a}$$

$$N_2^+ + O \rightarrow O^+ + N_2 \tag{8.7b}$$

$$N_2^+ + O \rightarrow NO^+ + N \tag{8.7c}$$

$$O^+ + N_2 \rightarrow NO^+ + N \tag{8.7d}$$

$$O^+ + O_2 \rightarrow O_2^+ + O \tag{8.7e}$$

$$O_2^+ + NO \rightarrow NO^+ + O_2 \tag{8.7f}$$

$$O_2^+ + N_2 \rightarrow NO^+ + NO \tag{8.7g}$$

In CO_2-rich atmospheres, such as those of Mars and Venus, the following reactions should be considered:

$$O^+ + CO_2 \rightarrow O_2^+ + CO \tag{8.7h}$$

$$N_2^+ + CO_2 \rightarrow CO_2^+ + N_2 \tag{8.7i}$$

$$CO_2^+ + O_2 \rightarrow O_2^+ + CO_2 \tag{8.7j}$$

$$CO^+ + CO_2 \rightarrow CO + CO_2^+ \tag{8.7k}$$

Electron attachment processes discussed in Chapter 2 result principally in the formation of negative oxygen ions although ion-shuffling reactions may consequently produce other negative ion species:

$$M + O_2 + e \rightarrow O_2^- + M \tag{8.8a}$$

$$O + e \rightarrow O^- + h\nu, \qquad \text{above} \sim 85 \text{ km} \tag{8.8b}$$

Electrons and ions recombine to form neutral species and this ultimately causes the disappearance of both ions and electrons. Recombination processes can be of several types depending, again, on the concentrations of species. Perhaps the most important of these is dissociative recombination of ions and electrons (which may also produce excited products), e.g.,

$$NO^+ + e \rightarrow N^* + O^* \tag{8.9a}$$

$$O_2^+ + e \rightarrow O^* + O^* \tag{8.9b}$$

$$CO_2^+ + e \rightarrow CO^* + O^* \tag{8.9c}$$

$$N_2^+ + e \rightarrow N^* + N^* \tag{8.9d}$$

A relatively slow and, therefore, minor recombination process is the radiative one, e.g.,

$$O^+ + e \rightarrow O + h\nu \tag{8.10}$$

Another important recombination process, particularly at lower altitudes is ion-ion recombination (or mutual neutralization), which is either a two- or three-body reaction depending upon the altitude, e.g.,

$$O_2^+ + O_2^- \rightarrow O_2 + O_2 \tag{8.11a}$$

or

$$O_2^+ + O_2^- + M \rightarrow O_2 + O_2 + M, \qquad \text{below} \sim 50 \text{ km} \tag{8.11b}$$

Negative ions can also be a source of free electrons via the processes of photodetachment or collisional detachment. Perhaps detachment reactions are better classified as electron source reactions than deionization reactions. However, since detachment reactions are secondary sources, it is convenient to group these reactions in this section. Negative molecular oxygen ions are particularly good sources of electrons when irradiated by visible solar radiation inasmuch as the electrons are not firmly attached. The detachment energy of 0.46 eV is easily supplied by sunlight. Hence,

$$O_2^- + h\nu \rightarrow O_2 + e \tag{8.12a}$$

is a fast reaction. Detachment can also occur by the reverse of the three-body attachment processes, i.e., collisional detachment,

$$O_2^- + M \rightarrow O_2 + M + e \tag{8.12b}$$

However, this reaction is endothermic and is very slow (unless, of course, M is a metastable species) in the terrestrial atmosphere and virtually non-existent in the Martian atmosphere. Other detachment reactions may also be important in special situations. An example is associative detachment:

$$O_2^- + O \rightarrow O_3 + e \tag{8.13a}$$

$$O_2^- + O \rightarrow O_3^- + h\nu; \qquad O_3^- + O \rightarrow O_2 + O_2 + e \tag{8.13b}$$

$$O^- + O \rightarrow O_2 + e \tag{8.13c}$$

8.3 Equilibrium

The partially ionized portion of the upper atmosphere which is called the ionosphere can be described by the following simple relationship which applies to any plasma

$$\frac{\partial N_e}{\partial t} = q - L + D$$

where q, L, and D are, respectively, the ionization rate, the ion and electron loss rate, and the diffusion rate; N_e is the free electron number density. Because ion diffusion is not important in the lower ionosphere, further consideration of this process is deferred to Chapter 9.

For the simple case of the single species ionosphere, the above equation may be written as

$$\frac{dN_e}{dt} = q - \alpha_D N_e N^+ - \beta N_e m + \gamma N^- n + \rho N^- \tag{8.14a}$$

where N^+ is the positive ion number density, N^- is the negative ion number density, n is the detaching species number density, m is the attaching species number density, α_D is the ion-electron recombination coefficient, β is the electron attachment coefficient, ρ is the photodetachment rate, and γ is the collisional detachment coefficient.

However, the number densities of the positive and negative ions are also changing with time and

$$\frac{dN^-}{dt} = -\alpha_i N^- N^+ + \beta N_e m - \gamma N^- n - \rho N^- \tag{8.14b}$$

$$\frac{dN^+}{dt} = q - \alpha_D N_e N^+ - \alpha_i N^- N^+ \tag{8.14c}$$

where α_i is the ion-ion recombination coefficient. The ionosphere is electrically neutral, hence,

$$N^+ = N_e + N^-; \tag{8.14d}$$

we assume that diffusion is not important at altitudes lower than 200 km.

Thus, even for this rather simple case, a set of nonlinear coupled differential equations must be solved. Equations (8.14a, b, c) are called the ionospheric continuity equations.

If we define $\lambda = N^-/N_e$ (negative ion to electron ratio), we can write the above equations as follows.[2]

$$\frac{dN_e}{dt} = \frac{q}{1+\lambda} - (\alpha_D + \lambda\alpha_i)N_e^2 - \frac{N_e}{1+\lambda}\frac{d\lambda}{dt} \tag{8.15a}$$

The equation obeyed by λ is[3]

$$\frac{1}{1+\lambda}\frac{d\lambda}{dt} = \beta m - \lambda\left[\rho + \gamma n + N_e(\alpha_i - \alpha_D) + \frac{q}{N_e(1+\lambda)}\right] \tag{8.15b}$$

The above equations implicitly assume that the ionosphere is composed of one species or can be described by one ionized species, one attaching species, etc., or that the reaction rates and concentrations are properly weighted composites of a multi-species atmosphere. For the general and more realistic case, Eqs. (8.14a, b, c) should be written to account for several species, ionization sources, and charge rearrangement reactions. Thus, the continuity equations are:

$$\frac{dN_e}{dt} = \sum_i q_i - \sum_i \alpha_i^D N_i^+ N_e - \sum_i \beta_i m_i N_e + \sum_i \gamma_i N_i^- n_i + \sum_i \rho_i N_i^- \tag{8.16a}$$

$$\frac{dN_i^+}{dt} = q_i - \alpha_i^D N_i^+ N_e - \sum_j \alpha_{ij}^i N_i^+ N_j^- - \sum_j k_{ij} N_i^+ n_j + \sum_{\substack{kl \\ (i \neq k)}} k_{kl} N_k^+ n_l \tag{8.16b}$$

$$\frac{dN_i^-}{dt} = -\sum_j \alpha_{ij}^i N_j^+ N_i^- - \sum_j l_{ij} N_i^- n_j + \sum_{\substack{kl \\ (i \neq k)}} l_{kl} N_k^- n_l$$

$$+ \beta_i m_i N_e - \sum_j \gamma_{ij} N_i^- n_j - \rho_i N_i^- \tag{8.16c}$$

$$N_e = \sum_i N_i^+ + \sum_i N_i^- \tag{8.16d}$$

where α^D is the ion-electron recombination coefficient, α^i is the ion-ion recombination coefficient, k and l are the charge rearrangement rate constants for positive and negative ions, respectively, and the subscripts denote the species.

Equation (8.16a, b, c, d) are obviously messy and to be avoided whenever possible. One of the advantages of considering the ionosphere in separate regions is that simplifications can be made which are valid in specific, though sometimes narrowly restricted, situations.

Useful forms of Eqs. (8.15a, b) can be derived if quasi-steady state conditions are assumed. Such conditions are reasonably valid during quiet solar conditions and periods of the day when q is nearly constant. They are not valid, for example, during sunrise or sunset periods.

Under most conditions, λ varies very slowly with time, i.e., $d\lambda/dt \approx 0$, and

$$\frac{dN_e}{dt} = \frac{q}{1 + \lambda} - (\alpha_D + \lambda\alpha_i)N_e^2 \tag{8.17a}$$

$$\lambda \cong \frac{\beta m}{\rho + \gamma n} \tag{8.17b}$$

Note: The last two terms in brackets on the right hand side of Eq. (8.15b) are frequently insignificant and are neglected; however, this assumption must be carefully evaluated.

Over most of the sunlit hemisphere $dN_e/dt \sim 0$ and $\rho \gg \gamma n$, hence,

$$q \cong (1 + \lambda)(\alpha_D + \lambda\alpha_i)N_e^2 = \alpha_{\text{eff}} N_e^2 \tag{8.18a}$$

$$\lambda \cong \frac{\beta m}{\rho} \tag{8.18b}$$

In the foregoing discussion, values for the various coefficients and rates (e.g., α_i^D, α_i^i, β_i, α_{eff}, ρ_i, k_{ij}, k_{kl}, l_{ij}, l_{kl}) were purposely not given. These values are continually being revised on the basis of improved laboratory measurements and ionospheric observations as well as improved theoretical interpretations. Thus, any set of values that is presented may well be outdated; reference to current literature is required for the investigation of specific problems. However, in order to provide the reader with a feeling for the magnitude of the values and some realistic numbers for illustrative examples and problems, a compilation of laboratory measurements is presented in Table 8.1. The values are rounded to one significant figure and error limits are not shown. Temperature dependence is indicated only where data are available. In the cases where no measurements have been made, currently popular estimates are given.

The concept and use of an effective recombination coefficient (α_{eff}) that represents all recombination reactions is confusing, but occasionally useful. A considerable literature exists on the measurement of this quantity, but the reports are too often vague concerning exactly what quantity was measured. If dN_e/dt and N_e are measured and q is calculated (as is often done), α_{eff} is computed by using the relationship

$$\frac{dN_e}{dt} = q - \alpha_{\text{eff}} N_e^2 \tag{8.17c}$$

Comparing with Eq. (8.17a), we see that (8.17c) can be valid only when $\lambda \ll 1$, or $\lambda \approx 0$; under such conditions α_{eff} must be α_D, or an effective dissociative recombination coefficient. If there is an appreciable negative-ion population, (and λ is significant), Eq. (8.17c) cannot be valid; nonetheless, values of α_{eff} computed by (8.17c) are frequently published. If the negative-ion population is measured or calculated (a very rare occurrence), then either of two versions of (8.17a) are used by the experimenters

$$\frac{dN_e}{dt} = \frac{q}{1+\lambda} - \alpha'_{\text{eff}} N_e^2 \tag{8.17d}$$

or

$$(1+\lambda)\frac{dN_e}{dt} = q - \alpha_{\text{eff}} N_e^2 \tag{8.17e}$$

In the first form (8.17d), $\alpha'_{\text{eff}} = \alpha_D + \lambda\alpha_i$; in the second form (8.17e) $\alpha_{\text{eff}} = (1+\lambda)(\alpha_D + \lambda\alpha_i)$. If the form used is stated clearly, no confusion exists; however, it is clear that $\alpha'_{\text{eff}} \neq \alpha_{\text{eff}}$, unless $\lambda \approx 0$. Inasmuch as this important point is often left to the reader's clairvoyance, even these validly obtained values have limited usefulness. When the common error of using (8.17c) is also considered, it is no wonder that the usefulness of α_{eff} has become

TABLE 8.1. ILLUSTRATIVE REACTION RATES

Reactions	Rate
Dissociative Recombination	
$NO^+ + e \rightarrow N + O$	$\alpha_D = 5 \times 10^{-7} \left(\frac{T}{300}\right)^{-1.2}$ cm^3 sec^{-1}
$O_2^+ + e \rightarrow O + O$	$\alpha_D = 2 \times 10^{-7} \left(\frac{T}{300}\right)^{-0.7}$ cm^3 sec^{-1}
$N_2^+ + e \rightarrow N + N$	$\alpha_D = 3 \times 10^{-7} \left(\frac{T}{300}\right)^{-0.2}$ cm^3 sec^{-1}
$CO_2^+ + e \rightarrow CO + O$	$\alpha_D = 5 \times 10^{-7}$ cm^3 sec^{-1}
Radiative Recombination	
$O^+ + e \rightarrow O + h\nu$	$\alpha_r = 4 \times 10^{-12} \left(\frac{T}{300}\right)^{-0.7}$ cm^3 sec^{-1}
Ion-ion Recombination	
$O_2^+ + O_2^- \rightarrow O_2 + O_2$	$\alpha_i = 2 \times 10^{-7}$ cm^3 sec^{-1}
$O_2^+ + O_2^- + M \rightarrow O_2 + O_2 + M$	$K = 3 \times 10^{-25} \left(\frac{T}{300}\right)^{-2.5}$ cm^6 sec^{-1}
Rearrangement	
$N_2^+ + O_2 \rightarrow N_2 + O_2^+$	$k = 2 \times 10^{-10}$ cm^3 sec^{-1}
$N_2^+ + O \rightarrow N_2 + O^+$	$k = 10^{-12}$ cm^3 sec^{-1} (est)
$N_2^+ + O \rightarrow NO^+ + N$	$k = 3 \times 10^{-10}$ cm^3 sec^{-1}
$N_2^+ + CO_2 \rightarrow CO_2^+ + N_2$	$k = 9 \times 10^{-10}$ cm^3 sec^{-1}
$O^+ + N^2 \rightarrow NO^+ + N$	$k = 2 \times 10^{-12}$ cm^3 sec^{-1}
$O_2^+ + NO \rightarrow NO^+ + O_2$	$k = 8 \times 10^{-10}$ cm^3 sec^{-1}
$O_2^+ + N_2 \rightarrow NO^+ + NO$	$k < 10^{-15}$ cm^3 sec^{-1} (upper limit only)
$O^+ + CO_2 \rightarrow CO + O_2^+$	$k = 10^{-9}$ cm^3 sec^{-1}
Electron Attachment	
$M + O_2 + e \rightarrow O_2^- + M, M = N_2$	$K = 10^{-31} \left(\frac{300}{T}\right) e^{-600/T}$ cm^6 sec^{-1}
$M = O_2$	$K = 10^{-29}$ cm^6 sec^{-1} $\beta = K[O_2]$
$O + e \rightarrow O^- + h\nu$	$\beta_r = 10^{-15}$ cm^3 sec^{-1}

TABLE 8.1. (CONTINUED)

Reactions	Rate
Electron Detachment	
$O_2^- + h\nu \rightarrow O_2 + e$	$\rho = 0.33\ sec^{-1}$
$O_2^- + M \rightarrow O_2 + M + e,\ M = N_2$	$\gamma = 2 \times 10^{-12} \left(\frac{T}{300}\right)^{1.5} e^{5\times 10^{-3}/T}$ $cm^3\ sec^{-1}$
$M = O_2$	$\gamma = 3 \times 10^{-10} \left(\frac{T}{300}\right)^{0.5} e^{6\times 10^{-3}/T}$ $cm^3\ sec^{-1}$
$O^- + O \rightarrow O_2 + e$	$k = 3 \times 10^{-10}\ cm^3\ sec^{-1}$
$O_2^- + O \rightarrow O_3 + e$	$k = 5 \times 10^{-10}\ cm^3\ sec^{-1}$

limited. However, it should be pointed out that the careful measurement of α_{eff} is nonetheless a very useful endeavor, because it is still one of the few ways to compare theory and laboratory measurements with actual ionospheric conditions. As more, and better, ion measurements are made, the usefulness of α_{eff} measurements will diminish; however, α_{eff} may continue to be a useful notion for studying the ionospheres of other planets.

8.4 Ionospheric Regions

Ionospheric regions were discovered, and subsequently studied, by radio reflection and propagation techniques. Even with increased probing of the upper atmosphere with rocket-borne instrumentation, most of our knowledge about these regions of the earth's atmosphere as well as those of other planets continues to be derived by radio techniques. Hence, it is not surprising that the characteristics that differentiate ionospheric regions are often described by their effects on radio propagation rather than by their physical and chemical uniqueness. Radio effects will be covered in Chapter 11; in this chapter, we shall be concerned with physical and chemical processes. It is important to note that the regions which were discovered and identified by radio techniques can also be characterized by aeronomic conditions other than electron number-density gradients. We will use ion-kinetic process regimes as the basis for classifying ionospheric regions; this scheme provides a logical basis for discussion of the terrestrial ionosphere and, in addition, provides a basis for classifying ionospheric regions on other planets.

The principal ionospheric regions are known as *D*, *E*, and *F* (see Chapter 1). Radio sounding techniques seemingly reveal that each of these regions contains one or more unique features of the electron number-density profile identified as layers. In the *D*-region, there is some ambiguity regarding the existence of any well-defined layer, whereas in the *F*-region, there are two, F_1 and F_2. Occasionally a *C*-layer is mentioned, but the existence of such a layer below the *D*-region is difficult to demonstrate. In this chapter, the *C*-layer, if any, will be considered to be part of the *D*-region; and the F_2 layer, because of rather unique characteristics involving geomagnetic control and diffusive effects, will be considered separately in Chapter 9.

Actually, the ionosphere contains only one well-defined layer, the F_2. The other so-called layers are actually regions with inflections or with large electron density gradients that reflect radio waves just as a well-defined layer would; in spite of this, the layer terminology is commonplace.

Although the designations *D*, *E*, F_1, and F_2 pertain to features of the terrestrial ionosphere, ionospheric regions of other planets can be designated by the same letters. In doing so, it is implied that the regions are formed by aeronomic conditions that are very analogous to those responsible for the formation of the same regions in the terrestrial atmosphere. For example, a region that is controlled by negative ion-kinetics can be designated a *D*-region whether it is formed on the earth or Mars. Photoionization and dissociative recombination are characteristic of the terrestrial *E*-region; a region in the Martian or Cytherean ionosphere that is controlled by similar processes would be similarly labeled. A region that is controlled by ion-interchange processes can be termed an F_1 region regardless of the planet. If a layer in the Martian or Cytherean ionospheres is controlled principally by diffusion, it would be designated F_2, just as it is in the terrestrial ionosphere.

8.4.1 D-region

The *D*-region is probably the most complex region in an ionosphere. There is considerable doubt that distinct layers exist in this region, but there is no doubt that perturbations of this region greatly affect the absorption of high-frequency radio signals and the reflection of low-frequency signals. A *D*-region is the lowest-lying region and hence is produced by the most penetrating of the ionizing radiations that impinge upon the planet; it is, characteristically, a region of weakly-ionized plasma and large neutral species number-density as well as complex ion-interchange and electron attachment and detachment reactions. The importance of the latter processes (electron attachment and detachment) is the most distinguishing feature of a *D*-region.

The responsible ionizing radiations are the very energetic (>BeV) charged particles that constitute the Galactic Cosmic Radiation (GCR),[1] the most energetic portion of the solar X-ray spectrum (<10A),[4,5] the intense solar hydrogen Lyman-α (Lα) emission line (1216 R),[6] and EUV (<1118 Å).[19] A secondary electron producing source is visible and ultraviolet solar emission which can provide the energy necessary to release weakly held electrons from negative ions. The very energetic GCR and solar X-radiations ionize all atmospheric constituents, whereas the Lα radiation can ionize only those constituents that have ionization potentials lower than ~10 eV; these include all of the known minor constituents, i.e., nitric oxide and metallic atoms; EUV (<1118 Å) will ionize the metastable molecular oxygen species O_2 $(a^1\Delta_g)$.

The relative importances of the ionizing radiations depend on their fluxes, the concentrations of ionizable species, and the recombination and charge transfer rates of the resulting ions (and also the attachment rates of the resulting electrons) as described by Eqs. (8.6) and (8.16). In terms of total energy, the Lα line radiation is the major source; it is approximately 5 ergs cm^{-2} sec^{-1} at the top of the earth's atmosphere, three to five orders of magnitude greater than the X-ray flux at wavelengths smaller than 10 Å and more than an order of magnitude greater than the flux in the EUV band 1027 to 1118 Å. The dissociative recombination coefficients for all diatomic gaseous molecules of interest in the terrestrial *D*-region are similar (within $\frac{1}{2}$ order of magnitude) and therefore do not affect the relative distributions of ions However, the relative abundances of ionizable species and the rates for charge transfer from ions to neutral constituents are critical in determining the ion distribution.

Thus, in the terrestrial *D*-region, the fraction of nitric oxide present below 90 km is critical because, although it is a minor constituent, it is ionized by the strongest of the ultraviolet sources (Lα). The flux of Lα radiation available for the ionization of NO is actually determined by molecular oxygen concentrations because oxygen molecules are very effective absorbers over most of the UV spectrum. Figure 8.3 shows, however, that a window (region of low absorption) exists in the oxygen absorption spectrum at the Lα wavelength; this fortunate coincidence allows a sufficient amount of Lα radiation to penetrate to *D*-region altitudes to make NO an important source of ions. The metastable oxygen molecule $O_2(^1\Delta_g)$ is also a minor species, but it is more plentiful than nitric oxide; hence, although the flux of the radiation (<1118 Å) that ionizes $O_2(^1\Delta_g)$ is less than $\frac{1}{10}$ that of Lα, it is sufficient to make $O_2(^1\Delta_g)$ a major *D*-region ion-source. This is illustrated in the following figures. Figure 8.4 shows the absorption as a function of altitude for several wavelengths of electromagnetic radiation. Figure 8.5 is an illustrative example of ionization rates produced by X-ray, GCR, Lα, and EUV (1027 to 1118 Å). In the case of Lα, two curves are shown; the first is based on theoretical estimates of NO concentrations[6] and the second is based on spectrometric measurements of nitric oxide airglow (see Chapters 4

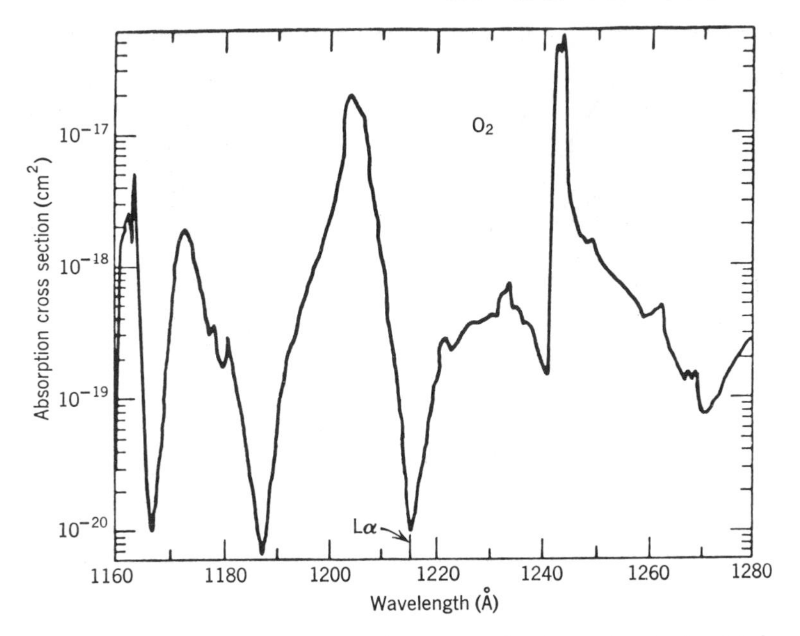

Figure 8.3 Absorption cross section of O_2 in the region 1160–1280 Å, obtained with 0.2 Å resolution. (After Watanabe, *Adv. in Geophysics*, **5**, 153 (1958) reprinted by permission of Academic Press, Inc.)

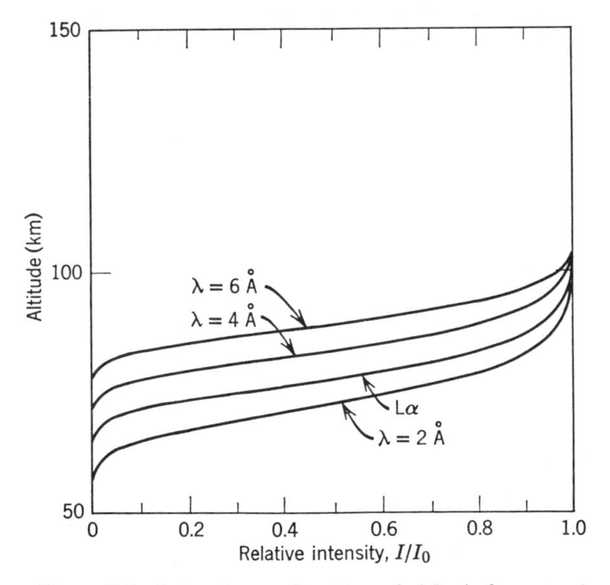

Figure 8.4 Intensity as a function of altitude for several wavelengths of ionizing electromagnetic radiation.

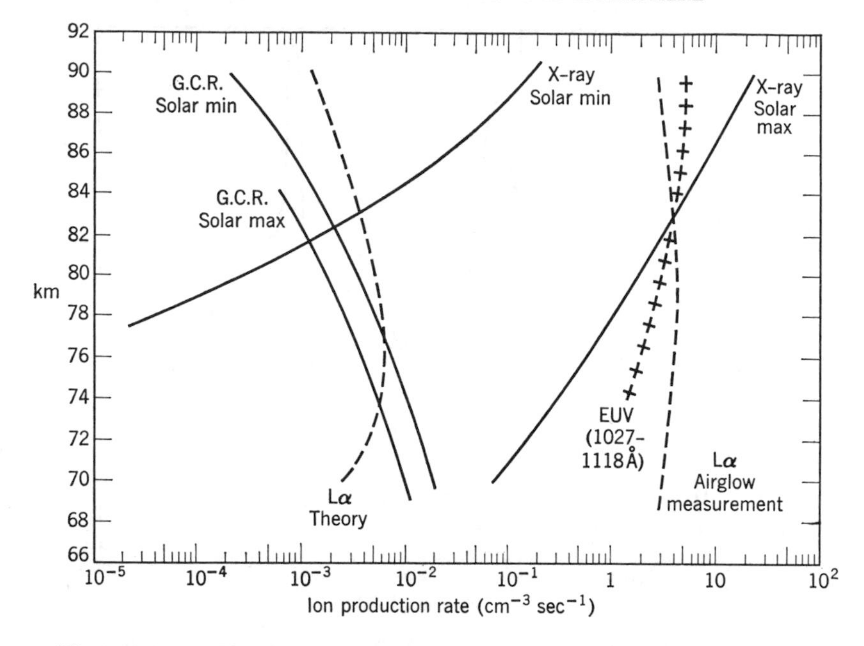

Figure 8.5 Illustrative example of ionization rates produced by X-ray, GCR, and hydrogen Lα radiation. The X-ray and GCR components are shown over a solar cycle; the Lα is shown for both theoretical and measured NO concentrations (NO measurement inferred from airglow observations); *see* Chapter 4. The EUV ionization of $O_2(^1\Delta)$ is from Hunten and McElroy.[19]

and 6). The curves of Figure 8.5 are computed through the use of Eq. (8.6a) and the assumption that the sun is directly overhead ($\chi = 0°$).

We should note the two principal reasons for not attempting to use the Chapman theory to compute q_m and z_m [Eqs. (8.4) and (8.5)]. In the case of Lα the ionizing radiation is strongly absorbed (and the intensity is, therefore, controlled) by molecular oxygen, but it ionizes only nitric oxide; hence the assumption of $Q_a = Q_i$ would not apply. If a sufficient number of simplifying assumptions are made to use the Chapman theory, the calculations force the conclusion that a layer exists; under most conditions, this would not be a valid conclusion for the *D*, *E*, or F_1 regions.

If we assume quasi-equilibrium conditions, the corresponding free electron number-densities can be calculated for each process according to Eqs. (8.18a, b). This is shown in Figure 8.6. In general, Eqs. (8.16a, b, c, d) should be used. The positive ions formed directly by ionizing radiation in this region are O_2^+, N_2^+, and NO^+. The N_2^+ ion charge-transfers to O_2 very rapidly (Eq. (8.7a)] and thus does not exist in significant quantities. The O_2^+ ion charge will transfer to NO ([Eq. (8.7f)] very rapidly; but because the concentration of NO is small, significant quantities of O_2^+ are expected to exist. If, however, the ion-interchange reaction (8.7g) is as important as it is thought to be in the *D*-region, O_2^+ concentrations may also become insignificant. Mass spectrometric analyses of ion species in the *D*-region reveal the presence of metallic ions and hydrated ions for which no satisfactory explanation has been advanced. See Figure 8.7.

As can be seen easily from 8.17b, the negative-ion to electron ratio, λ, increases rapidly with density (i.e., with decreasing altitude). This is illustrated by Figure 8.8. At the lower end of the altitude range, $\beta m \gg \rho + \gamma n$ and essentially no free electrons are found under normal conditions; therefore, ion-ion recombination [such as Eq. (8.11)] is the principal ion-loss process. At intermediate altitudes where λ is very important in electron-density calculations, considerable uncertainty exists regarding ionospheric conditions. If it is assumed (as in the calculation of the profiles of Figure 8.8) that the electron attachment and detachment processes involve only molecular oxygen, the problems are relatively straightforward and solvable. Actual measurements in this region are very difficult, and to date the composition, ionic or neutral, is not known. More to the point, the actual

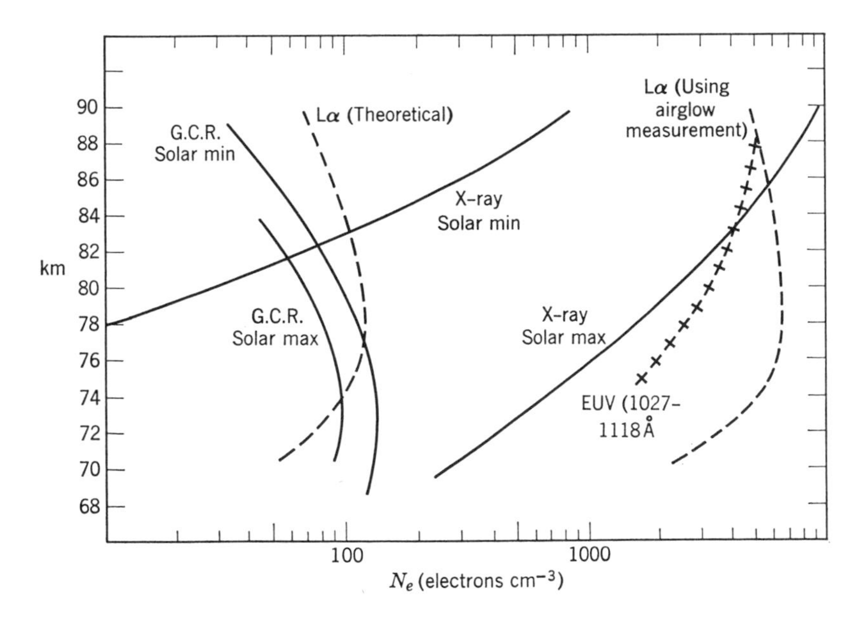

Figure 8.6 Calculated electron concentrations corresponding to ionization rates of Figure 8.5. A simple attachment-detachment scheme involving only O_2 was used for this illustration.

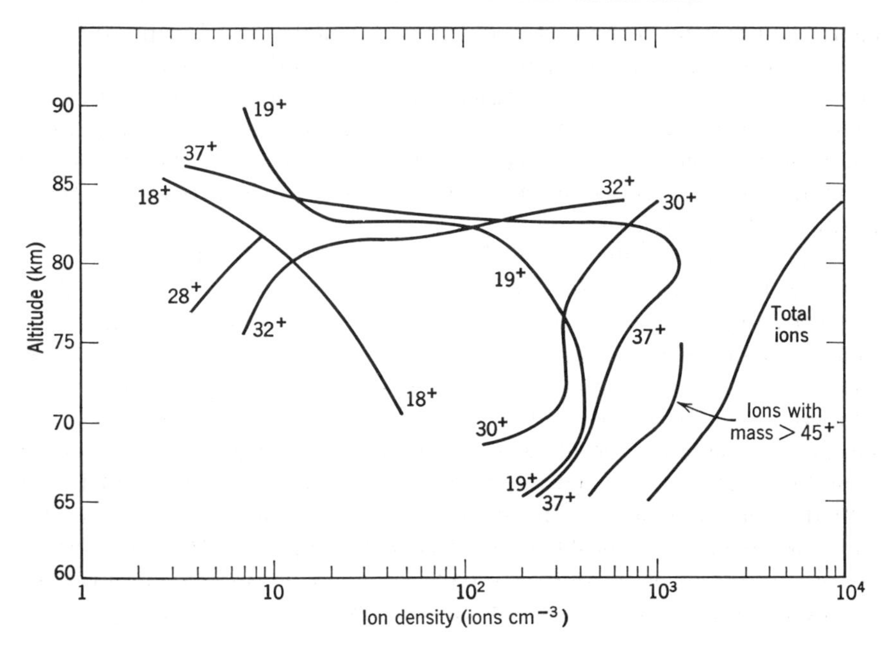

Figure 8.7 Concentrations of positive ions detected by the mass spectrometer within the altitude range 64 to 83 km. (After Narcisi and Bailey, *JGR* **70**, 3687–3700 (1965); reprinted by permission of the American Geophysical Union.)

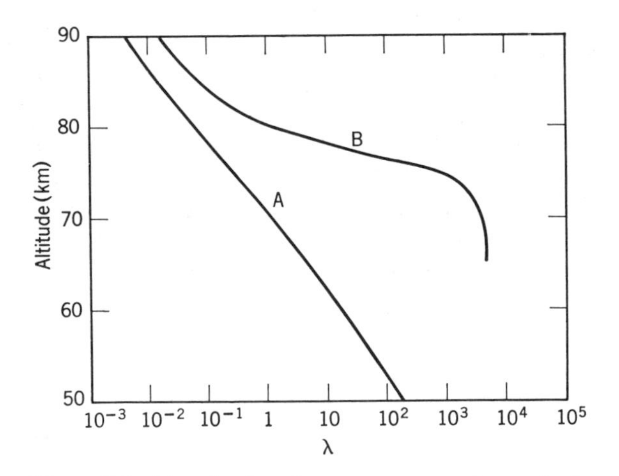

Figure 8.8 Negative ion-electron concentration ratio λ for altitudes below 90 km. A—Daytime B—3 hours after sunset. This illustration is calculated for a simple attachment-detachment model involving only molecular oxygen.

negative-ion species in the *D*-region are unknown and total negative-ion measurements are doubtful at best. Also ion-ion recombination rates are not well known. Hence this region is still a mystery and provides grounds for countless speculations. Fragmentary observations of electron number-densities and their variation have given rise to paradoxes that virtually defy possibilities of solution. Current speculation[7–9] involves negative-ions formed by attachment to species other than molecular oxygen and/or a series of negative-charge "shuffling" reactions. Ozone, nitrogen dioxide, hydroxyl radicals, and carbon trioxide have been advanced as candidates for attachment or for intermediaries in the series. Except for ozone, the theories regarding the production and abundance of the candidate agents are not well developed. In addition, associative detachment schemes[8,10] such as Eq. (8.13a, b, c) have become popular; but, of course, the abundances and sources of species taking part in these reactions have yet to be experimentally determined with adequate reliability. Hypothesized negative-ion processes are summarized in Figure 8.9.

Certain implications of the preceding comments on the conditions thought to exist in the Martian lower atmosphere should be noted. Current suppositions are that the lower atmosphere is a mixture of CO_2 with small amounts of argon, nitrogen, or neon in various proportions; photolysis of CO_2 is thought to be minor. Most of the X-ray ionization occurs at altitudes over 100 km; but galactic cosmic radiation should furnish a relatively abundant and constant source of ion-electron pairs in the region well below 100 km (in fact down to the surface) just as it does in the terrestrial atmosphere. It follows, however, that Mars may not have a *D*-region because there are no attaching species and hence the processes responsible for a lower ionospheric region are simply ionization of a molecular species and molecular-ion recombination with electrons. Thus, one might indeed speculate that on Mars the *E*-region extends down to the surface of the planet. If, on the other hand, appreciable quantities of oxygen, or hydrocarbons exist a true *D*-region could be formed. Figures 8.10a, b, c, d illustrate the consequences of a GCR-formed lower ionosphere on Mars with oxygen as a candidate electron-attaching species (*D*-type region) and without oxygen (low-lying *E*-type region). Similar speculations could be made about Venus. Models constructed by the authors (*Planet. Space Sci*, in press 1970) indicate the existence of a very definite D-region on Mars.

8.4.2 E-region

An *E*-region is somewhat easier to describe inasmuch as the ionization-deionization processes are straightforward. Ionization of molecular species is dominant and electron-ion losses are solely dissociative recombination

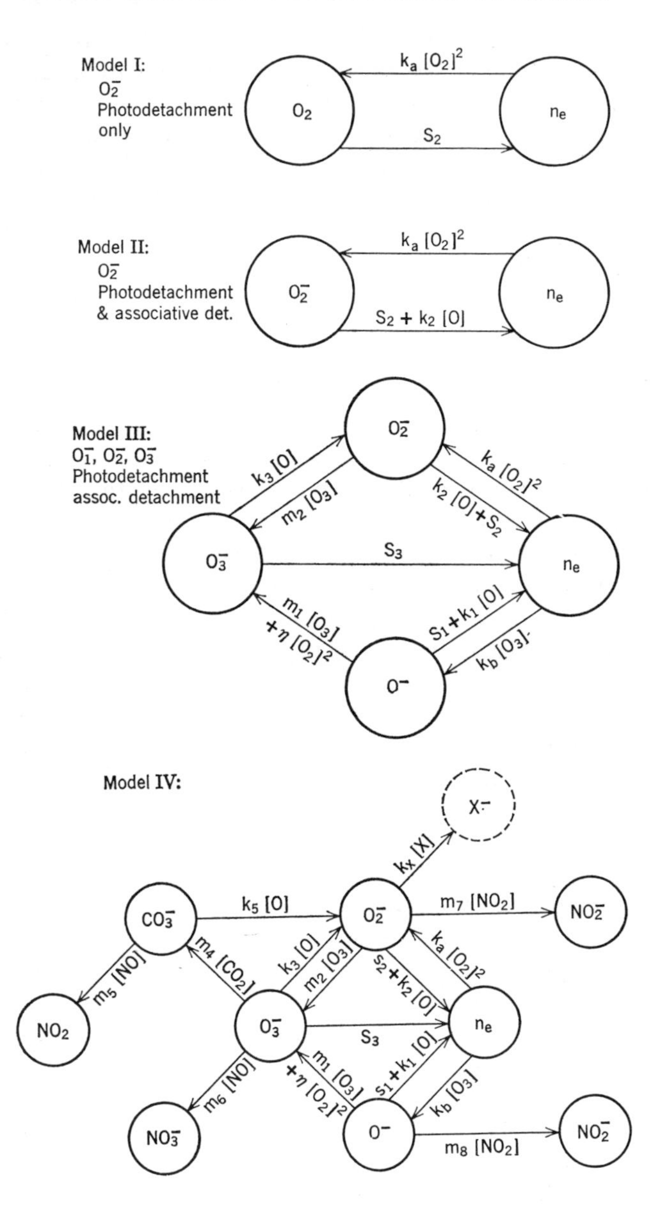

Figure 8.9 Four hypothesized negative ion attachment, detachment, and ion-shuffling schemes. (After LeLevier and Branscomb, *JGR* **73**, 27 (1968); reprinted by permission of the American Geophysical Union.)

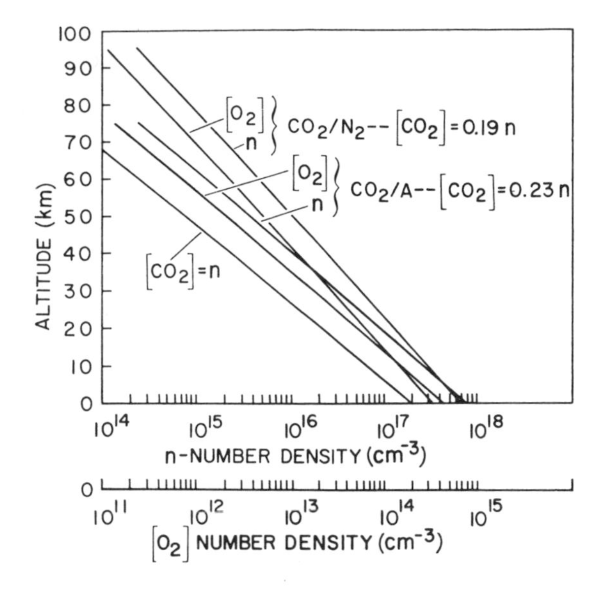

Figure 8.10a Several hypothesized models of the lower Mars neutral atmosphere.

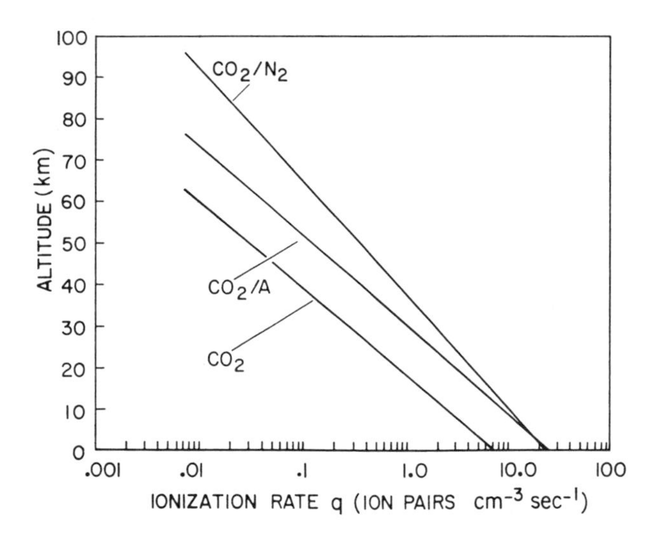

Figure 8.10b Ionization rate profiles for the neutral atmosphere models illustrated in (a). It is assumed that GCR ionization is the only significant ion-electron production mechanism in this region.

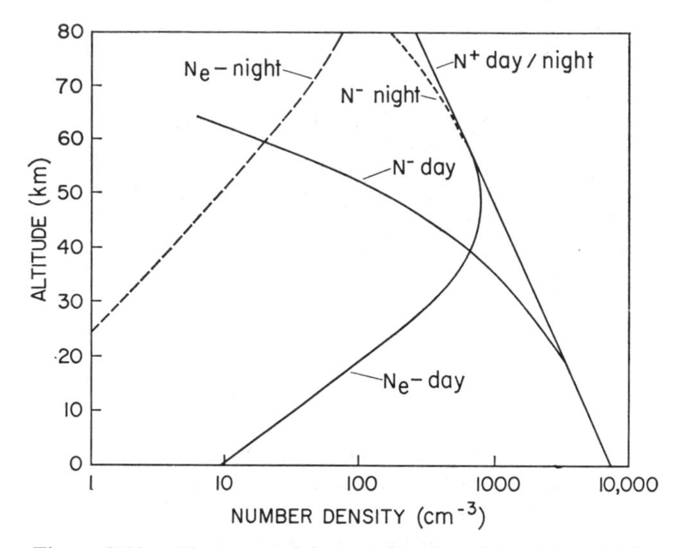

Figure 8.10c Electron and ion profiles based on (a) and (b) for the case of 15 cm-atm of oxygen. This illustrates the possibility of a *D*-region on Mars; the existence of a *D*-region depends on the presence of O_2 or some other attaching species. If no attaching species exists in the Mars atmosphere, the electron profile of the ionosphere would extend to the surface (or to a boundary layer that would provide dust for electron attachment) and would exist day or night; it would be a permanent *E*-region as illustrated by profile labeled N^+ day/night.

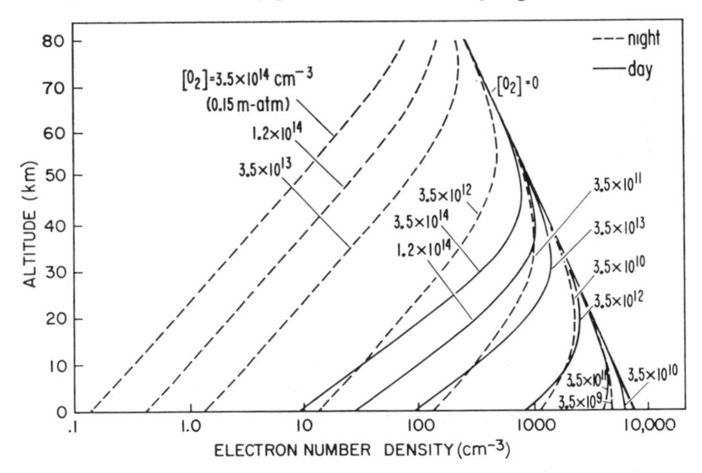

Figure 8.10d Illustrative profiles showing that the electron profile would be very sensitive to the actual concentrations of O_2 in the Mars atmosphere.

processes such as Eqs. (8.9a, b, c, d). Negative-ion and ion-ion recombination processes are insignificant. For neutral species in an E-region, photoequilibrium becomes more important and turbulent processes become less important.

For an E-region, Eqs. (8.16a, b, c, d) can be simplified by eliminating the attachment, detachment, and ion-ion recombination terms. Hence

$$\frac{dN_e}{dt} = \sum_i q_i - \sum_i \alpha_i^D N_i^+ N_e \tag{8.16a$'$}$$

$$\frac{dN_i^+}{dt} = q_i - \alpha_i^D N_i^+ N_e - \sum_j k_{ij} N_i^+ n_j + \sum_{\substack{kl \\ (i \neq k)}} k_{kl} N_k^+ n_l \tag{8.16b$'$}$$

$$N_e = \sum_i N_i^+ \tag{8.16d$'$}$$

under quasi-equilibrium conditions, $dN_e/dt = dN_i^+/dt = 0$,

$$q_i = \alpha_i^D N_i^+ N_e + \sum_j k_{ij} N_i^+ n_j - \sum_{\substack{kl \\ (i \neq k)}} k_{kl} N_k^+ n_l \tag{8.19}$$

Equilibrium concentrations of N_e and N_i^+ can be determined by solving the set of Equations (8.19) and remembering the condition of (8.16d′).

For a particular ionosphere, the procedure is simply to determine the responsible ionizing radiations, the ionizable species, and the applicable ion interchange reactions. We assume, of course, that pertinent cross sections and rates are available.

In the terrestrial ionosphere, the principal ions formed directly are N_2^+ and O_2^+; to a lesser extent O^+ and NO^+ are formed directly. Na^+ and other metallic ions may also be formed directly but at insignificant rates and are not ordinarily important; however, metallic ions may play an important role in sporadic E layer formation (see Chapter 10). Important ionizing radiations in this region are solar hydrogen Lyman β at λ 1025.7 Å, the EUV spectrum below λ 1000 Å (He II, He I lines), and soft X-rays ($\lambda > 10$ Å) Molecular nitrogen and atomic oxygen are ionized by EUV and X-radiations at $\lambda <$ 900 Å. Molecular oxygen is ionized by EUV and X-radiation and also by radiations at wavelengths as long as the Lyman-β line; because of this Lyman β is considered to be the controlling terrestrial E-region ionization source by some, though not all, aeronomers. The key to the controversy is obviously the differences in both the assumed model atmospheres and the assumed values for solar spectral irradiance. Table 8.2 illustrates relative ionization efficiencies of O_2, N_2, and O as a function of wavelength. Table 8.3 and Figures 8.11 and 8.12 illustrate the relative photoionization rates as functions of both ionizing radiation and ionizable species for the two model

TABLE 8.2. PHOTOIONIZATION DATA

λ (Å)	$F_\lambda(\infty)$, 10^8 photons cm^2 sec^{-1}	Q_{N_2}, 10^{-18} cm^2	Q_{O_2}, 10^{-18} cm^2	Q_O, 10^{-18} cm^2	Ionization Efficiency for N_2†	Ionization Efficiency for O_2†	Ionization Efficiency for O†
1025.7	25	0.005	1.9			0.58	
1000–1027	18	0.1	1.3			0.79	
977	40.3	1.0	4.7			0.73	
911–1000	52.0	1.5	8.0			0.69	
850–911	75.7	7.0	8.0	3.2		0.70	1.0
796–850	39.3	3.0	16	3.2		0.31	1.0
700–796	50.0	20	25	4.5	0.75	0.55	1.0
400–700	111	25	28	10	0.9	0.9	1.0
230–400	114	8	16	7	1.0	1.0	1.0
170–230	54.4	6	6	3	1.75	1.75	1.75
80–170	~15	2.0	2.0	1.0	4.0	4.0	4.0
40–80	~2.4	0.4	0.6	0.3	6.0	6.0	6.0
10–40	~0.5	0.1	0.2	0.1	12	12	12

† We define the ionization efficiency as the ratio of ionization to absorption cross section, or

$$\text{ionization efficiency} = \frac{\text{no. of ion-pairs formed per sec}}{\text{no. of photons absorbed per sec}}$$

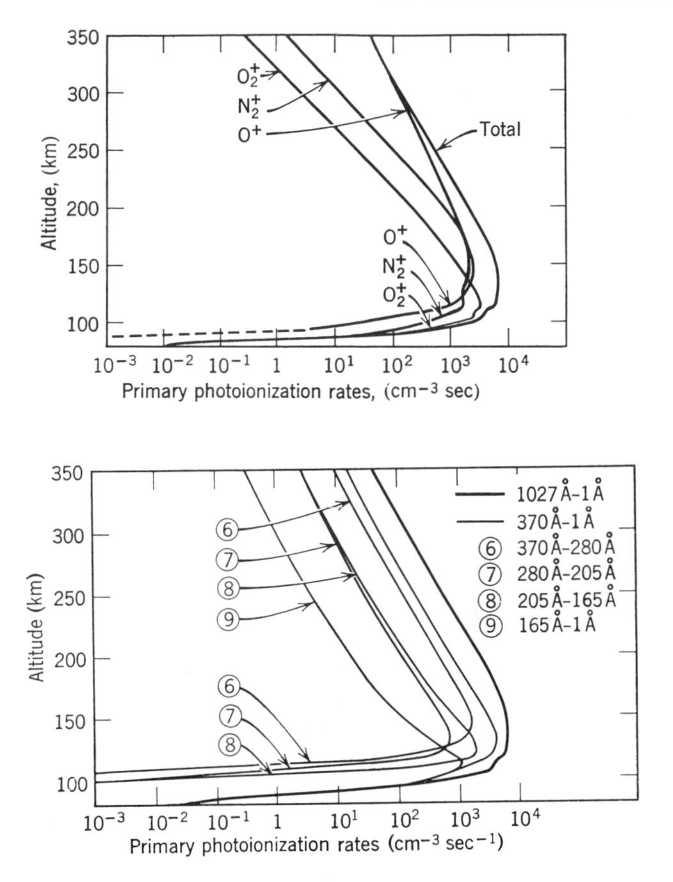

Figure 8.11 Calculated photoionization rates in the *E* and *F*-regions. (After Hinteregger, Hall, and Schmidtke, *Space Research* V, 1175 (1965); reprinted by permission of the North-Holland Publishing Co.)

atmospheres in Table 8.3. Note how the relative importances of Lyman β and X-rays depend on the assumed ratios of O/O_2.

Although N_2^+ is copiously produced in the earth's *E*-layer, it is not a major ionic constituent because of rapid charge-transfer and ion-interchange reactions. Most of the ions formed directly will undergo charge rearrangements depending on rates and concentrations. The important reactions are listed as Eqs. (8.7a) to (8.7i). The resultant free electron densities will, of course, depend on the equilibrium distribution of ion-species and their corresponding recombination rates.

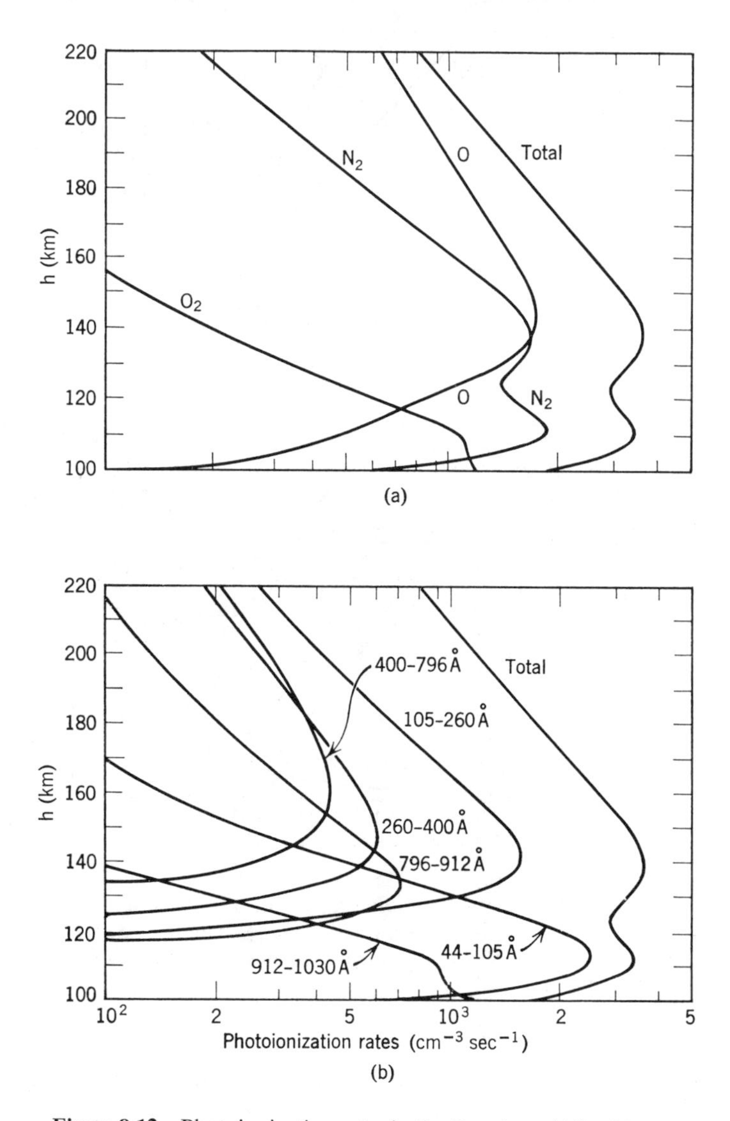

Figure 8.12 Photoionization rates in the *E*-region. (After Norton, Van-Zandt, and Denison, *Proceedings of the International Conference on the Ionosphere*, London, 1962; reprinted by permission of the Institute of Physics and the Physical Society.)

TABLE 8.3. COMPARISON OF MODEL ATMOSPHERES USED IN *E*-REGION MODELS

Altitude km	Hinteregger, Hall, and Schmidtke (1965)			Norton, Van Zandt, and Denison (1963)		
	[O] cm^{-3}	[O_2] cm^{-3}	[N_2] cm^{-3}	[O] cm^{-3}	[O_2] cm^{-3}	[N_2] cm^{-3}
100	5.0×10^{11}	2.0×10^{12}	1.0×10^{13}	2.7×10^{12}	6.5×10^{11}	7.4×10^{12}
105	2.5	7.0×10^{11}	4.0×10^{12}	1.2	3.0	3.4
110	1.7	3.0	2.0	5.9×10^{11}	1.5	1.7
120	8.0×10^{10}	6.0×10^{10}	3.0×10^{11}	2.4	2.9×10^{10}	3.9×10^{11}
130	5.0	2.0	1.0	1.1	7.4×10^{9}	1.2
140	3.0	8.0×10^{9}	4.0×10^{10}	5.3×10^{10}	2.4	4.1×10^{10}
150	2.0	4.2	2.0	3.0	9.6×10^{8}	1.8
160	1.2	2.1	1.1	1.9	4.7	9.5×10^{9}
170	8.0×10^{9}	1.0	8.0×10^{9}	1.3	2.6	5.6
180	6.0	9.0×10^{8}	5.0	9.9×10^{9}	1.6	3.5
200	3.3	3.0	2.0×10^{9}	6.0	6.5×10^{7}	1.6

Because the atmosphere of Mars appears to be similar to the atmosphere of Venus (above the cloud layer), it is speculated that the ionospheres of both planets should be similar. And, indeed, observations[11,16] of the ionospheres of both planets by radio occultation techniques would seem to confirm such a hypothesis. The observations of each planet reveal a large peak at an altitude where the neutral particle concentration is 10^{10}-10^{11} cm^{-3} and a smaller peak or inflection a few kilometers lower (see Figures 8.13 and 8.14). These have been interpreted[12–15] as both *E*-region and *F*-region type ionospheres, but the most plausible is the *E*-region interpretation. The theoretical models[20] that come closest to explaining the observed features all require that the major peak in each planet's ionosphere is due to the photoionization of CO_2 by EUV, that the lower peak or inflection is due to ionization of CO_2 by X-rays, and that the principal ion-electron loss process is the dissociative recombination of CO_2^+ with thermal electrons. However, inspection of the observed ionospheres and the model calculations illustrated in Figures 8.13 and 8.14 show that agreement is far from perfect in the case of Mars; although the agreement for Venus may be much more satisfactory, it is still not complete. The inclusion of other hypothesized ionic processes in the models simply makes the agreement worse; hence, these models are the best available and probably will not be improved until much more detailed information is obtained regarding the atmospheres of both planets and the details of relevant photochemical and ionic reactions.

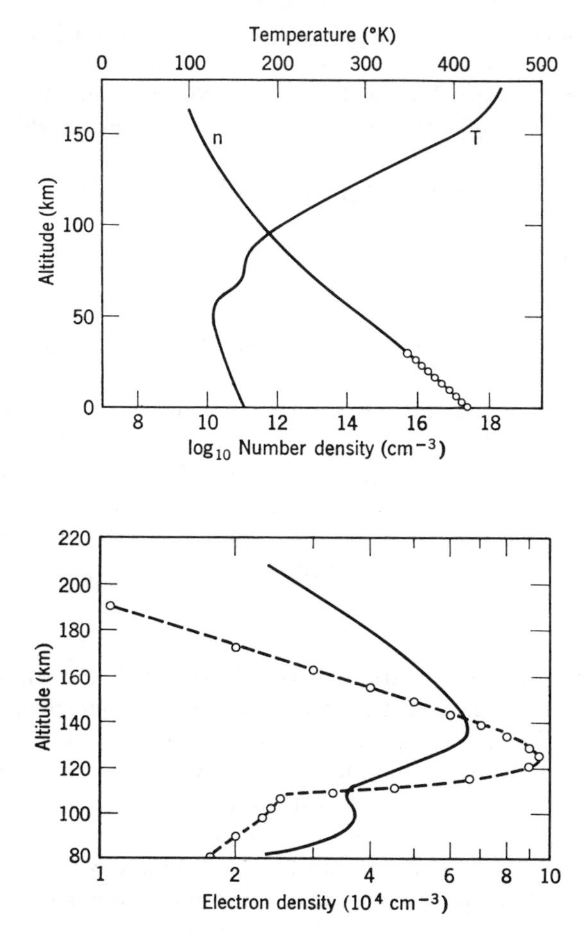

Figure 8.13 (a) A pure CO_2 model of the Martian atmosphere. This model should be appropriate for solar minimum conditions. (b) The Mars ionosphere. The dashed line is the profile derived from Mariner 4 radio occulation measurements. The solid line is a model based on the pure CO_2 neutral atmosphere model above and the assumption that photoionization of CO_2 and the dissociative recombination of CO_2^+ and electrons are the controlling production and loss mechanizations. (After McElroy, *JGR* **74,** 29 (1969); reprinted by permission of the American Geophysical Union.)

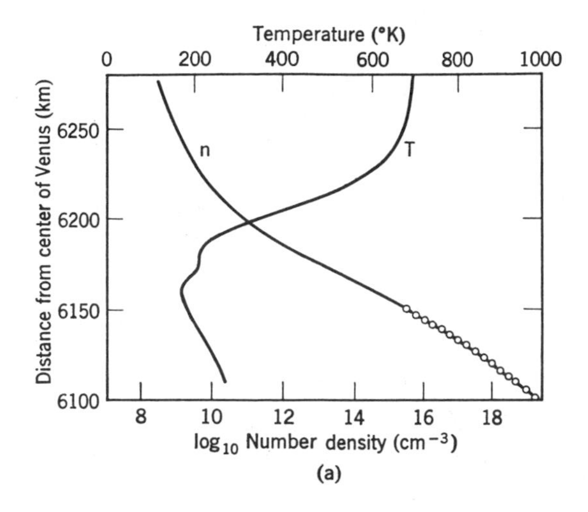

Figure 8.14 (a) Model of the Venus upper atmosphere based on an assumed atmosphere of 90% CO_2 and 10% N_2 with diffusive separation above 6190 km. (b) Model of ionosphere (solid line) based on the neutral atmosphere model represented in (a). It is assumed that the ion-electron production mechanism is photoionization of CO_2 and loss is by dissociative recombination of CO_2^+. Profile derived from radio measurements (dashed line) is also shown. (c) Two model atmospheres with diffusive separation above 6185 km. Model D assumes a mixture of CO_2 (80%), CO (10%), and O (10%) at the turbopause. In model E, the assumed mixture is CO_2 (95%), CO (2.5%), and O (2.5%). (d) Ionospheric model based on neutral model D. Note lack of agreement of scale heights between model (solid line) and observed (dashed line) profiles. (e) Ionospheric model based on neutral model E. Note much better agreement between model (solid line) and observed (dashed line) profiles. The profiles of (e) and (b) demonstrate that it appears to be unlikely that atomic oxygen ions can play an influential role in the formation of the Venus ionosphere. (After McElroy, *JGR* **74**, 29 (1969); reprinted by permission of the American Geophysical Union.)

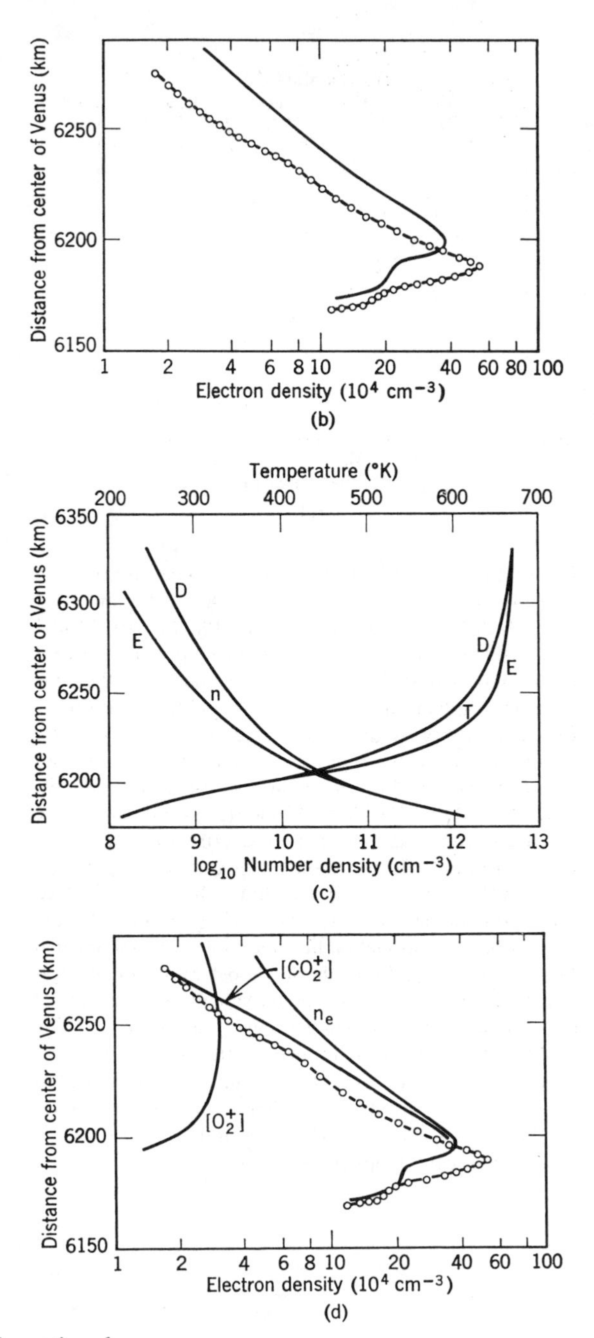

Figure 8.14 continued

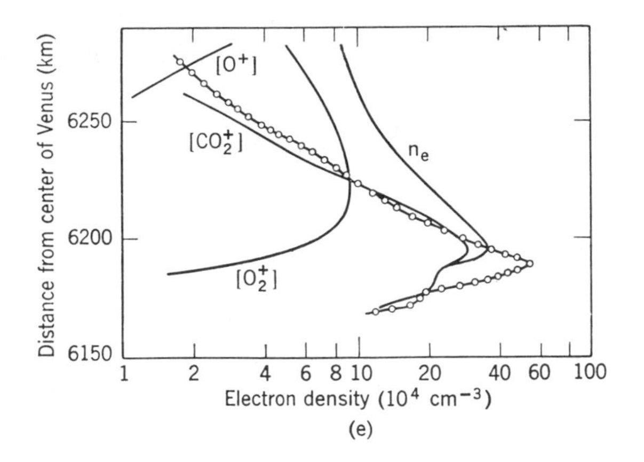

Figure 8.14e continued

8.4.3 F_1-Region

The unique characteristics of an F_1 region are that the principal ion produced is atomic, whereas the principal electron disappearance process is dissociative recombination with a molecular ion. Thus, ion-interchange is the key process that characterizes an F_1 region. On the other hand, it should be noted that ambipolar diffusion effects are not yet important. When they become important, the region becomes an F_2 region which is discussed in the following chapter. For neutral species in an F_1 region, photodissociation and diffusive equilibrium are dominant and mass transport and turbulent processes are not significant.

In an F_1 region the ions produced by solar radiation are principally atomic (or are molecular and transfer charge so quickly to atoms that the effective primary ion is atomic); hence, the continuity equations (8.16) become:

$$\frac{dN_e}{dt} = \sum_i q_i - \sum_k a_k^D N_k^+ N_e \qquad (8.16a'')$$

$$\frac{dN_i^+}{dt} = q_i - \sum_j k_{ij} N_i^+ n_j \qquad (8.16b'')$$

$$\frac{dN_k^+}{dt} = \sum_{\substack{ij \\ (i \neq k)}} k_{ij} N_i^+ n_j - \alpha_k^D N_k^+ N_e \qquad (8.16c'')$$

$$N_e = \sum_i N_i^+ + \sum_k N_k^+ \qquad (8.16d'')$$

where the subscript i refers to atomic-ion production and loss processes and subscript k refers to molecular-ion loss processes. For quasi-equilibrium solutions,

$$\frac{dN_e}{dt} = \frac{dN_i^+}{dt} = \frac{dN_k^+}{dt} = 0.$$

In the terrestrial ionosphere, the dominant F_1 region ionizable constituent is considered to be atomic oxygen; if Mars and Venus have F_1 regions, the ionizable constituent would, presumably, also be atomic oxygen. Reference to Figures 8.11 and 8.12 shows that in the terrestrial ionosphere, X-radiation is not important. Ionization is by EUV in the $\lambda\lambda$100–800 Å region.

Recombination between atomic ions and electrons occurs via a radiative mode

$$O^+ + e \rightarrow O + h\nu \tag{8.20}$$

Recombination coefficients for radiative processes are typically on the order of 10^{-12} cm^3 sec^{-1}, perhaps 5 orders of magnitude slower than dissociative recombination. Ion-interchange reactions with molecular constituents are also more rapid than radiative recombination; therefore, in the terrestrial F_1 region, the following processes probably control the electron loss rate:

$$O^+ + N_2 \rightarrow N + NO^+ \tag{8.7d}$$

$$O^+ + O_2 \rightarrow O + O_2^+ \tag{8.7e}$$

The rate-controlling processes are followed by:

$$NO^+ + e \rightarrow N + O \tag{8.9a}$$

$$O_2^+ + e \rightarrow O + O \tag{8.9b}$$

The ion-interchange reaction is slower and controls the electron disappearance rate.

In a Martian or Cytherean F_1 region, the interchange would probably involve CO_2 (a very rapid process[17]):

$$O^+ + CO_2 \rightarrow CO + O_2^+ \tag{8.7h}$$

followed by:

$$O_2^+ + e \rightarrow O + O \tag{8.9c}$$

When CO_2 is photolyzed (see Chapter 4), one of the products is atomic oxygen. Hence, one might reasonably expect that an F_1 type of region would be found in the CO_2-rich atmospheres of Mars and Venus; this would be a direct analogy to the terrestrial F_1 region which is made possible by the photolysis of O_2. However, a distinct F_1 peak is obviously not seen in Figures 8.13 and 8.14. Attempts to hypothesize significant concentrations of O atoms (and, hence, a supply of O^+ ions) for an F_1 explanation of the observed electron profiles simply make the agreement of models and observations worse. We must, therefore, conclude that a region directly analogous to the terrestrial F_1 region does not exist on Mars or Venus. A full explanation of why this is the case must await further details of Mars and Venus atmospheric compositions as well as a definitive unraveling of CO_2 photochemistry. This does not imply, however, that some kind of F-region does not occur at high altitudes on Mars and Venus. Processes involving He and deuterium ions have been hypothesized and appear to be reasonably successful for models of the Venus ionosphere at altitudes well above the observed ionization peak; however, the existence of the large (compared with terrestrial experience) abundance of helium necessary for these models has yet to be confirmed experimentally or satisfactorily explained by neutral atmosphere models. This is considered in more detail in Chapter 9.

8.5 Variations

From the foregoing discussions, it is evident that variations in the ionizing flux and spectral distribution or in atmospheric density or composition will affect ionospheric characteristics. Such variations do indeed occur, some regularly and some sporadically. The sporadic variations will be considered in Chapter 10. Regular variations can be classified in accordance with the time scale of occurrence; these are principally diurnal, seasonal, and solar-cyclic.

8.5.1 Diurnal Variations

Even a cursory examination of the continuity equations reveals that large differences should be found between the daytime and nighttime ionosphere. When the principal ionizing source, the sun, drops below the horizon, Σq_i effectively becomes zero except for a small component due to GCR. Photodetachment processes also cease and photochemical products begin to disappear.

In the D-region, electrons disappear through attachment processes, although a residual ionosphere may be found at night because of collisional and associative detachment processes and because of a small Lα flux which is scattered by the hydrogen corona into the dark hemisphere and ionizes nitric oxide. Ions will disappear via ion-ion recombination processes. Figure 8.8 illustrates the magnitude of change of the negative-ion to electron ratio, λ, from day to night (neglecting associative attachment and nighttime ionization sources). E-region electron concentrations also diminish rapidly at night mainly because of ion-electron recombination, although here again a small nighttime ionization source (Lβ) is provided by scattering from the hydrogen corona. The F_1 region will be affected not only by eliminating the photoionization source, but also because the photochemical production of the principal atomic species ceases. In all cases, of course, the disappearance

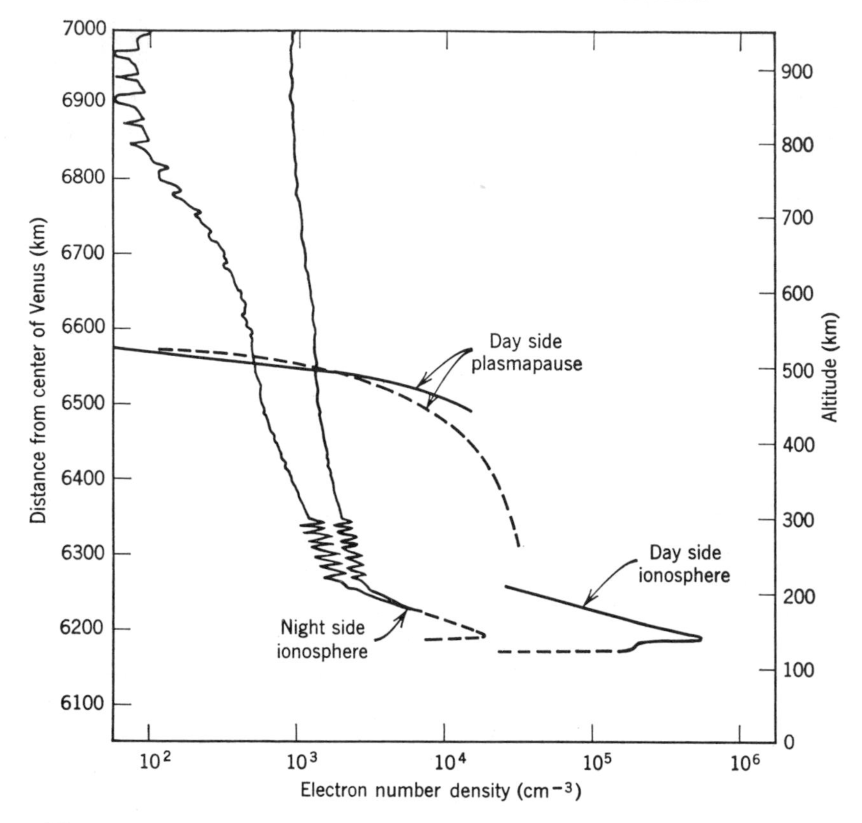

Figure 8.15 Daytime and nightime ionospheres of Venus. Electron concentration profiles derived from observations of the occulation of radio signals from the Mariner 5 spacecraft. (After Fjeldbo and Eshleman, *Radio Science* **4,** 879 (1969); reprinted by permission of the American Geophysical Union.)

rate diminishes with time inasmuch as the concentration of recombining species decreases.

The onset of dusk varies with altitude; hence, electron disappearance begins later at F_1 region altitudes than at D-region altitudes. Electron loss rates are also slower at F-region temperatures and particle number densities. Quasi-equilibrium assumptions are more difficult to justify at night; however, they can be justified when the ion disappearance rates become insignificant in comparison with ion-interchange rates. At twilight or dawn, Eqs. (8.16) are difficult to solve because several components are time-dependent and the functional relationships are not known. This is especially true in the lower D-region where attachment and detachment processes are important but the important species and reactions have not been identified.

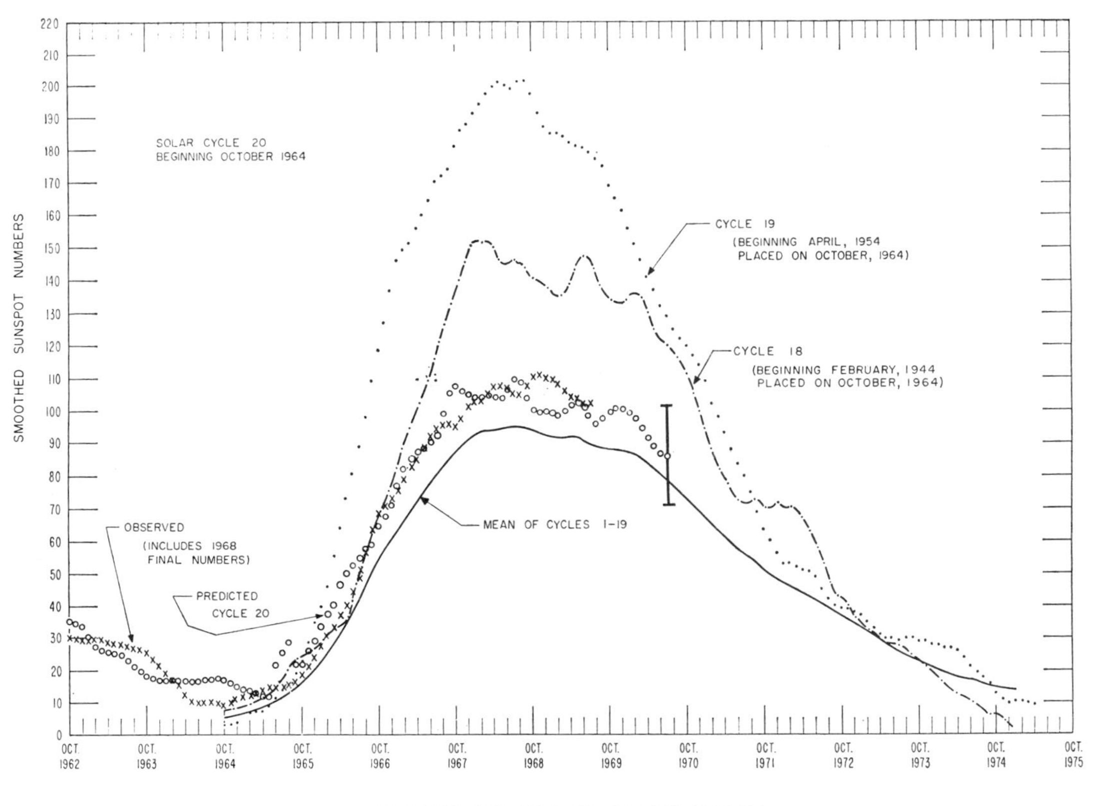

Figure 8.16 Predicated and observed sunspot numbers. (Reprinted from *Solar Geophysical Data*, No. 306, Part I, Feb. 1970, ESSA Research Laboratories by permission of the World Data Center A.)

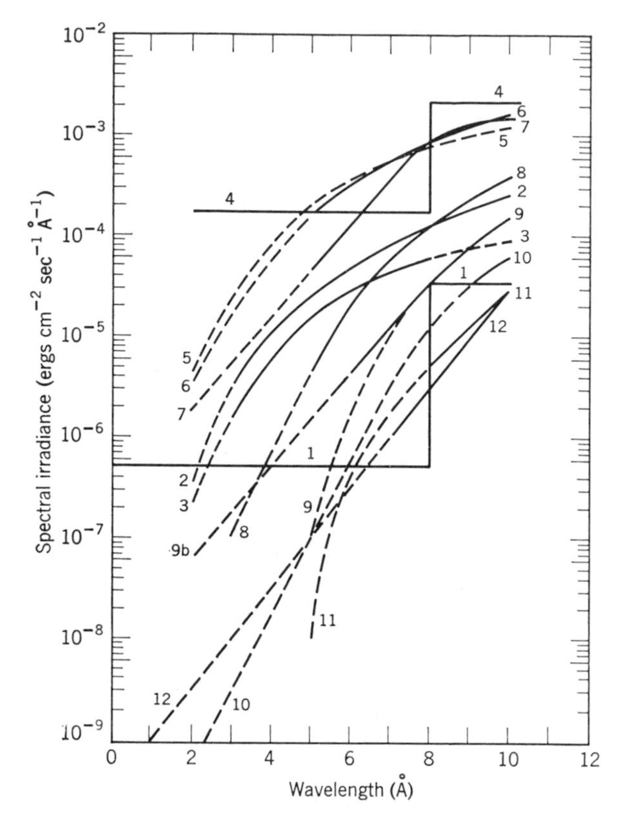

Figure 8.17 Non-flare solar X-ray spectra over a solar activity cycle. (After Poppoff, Whitten, and Edmonds, *JGR* **69**, 4081 (1964); reprinted by permission of the American Geophysical Union.)

A small but definite ionization peak is observed on the night side of Venus at almost the same altitude as the day side ionosphere (Figure 8.15). Several explanations have been tendered; these include:[21] charged particle ionization, ionization by Lyman α radiation scattered into the night side by the hydrogen corona, direct transport of ions from the day side to the night side at the level of the ionization peak, and a two-step process that involves transport from the day side to night side at a high altitude and vertical transport down to the ionization peak. Although none of these can be definitely ruled out, the most plausible explanation is the last one; again, as with the *F*-region

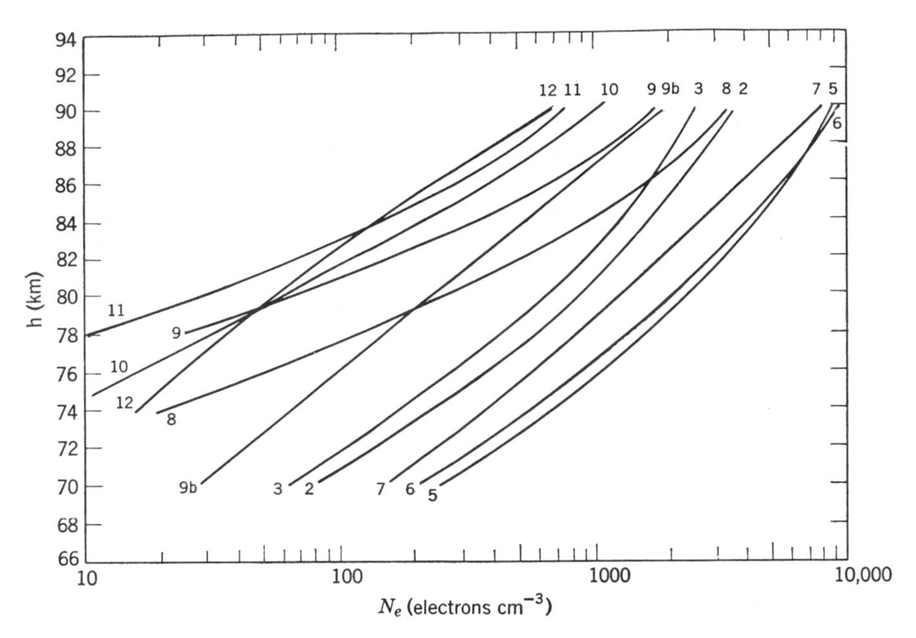

Figure 8.18 Electron concentrations corresponding to fluxes illustrated in Figure 8.15. (After Poppoff, Whitten, and Edmonds, *JGR* **69,** 4081 (1964); reprinted by permission of the American Geophysical Union.)

hypothesis (Section 8.5.3), the explanation requires a large abundance of helium.

8.5.2 Seasonal Variations

Normal seasonal variations are caused by two effects. The zenith angle of the sun is a function of the season and this is reflected through the secant term in Eq. (8.6a). In addition, temperature and density of the upper atmosphere vary with the season; these are discussed in Chapter 3.

Anomalous increases in radio absorption are observed during the winter. These "winter anomalies" are attributed to temporary increases in *D*-region electron concentrations. The most plausible cause of such an electron concentration enhancement is a corresponding enhancement of the nitric oxide concentration, which in turn can be explained by the effect of *D*-region warming on the equilibrium photochemistry of nitric oxide and by downward transport processes. Both the warming and the vertical transport are postulated to be the results of planetary wave action.

8.5.3 Solar Cyclic Variations

The activity of the sun as evidenced by changes in sunspot number varies over an 11-year cycle. (See Figure 8.16.) The radiation of the sun in the shorter wavlengths of interest in ionospheric research is particularly affected by this cyclic variation. Fluxes of X-rays, which are important to the lower ionosphere, vary by orders of magnitude. This is illustrated in Figure 8.17 and the effect on *D*-region electron concentration is illustrated in Figure 8.18. The GCR component varies with solar activity; this effect is also illustrated in Figure 8.18. The relationship between 10.7 cm radio noise, which varies with the solar cycle, and upper atmosphere heating is discussed in Chapter 3; consideration of ionospheric heating is deferred to Chapter 9.

A consequence of increased upper atmospheric temperature with increased solar activity is that scale heights of atmospheric species also increase. The increase in scale height results in the upward movement of heavier species; hence, the relative concentrations of O, N_2, and O_2 change such that the O_2 and N_2 molecules become more important in the *F*-region.

PROBLEMS

8.1. Derive expressions for the maximum ionization rate q_{max} [Eq. (8.5)] and for z_{max} [Eq. (8.4)].

8.2. The Chapman equations derived in this chapter (8.3a, b) assume a constant scale height. For a more realistic atmosphere in which the scale height varies with altitude as $H = H_0 + \alpha z$, write expressions for (a) ionization rate $q(z)$, (b) maximum ionization rate q_{max}, and (c) height of maximum ionization z_{max}. *Hint:* See Chapter 3.

8.3. a. Derive Eqs. (8.18a) and (8.17b) from steady state ($dN_e/dt = dN^+/dt = dN^-/dt = 0$) forms of Eq. (8.14a, b, c).
b. Assume $\alpha_D = \alpha_i$ and derive a form of (8.15b) that does not involve q.
c. Derive Eq. (8.15a, b) from (8.14a, b, c).

8.4. Summarize in tabular form the distinctive features of the *D*, *E*, and F_1 regions.

8.5. Using the concept of an effective recombination coefficient as in Eq. (8.18a), (a) convert Chapman equation for $q(z)$ to $N_e(z)$ assuming quasi-equilibrium conditions, (b) find an expression for the altitude for maximum N_e, and (c) find an expression for maximum N_e.

8.6. For an *E*-region, assume $\sum_{ij} k_{ij} N_i^+ n_j = \sum_{kl} k_{kl} N_k^+ n_l$.
a. For quasi-equilibrium conditions, find $\sum q_i$ necessary to maintain a positive ion concentration of 10^4 cm^{-3} at 300°K. Assume that ion species is O_2^+. See Eq. (8.16b).

b. Assume that nightfall in the E-region is instantaneous and independent of wavelength of incident radiation and that $N_e = 10^5$ cm^{-3} at nightfall. How long will it be before the concentration N_e drops to 10^4, 10^3, 10^2 cm^{-3}? Assume that positive ion species is NO^+ and temperature is 300°K.

8.7. Derive the equation for the altitude at which the optical depth is unity, i.e., the altitude at which $I(z)/I_\infty = 1/e$ Assume an isothermal atmosphere. What would that altitude be for a Mars-like atmosphere ($H = 10$ km, $n_0 = 2 \times 10^{17}$ cm^{-3}), 1 Å X-rays ($Q_a = 10^{-22}$ cm^2), and $\chi = 0°$?

8.8. What would be the principal reason for expecting to find much higher electron concentrations in a CO_2-N_2 Martian lower atmosphere than in a O_2-N_2 terrestrial atmosphere? Compare recombination rates (not coefficients) for CO_2-N_2 and O_2-N_2 atmospheres with attachment rates for O_2-N_2 atmosphere and assume O_2 concentration of 10^{16} molecules cm^{-3} (typical of 50 km terrestrial) and N_e of 10^3 cm^{-3}.

8.9. For a Martian F_1 region:

a. Assume that at an altitude of 125 km, there is an atomic oxygen concentration of 10^9 cm^{-3}, and an EUV flux (in the range from 200–900 Å) of 5×10^{10} photons cm^{-2} sec^{-1}. If Q_i is 5×10^{-18} cm^2, find $\sum_i q_i$.

b. If the neutral CO_2 concentration is 10^5 cm^{-3}, find N^+ when the ionosphere is in equilibrium.

c. Assume that the concentration of O is 10^{10} cm^{-3} and CO_2 is 10^{12}. Find concentration of O^+ at equilibrium. Would these conditions really describe an F_1 region?

d. Assume that dusk occurs instantaneously, how long would it take N^+ to drop to 10^{-1}, 10^{-2}, 10^{-3} of pre-dusk values? Use data computed above.

8.10. Construct the electron concentration profile due to GCR in the Martian atmosphere for altitudes 0–60 km, assuming that the ionization rate for GCR in the Martian atmosphere is the same as it is in the terrestrial atmosphere for altitudes with equal neutral particle density. Assume that the Martian atmosphere is pure CO_2, has a scale height of 10 km, and a surface concentration of 2×10^{17} molecules cm^{-3}, and that $\alpha_D = 4 \times 10^{-7}$ cm^3 sec^{-1} for CO_2^+ ions.

8.11. In Problem 10, assume that the Mars atmosphere is composed of CO_2 and O_2 in the ratio $1{:}10^{-3}$. Construct profiles of λ, N_e, N^-, and N^+.

8.12. a. Derive the following expression for the negative ion-electron number density ratio λ which is sometimes useful in D-layer analysis:

$$\lambda = -\frac{1}{2}\left(1 + \frac{D}{\alpha_i N_e}\right) + \frac{1}{2}\sqrt{\left(1 + \frac{D}{\alpha_i N_e}\right)^2 + \frac{4A}{\alpha_i N_e}}$$

b. Assume that $\alpha_i = \alpha_D$ and derive a formula for λ that is independent of N_e.

Hint: D is the summation of all detachment rates and A is the summation of all attachment rates; the above is the root of a quadratic equation.

8.13. By comparing ion species lifetimes, show that:
a. Electron attachment is the controlling electron loss process in the lower terrestrial *D*-region,
b. Recombination with N_2^+ cannot be an important electron loss mechanism in the terrestrial *E*-region, and
c. Ion-interchange controls the electron loss rate in the F_1 region.

REFERENCES

1. W. R. Webber, *J. Geophys. Res.* **67,** 5091 (1962).
2. S. K. Mitra, *The Upper Atmosphere*, 2nd Ed., The Asiatic Society, Calcutta, 1952.
3. D. R. Bates and H. S. W. Massey, *Proc. Roy. Soc.* **A192,** 1 (1947).
4. I. G. Poppoff and R. C. Whitten, *J. Geophys. Res.* **67,** 2986 (1962).
5. I. G. Poppoff, R. C. Whitten, and R. S. Edmonds, *J. Geophys. Res.* **69,** 4081 (1964).
6. M. Nicolet and A. C. Aikin, *J. Geophys. Res.* **65,** 1469 (1960).
7. R. C. Whitten and I. G. Poppoff, *Physics of the Lower Ionosphere*, Prentice-Hall, Inc., Englewood Cliffs, N.J., 1965.
8. F. C. Fehsenfeld, A. L. Schmeltekopf, H. I. Schiff, and E. E. Ferguson, *Planet. Space Sci.*, **15,** 373 (1967).
9. R. E. LeLevier, and L. M. Branscomb, *J. Geophys. Res.* **73,** 27 (1968).
10. R. C. Whitten and I. G. Poppoff, *J. Geophys. Res.* **67,** 1183 (1962).
11. A. Kliore, D. L. Cain, G. S. Levy, Von R. Eshleman, G. Fjeldbo, F. D. Drake, *Science* **149,** 1243 (1965).
12. G. W. Fjeldbo, W. C. Fjeldbo, and Von R. Eshleman, *J. Geophys. Res.* **71,** 2307 (1966).
13. G. Fjeldbo, W. C. Fjeldbo, Von R. Eshleman, *Science* **153,** 1518 (1966).
14. J. W. Chamberlain and M. B. McElroy, *Science* **152,** 21 (1966).
15. Michael B. McElroy, *Ap. J.*, **150,** 1125 (1967).
16. Mariner Stanford Group, *Science* **158,** 1678 (1967).
17. R. B. Norton, E. E. Ferguson, F. C. Fehsenfeld, and A. L. Schmeltekopf, *Planet. Space Sci.* **14,** 969 (1966).
18. S. Chapman, *Proc. Phys. Soc.* **43,** 26 and 484 (1931).
19. D. M. Hunten and M. B. McElroy, *J. Geophys. Res.* **73,** 2421 (1968)
20. M. B. McElroy, *J. Geophys. Res.* **74,** 29 (1969).
21. M. B. McElroy and D. F. Strobell, *J. Geophys. Res.* **74,** 1118 (1969).

GENERAL REFERENCES

Alex E. S. Green, Ed., *The Middle Ultraviolet: Its Science and Technology*, John Wiley & Sons, New York, 1966. A good source of information on solar UV emission and interactions with atmospheric species.

C. O. Hines, Irvine Paghis, Theodore R. Hartz, and Jules A. Fejer, *Physics of the Earth's Upper Atmosphere*, Prentice-Hall, Englewood Cliffs, N.J., 1965. Somewhat outdated compendium of articles on the ionosphere. Good for a general survey.

S. K. Mitra, *The Upper Atmosphere*, 2nd Ed, The Asiatic Society, Calcutta, 1952. This is the classic work, much of it is out-of-date, but it should be pursued for historical interest.

P. J. Nawrocki, and R. J. Papa, *Atmospheric Processes*, Prentice-Hall, Englewood Cliffs, N.J., 1963. A collection of interesting and sometimes useful data, but again much of it is obsolete.

J. A. Ratcliffe, Ed., *Physics of the Upper Atmosphere*, Academic Press, New York, 1960. This was the authoritative work for many years. As with many of these books, some of this discussion is quite pertinent, but much is out-of-date.

K. Rawer, *The Ionosphere*, Ungar Publishing Co., New York, 1952. Written by one of the pioneers in ionospheric physics, it should also be pursued for historical interest.

R. C. Whitten, and I. G. Poppoff, *Physics of the Lower Ionosphere*, Prentice-Hall, Englewood Cliffs, N.J., 1965. A monograph dealing specifically with *D* and lower *E*-region processes, it is also becoming outdated.

T. Yonezawa, Theory of formation of the ionosphere, *Space Science Reviews* **5,** 3 (1966). An excellent treatment of *F*-region ionospheric processes.

Henry Rishbeth, and Owen K. Garriott, *Introduction to Ionospheric Physics*, Academic Press, New York and London, 1969. A very good and thorough monograph on the ionosphere, this book treats the upper terrestrial ionosphere in considerable detail. It is highly recommended for those readers who wish to learn more about the complexities of ionospheric observations.

9

Structure of the Upper Ionosphere

In the preceding chapter we investigated in considerable detail the processes which are responsible for the formation and maintenance of the earth's ionosphere below an altitude of ~200 km. Under quiet daytime conditions all charged constituents are in, or nearly in, a condition of photochemical equilibrium. In fact, photochemical processes determine the structure at all times of the day and night. At greater altitudes, i.e., above ~250 km, this is not the case; at these heights the chemical loss rate of the ions is comparable to their diffusion rate through the neutral gas, and transport mechanisms (diffusion) begin to influence the ion and electron number densities. With ascending altitude, diffusion becomes more and more important, finally dominating the picture at great heights. The competition of chemical loss (recombination) and diffusion leads to the formation of a peak in the number density profile of the ions and electrons, the well-known F_2 peak. In the F_2 region itself the dominant ion is O^+. However, at altitudes of the order ~1000 km, H^+ is present in greater concentrations owing to charge transfer. At first thought, it may seem that this region, the protonosphere, should have many of the characteristics of the exosphere. In fact, because of the long range nature of the Coulomb force between ions, this is not true.

Chemical loss and diffusion are processes which, for an isothermal ionosphere, are relatively easy to treat analytically. However, the ionospheric models constructed by simple solution of the continuity equation do not predict many of the peculiarities or anomalies of F_2 region behavior. These are for the most part associated with the influence of the geomagnetic field on ionic diffusion. As we saw in Chapter 7, ions and electrons can move freely along the lines of force in regions of low neutral particle density, but not across the field lines unless electric fields are present. Such fields are usually present and cause the drifts studied previously. Electrodynamic drifts and geomagnetic effects are thought to be closely associated with most of the anomalies in F_2 region behavior.

In the ionosphere below ~100 km all the charged species including electrons are in or very close to thermal equilibrium with the neutral gas. At higher altitudes this is not necessarily true, especially for electrons whose small mass requires a large thermal equilibration time. Of course, the energy distribution of the electrons is very nearly Maxwellian because *e-e* collisions occur via the very long range Coulomb force. However, the temperature of the electron gas is expected to be substantially larger than that of the ambient atmosphere at certain altitudes. This prediction is confirmed by observation. In this chapter we shall discuss the electron temperature at altitudes above 120 km, thus intruding upon the ground of the previous chapter. We take this course because thermal transport is significant as low as 120 km. Hence, dividing our study into two parts at some higher altitude (e.g., 250 km, where ambipolar diffusion sets in) would be artificial and probably confusing to the reader.

In addition to the so-called "anomalies" the F_2 region is also characterized by the occurrence of irregularities, such as large scale irregularities, spread F, and traveling ionospheric disturbances. Of these, spread F has been most intensively investigated, although the exact nature of the causative mechanism is not known yet. All of these aspects of the upper ionosphere will be discussed in this chapter.

In the preceding chapter we also discussed the lower ionospheres of Mars and Venus. We found that unlike the ionosphere of the earth, F_1 or E-layers characterize the most prominent portions of the ionospheres of those planets; there is no prominent F_2 layer as there is above the earth. Nevertheless, there does exist a "ledge" of ionization in the "topside" ionosphere of Venus (we do not yet know whether such a ledge exists above Mars). Here ion and electron diffusion is significant, giving the region many of the characteristics of the earth's F_2 region. The character of the ionosphere of Jupiter is so speculative that it does not seem worthwhile to discuss it here.

9.1 Formation of the F_2 Layer

In this section we shall first briefly discuss photoionization in the upper ionosphere after which a more complete treatment of electron loss processes and ambipolar diffusion will be given.

9.1.1 Photoionization

Photoionization of atmospheric constituents at E and F_1 region heights was discussed in considerable detail in the preceding chapter and it is only necessary to make a few remarks here about the situation above 200 to 250 km. It is evident from Figure 8.5 that radiation of wavelength $\lambda \lesssim 800$ Å is mainly responsible for ionization in the F_2 region. The threshold wavelength at 800 Å results from the large ionization potentials of the dominant constituents, atomic oxygen (13.614 eV) and molecular nitrogen (15.58 eV). The minor constituent, molecular oxygen, has a lower threshold (12.08 eV) and is thus ionizable by radiation of wavelength $\lambda < 1029$ Å. However, its relative concentration is low enough that it does not contribute significantly to the total ionization rate.

The ionization rate profile can be approximated by an appropriate Chapman function discussed in previous chapters. Reference to Figure 8.5, reveals, however, that the profile is quite straight (on a semi-log plot) down to nearly 200 km. In other words, in the optically thin region, where attenuation is not important, the ionization rate of a constituent is proportional to its number density. Therefore, it is unnecessary to employ Chapman functions when constructing models of the F_2 layer although some authors have done so. Computations are greatly simplified by using exponential functions and we shall follow this prescription in the present chapter. Of course, as the solar zenith angle is increased, the height of the ionization maximum increases and it may be necessary to employ a Chapman function at very large zenith angles.

9.1.2 Electron Loss

The ionic species formed initially in the F_2 region are the same as at lower altitudes, O^+, N_2^+, and O_2^+. Of these, the first two are the most important. The molecular ions can recombine directly via the usual dissociative processes

$$e + N_2^+ \rightarrow N + N \tag{9.1}$$

$$e + O_2^+ \rightarrow O + O \tag{9.2}$$

which occur rapidly because of the large recombination coefficients (see Section 2.6.1). Atomic oxygen is a different case. The mechanism of direct recombination is necessarily radiative and is thus very slow (see Section 2.6.1). A more rapid means of electron removal is offered by the chain of processes

$$O^+ + N_2 \rightarrow NO^+ + N \tag{9.3}$$

and

$$O^+ + O_2 \rightarrow O_2^+ + O \tag{9.4}$$

followed by

$$NO^+ + e \rightarrow N + O \tag{9.5}$$

and process (9.2).

One can easily obtain the electron loss rate under quasi-chemical equilibrium conditions. Neglecting photoionization of N_2 and O_2, the pertinent rate equations are

$$q(O^+) = (k_3[N_2] + k_4[O_2])[O^+] \tag{9.6}$$

$$\alpha_2[O_2^+]N = k_4[O_2][O^+] \tag{9.7}$$

$$\alpha_5[NO^+]N = k_3[N_2][O^+] \tag{9.8}$$

where

$$N = [O^+] + [NO^+] + [O_2^+] \tag{9.9}$$

is the total ion (and electron) number density, $q(O^+)$ is the photoionization rate of atomic oxygen, and the subscripts on the rate coefficients refer to processes (9.2) to (9.5). By eliminating $[(O^+]$ and $[NO^+]$ from Eq. (9.9) with the aid of Eqs. (9.7) and (9.8), and substituting the resulting equation for $[O^+]$ into Eq. (9.6), one obtains a relation between $q(O^+)$ and the total ion number density

$$q(O^+) = \left(\frac{(k_3[N_2] + k_4[O_2])N}{1 + \dfrac{k_4[O_2]}{\alpha_2 N} + \dfrac{k_3[N_2]}{\alpha_5 N}} \right) = \beta N \tag{9.10}$$

We call β the electron loss rate. As we saw in Section 2.6.1, the rate constants corresponding to reactions (9.3) and (9.4) are, respectively, about 2 to 4 $\times$ 10^{-12} and 2 to 4 $\times$ 10^{-11} cm^3 sec^{-1}. For typical values of α_2 and α_5 ($\sim 10^{-7}$ cm^3 sec^{-1}), and ion number density $N(\gtrsim 2 \times 10^5$ cm$^{-3})$ one can show that

$$\left(\frac{k_4[O_2]}{\alpha_2} + \frac{k_3[N_2]}{\alpha_5} \right) \frac{1}{N} \ll 1 \tag{9.11}$$

at altitudes above 250 km. Then we have

$$\beta \approx k_3[N_2] + k_4[O_2], \tag{9.12}$$

which, because N_2 and O_2 have nearly the same scale height H, behaves almost exponentially in an isothermal atmosphere;

$$\beta = \beta_0 \exp \left[- \left(\frac{z - z_0}{H} \right) \right] \tag{9.13}$$

9.1.3 Ambipolar Diffusion

At an altitude of $\sim$250 km or somewhat higher where the diffusion rate of the ions through the neutral gas becomes comparable to the electron loss rate, ambipolar diffusion is a highly important process. That is, the charged particles can move long distances before they are removed by one of the loss processes. Before launching into a mathematical treatment it is worth briefly repeating the remarks in Chapter 7 concerning the effect of the geomagnetic field. Diffusion of ions across field lines is greatly inhibited if the neutral particle number density is low and electric fields are sufficiently weak. Hence, at middle and high latitudes we shall assume that all charged particles diffuse in the direction of the geomagnetic field. In the equatorial region, where the lines of force are nearly horizontal, it is necessary to modify the simple treatment developed in the following paragraphs; the modification will be briefly discussed later in the chapter.

The starting point for discussing ambipolar diffusion is the appropriate Boltzmann transport equation [see Eqs. (2.62) and (7.2)] for each species of charged particles. In Chapter 7 we neglected the gravitational and pressure terms which are unimportant at relatively low altitudes. However, in the F_2 region all the terms must be retained.

$$m_k \frac{d\bar{\mathbf{v}}_k}{dt} = \pm e\mathbf{E} - m_k\mathbf{g} - \frac{1}{N_k}\nabla p_k - m_k\nu_{kn}(\bar{\mathbf{v}}_k - \bar{\mathbf{v}}_n) \tag{9.14}$$

where $\bar{\mathbf{v}}_k$ is the average velocity of the kth species, $\mathbf{g}$ is the acceleration due to gravity ,p_k is the partial pressure exerted by the kth species, ν_{kn} is the collision frequency of the kth species with neutral particles, and the sign of the first term on the right hand side is positive for positively charged ions and negative for electrons. Under conditions of equilibrium the acceleration $d\mathbf{v}_k/dt$ vanishes and, with the aid of the equation of state of an ideal gas

$$p_k = N_k k T_k \tag{9.15}$$

we can write, for the major species of positive ions,

$$\nabla(N_+ kT_+) = N_+ e\mathbf{E} - m_+ N_+ \mathbf{g} - m_+ N_+ \nu_+ (\bar{\mathbf{v}}_+ - \bar{\mathbf{v}}_\mathbf{n}) \tag{9.16}$$

and, for electrons

$$\nabla(N_e kT_e) = -N_e e\mathbf{E} - m_e N_e \mathbf{g} - m_e N_e \nu_e (\bar{\mathbf{v}}_\mathbf{e} - \bar{\mathbf{v}}_\mathbf{n}) \tag{9.17}$$

Since charge neutrality demands that $N_e = N_+$ in the F_2 region, the sum of Eqs. (9.16) and (9.17) yields

$$\nabla(NkT_+ + NkT_e) = -m_+ N\mathbf{g} - m_+\nu_+ N\mathbf{w}_\mathbf{D} \tag{9.18}$$

where the approximations $m_e \ll m_+$, $m_e\nu_e \ll m_+\nu_+$ have been made. Furthermore, we have assumed an atmosphere at rest ($\mathbf{v}_\mathbf{n} = 0$), and that convective drifts are absent; equilibrium demands that $\bar{\mathbf{v}}_\mathbf{e} = \bar{\mathbf{v}}_+ = \mathbf{w}_\mathbf{D}$, the charged particle drift velocity. In a more rigorous treatment, electron-ion terms would have been included in Eqs. (9.16) and (9.17), but they cancel when the two equations are added because of Newton's third law of motion. Rearrangement of Eq. (9.18) leads to an equation for the diffusion velocity of the major species ion

$$w_D = -\frac{1}{m_+\nu_+ N}\left[\frac{\partial}{\partial z}(Nk(T_+ + T_e)) + m_+ gN\right] \tag{9.19}$$

If one wishes to include the effects on diffusion of geomagnetic control, it is necessary to use the component of the pressure gradient and gravitational force in the direction of the field lines:

$$\sin I \frac{\partial}{\partial z}[Nk(T_+ + T_e)] \quad \text{and} \quad m_+ g \sin I\, N$$

where I is the angle between the field line and the horizontal (dip angle). Then Eq. (9.19) is written

$$w_D = -\frac{\sin I}{m_+\nu_+}\left\{\frac{1}{N}\frac{\partial}{\partial z}[Nk(T_+ + T_e)] + m_+g\right\} \tag{9.20}$$

We are now in a position to discuss the continuity equation.

9.2 The Continuity Equation

The continuity equation, which is simply a statement of the law of conservation of matter, was introduced earlier in Sections 2.3.4 and 3.4. In fact, the whole discussion of molecular diffusion in a neutral atmosphere which was presented in Chapter 3 is very similar to the present treatment of ambipolar diffusion. In the notation used in the preceding sections the continuity equation is

$$\frac{\partial N}{\partial t} = q - \sin I\frac{\partial}{\partial z}(Nw_D) - \beta N \tag{9.21}$$

if the geomagnetic constraint is included. Written out in full, it is

$$\frac{\partial N}{\partial t} = q + \frac{\sin^2 I}{m_+}\frac{\partial}{\partial z}\left[\frac{1}{\nu_+}\frac{\partial}{\partial z}(Nk(T_e + T_+)) + m_+\frac{gN}{\nu_+}\right] - \sin I\frac{\partial}{\partial z}(Nw_{DE}) - \beta N \tag{9.22}$$

where w_{DE} is the convective drift velocity arising from induced electric fields.

9.2.1 *Solutions for Simple Models of the F_2 Region*

Equation (9.22) is rather intractable unless we can suitably simplify it. The most useful simplification and one that is under some conditions qualitatively justifiable is the assumption of an isothermal atmosphere with $T_+ = T_n$. As we shall see later, this is not usually valid. Secondly, we neglect w_{DE}; this is probably not strictly valid either, particularly at night. In spite of these limitations it is instructive to solve the simplified equation because it demonstrates the origin of the most outstanding feature of the F_2 region, the F_2 peak and the resulting profile represents the electron number density fairly well near the peak. We write

$$\frac{\partial N}{\partial t} = q - \beta N + \sin^2 I\frac{\partial}{\partial z}\left[D\left(\frac{\partial N}{\partial z} + \frac{m_+g}{(1 + c)kT}\right)\right] \tag{9.23}$$

where $D = (1 + c)kT/m_+\nu_+$ is the diffusion coefficient and $c = T_e/T_n$. In the F_2 region the dominant ion is O^+ which must diffuse through a mixture of O and N_2. Because of symmetry effects diffusion of an ion through its parent gas is slower than through other neutral species.[1] At F_2 layer heights we need consider only the diffusion of O^+ through O since O is the major neutral constituent, then we can replace the quantity kT/m_+g by H, the scale height of atomic oxygen. Furthermore, the collision frequency ν_+ in an isothermal atmosphere can be expressed as

$$\nu_+ = \nu_0 \exp\left(-\frac{z - z_0}{H}\right) \quad \text{or} \quad D = D_0 \exp\left(\frac{z - z_0}{H}\right) \tag{9.24}$$

where z_0 is some reference altitude. Equation (9.23) thus takes the form

$$\frac{\partial N}{\partial t} = q - \beta N + D \sin^2 I \left[\frac{\partial^2 N}{\partial z^2} + \left(\frac{c + 2}{c + 1}\right) \frac{\partial N}{\partial z} + \frac{N}{(c + 1)H^2}\right] \tag{9.25}$$

In order to obtain solutions to it, one must make some assumptions concerning the altitude dependence of q and β. According to our earlier discussion concerning photoionization and electron loss, O^+ is the dominant species present at F_2 region altitudes, and so our ionization rate is expected to be proportional to the concentration of atomic oxygen:

$$q = q_0 \exp\left(-\frac{z - z_0}{H}\right) \tag{9.26}$$

where H is the scale height of O. Assuming the electron loss rate given by Eq. (9.13), we can express β as

$$\beta = \beta_0 \exp\left[-\frac{7}{4}\left(\frac{z - z_0}{H}\right)\right] \tag{9.27}$$

the derivation of which is left as an exercise. Using the foregoing expressions for q, β, and D, solutions of the continuity equation can be obtained analytically.

Before discussing such solutions further, it is necessary to investigate their asymptotic character, that is, their form at great heights where q and β can be neglected and at low altitudes where D is negligible. At equilibrium Eq. (9.25) can be approximated by

$$\frac{d^2 N}{dz^2} + \left(\frac{c + 2}{c + 1}\right) \frac{dN}{dz} + \frac{N}{(c + 1)H^2} = 0 \tag{9.28}$$

for which the solution is

$$N \sim C_1 e^{-(z - z_0)/H} + C_2 e^{-(z - z_0)/(c+1)H} \tag{9.29}$$

when production and loss are neglected. In this context the "$\sim$" sign signifies "asymptotic to." For the combination of C_1 and C_2 which is needed, one must resort to the appropriate boundary conditions. One possibility is that the upper ionosphere "tails off" to zero at an infinite height such that there is no influx of charged particles from outer space. In this case the drift velocity at great heights is zero and Eq. (9.19) predicts a scale height of

$$H_+ = \frac{(c + 1)kT}{mg} \tag{9.30}$$

for the ions and electrons. This condition corresponds to the second term on the right hand side of Eq. (9.29). The first term corresponds to a flow of charged particles sufficient to insure that the diffusive current (Nw_D) is divergenceless

$$\frac{\partial}{\partial z}(Nw_D) = 0 \tag{9.31}$$

Most of our knowledge gained from investigation of the upper ionosphere at midlatitude indicates that the asymptotic form given by the second term is the dominant one. We shall assume this to be the case in the remainder of the chapter and shall therefore choose the upper boundary condition for all F_2 region models to be

$$N \sim \exp\left(-\frac{z - z_0}{(c + 1)H}\right) \tag{9.32}$$

For the lower boundary condition we choose

$$\lim_{z \to -\infty} N(z) = 0 \tag{9.33}$$

At low altitudes, $D \sim 0$ and

$$q \approx \beta N \tag{9.34}$$

under quasi-equilibrium conditions. If q is due to ionizing radiation of a single wavelength, we have the famous "Chapman β" profile[2] for which a linear loss rate is assumed.

Since q decreases more slowly with altitude than does β, the electron number density must grow with altitude. On the other hand, in the asymptotic region at much higher altitudes, N decreases with increasing height. One concludes that N must have a maximum at some intermediate altitude. At this peak the lifetime of an electron-ion pair against recombination, τ_R, must be approximately equal to the time required for diffusive transport through one scale height, τ_D. By definition

$$\tau_D = \frac{H}{w_D} \tag{9.35}$$

or, from Eq. (9.20),

$$\tau_D = H^2(1 + c)/D \tag{9.36}$$

at the peak. Since

$$\tau_R = 1/\beta \tag{9.37}$$

we have[3]

$$\beta_{\text{peak}} = D_{\text{peak}}/H^2(1 + c) \tag{9.38}$$

Returning to Eq. (9.25), we wish to solve it for various conditions of ionization and electron loss rates. In order to make the form of the equation more tractable, it is convenient to transform the independent variable:

$$x = Ae^{\nu(z-z_0)/H} \tag{9.39}$$

where ν is a parameter to be chosen such that the continuity equation assumes its simplest form. As an illustrative example, let us assume equilibrium ($\partial N/\partial t = 0$) and

$$\begin{aligned} D &= D_0 \exp\left(\frac{z - z_0}{H}\right) \\ \beta &= \beta_0 \exp\left(-\frac{z - z_0}{H}\,a\right); \\ q &= q_0 \exp\left(-\frac{z - z_0}{H}\right) \end{aligned} \tag{9.40}$$

Equation (9.25) is recast as

$$\nu^2 \frac{d^2N}{dX^2} + \nu\left(\nu + \frac{c + 2}{c + 1}\right)\frac{1}{x}\frac{dN}{dx} - \nu^2 x^{-(1+a)/\nu-2}N + x^{-2}\frac{N}{c + 1} + \nu^2\frac{q_0}{\beta_0}\left(\frac{x}{A}\right)^{-(2/\nu)-2} = 0 \tag{9.41}$$

where

$$A = \frac{1}{\nu}\left(\frac{D_0}{\beta_0 H^2}\right)^{\nu/(1+a)}$$

The solution of Eq. (9.41) for various values of a and the boundary conditions (9.32) and (9.33) has been carried out, for example, by Yonezawa[4]† using the standard mathematical techniques such as the method of variation of parameters.[5] A plot of N for $a = 1$ is shown in Figure 9.1; the F_2 peak, which we discussed earlier in this section, is its most striking feature. Comparison of this profile with the observed profiles shown in Figure 9.2 verifies that our model is qualitatively correct. Computation of electron number density profiles for various values of a is left as an exercise.

† Yonezawa employed Chapman functions to describe q, but the results are not much different from an exponential dependence.

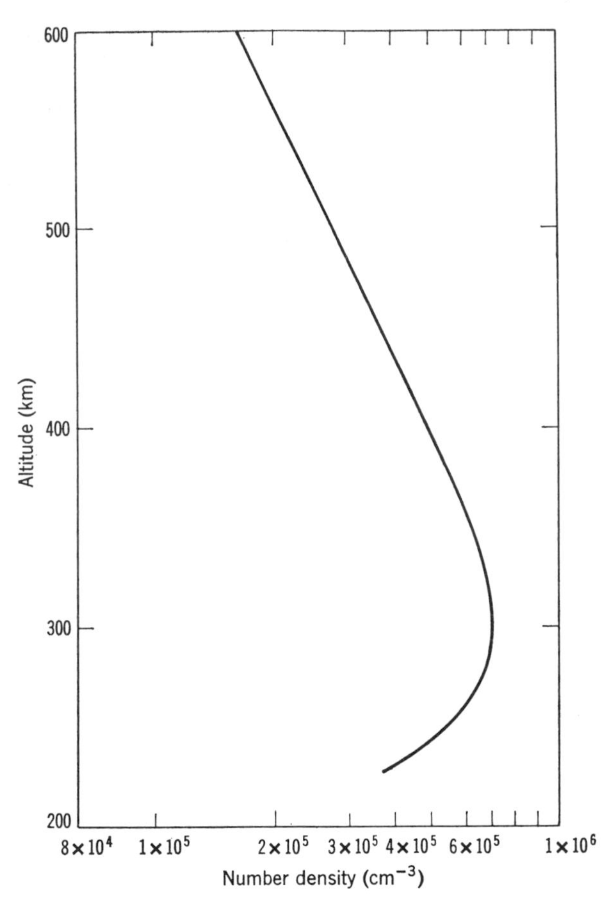

Figure 9.1 Electron number density profile computed from Eq. (9.41) with $a = 1$. The values of other parameters were $\beta = 2 \times 10^{-5}\ \text{sec}^{-1}$, $D = 2 \times 10^{10}\ \text{cm}^2\ \text{sec}^{-1}$, $H = 60$ km, all at 300 km altitude; also $T_e = 2200°$K; $T_+ = T_n = 1100°$K.

9.2.2 A Multicomponent Topside Ionosphere[11]

Observations by rocket- and satellite-borne mass spectrometers[12] have revealed the occurrence of many minor species of ions such as H^+, He^+, N^+, O^{++}, etc. In the altitude range in which the vertical distributions of all the ionic species are in diffusive equilibrium, the number density N_i of each species obeys an equation like Eq. (9.16) with $v_+ - v_n$ set equal to zero

$$\frac{kT_+}{N_i}\frac{dN_i}{dz} = eE - m_i g \tag{9.42}$$

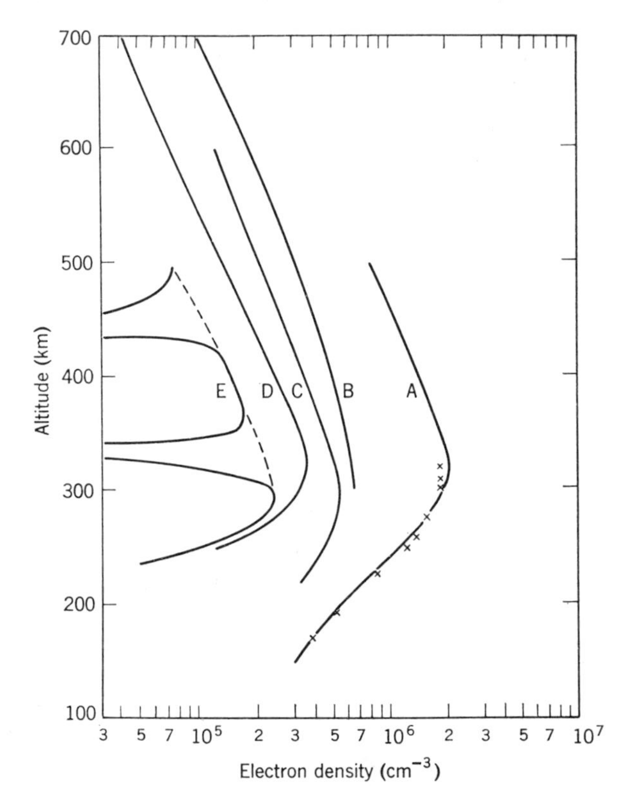

A Rocket Measurement of Posititive Ion. Densities at Latitute 28° N, time 1300 LMT, 22 Sept. 1959. The cross represents simultaneous ionosonde data for the same geographical area.

B Alouette topside sounder data, position 16 S, 74.3°W, time 2030 UT, 19 Feb. 1963.

C From rocket-to-ground CW propagation experiment at Wallops 1s., Va., time 1502 EST, 27 April 1961.

D Rocket measurement of positive ion densities above Wallops 1s., Va., time 2044 LMT, 9 Nov. 1960.

E Rocket measurement of positive ion densities above Woomera, Australia, time 2242 LMT, 1 May 1962. The troughs are associated with rocket outgassing; the broken line represents interpolated densities.

Figure 9.2 *F*-region ion and electron number densities.

The corresponding equation for the total ion number density N_+ is obtained by summing over i

$$\frac{kT_+}{N_+}\frac{dN_+}{dz} = eE - m_+ g \tag{9.43}$$

where m_+ is the *mean* ionic mass defined by

$$m_+ = \left(\sum_i m_i N_i\right) \Big/ N_+ \tag{9.44}$$

Use of a similar equation for the electron number density N_e

$$\frac{kT_e}{N_e}\frac{dN_e}{dz} = -eE - m_e g \approx -eE \tag{9.45}$$

enables one to eliminate the electric field E which results from charge separation. We obtain for N_+

$$\frac{1}{N_+}\frac{dN_+}{dz} = -\frac{m_+ g}{(1+c)kT_+} = -\frac{eE}{kT_e} \tag{9.46}$$

where $c = T_e/T_+$. Substituting this result for eE into Eq. (9.42), we have

$$\frac{1}{N_i}\frac{dN_i}{dz} = -\frac{m_i g}{kT_+}\left(1 - \frac{m_+}{(1+c)m_i}\right) = \frac{1}{H_i}\left(1 - \frac{m_+}{(1+c)m_i}\right) \tag{9.47}$$

which describes the vertical distribution of ionic species i. Thus, we see that in an ionosphere in diffusive equilibrium, the scale height of any ion species is related to the scale height H_i of the corresponding neutral species by

$$H_{\text{ion}_i} = H_i\left(1 - \frac{m_+}{(1+c)m_i}\right)^{-1} \tag{9.48}$$

9.2.3 The F_2 Layer at Equatorial Latitudes

At equatorial latitudes electrons and ions cannot diffuse vertically because the geomagnetic field lines are nearly horizontal. Hence the continuity equation must be suitably modified. The derivation of the diffusion term [i.e., the third term on the right hand side of Eq. (9.25)] under these conditions has been carried out by Kendall.[13] Because it is a rather tedious affair we shall content ourselves by merely writing it down:

$$D\sin^2 I\left(\frac{d^2N}{dr^2} + \frac{3}{2H}\frac{dN}{dr} + \frac{N}{2H^2}\right) + D\left(\frac{15\mu^4 + 10\mu^2 - 1}{(1+3\mu^2)^2}\right)\left(\frac{2}{r}\frac{dN}{dr} + \frac{N}{Hr}\right) \tag{9.49}$$

Here r is distance from the center of the earth and $\mu = \cos\theta$; θ is the magnetic co-latitude, which is related to the dip angle by $\sin I = 2\cos\theta/(1 + 3\cos^2\theta)^{1/2}$. Further treatment of the equatorial case is left for an exercise.

9.2.4 Time-Dependent Model of the F_2 Region

Time-dependent solutions of Eq. (9.25) are easy enough to obtain by standard methods such as the separation of variables (see Problem 9.3). At first glance it would seem that such solutions are appropriate to the condition immediately after sunset. However, the situation is much more complicated than has been indicated, principally because of three factors: the effects of electrodynamic drifts; thermal contraction of the electron-ion gas and thermally induced changes in the electron loss rate. For example, the cooling of the gas after the solar heat source is removed results in its contraction. The complete solution of the time-dependent continuity equation with drifts and thermal effects included has never been attempted. However, Rishbeth[14] has developed a time-varying model of the mid-latitude F_2 region that does partially include thermal changes. He found that many of the grosser features of the time variation of the F_2 region were reproduced by the model but that different diffusion rates had to be assumed for day and night in order to obtain agreement with observations. In particular, the daytime model required slow diffusion rates while the nighttime model required fast ones. Although atmospheric temperature changes were taken into account, the observed increase of electron density at sunset did not appear in the model. This increase is thought to be associated with cooling of the electron gas. Some of Rishbeth's results for conditions of low solar activity are presented in Figure 9.3 which is to be compared with some experimental data shown in Figure 9.4.

Neglect of thermal changes in the electron loss rate may have caused or contributed to the apparent change in diffusion rate. It has been shown by Ferguson and collaborators[15] that the rate of reaction (9.3) is increased substantially when the N_2 molecules are vibrationally excited. If the rate is much larger in the daytime due to collisions with energetic thermal electrons, the loss rate β can be expected to decrease after sunset when the electron gas cools. However, the magnitude of the change in β is unknown and we can do little more than offer "hand waving" arguments at the present time.

9.3 Anomalies

From what has already been said concerning F_2 region behavior, it should be obvious that it can not be adequately described by the simple continuity equation of the type Eq. (9.25). Departures from the structure predicted by this approach were originally called "anomalies" mainly to conceal ignorance of the true state of affairs. Some of these have since yielded to the efforts of various investigators; some have not. In the following subsections characteristics and (where known) the causes of these peculiarities will be discussed. Before doing so, however, we shall briefly consider two phenomena: electrodynamic drifts and nocturnal ionization which are undoubtedly associated with some of the "anomalies."

9.3.1 Electrodynamic Drifts

Electrodynamic drifts at E-region heights were discussed at length in Chapter 7. There we pointed out that electric fields can also cause drifts across the geomagnetic field lines in the F_2 region. In addition, motions of the neutral atmosphere can induce currents by viscous drag (see Chapter 5). Such convection currents, which should be added (vectorially) to the diffusion

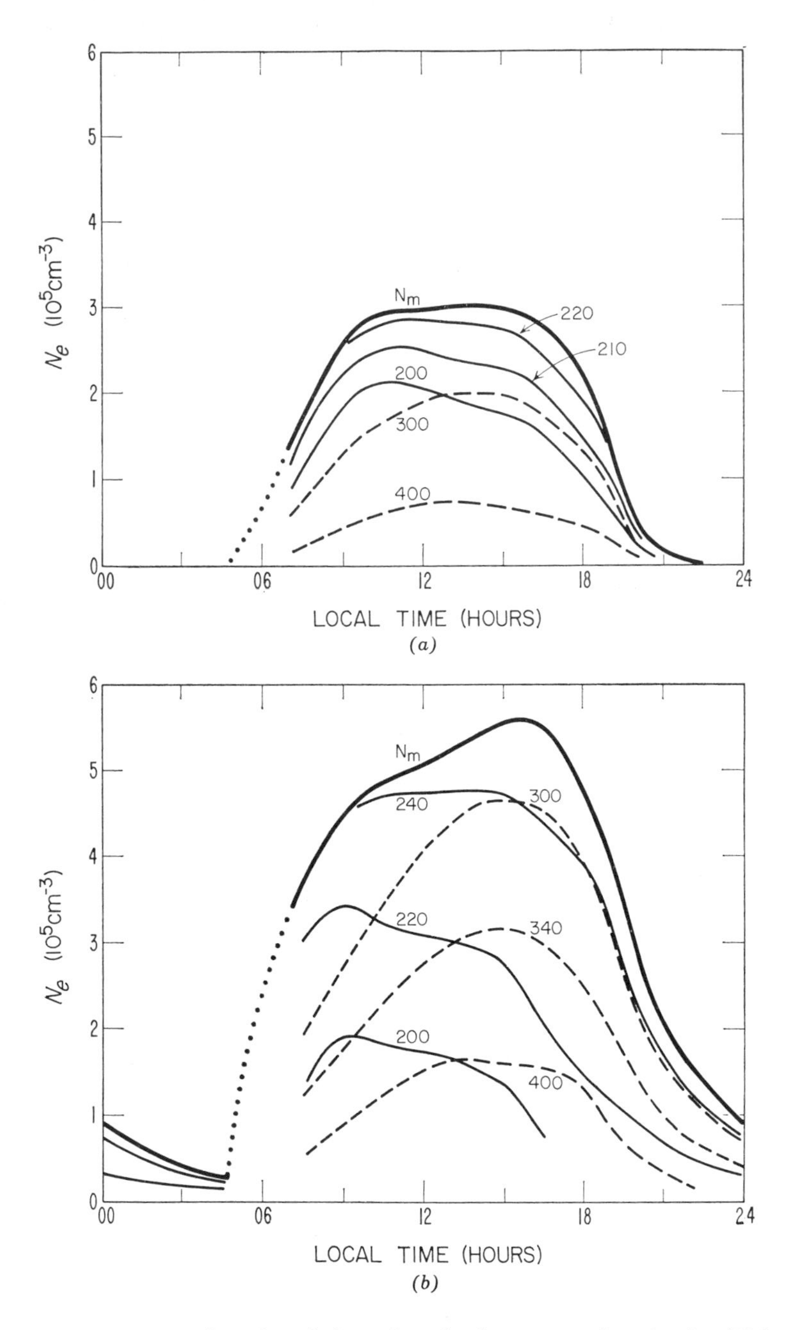

Figure 9.3 Diurnal variation of peak electron number density (N_m) (heavy solid line) and of electron number densities at several fixed heights (thin solid line) and above the peak (broken lines); (a) fast diffusion, low solar activity (S = 70); (b) slow diffusion, low solar activity (S = 70). (After H. Rishbeth, *J. Atmos. Terr. Phys.* **26,** 657 (1964); reprinted by permission of Pergamon Press.)

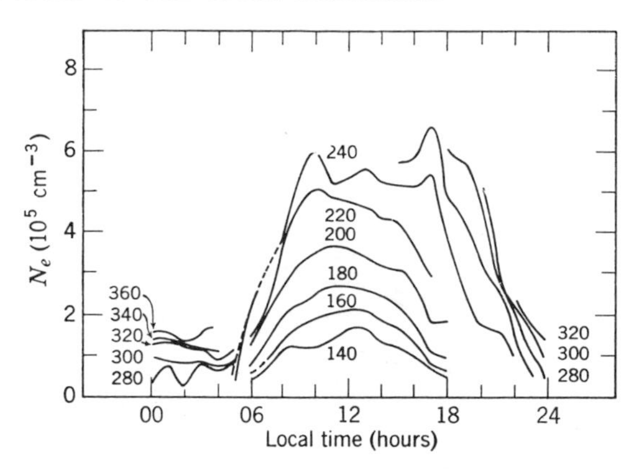

Figure 9.4 Mean quiet $N_e(t)$ curves for Slough, United Kingdom, September 1950. (After H. Rishbeth, *J. Atmos. Terr. Phys.* **26,** 657 (1964); reprinted by permission of Pergamon Press; data originally due to Croom, Robbins, and Thomas, Cavendish Laboratory, Cambridge).

drift velocity given in Eq. (9.20), can have a pronounced effect on the diurnal variation of the maximum electron density, particularly in the equatorial F_2 layer. Hirono and Maeda[16] solved the time-dependent continuity equation with a vertical drift term and obtained two peaks, one in the forenoon and the other in the afternoon. Since the resulting noontime "bite out" in the peak electron number density has been observed (see Figure 9.5), vertical drifts appear to be quite important at low geomagnetic latitudes. Vertical drifts due to heating or cooling of the electron gas just before sunrise or after sunset can also be expected to produce observable effects in the F_2 layer. The predicted evening increase at mid-latitudes has apparently been observed.[18]

Typically, the horizontal drift speeds are of the order 50 to 200 m sec^{-1}, while vertical drifts are much smaller, $\sim$20 m sec^{-1}. During the daytime the direction of the east-west drift is eastward, shifting to westward near sunset; the north-south drift is directed toward the equator at night and away during the day.

The few remarks concerning F_2 region drifts of ionized particles are about all that we can legitimately make at this point. The quantitative work such as that of Hirono and Maeda is too speculative to warrant a more detailed discussion.

9.3.2 Nocturnal Ionization

It is apparent from Figure 9.4 that the F_2 region does not disappear at night, but rather falls to some quasi-equilibrium profile. That is, it varies in electron

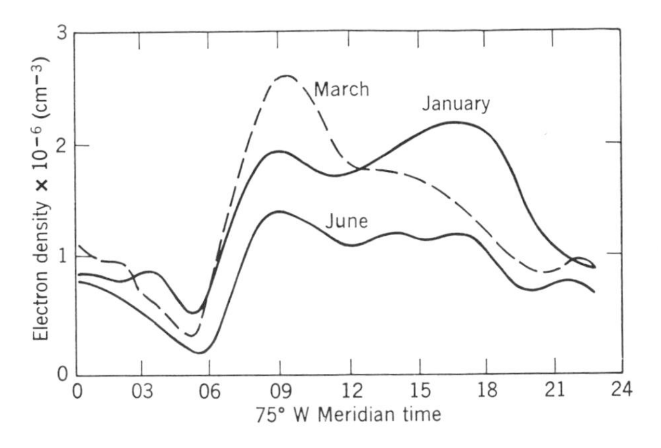

Figure 9.5 Mean quiet day variation of N_mF at Huancayo for three IGY months. (After Y. V. Somayajulu, *J. Geophys. Res.* **69**, 1392 (1964) 17; reprinted by permission of the American Geophysical Union).

number density only slightly following the post-sunset decrease. As may be expected, thermal contraction of the atmosphere and downward diffusion of electrons and ions will help to maintain the F_2 region at night. This was, in fact, the basis of Rishbeth's model[14] of the temporal variations of the upper ionosphere and its importance is evident in Figure 9.3. However, downward diffusion apparently cannot by itself account for the maintenance of the nocturnal F_2 layer. Alternatively, it has been suggested that there is some source of ionization at night. As will be discussed at some length in Section 9.4, the electron gas is considerably warmer, even at night, than is the ambient atmosphere. Whether the thermal input to the former arises from ionization processes or from influx of heat from very high altitudes (protonosphere) is still an open question. It is also possible that winds in the neutral atmosphere raise the ionized layer enough to substantially decrease the electron loss rate.

9.3.3 Geographical Anomalies

Of the various geographical anomalies, the most striking is the equatorial anomaly which is undoubtedly a consequence of geomagnetic control of the F_2 region. The equatorial anomaly can be described as follows: If one plots the electron number density at the F_2 peak as a function of geomagnetic latitude, one finds a trough that is roughly symmetric about the geomagnetic equator with maxima occurring at latitudes of about 15° magnetic north and south. This is illustrated in Figure 9.6. Typically, the F_2 maximum electron density at the equator is about 20 to 50 per cent less than at 15° latitude. In

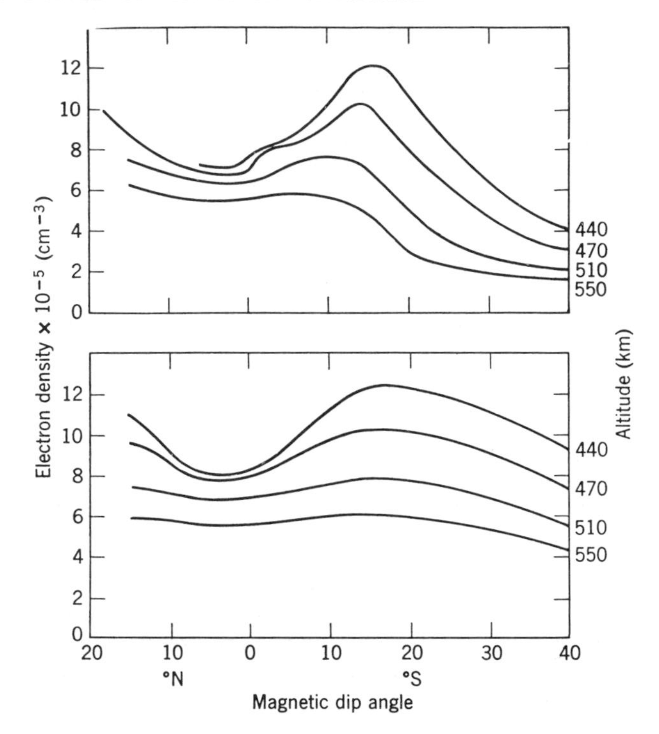

Figure 9.6 Electron number density as a function of magnetic dip angle for various altitudes. The upper curves represent measurements made by the Alouette topside sounder satellite at 1000 LMT, 30 Oct. 1962. (King, Smith, Eccles, and Helm, unpublished report, DSIR Radio Research Station, England, 1963). The lower curves represent theoretical calculations of the electron number density (Goldberg *et al.*, 1964[19]). Magnetic dip angle I is related to geomagnetic colatitude θ by $\sin I = 2\cos\theta/(1 + 3\cos^2\theta)^{1/2}$.

addition, it has been observed that the height of the peak electron density is a maximum at the geomagnetic equator. Recent attempts[19] to reconstruct this effect by numerical solution of the appropriate continuity equation have met with partial success. The trough was present in the theoretical results, but the variation of the height of peak electron number density did not entirely agree with observations. This may arise from the omission of some important electrodynamic transport terms or the use of incorrect electron production and loss terms. At the present time the situation is far from clear and all computational attempts along these lines must be regarded with some suspicion.

There are various other geographic anomalies which are less well understood than the behavior of the F_2 layer near the equator. Among these are the "Ottawa depression" and the morphology of the high latitude F_2 region where the occurrence of unusual geographic variations are unexplained. Probably they are due to transport processes, but the relationship has not been established. The depression in the F_2 region in the vicinity of Ottawa is undoubtedly associated with local geomagnetic peculiarities, but beyond this we can say no more.

9.3.4 Temporal Anomalies

There are three temporal anomalies, one of which is diurnal, one seasonal and the third annual. Since their causes are unknown we can do little more here than to describe them. The diurnal anomaly is characterized by changes in the time of occurrence of the peak electron number density with season and epoch of the solar cycle. Near the equator the previously described "bite-out" (see Figure 9.5) occurs, mainly during the daylight hours near sunspot minimum; at sunspot maximum the effect lasts until after midnight. At somewhat higher magnetic latitudes (10°–15°) the maximum electron concentration at the F_2 peak occurs in late afternoon during summer and equinoctial months; however, in winter the peak occurs near noon.

The seasonal effect is manifested in larger F_2 region electron number densities during winter than during summer. The effect is particularly strong near sunspot maximum. If the electron concentration were controlled entirely by the ionization rate, we should expect it to reach a daytime maximum in the summer. Since this is not the case, we conclude either that it is due to atmospheric "settling" in the winter or to geomagnetic effects. The former is thought to result from a decrease in the ratio $[N_2]/[O]$ as a consequence of atmospheric cooling. This, in turn, is manifested as an increase in the ratio of the ionization rate to the electron loss rate. On the other hand, the earth's magnetic field is believed to guide plasma from the summer to the winter hemisphere. It is well known, for example, that at low latitudes near sunrise photoelectrons move along the field lines from the newly-illuminated hemisphere to the dark one, producing a pre-sunrise increase in electron number density. Whether or not a similar effect lies behind the seasonal anomaly is unknown.

Superposed on this seasonal peculiarity is the "December" anomaly which is characterized by an increase in both hemispheres (mainly at mid-latitudes) of the electron number density at the F_2 peak during the months of November through January. The exact reason for its occurrence is unknown but we can be reasonably certain that it is geomagnetic in character.

The foregoing discussion, brief as it is, is actually a rather complete description of our current knowledge about the anomalies. Hence, we shall not pursue the topic further.

9.4 Thermal Properties of the F_2 Region

The rather strong absorption of solar ultraviolet radiation in the F_2 layer results in intense heating of the upper ionosphere. In fact, as we saw in Chapter 3, it constitutes the principal heat source at ionospheric altitudes. Until relatively recently it was assumed that the energy so absorbed equilibrated quite rapidly with the neutral atmosphere. However, Hanson and Johnson[20] and later Dalgarno and coworkers[21] showed that even a small heat input to the electron gas must result in substantial departure from thermal equilibrium in the ionosphere above ~150 km. Measurements[22] of the temperature of the electron and ion gases have verified their predictions.

Because the photoelectrons rapidly thermalize with the ambient electron gas, they constitute a heat source for this gas. Energy is lost from the ambient electrons to the neutral particles and ions by elastic and inelastic collisions; it is also spatially redistributed by transport (heat conduction) so that the equilibrium temperature at any given height is set by the rates of heat input, loss, and conductive transfer. The temperature of the ion gas is established in a similar manner; the heat input occurs via Rutherford scattering of hot electrons, loss through elastic scattering of neutral particles, and transport by means of conduction. Hence, we find the electron temperature to be substantially higher than that of the neutral atmosphere, and the ion temperature to be intermediate between the two.

9.4.1 *Energy Spectrum of the Photoelectrons*

The energy spectrum of the photoelectrons is dependent upon the ultraviolet spectrum of the sun, and upon the ionization cross sections and number densities of the atmospheric constituents. Since O and N_2 are the principal species at F_2 region altitudes, the ionization of these two species can be expected to be the means by which most of the energy is deposited. In order to arrive at estimates of energy deposition, we need to know more than just the energy and cross section for transition to the ionic ground states. We must have similar information about transitions to their various excited states. For example, the processes

$$O(^3P) + h\nu \rightarrow O^+(^4S) + e \tag{9.50a}$$

$$O(^3P) + h\nu \rightarrow O^+(^2D) + e \tag{9.50b}$$

$$O(^3P) + h\nu \rightarrow O^+(^2P) + e \tag{9.50c}$$

have ionization energies of 13.6, 16.9, and 18.7 eV, respectively (or equivalently "spectral heads" of 910 Å, 732 Å, and 663 Å), and the formation of the states of O^+ indicated will have a pronounced effect on the electron energy spectra, depending upon the UV spectrum. Energetically higher excited states

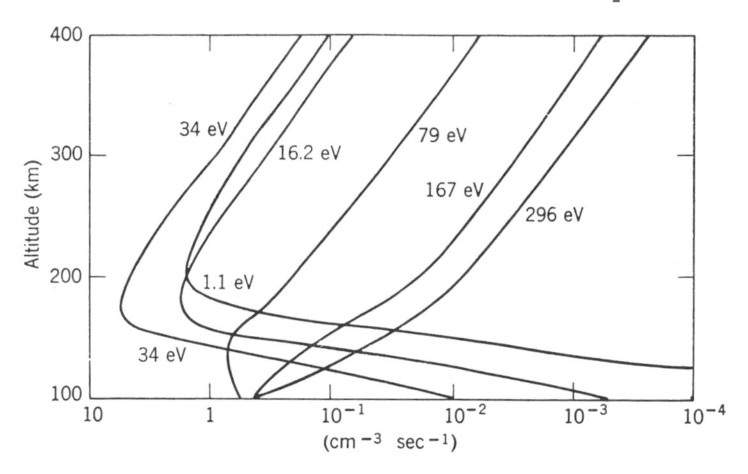

Figure 9.7 Distribution of photoelectrons produced by reaction 9.50a if $T(\infty) = 1000°K$. (After Dalgarno, McElroy, and Moffett, *Planet. Space Sci.* **11,** 451 (1963); reprinted by permission of Pergamon Press).

of O^+, which may also be formed, must be similarly considered in computations of energy deposition.

In order to compute the energy spectrum (i.e., energy per unit energy interval per unit volume) of the electrons produced, say, by process (9.50b), it is only necessary to evaluate the product

$$E_E = [O]F_E Q_E(O(^3P) \rightarrow O^+(^4S))(h\nu - 13.6) \tag{9.51}$$

where [O] is the number density of atomic oxygen, Q_E is the appropriate ionization cross section (a function of the photon energy) and F_E is the photon spectral irradiance (photons cm^{-2} sec^{-1} eV^{-1}). It must, of course, be remembered that photons with energy 20 eV, for example, will produce electrons of 6.4 eV. Dalgarno and co-workers[21] have carried out such a program for atmospheres characterized by thermospheric temperatures of 1000° and 2000°K. The distribution of photoelectrons produced by process (9.31a) is presented in Figure 9.7 for a thermospheric temperature of 1000°K. Note that the energy spectrum appears to peak at about 30 to 50 eV.

9.4.2 Energy Loss Mechanisms

A number of energy loss mechanisms are involved in thermalization of the photoelectrons. These include ionization of the neutral species by the very energetic primary electrons, electronic excitation, vibrational and rotational excitation of N_2, coulomb scattering by ambient electrons, and elastic collisions with neutral species and ions (see Section 2.5). At energies of the order of 20 to 50 eV the principal mechanism is excitation of neutral species to excited states which return to the ground state via allowed transitions; some ionization also occurs. The photons produced can themselves be absorbed by atmospheric components. At energies of a few to ~20 eV, excitation to metastable states of atomic oxygen, molecular oxygen and molecular nitrogen is more significant. The metastable species may then disappear by photon emission or by some sort of quenching. A very prominent spectral line observed during electron bombardment of the upper atmosphere is the atomic oxygen red line (6300 Å) produced by the transition

$$O(^1D) \rightarrow O(^3P) + h\nu \tag{9.52}$$

At still lower energies vibrational excitation of N_2 is quite significant and is, in fact, the principal mode of energy loss (with the exception of electron-electron collisions) in the energy range 2 to 4 eV. Actually there is some question as to the dissipation of the energy stored in the vibrational mode. It may be transferred to other neutral particles or ions as kinetic energy (a very slow process) or it may be returned to the electrons as kinetic energy via superelastic collisions. Finally, at thermal energies, inelastic collisions with nitrogen molecules and ground state oxygen atoms are the dominant energy loss mechanisms for electrons. Energy is lost to the N_2 by rotational excitation due mainly to the permanent quadrupole moment of the N_2 molecule, and to a lesser extent to the induced dipole interaction. Until very recently, it was assumed that thermal electrons do not collide inelastically with $O(^3P)$ atoms. However, Dalgarno and co-workers[23,24] have shown that the excitation of the fine structure levels of $O(^3P)$ by electron impact

$$e + O(^3P_J) \rightarrow e + O(^3P_{J'}); \qquad J' < J \tag{9.53}$$

is a very efficient cooling mechanism for the electron gas. (Recall that J is the total angular momentum, which can be 0, 1, or 2.) After the O atoms are excited, they can return to the ground state by radiation (see Section 3.2.3) or by superelastic collisions with electrons. The latter mechanism returns energy to the electron gas, but the former leads to permanent loss by radiation into space.

Before discussing the approach to thermal equilibrium of the ionospheric plasma, we must say a few words about elastic collisions. Electron-electron collisions via the coulomb interaction constitute the most efficient means of energy removal from electrons below ~30 eV. The pertinent equation for energy loss per unit path length, which was presented in Section 2.5.3 [Eq. (2.146)], is of the form

$$\frac{dE}{dx} = -N_e\left(\frac{K}{E}\right) \ln E \tag{9.54a}$$

where E is the electron energy, N_e is electron number density, and K is a constant. This expression may be simplified to

$$\frac{dE}{dx} \approx -\frac{1.95 \times 10^{-12}}{E} N_e(\text{eV cm}^{-1}) \tag{9.54b}$$

for $E \gtrsim 3$ eV. Elastic collision of energetic electrons with heavy particles is a much less significant mode of energy loss, but such collisions involving slow electrons constitute an important mechanism for cooling of the heated electron gas.

In finding the temperature of the electron gas, one must know the fraction of the photoelectron energy which is ultimately converted into electron heat energy. Quantitatively, the heating rate $Q(z)$ is expressed in terms of the photoionization rate $q(z)$ and the heating efficiency $\epsilon(z)$

$$Q(z) = q(z)\epsilon(z) \tag{9.55}$$

Dalgarno and coworkers[21] have shown that one can obtain a very good approximation to the heating efficiency at a given altitude by finding the electron energy at which the rate of energy transfer to other species (ions and neutrals) is equal to the rate of transfer to the electron gas. This energy is taken as the heating efficiency.

Figures 9.8 through 9.11 show as a function of altitude the estimated rates of energy loss of electrons of various energies up to 10 keV for an atmosphere characterized by a thermospheric temperature of 1000°K. Note the importance of the ambient electron gas as an energy absorber. Figure 9.12 shows the heating efficiency (for a representative mode atmosphere) determined by Dalgarno and coworkers.[24]

9.4.3 Cooling of the Electron and Ion Gases; Heat Conduction

In the preceding section we discussed the various mechanisms responsible for cooling of the electron gas. The appropriate collision cross sections for most of these processes have been measured or calculated, thus permitting one to write fairly precise expressions for the rate of heat transfer. The following are those proposed by Banks[25] and by Dalgarno and Degges[23] given in terms of the electron temperature T_e, neutral gas temperature T_g, and ion temperature T_+:

$$L_{e_1}(\text{O}) = 2.47 \times 10^{-18} N_e[\text{O}](T_e - T_g)T_e^{1/2}(\text{eV cm}^{-3}\,\text{sec}^{-1}) \tag{9.56a}$$

for the energy loss in *elastic* collisions with atomic oxygen;[25]

$$L_{e_2}(\text{O}) = 4 \times 10^{-12} N_e[\text{O}](T_e - T_g)T_g^{-1}(\text{eV cm}^{-3}\,\text{sec}^{-1}) \tag{9.56b}$$

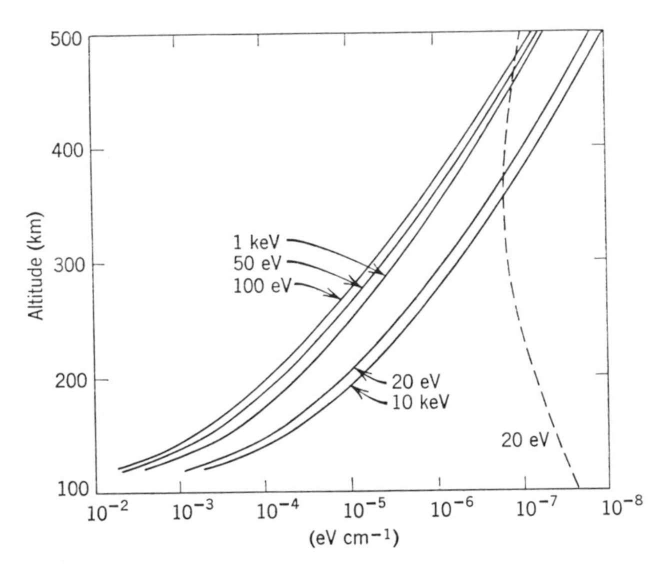

Figure 9.8 Rates of energy loss to neutral particles (solid lines) and to electrons (dashed line) for electrons of various energies above 20 eV moving in a model atmosphere with $T(\infty) = 1000^{\circ}$K. (After Dalgarno, McElroy, and Moffett, *Planet. Space Sci.* **11,** 451 (1953); reprinted by permission of Pergamon Press.)

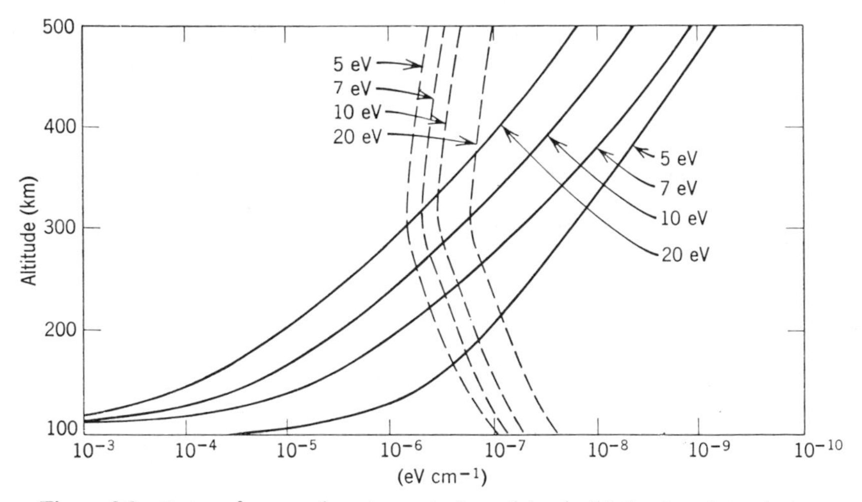

Figure 9.9 Rates of energy loss to neutral particles (solid lines) and to electrons (dashed lines) for electrons of various energies below 20 eV moving in a model atmosphere with $T(\infty) = 1000^{\circ}$K. (After Dalgarno, McElroy, and Moffett, *Planet. Space Sci.* **11,** 451 (1963); reprinted by permission of Pergamon Press.)

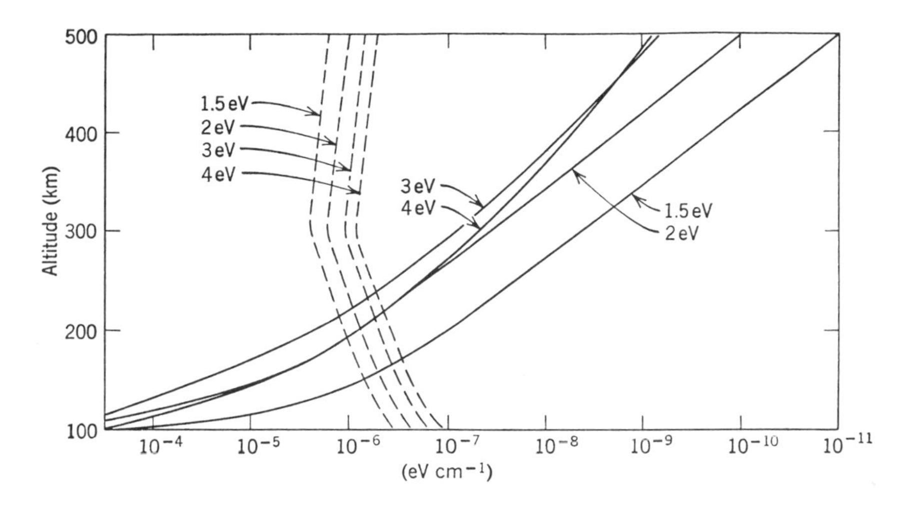

Figure 9.10 Rates of energy loss to neutral particles (solid lines) and to electrons (dashed lines) for electrons with energies between 1.5 eV and 4 eV moving in a model atmosphere with $T(\infty) = 1000°K$. (After Dalgarno, McElroy, and Moffett, *Planet. Space Sci.* **11,** 451 (1963); reprinted by permission of Pergamon Press.)

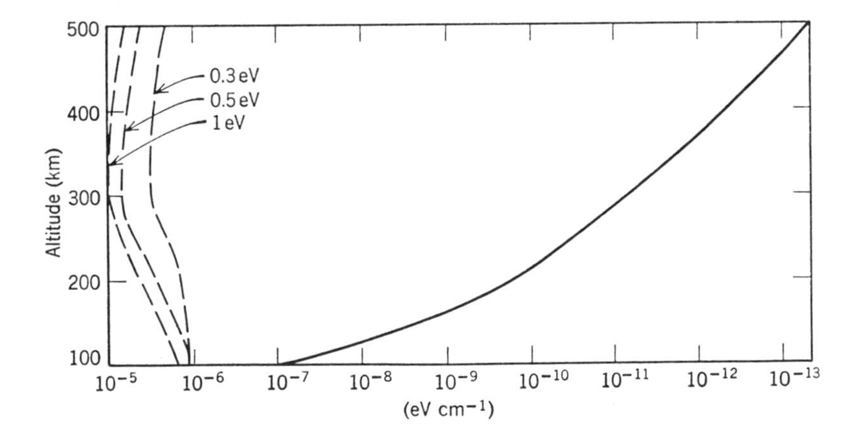

Figure 9.11 Rates of energy loss through rotational excitation of N_2 (solid lines) and through elastic collisions (dashed lines) with the ambient electrons for $T(\infty) = 1000°K$. (After Dalgarno, McElroy, and Moffett, *Planet. Space. Sci.* **11,** 451 (1963); reprinted by permission of Pergamon Press.)

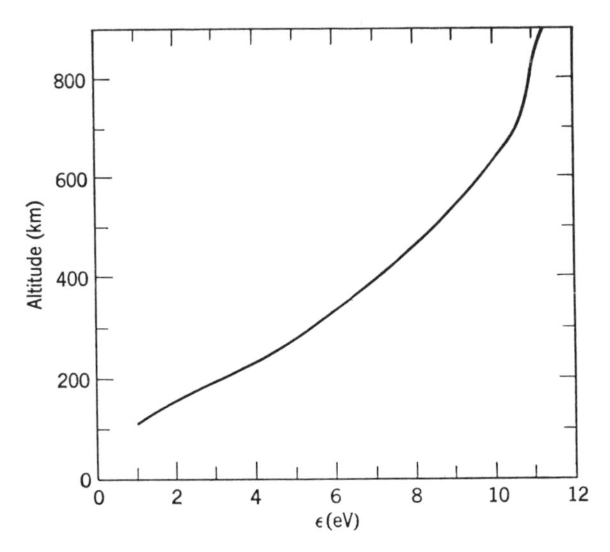

Figure 9.12 The altitude profile of the electron heating efficiency at noon.[24]

for the energy loss in *inelastic* collisions (i.e., fine structure excitation) with atomic oxygen[23];

$$L_e(N_2) = 1.41 \times 10^{-18} N_e[N_2][1 - 3.84 \times 10^{-2} T_e^{1/2}](T_e - T_g)T_e^{1/2}$$
$$+ 3.1 \times 10^{-14} N_e[N_2](T_e - T_g)T_e^{-1/2}(\text{eV cm}^{-1}\text{ sec}^{-1}) \quad (9.56c)$$

for energy loss to N_2 by elastic (first term) and rotational excitation collisions[25];

$$L_e(O_2) = 1.0 \times 10^{-13} N_e[O_2](T_e - T_g)T_e^{-1/2}(\text{eV cm}^{-3}\text{ sec}^{-1}) \quad (9.56d)$$

for energy loss to O_2 by inelastic collisions[25]; and

$$L_e(+) = 7.7 \times 10^{-6} \frac{N_e}{A} N_+(T_e - T_g)T_e^{-3/2}(\text{eV cm}^{-3}\text{ sec}^{-1}) \quad (9.56e)$$

for energy loss to ions of mass number A and concentration N_+ by elastic collisions.[25] In the case of electrons in a carbon dioxide atmosphere, the energy loss occurs principally by vibrational excitation. Although the functional dependence of the cross section on electron energy is extremely complicated, one can express the cooling rate as

$$L_e(\text{CO}) = 4.5 \times 10^{-13} N_e[CO_2](T - T_g)(\text{eV cm}^{-3}\text{ sec}^{-1}) \quad (9.56f)$$

to a very good approximation.[26]

The term $L_e(+)$ which expresses the rate of energy loss of the electrons to the ion gas is a heat *source* function for the ions. In fact virtually all of the ion heating is caused by elastic collisions with electrons. Because of the large mass of the ions they acquire very little recoil energy during the photoionization process. The ion gas is cooled principally by elastic collisions (and "symmetric" charge transfer in the case of $O^+ - O$ collisions) with neutral atoms and molecules. Although energy transfer via inelastic collisions is indeed possible, the near-equality of ion and neutral particle masses makes energy transfer via elastic collisions so efficient that it dominates. Banks[27] has employed the following ion cooling rate $L_+(O^+)$ in an atmosphere of N_2 and O

$$L_+(O^+) = 6.6 \times 10^{-14}[O^+][N_2](T_+ - T_g) + 2.1 \times 10^{-15}[O^+][O](T_+ + T_g)^{1/2}(T_+ - T_g)(\text{eV cm}^{-3}\text{ sec}^{-1}) \quad (9.57)$$

As in the earlier discussion concerned with the thermal structure of neutral atmospheres (Section 3.2.2), heat transport by conduction plays a very important role in the electron and ion temperature profiles. For the electrons heat conduction is already significant at altitudes below 200 km, while for ions, it can be neglected below ~500 km. Thermal conductivity in a gas was discussed in Section 2.3.4 where it was shown how to derive its relationship of collision frequency: the thermal conductivity is inversely proportional to the *momentum transfer* collision frequency. Since the rate of heat flow is dependent upon the *total* collision frequency with *all* species, the overall conductivity K must be related to the individual conductivities K_i for each component gas: by

$$\frac{1}{K} = \sum_i K_i^{-1} \quad (9.58)$$

Spitzer[28] has derived the thermal conductivity for electrons in a gas of singly charged ions by assuming screened Rutherford scattering. We shall not go into the details of the derivation but merely write down the numerical result

$$K_{e,+} = 7.7 \times 10^5 T_e^{5/2}\text{ eV cm}^{-1}\text{ sec}^{-1}\ (°K)^{-1} \quad (9.59)$$

One can use the same approach to evaluate K_i for neutral species; the result is[25]

$$K_{e,i} = \frac{23.9\ N_e}{N_i \overline{Q_{D,i}}} T_e^{1/2}\text{ eV cm}^{-1}\text{ sec}^{-1}\ (°K)^{-1} \quad (9.60)$$

where N_i is the number density of species i, and $\overline{Q_{D,i}}$ is the momentum transfer cross section *averaged* over the energy distribution of the electrons.

The thermal conductivity for ions of mass number A in a neutral atmosphere is[27]

$$K_{+,i} = \frac{4.6 \times 10^4}{A^{1/2}} T_+^{5/2} (\text{eV cm}^{-1} \text{sec}^{-1} {}^\circ K^{-1}) \tag{9.61}$$

9.4.4 *Nonlocal Energy Deposition*

In addition to transfer of the thermal energy of the electron gas from one level to another, nonlocal energy deposition by the energetic photoelectrons affects the spatial distribution of energy in the ionosphere. By "local" we mean that the electron deposits its energy at altitudes within a scale height H of the point of formation; here "H" refers to the *neutral constituent* from which the photoelectrons are formed. "Nonlocal," on the other hand, refers to the deposition of a major portion of the energy at altitudes $>H$ from the point of formation. One can easily show that if the scale height of the ionosphere is also equal to H, the energy deposition is "local."

The rate of energy deposition Q at height z' in the electron gas is expressed by the equation

$$Q(z') = \frac{1}{2}\left[\int_{z_1}^{z'} q(z) \left|\frac{dE}{dz}\right|_{z',z} dz + \int_{z'}^{z_2} q(z) \left|\frac{dE}{dz}\right|_{z',z} dz\right] \tag{9.62}$$

where $q(z)$ is the ionization rate at altitude z, and $|dE/dz|_{z',z}$ is the stopping power at altitude z' of a photoelectron which was formed at z. The integration is carried out over the path of the photoelectrons; z_1 and z_2 are, respectively, the lowest and highest altitudes from which electrons can reach height z'. The photoelectrons are, of course, constrained to move along geomagnetic field lines with "pitch angle" distributions characteristic of the ionization process and the field line orientation. In order to simplify the approach, we have assumed that half of the photoelectrons have pitch angles of 0° and half 180°. It is not indeed necessary to make this assumption, although the computations are more complicated, if it is not.[29,30]

It has been shown[31] that nonlocal energy deposition in the electron gas apparently has a negligible effect on the electron temperature because of the "fine structure" cooling by atomic oxygen. Were it not for the presence of the term given by Eq. (9.56b), the predicted temperature difference $T_e - T_g$ at altitudes above 250 km would be raised by about ten percent when nonlocal deposition is included in the calculations.

9.4.5 *The Heat Equations for the Ionosphere*

The heat conduction equation for the electron gas is (assuming no horizontal stratification)

$$C_{v,e}\frac{\partial T_e}{\partial t} = \sin^2 I \frac{\partial}{\partial z} K_e \frac{\partial T_e}{\partial z} + Q - [L_e(\text{O}) + L_e(\text{O}_2) + L_e(N_2) + L_e(+)] \tag{9.63}$$

where the L_e are given by Eqs. (9.56a) to (9.56e), respectively, K_e by Eqs. (9.58) to (9.60), and Q by Eq. (9.55); I is the magnetic dip angle, $C_{v,e}$ is the specific heat capacity at constant volume of the electron gas

$$C_{v,e} = \tfrac{3}{2} N_e k \tag{9.64}$$

in which k is Boltzmann's constant and N_e is the electron number density. There is a similar equation for the ions

$$C_{v,+} \frac{\partial T_+}{\partial t} = \sin^2 I \frac{\partial}{\partial z} K_+ \frac{\partial T_+}{\partial z} + L_e(+) - L_+ \tag{9.65}$$

where K_+ is given by Eq. (9.61), $L_e(+)$ by Eq. (9.56e), and L_+ by Eq. (9.57); $C_{v,+}$, the specific heat capacity at constant volume, is given by

$$C_{v,+} = \tfrac{3}{2} N_+ k \tag{9.66}$$

where N_+ is the ion number density.

It is apparent that Eqs. (9.63) and (9.65) are coupled by the term $L_e(+)$, thus requiring their simultaneous solution in order to obtain T_e and T_+ rigorously. Fortunately, this quite difficult mathematical task can be simplified considerably by using the (very good) approximation in Eq. (9.63) that $T_e = T_+$.

Since the two heat equations are of second order, their solutions must satisfy two boundary conditions. The lower condition usually takes the form of holding T_e or T_+ fixed at some given altitude, while the upper one is established by the heat flux Φ at some very high altitude

$$\Phi = -K \sin I \frac{\partial T}{\partial z} \tag{9.67}$$

where Φ, K, and T refer either to the electron or the ion gas. The heat conduction equations are, of course, highly nonlinear, and thus are not amenable to analytic solution. They can be readily solved on a digital computer using finite difference methods.[27] Even in equilibrium situations it is almost mandatory to solve the time-dependent equation in order to avoid mathematical stability problems. In such a "relaxation" method, we establish the initial condition by setting T_e or T_+ equal to T_g at all altitudes and then calculate solutions until T_e or T_+ approaches equilibrium at each altitude in the time-space grid. We show some typical computed T_e and T_+ profiles[31] in Figures 9.13. In obtaining these solutions, influence of electric fields were ignored principally because we do not know them in sufficient detail to include them in the computations. The main difference between Figures 9.13a and 9.13b is caused by the exclusion of the term L_{e_2} [Eq. (9.56b)] from the calculations which yielded profile (b). Figure 9.13b also shows a computed ion temperature profile.

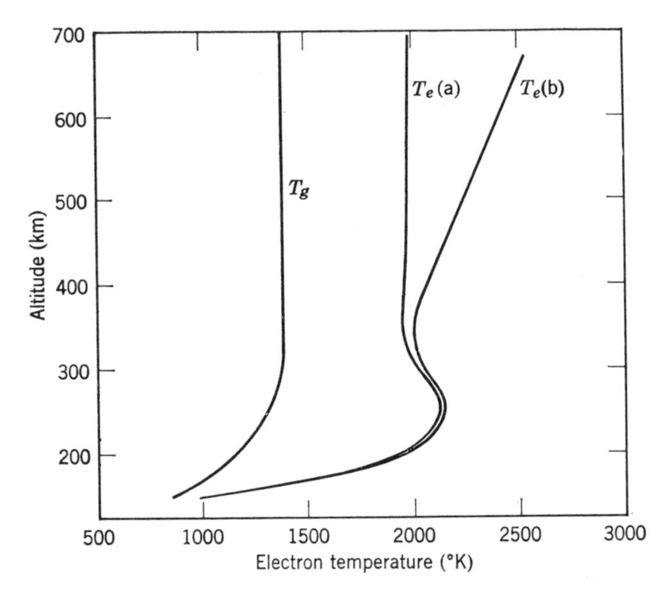

Figure 9.13d Computed electron temperature profiles using the ϵ profile in Figure 9.12 and a Harris and Priester model atmosphere[32] (S = 150, 1400 local time). The ionization rates at 300 km were taken as $q(\text{O}) = 250\ \text{cm}^{-3}\ \text{sec}^{-1}$, $q(\text{N}_2) = 80\ \text{cm}^{-3}\ \text{sec}^{-1}$, and the boundary conditions are $T_e = T_g$ at 120 km, $\Phi(a) = 0$, $\Phi(b) = 3 \times 10^9\ \text{eV cm}^{-2}\ \text{sec}^{-1}$.

At low latitudes the factor $\sin^2 I$ which appears in Eqs. (9.64) and (9.65) becomes extremely small, and the heat equations must be modified in much the same way as the continuity equation [Eq. (9.49)].

9.4.6 Observations of Electron and Ion Temperature

Observations of electron and ion temperatures have been obtained by two techniques: rocket and satellite-borne Langmuir probes[33] and Thomson scatter radar.[34] We shall discuss both devices briefly in Chapter 11, although the theoretical basis for deducing temperatures from the radar results is so complicated that we do not have the space to develop it here.

Some observed profiles are shown in Figure 9.14. Figure 9.14a shows two electron temperature profiles which were measured simultaneously by rocket (NASA 6.07) and satellite (Explorer 17) at mid-latitude in April 1963, near sunspot minimum. Figure 9.14b shows a profile of electron temperature obtained simultaneously by Javelin rocket probes and satellite (Explorer 31) during a much greater degree of solar activity in June 1967. The principal difference in the profiles is the much larger electron temperature at 300 km in

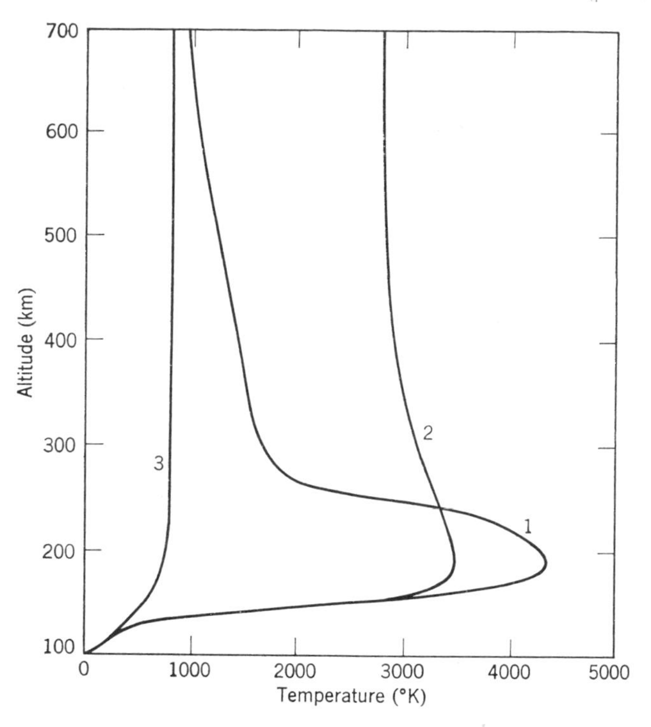

Figure 9.13b Electron and neutral gas temperatures typical of quiet solar conditions. Curve 1 does not include the effects of heat conduction, but curve 2 does. Curve 3 is the assumed natural gas temperature. [After P. M. Banks, *Ann. Geophys.* **22,** 577 (1966); reprinted by permission of Service des Publications du C.N.R.S., Paris.]

1967 then in 1963, reflecting a change in the emission rate of solar EUV radiation. Although the 1963 profile does not extend high enough to be certain, it appears that the vertical gradient of T_e is not likely to be very large at heights above 500 km. On the other hand, the temperature gradients in Figure 9.14b are quite large; Hanson and coworkers[22] estimated a downward heat flux from the protonosphere of 1.9×10^{10} eV cm^{-2} sec^{-1}. Consistently higher temperatures have been obtained by Thomson scatter radar than by probe; attempts to account for this difference have proved fruitless so far. A typical radar profile of T_e is shown in Figure 9.14c; unfortunately, we do not have a profile obtained simultaneously by Langmuir probe.

Figures 9.14b and c also show observations of the ion temperature. Both profiles agree qualitatively with the theoretical curve of Banks shown in Figure 9.13b.

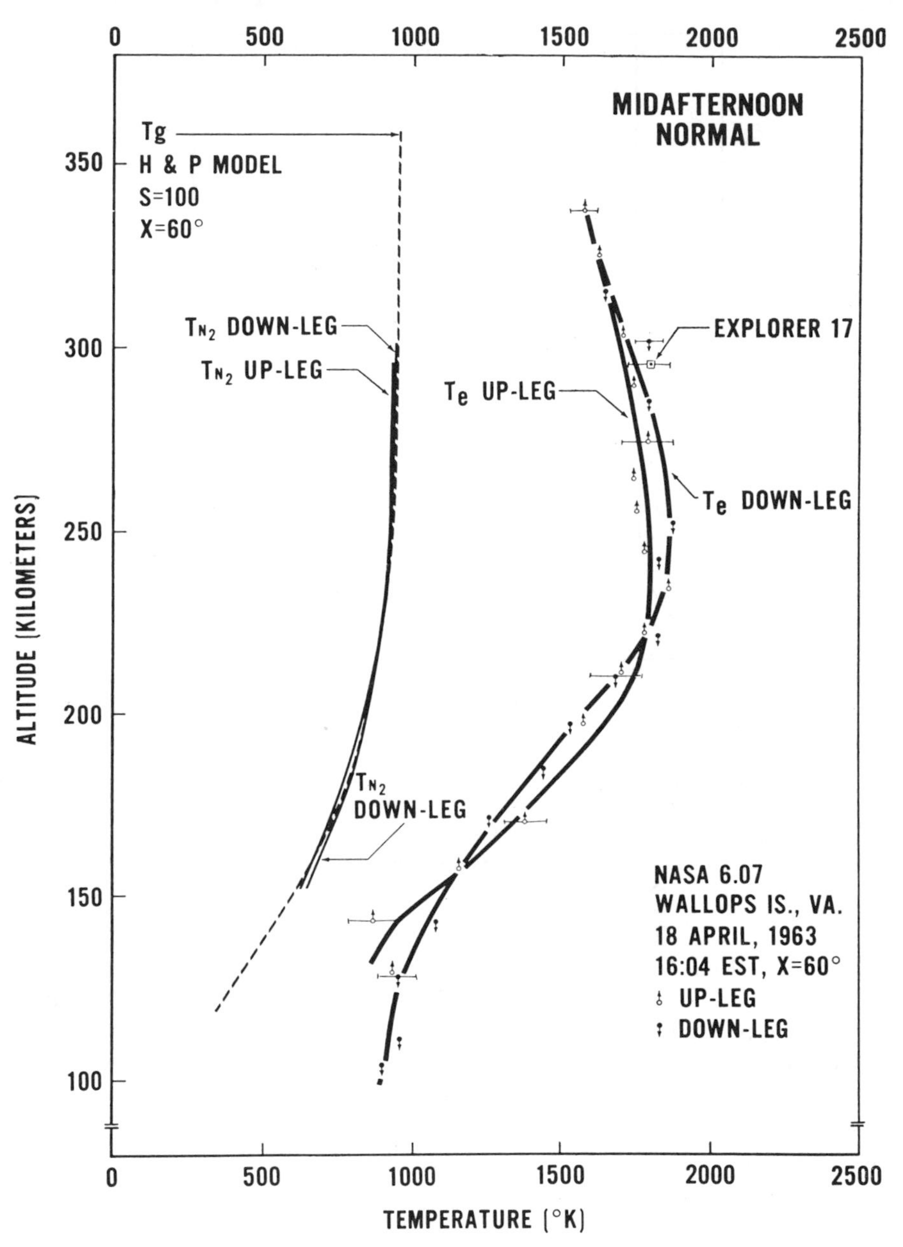

Figure 9.14a T_{N_2} and T_e measurements from the rocket NASA 6.07 with simultaneous Explorer 17 T_e measurement and Harris and Priester model. [After Spencer, Brace, Carignan, Taeusch, and Nieman, *J. Geophys. Res.* **70**, 2665 (1965); reprinted by permission of the American Geophysical Union.]

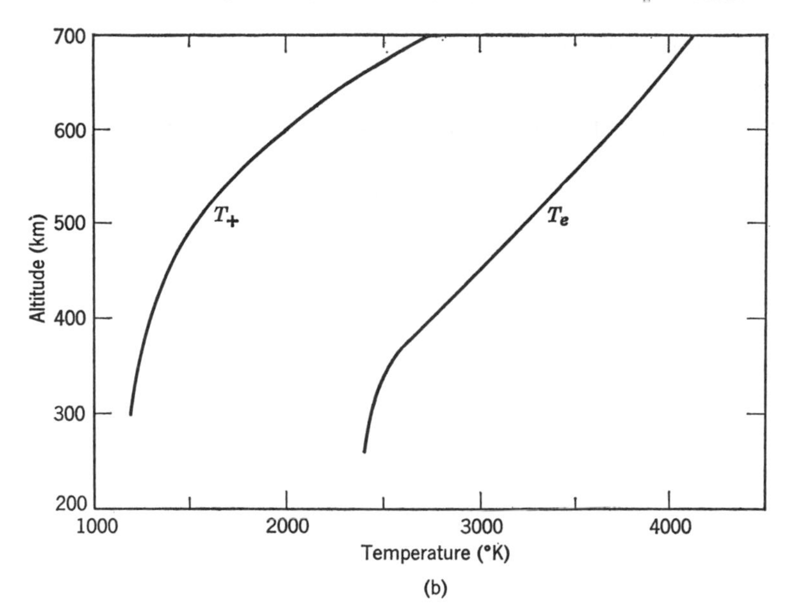

Figure 9.14b Mid-latitude ion and electron temperature, T_+ and T_e, obtained in June 1967 by probes carried in Javelin rockets and Explorer 31.[22]

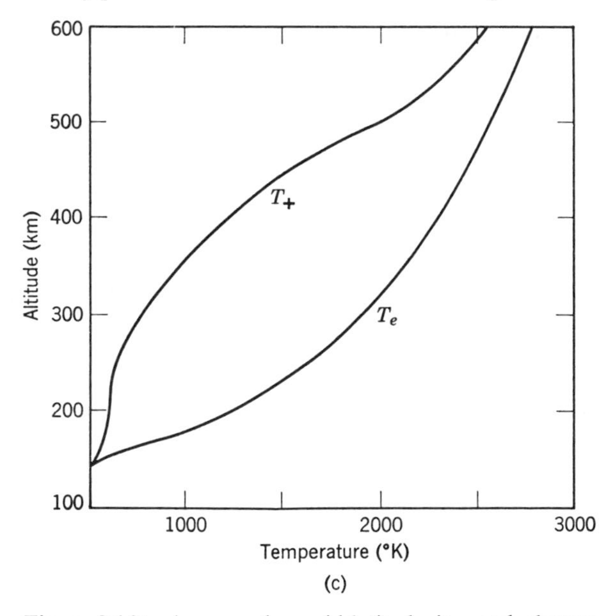

Figure 9.14c Average day mid-latitude ion and electron temperatures, T_+ and T_e, obtained in November 1964 by Thomson scatter radar.[35]

9.5 The Protonosphere

We have seen in Chapter 3 (e.g., Table 3.1) that as one ascends in altitude, helium and atomic hydrogen comprise larger and larger fractions of the neutral atmosphere. Both species are photoionized by solar radiation and both types of ions are lost by reaction with neutral species

$$He^+ + N_2 \rightarrow He + N + N^+ \tag{9.68}$$

$$H^+ + O(^3P) \rightarrow H(^1S) + O^+(^4S) \tag{9.69}$$

Here, the similarity ends, however, because reaction (9.69) proceeds very nearly as rapidly in the reverse as in the forward direction (for equal concentrations of reactants) due to the near equality of the ionization potentials of O ($IP = 13.614$ eV) and H ($IP = 13.595$ eV). The slight difference in ionization potentials (0.019 eV) is much less than the thermal energies of the ion and neutral gases, leading to a case of *asymmetric resonant* charge transfer. The equilibrium constant for reaction (9.69) is usually taken as

$$\frac{[H^+][O]}{[H][O^+]} = \frac{9}{8} \tag{9.70}$$

where we have assumed that statistical factors μ due only to angular momenta J are involved:

$$\mu = 2J + 1 \tag{9.71}$$

(see Section 2.4.2). The rate of the reaction is then proportional to the product of the statistical weights of the reactants. For example, H^+ has $J = 0$, $O(^3P)$ has $J = 0$, 1, or 2, $H(^1S)$ has $J = \frac{1}{2}$ and $O^+(^4S)$ has $J = \frac{3}{2}$. Thus the rate of reaction (9.69) must be proportional to $(2 \times \frac{1}{2} + 1)\,(2 \times \frac{3}{2} + 1) = 8$, while its reverse is proportional to 1×1 for $O(^3P_0)$, to 1×3 for $O(^3P_1)$, and to 1×5 for $O(^3P_2)$. The total rate for the reverse reaction is thus proportional to $1 + 1 \times 3 + 1 \times 5 = 9$. The magnitude of the rate coefficient of reaction (9.69) is not known for thermal energies but Rapp[36] argues that it should be expected to be as large as 10^{-9} cm^3 sec^{-1}. The fact that the rate coefficients for reaction (9.69) and its reverse are nearly equal together with the small mass of H^+ relative to O^+ is responsible for the existence of a barrier to the diffusion of H^+ and O^+. That is, H^+ ions which are transported downward are changed to O^+ at a rate fast enough to maintain chemical equilibrium of H^+ at altitudes below $\sim$1000 km. Also, the charge separation electric field causes the light ions to be "buoyed up" over the heavier ions. Hence, in the region where there are large concentrations of O^+, there is little H^+ and vice versa. The region where H^+ dominates is the *protonosphere*.

Although we shall not do so here, it is quite straightforward to apply the techniques discussed in Section 9.2 to compute number density profiles of H^+ and O^+ in the topside ionosphere. Figure 9.15 shows a pair of such profiles due to Geisler.[37]

The results of several mass spectrometric determinations of the number densities of He^+, H^+, and O^+ are shown in Figure 9.16. They obviously bear out our qualitative remarks concerning relative abundances. The difference between profiles n_{t_1} and n_{t_2} in Figure 9.16 probably reflects thermal and diffusion effects not included in the simple theory which predicts coincidence of n_{t_1} and n_{t_2}.

9.6 Spread-*F* and Other Irregularities

There are several types of irregularities found in the upper ionosphere, but the one which is the most interesting and has received the most attention is "spread *F*." This name derives from the "spread" appearance of an echo trace on an ionogram (see Section 11.9.1); that is, the reflection of a vertically incident electromagnetic wave appears to occur at a continuum of altitudes. Several examples of this effect are shown in Figure 9.17. Spread-*F* is not confined to any particular time or place, but does occur with greatest frequency at equatorial and high latitudes. Evidently, the causative mechanisms are different for these two basic types, although we do not yet know what they are. The high latitude variety, an example of which is shown in Figure 9.17a,

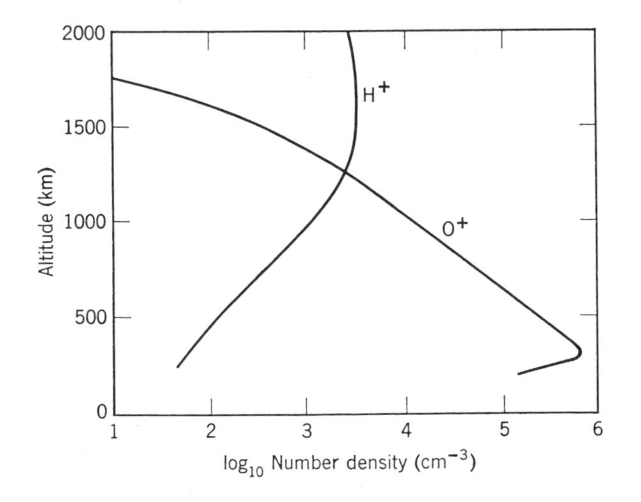

Figure 9.15 Typical equilibrium profiles of H^+ and O^+ number densities.[37]

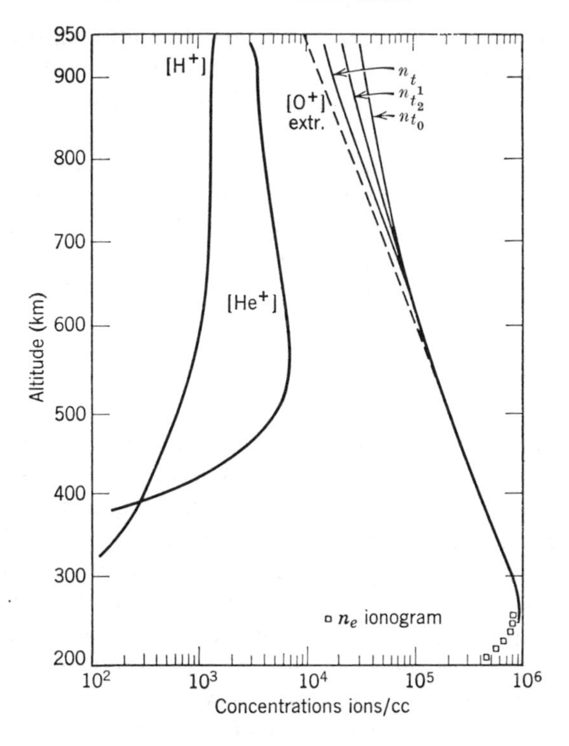

Figure 9.16 Composite ion concentration profiles including total ion number density, n_{t_1}, obtained by adding [H$^+$], [He$^+$], and [O$^+$]; and n_{t_2} derived by correction of the measured total ion number density, n_{t_0}, for the presence of lighter ions. [After Taylor, Brace, Brinton, and Smith, *J. Geophys. Res.* **68,** 5397 (1963); reprinted by permission of the American Geophysical Union.]

is usually characterized by frequency spreading in which the frequencies of the reflected signals appear to be shifted by varying amounts. Doppler shift seems to be an obvious suggestion in this regard, but it is not really known if the irregularities are moving or not. At equatorial latitudes, the ionograms (Figure 9.17b) usually appear to be dominated by range spreading in which apparent variations in the reflection height "smear out" the traces. Figure 9.17b shows range spreading quite clearly although the echo traces have not been completely obliterated. A comparison of the characteristics of the two types is shown in Table 9.1. Because of the great differences between the two types, the conclusion that they have different origins seems inescapable.

TABLE 9.1. CHARACTERISTICS OF SPREAD-F

Equatorial Spread-F	High Latitude Spread-F
(1) Occurs within the geomagnetic latitude zone $\lambda_m = 0°$ to $30°$ ~0 incidence at $\lambda_m = 40°$	First evident at $\lambda_m \approx 40$ to $60°$; Maximum incidence occurs just above auroral zone
(2) Negatively correlated with magnetic activity	Positively correlated with magnetic activity
(3) Maximum incidence at equinoxes	Maximum incidence in winter
(4) Occurs at night only	Can occur throughout the day, but usually at night
(5) Range-spreading type occurs most frequently	Frequency-spreading type occurs most frequently

Investigations of the angular distribution of radiation scattered by the irregularities associated with the spread-F condition have definitely established that the irregularities are aligned with the geomagnetic field. Dimensions are up to 10 km in length and as little as 10 m in width. Most theories of the scattering mechanism have proposed under-dense field aligned regions in which the incident electromagnetic wave is trapped for a short period of

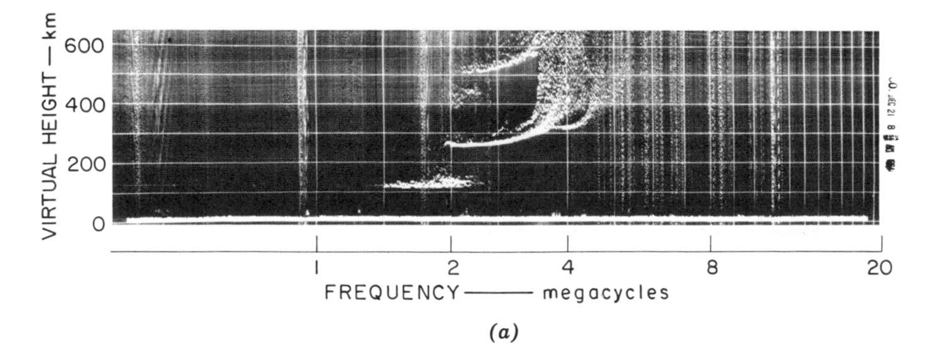

(a)

Figure 9.17 Ionograms showing spread-F conditions (courtesy J. B. Lomax) (a) At College, Alaska, 0855 local time, Dec. 21, 1964;

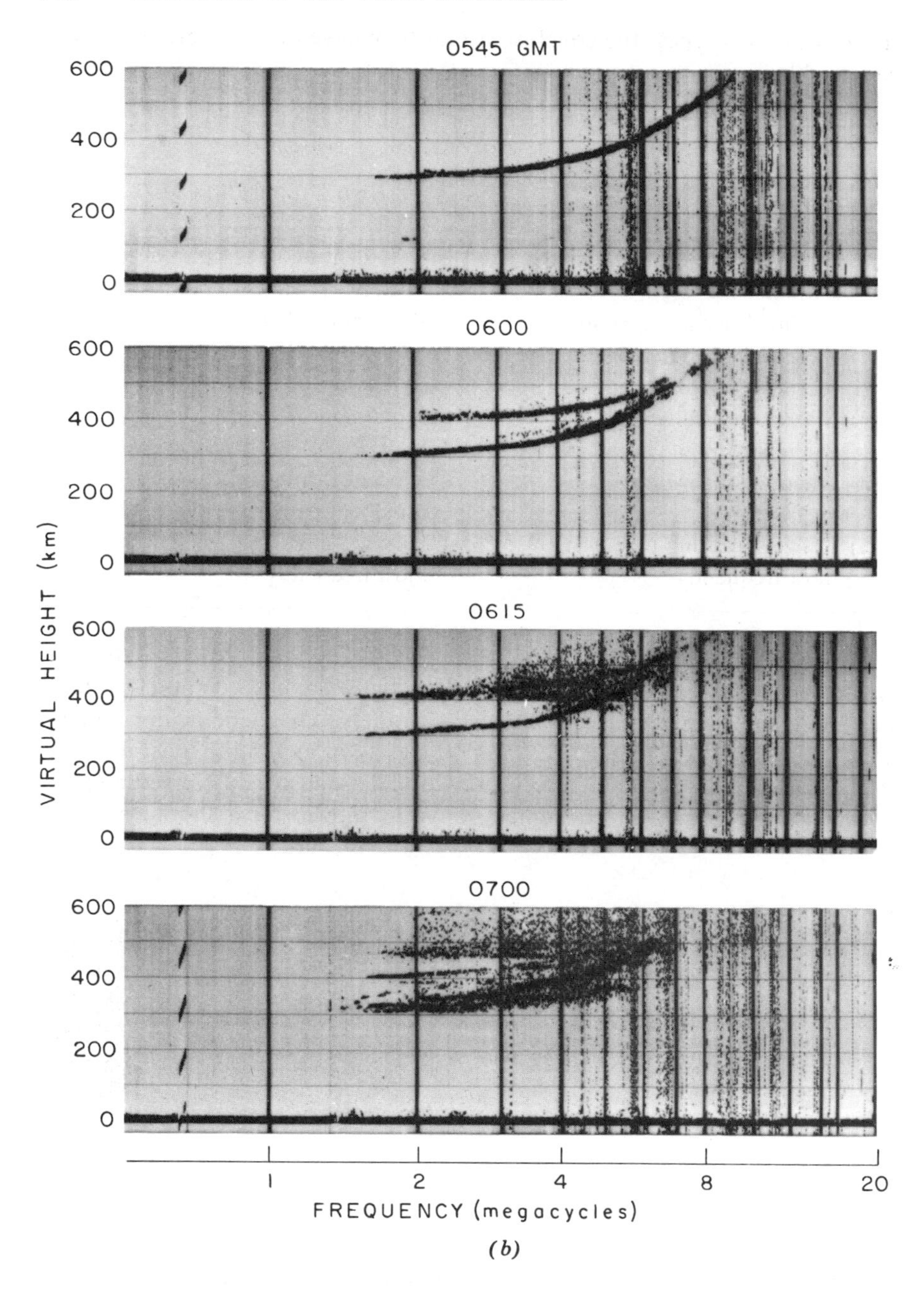

(b)

(b) At Canton Island, October 27, 1962; local time for the top ionogram is 1645. Note the onset of spread-*F* just after sunset.

time. The resultant time delay is associated with the range spreading mentioned in the preceding paragraph. At present, some sort of motion of the irregularities appears to be required to explain the frequency spreading.

Until the development of topside sounder satellites such as Alouette, it was thought that spread-F was associated with the bottomside of the F_2 region only. However, such soundings have observed topside spread-F to be strongly correlated with its bottomside counterpart. In general, the former persists well after sunrise while the latter is rapidly smoothed out during those hours. This may be associated with amplification mechanisms which have been proposed as the cause of spread-F.

The theories as to the origin of spread-F, which have been proposed in recent years, are almost unanimous in ascribing the equatorial and high latitude types to different causes. Theories of the equatorial spread-F may be divided into two groups: those which postulate amplification of weak irregularities formed directly in the F_2 layer, and those which postulate transfer of irregularities into the F_2 along magnetic field lines. The amplification mechanisms could, for example, be associated with hydromagnetic waves propagating downward from the magnetosphere or perhaps with ion-acoustic waves (cf. Chapter 11). At high latitudes three possible mechanisms have been suggested: coupling of the E-layer or magnetosphere with the F_2 region, downward flow of heat flux from the protonosphere at night, and ionization by charged particles. There does seem to be some evidence that equatorial spread-F is associated with the equatorial electrojet[38] (see Section 7.2), but the character of the relationship, if it does exist, is not understood. Of the proposals with respect to high latitudes, Bowhill's suggestion[39] that spread-F is caused by the nocturnal downward flow of heat from the protonosphere is the most intriguing. Such an effect would produce field aligned regions of higher-than-ambient electron number density in the topside ionosphere and lower-than-ambient density in the bottomside. These effects have been observed by satellite, but like the other proposals there are several defects. Not the least of these is failure to predict correlation between spread-F and magnetic activity. The large number of spread-F theories and their speculative nature forbids more detailed discussion here; the interested reader is referred to the various review papers such as that by Herman.[40]

In addition to spread-F there are other irregularities present in the upper ionosphere. For example, there are large scale irregularities (5 to 500 km in size) which are associated with variations in the fading period of radio wave from a satellite. This variety, which is not magnetic field-aligned, may be associated with traveling disturbances. The smaller of these is probably the cause of so-called "radio star scintillation," or fluctuations in the observed amplitude of the radio waves from a strongly emitting star. Actual analysis of the details and characteristics is quite complex and depends upon a statistical treatment of the data. The topic will not be pursued further here.

9.7 Traveling Ionospheric Disturbances

Quasi-periodic disturbances with periods ranging from 15 minutes to one hour have been observed in the upper ionosphere for a number of years. The obvious inference is that they are wavelike phenomena produced by some sort of motion in the neutral atmosphere. Hines[41] has suggested gravity waves (see Section 5.2) as the most likely causative mechanism, and indeed the magnitude of the periods associated with the latter lie in the same range as those of the traveling ionospheric disturbances. As we saw in Chapter 5, atmospheric gravity waves apparently propagate in the region below the mesopause rather than at F_2 region altitudes. It is conjectured that the coupling occurs through upward leakage of energy from the gravity wave duct. Because of the low densities which are characteristic of the F_2 layer, leakage of a very small fraction of the energy of the wave can cause significant fluctuations in the upper ionosphere.

Traveling ionospheric disturbances have been detected following high altitude nuclear explosions and sudden commencements of magnetic storms. If the phenomenon is indeed associated with internal gravity waves in the atmosphere, the latter must be excited directly by pressure fluctuations or perhaps by hydromagnetic waves. Due to the very speculative nature of the whole topic of traveling ionospheric disturbances, it is not worthwhile to consider them at greater length.

9.8 Upper Ionospheres of Other Planets

The Mariner series of spacecraft have yielded valuable information on the upper ionospheres of both Venus and Mars, especially Venus. Presumably the general features of the upper ionosphere of Jupiter will be revealed by space probes which are yet to be launched. At the moment, though, the structure of the Jovian ionosphere is so speculative that it is not worthwhile discussing it here. Because of space limitations we shall also omit the case of Mars and confine our treatment to the daytime upper ionosphere of Venus.

9.8.1 The Daytime Upper Ionosphere of Venus

The bistatic radar occultation measurements[42] made by Mariner 5, which were discussed previously (see Sections 3.7.3, 3.8.3, and 8.5) revealed the existence of a "ledge" of ionization between 200 and 450 km altitude. Above 450 km the electron number density was observed to fall rapidly with increasing altitude to that characteristic of the solar wind. Unfortunately, no direct measurements were made because of the formation of "caustics;" that is, successive rays from Earth to the spacecraft crossed each other. Unambiguous "unfolding" of the electron number density from the radar data requires that the ray paths *not cross each other.* Nevertheless, we do know the order of magnitude of the electron number density for which a "best guess" profile is shown in Figure 9.18.

The formation of the upper ionosphere of Venus has been considered in detail by one of the authors,[43] who proposed that it may be due to the photoionization of a layer of helium of concentration ~3 to 5 × 10^7 cm^{-3} at 200 km altitude. It is indeed unfortunate that no attempt was made by either Mariner 5 or Venera 4 to detect helium.

One can compute an electron number density profile by using the methods described earlier in the chapter. The photoionization rate per neutral per helium atom is approximately

$$\frac{q(\mathrm{He})}{[\mathrm{He}]} \approx 1 \times 10^{-7}\ \mathrm{cm}^3\ \mathrm{sec}^{-1} \tag{9.72}$$

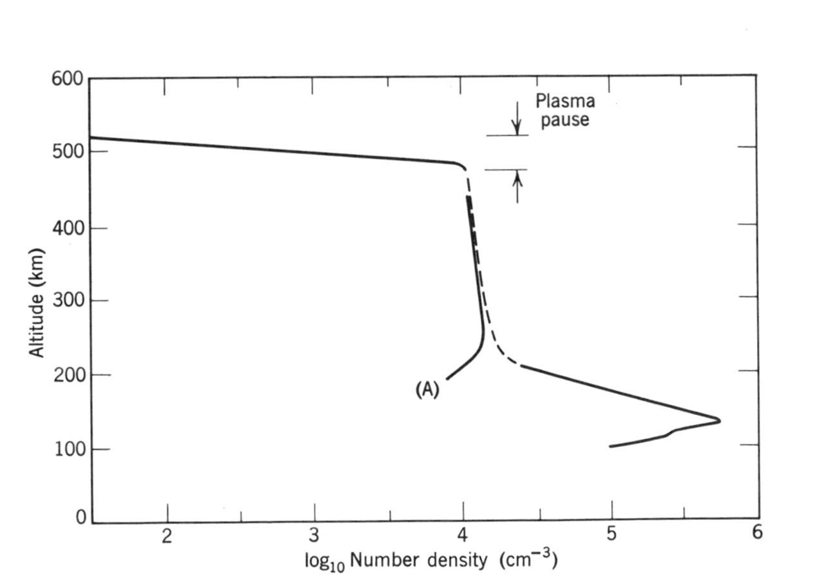

Figure 9.18 Profile of the electron concentration in the daytime ionosphere of Venus as measured by Mariner 5. The broken line is in the area where no direct measurements were made because of the formation of caustics.[42] Profile (A) is a theoretical curve based on a concentration of neutral helium of 5 × 10^7 cm^{-3} at 250 km altitude.[43]

Loss of He^+ ions probably occurs through the sequential reactions

$$He^+ + CO_2 \rightarrow \begin{cases} CO + O^+ + He & (9.73a) \\ CO^+ + O + He & (9.73b) \end{cases}$$

with

$$O^+ + CO_2 \rightarrow CO_2^+ + O \qquad (9.74a)$$

$$CO_2^+ + e \rightarrow CO + O \qquad (9.75a)$$

$$CO^+ + e \rightarrow C + O \qquad (9.74b)$$

or

$$CO^+ + CO_2 \rightarrow CO + CO_2^+ \qquad (9.74b')$$

followed by reaction (9.75a). The rate coefficient[44] for reactions (3) is about 10^{-9} cm^3 sec^{-1}, while that for the dissociative recombination processes is probably $\sim 10^{-7}$ cm^3 sec^{-1}. For the concentrations of CO_2 and electrons found at 200 km altitude, where the transition from the lower to the upper ionosphere occurs, the electron loss rate is controlled by dissociative recombination. The electron loss term which appears in the analogue of Eq. (9.25) is thus nonlinear. However, one can assume the term to be linear without introducing appreciable error. We linearize β [see Eq. (9.25)], where

$$\beta = \alpha_D N \qquad (9.76)$$

by setting N equal to the *average* electron number density in the region near 200 km.

Because of the small neutral particle number densities at heights above 200 km, ambipolar diffusion is expected to be quite significant. One must, therefore, use the methods introduced in Section 9.1.3 in order to compute the electron and ion number density. Due to the absence of a strong planetary magnetic field, the Cytherian ionosphere is almost in direct contact with the solar wind. However, because of the magnetic field lines carried by the solar wind, a boundary layer is formed, which probably isolates the ionosphere from the solar wind, at least under quiescent conditions. The theoretical electron number density profile (A) shown in Figure 9.18 corresponds to (1) no influx of solar wind plasma and (2) [He] $= 5 \times 10^7$ cm^{-3} at 250 km.

We do not have the space to discuss either the theoretical work which has been carried out for the nighttime ionosphere of Venus or the daytime topside ionosphere of Mars, but strongly urge the reader to consult the original papers.[45,46]

9.8.2 *Thermal Structure of the Ionosphere of Venus*

To a large extent the problem of heat production, loss and flow in the Cytherean ionosphere is much like that of the earth. That is, one must solve the heat conduction equations (9.63) and (9.65), subject to appropriate

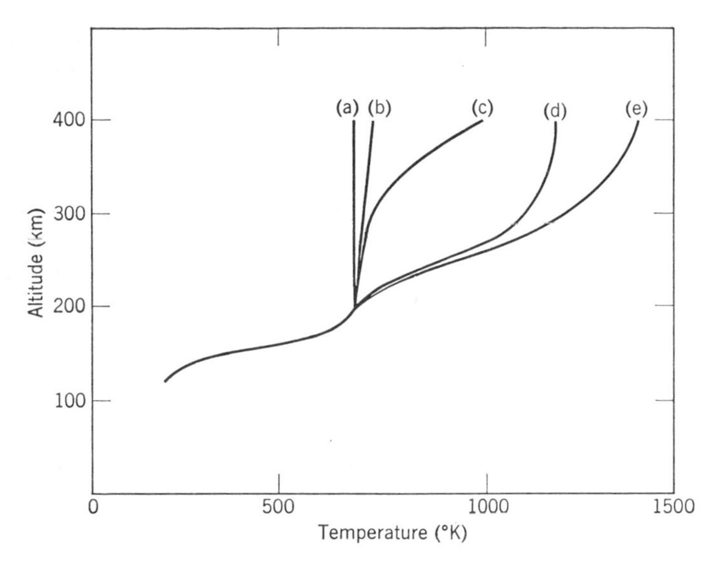

Figure 9.19 Temperature profiles in the upper atmosphere of Venus. (a) the neutral atmosphere, (b) the ion gas, no heat flux from the solar wind, (c) the ion gas, heat flux = 10^9 eV cm^{-2} sec^{-1}, (d) the electron gas, no heat flux, (e) the electron gas, heat flux = 10^9 eV cm^{-2} sec^{-1}.

boundary conditions. The electron and ion cooling terms are somewhat different than in the case of the earth [e.g., Eq. (9.56f) for electron energy loss to CO_2], as is the heating efficiency of photoelectrons in CO_2. In addition to energy deposition by photoelectrons, there is probably some energy transmission across the ionosphere-solar wind boundary layer by plasma waves, which are dissipated in the high ionosphere. Unfortunately, we do not know how to incorporate this heating mechanism into the conduction equations. However, if the waves are dissipated within ~50 to 100 km below the plasmapause (Figure 9.18), they can be treated as a surface source or heat flux, thus establishing the upper boundary condition. Figure 9.19 shows some theoretical profiles of the electron and ion temperatures.[26]

PROBLEMS

9.1. Obtain the analogue of the Chapman function for an atmosphere in which the scale height is a linear function of altitude z: $H = H_0 + \alpha z$.

9.2. Obtain an equation for the diffusion drift velocity w_D in terms of the diffusion coefficients and mobilities of the ions and electrons. Do not use the approximation that $m_e \nu_e \ll m_+ \nu_+$.

9.3. Solve Eq. (9.25) for the case $N(z, t = 0) = N_0 \exp[-(z - z_0/2H) - \delta \exp(-(z - z_0)/H)]$, $\beta = \beta_0 \exp[-(z - z_0/H)]$, $D = D_0 \exp(z - z_0/H)$ with the lower boundary condition $N(-\infty, t) = 0$ and upper boundary conditions corresponding to (a) $N \sim e^{-(z-z_0)/(c+1)H}$, (b) $N \sim e^{-(z-z_0)/H}$. *Hint:* Use transformation $x = Ae^{-\nu(z-z_0)/H}$.

9.4. If $q = q_0 \exp[-(z - z_0/H)]$ where H is the scale height of atomic oxygen and the electron loss rate is controlled by process (9.3), prove Eq. (9.27).

9.5. Prove that $w_D(z = \infty) = 0$ corresponds to upper boundary condition (a), and div $(N\mathbf{w}_D) = 0(w_D > 0)$ to upper boundary condition (b) of problem 9.3. To what physical conditions do (a) and (b) correspond? Assume an isothermal atmosphere.

9.6. Solve Eq. (9.41) for $a = 2$.

9.7. Obtain a solution to the continuity equation for the F_2 region at the equator.

9.8. The F_2 region of the earth's ionosphere contains doubly charged oxygen ions (O^{++}) in concentrations of about 2 to 3 orders of magnitude less than O^+. Find a relation between the scale heights of O^+ and O^{++}, assuming the ionosphere to be isothermal and in diffusive equilibrium. *Hint:* The electric force action on O^{++} is 2e times the charge separation electric field.

9.9. Compute the spectrum of photoelectrons from the transition $O(^3P) + h\nu \rightarrow O^+(^4S) + e$ from threshold to 280 Å. Neglect the solar continuum. (Refer to *Physics of the Lower Ionosphere* by Whitten and Poppoff, Chapters 2 and 4).

9.10. Develop a numerical method for solving the time-dependent heat equation [Eq. (9.63)] for the electron gas. *Hint:* See P. M. Banks, *J. Geophys. Res.* **72,** 3365 (1967), p. 3372.

9.11. Show that at height z the energy E of a photoelectron which is traveling parallel to the local geomagnetic field is given by

$$E = E_0\left\{1 - 4 \times 10^{-12}\frac{HN(z_0)}{E_0 \sin I}\left[1 - \exp\left(\frac{z_0 - z}{H}\right)\right]\right\}^{1/2}$$

where E_0 is the initial energy, z_0 is the altitude of formation, H is the plasma scale height, N is the electron number density, and I is the magnetic dip angle.

9.12. In an isothermal protonosphere of $T = 2000°K$ $[H^+] = 10^3\ cm^{-3}$ at 1000 km altitude. What is the number density at 3000 km?

9.13. Obtain a solution to the continuity equation for H^+ in the protonosphere. Assume that the electron, ion, and neutral gases are isothermal and in thermal equilibrium with each other, and that the O^+ profile is known.

REFERENCES

1. A. Dalgarno, *J. Atmos. Terr. Phys.* **26,** 939 (1964).
2. S. Chapman, *Proc. Phys. Soc.* **43,** 26 (1931).
3. H. Rishbeth and D. W. Barron, *J. Atmos. Terr. Phys.* **18,** 234 (1960).

4. T. Yonezawa, *Space Sci. Revs.* **5,** 3 (1966).
5. *See*, for example, P. M. Morse, and H. Feshbach, *Methods of Theoretical Physics*, Vol. I, McGraw-Hill, New York, 1953.
6. W. B. Hanson and D. D. McKibbin, *J. Geophys. Res.* **66,** 1667 (1961).
7. K. Norman and A. P. Willmore, *Planet. Space Sci.* **13,** 1 (1965).
8. R. C. Sagalyn, M. Smiddy, and J. Wisnia, *J. Geophys. Res.* **68,** 199 (1963).
9. A. K. Paul and J. W. Wright, *J. Geophys. Res.* **69,** 1431 (1964).
10. J. E. Jackson and S. J. Bauer, *J. Geophys. Res.* **66,** 3055 (1961).
11. P. Mange, *J. Geophys. Res.* **65,** 3833 (1960).
12. *See* e.g., J. H. Hoffman, C. Y. Johnson, J. C. Holmes and J. M. Young, *J. Geophys. Res.* **74,** 6281 (1969).
13. P. C. Kendall, *J. Atmos. Terr. Phys.* **24,** 805 (1962).
14. H. Rishbeth, *J. Atmos. Terr. Phys.* **26,** 657 (1964).
15. A. L. Schmeltekopf, E. E. Ferguson, and F. C. Fehsenfeld, *J. Chem. Phys.* **48,** 2966 (1968).
16. M. Hirono and H. Maeda, *J. Geomag. Geoelec.* **6,** 127 (1954).
17. Y. V. Somayajulu, *J. Geophys. Res.* **69,** 1392 (1964).
18. J. V. Evans, *J. Geophys. Res.* **70,** 1175 (1965).
19. For example, H. Rishbeth, A. J. Lyon, and M. Peart, *J. Geophys. Res.* **68,** 2559 (1963); R. A. Goldberg and E. R. Schmerling, *J. Geophys. Res.* **68,** 1927 (1963); R. A. Kendall, P. C. Kendall, and E. R. Schmerling, *J. Geophys. Res.* **69,** 417 (1964).
20. W. B. Hanson, and F. S. Johnson, *Mem. Soc. Liege Serie* 5, **4,** 390 (1961); also W. B. Hanson, "Electron temperatures in the upper atmosphere," in *Space Research III*, North-Holland Publishing Co., Amsterdam, 1963.
21. A. Dalgarno, M. B. McElroy, and R. J. Moffett, *Planet. Space Sci.* **11,** 451 (1963).
22. N. W. Spencer, L. H. Brace, G. R. Carignan, D. R. Taeusch, and N. Nieman, *J. Geophys. Res.* **70,** 2665 (1965); also W. B. Hanson, S. Sanatani, L. H. Brace, and J. A. Findlay, *J. Geophys. Res.* **74,** 2229 (1969).
23. A. Dalgarno and T. C. Degges, *Planet. Space Sci.* **16,** 125 (1968).
24. A. Dalgarno, M. B. McElroy, M. H. Rees, and J. C. G. Walker, *Planet. Space Sci.* **16,** 1371 (1968).
25. P. M. Banks, *Ann. Geophys.* **22,** 577 (1966).
26. R. C. Whitten, *J. Geophys. Res.* **74,** 5623 (1969).
27. P. M. Banks, *J. Geophys. Res.* **72,** 3365 (1967).
28. L. Spitzer, *Physics of Fully Ionized Gases*, Interscience (John Wiley), New York, 1956.
29. J. E. Geisler and S. A. Bowhill, *J. Atmos. Terr. Phys.* **27,** 1119 (1965); also see their *Aeronomy Report No. 5*, University of Illinois, January 1965.
30. R. C. Whitten, *J. Atmos. Terr. Phys.* **30,** 1523 (1968).
31. R. C. Whitten, *J. Atmos. Terr. Phys.* **32,** 1143 (1970).
32. See Reference 24 of Chapter 3.
33. *See*, e.g., N. W. Spencer, L. H. Brace, and G. R. Carignan, *J. Geophys. Res.* **67,** 157 (1962).
34. *See*, e.g., J. V. Evans, and M. Loewenthal, *Planet. Space Sci.* **12,** 915 (1964).

35. J. V. Evans and G. P. Mantas, *J. Atmos. Terr. Phys.* **30,** 563 (1968).
36. D. Rapp, *J. Geophys. Res.* **68,** 1773 (1963).
37. J. E. Geisler, Ph.D. dissertation, University of Illinois, 1965.
38. J. B. Lomax, *Spread F in the Pacific*, paper presented at AGARD-NATO Meeting, Copenhagen, August 1964.
39. S. A. Bowhill, *Origin of field-aligned irregularities in the F_2 layer*, paper presented at AGARD-NATO Meeting, Copenhagen, August 1964.
40. J. R. Herman, *Revs. of Geophys.*, **4,** 255 (1966).
41. C. O. Hines, Ionospheric movements and irregularities, in *Research in Geophysics*, Vol. 1, Ed. by H. Odishaw, MIT Press, Cambridge, 1964.
42. Mariner Stanford Group, *Science* **158,** 1678 (1967); also article by Kliore et al. in the same issue.
43. R. C. Whitten, *J. Geophys. Res.* **75,** 6215 (1970).
44. F. C. Fehsenfeld, A. L. Schmeltekopf, P. D. Goldan, H. I. Schiff, and E. E. Ferguson, *J. Chem. Phys.* **44,** 4087 (1966).
45. M. B. McElroy and D. F. Strobel, *J. Geophys. Res.* **74,** 1118 (1969).
46. P. A. Cloutier, M. B. McElroy, and F. C. Michel, *J. Geophys. Res.* **74,** 6215 (1969.)

GENERAL REFERENCES

S. K. Mitra, *The Upper Atmosphere*, 2nd Ed. Asiatic Society, Calcutta, 1952 is out of date in some respects but is still a valuable reference.

More recent reviews are contained in J. Belrose, The Ionospheric F Region, in *Physics of the Earth's Upper Atmosphere*, Ed. by Hines, Paghis, Hartz, and Fejer, Prentice-Hall, Inc., Englewood Cliffs, N.J., 1965 and

T. Yonezawa, Theory of Formation of the Ionosphere, *Space Science Reviews*, **5,** 3 (1966).

H. Rishbeth and O. K. Garriott, *Introduction to Ionospheric Physics*, Academic Press, New York, 1969. Chapters 4 and 5 give an excellent introduction to the physics of the upper ionosphere.

Hexcellent discussion of ambipolar diffusion is given in E. H. Holt and R. E.
An askell, Foundations of Plasma Dynamics, Macmillan, New York, 1965; sections 7.11–7.13, 10.23–10.24.

10

Disturbances in the Ionosphere

In the discussions of Chapters 8 and 9, it was assumed that ionospheric conditions were nearly constant with respect to time, or at least slowly varying; this is a reasonably valid assumption during solar quiescent periods. However, ionospheric conditions change rather rapidly during and following solar disturbances. The frequency of these disturbances varies throughout the 11-year solar sunspot cycle, becoming more frequent as sunspot number increase. A solar disturbance can take the form of increased radiation, corpuscular or electromagnetic, or both, and/or magnetic variations. The disturbances may vary from barely perceptible perturbations occurring at less than monthly intervals during solar minimum periods to dramatic perturbations lasting hours or days and occurring so often that they overlap one another during solar maximum periods. It is something of an understatement to say that the causes and effects of these perturbations are not well understood. Attempts thus far to calculate the electron and ion variations in any detail have proven to be unsuccessful because of the many processes involved and the lack of knowledge regarding both the processes and boundary conditions. Computer programs have so far been unable to even approximate observed effects. *In situ* measurements have not yet added significant unambiguous information.

Although the primary sources of the disturbing phenomena have not been satisfactorily described in detail, a considerable amount of knowledge has been catalogued in recent years regarding the causes of the observed ionospheric effects and of the qualitative aspects of the interactions between disturbing agents and the atmosphere.

Many ionospheric perturbations can be attributed directly to solar disturbances called flares; they have been shown to be impulsive sources of both electromagnetic and corpuscular radiation. Another set of ionospheric perturbations caused by corpuscular radiation is confined mainly to the auroral regions and probably is caused indirectly by solar disturbances, but the web of events leading to auroral disturbances is complex and not likely to be untangled for some time. "Ionospheric storms" and aurora may be linked with magnetic regions of the sun but again the link is not well understood.

Perturbations caused by solar eclipses are more straightforward; the causes are evident but because of the difficulty of accounting for all sources of ionizing radiations on the solar disc, the analyses are not definitive. Sporadic-*E* events are not thought to be connected with solar events but are included in this chapter for convenience.

Aside from the phenomenological details, the key analytical problems are related to the simplifying assumptions invoked in Chapters 8 and 9. Assumptions of constant q, or $dN/dt = 0$, or $d\lambda/dt = 0$ must be treated with caution and can be used in only very carefully qualified conditions, if at all. In addition, many of the perturbing radiations are very energetic and penetrate to much lower altitudes than do the normal "background" radiations. Hence, the general lack of knowledge of terrestrial atmospheric conditions and processes below 80 km becomes a very real obstacle.

10.1 Perturbations by Electromagnetic Radiation

In this section we shall discuss in some detail the nature of the electromagnetic radiation emanating from a solar flare and the resulting effects in the ionosphere.

10.1.1 Solar X-ray Flare

In Chapter 8 we noted briefly that as the sunspot activity increases during a solar cycle, the average background X-ray flux emitted by the sun increases especially in the shorter wavelength region, i.e., the flux is said to "harden." The X-ray emission is thought to originate over the very bright solar regions (observed in hydrogen Hα light) called plages, and the area of the sun covered by plages increases with the sunspot number. Sunspots are found within the plage areas.

Occasionally, a dramatic eruption occurs in a plage area; this is called a solar flare. It has been observed for many years that the flares cause ionospheric disturbances. The disturbances were observed as fading of short wave communication and called the "Mögel-Dellinger Effect," after the observers who first reported the phenomena; the term is seldom used nowadays. The causes of solar flares have been variously thought to be connected with intense electric fields, hydromagnetic waves, and/or electron jets caused by local high compression. In any case, flares are observed to be intimately associated with complex sunspot groups and are observed as arcs of glowing gases which originate in one sunspot and terminate in another; however, exceptions to such observations have been reported.

It was assumed for some time that the flares emitted large fluxes of hydrogen Lα radiation which ionized the nitric oxide in the lower ionosphere, and, hence, caused increased free electron concentrations which absorbed short wave radio signals propagated through the *D*-region. However, a series of rocket flights by the Solar Physics Group at the Naval Research Laboratory[1] showed conclusively that the total solar Lα radiation did not increase significantly during a flare, but that the X-ray radiation did. This has since been confirmed by satellite observations.

Two important parameters are required in order to properly assess the effect of solar X-ray flares on the ionosphere. These are the spectral irradiance (intensity per unit wavelength as a function of wavelength) and the temporal variations of the spectral irradiance. To date, we have only fragments of direct knowledge of these parameters. Figure 10.1 shows the change in spectral irradiance of the X-radiation emitted by the sun as observed with the Ariel I Satellite[2] during two small flares. Figure 10.2 shows the temporal variation of the 2–8 Å component of solar radiation during a flare.[3] It is clear that all components of solar X-radiation do not increase at the same time; this is illustrated very well[4] by Figure 10.3. Higher energy electromagnetic radiations (up to 500 keV) have also been measured during flares with rocket-borne and balloon-borne detectors.

10.1.2 Observed Effects

Ionospheric disturbances caused by pulses of electromagnetic radiation are observed as perturbations in the propagation of radio signals and are commonly described by these radio-observed effects. The terms commonly used to describe ionospheric events are summarized below. Some of these effects will be considered further in Chapter 11 on radio waves in the ionosphere.

Observed ionospheric events of an irregular, impulsive character are known, collectively, as Sudden Ionospheric Disturbances (SID). Specific events affecting particular aspects of radio propagation are known as:

a. Short Wave Fadeout (SWF). At frequencies above 500 kHz, the observed effect is one of signal absorption (or signal fading); it is most pronounced at frequencies around 1 MHz. Large flares can cause complete "blackout" of long-distance short-wave radio communications. Sudden Short Wave Fadeouts are designated SSWF.

b. Sudden Enhancement of Atmospherics (SEA). This is a low frequency (10–500 kHz) phenomenon. The increased ionization causes increased reflectivity of low frequencies by the *D*-region. Hence, intensity increases are observed of the natural low-frequency radio sources known as atmospherics.

c. Sudden Cosmic Noise Absorption (SCNA). High frequency signals of extraterrestrial origin (Cosmic Noise) are attenuated by flare-induced effects (principally by absorption in the *D*-region). This effect is observed regularly around the world by stations equipped with riometers (Relative Ionospheric Opacity Meters). Studies of the characteristics of SCNA's have been useful in deducing ionospheric processes.

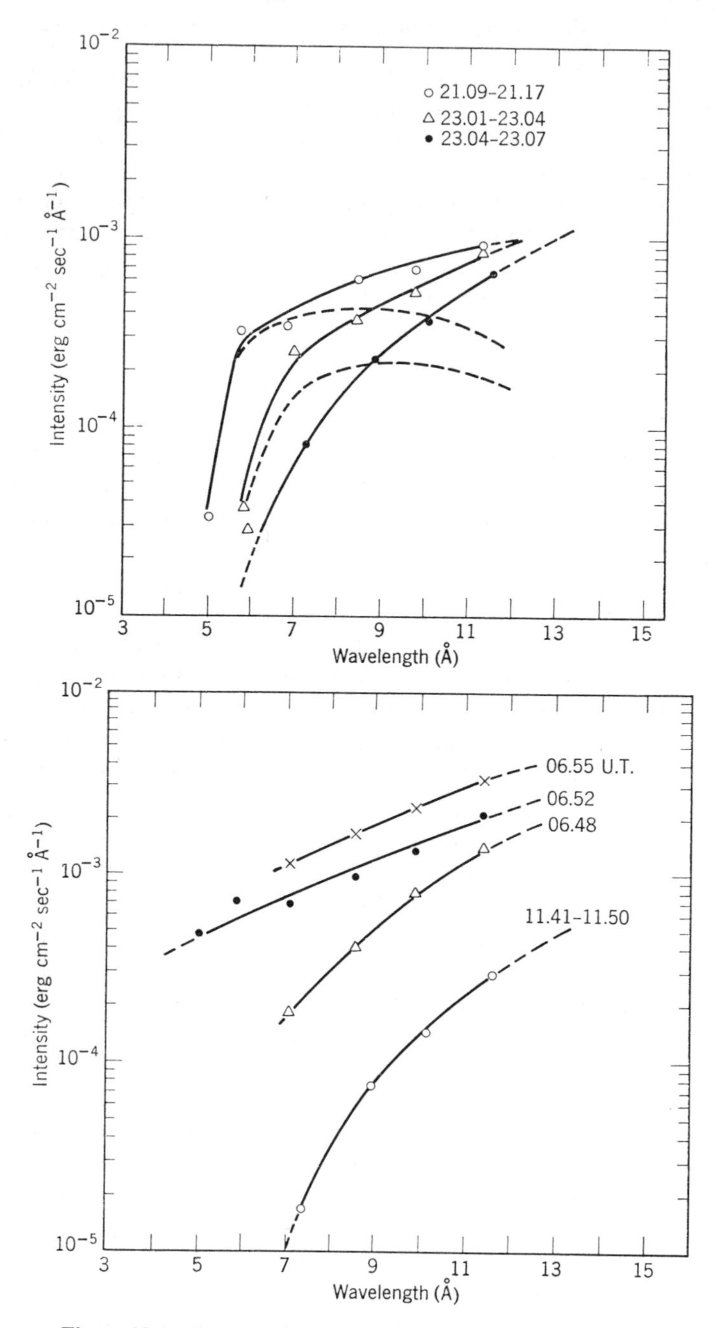

Figure 10.1 Spectra of small solar flares. [After Bowen, Norman, Pounds, Sanford, and Willmore, *Proc. Roy. Soc.* **A281,** 538 (1964); reprinted by permission of the Royal Society.]

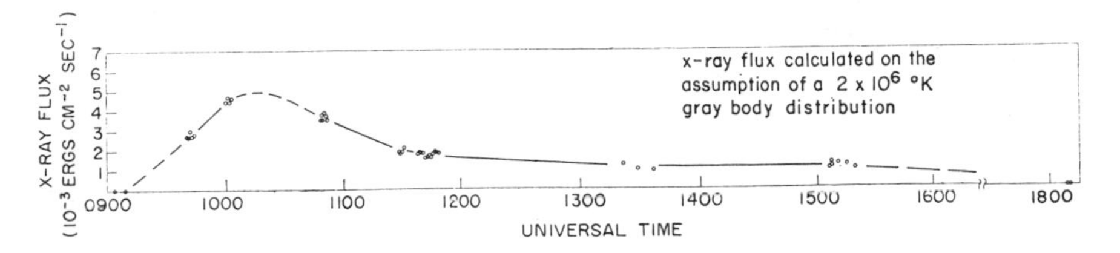

Figure 10.2 Temporal variation of 2–8 Å solar flare radiation. [After Kreplin, Chubb and Friedman, *JGR* **67**, 2231 (1962); reprinted by permission of the American Geophysical Union.]

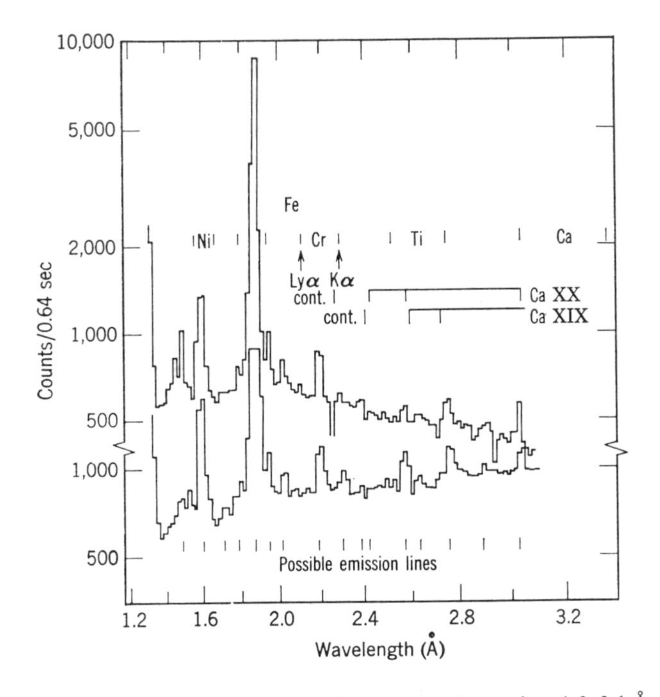

Figure 10.3a Two spectral scans in the region 1.3–3.1 Å obtained during the increasing phase of a solar X-ray burst on March 22, 1967. Apparent differences in spectral distribution are due to the increase in intensity of the X-ray burst in the time (5 min) required to make the two scans. The onset of increased count rate at 1.34 Å coincides with the detector position at which it begins to be illuminated directly by the sun.

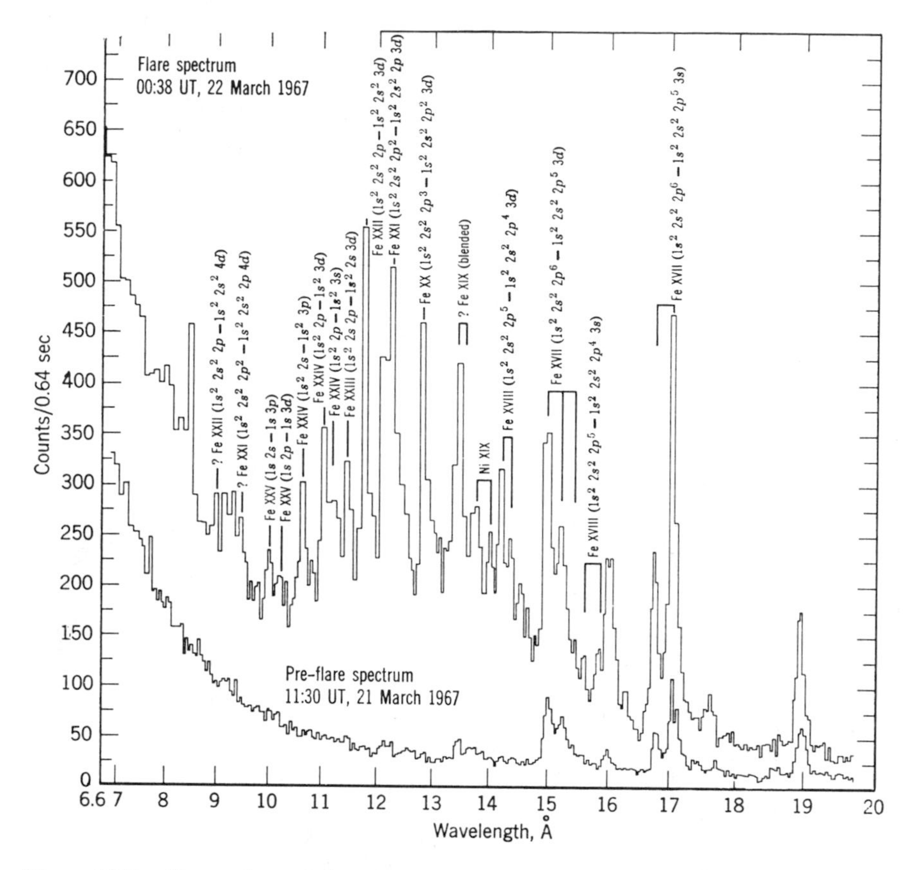

Figure 10.3b Comparison of the solar spectrum between 6.3 and 20.0 Å obtained during a flare on March 22, 1967, with a spectrum obtained on the previous day when no flares were in progress. Emission of line spectra during solar flares. [After Neupert, Gates, Schwartz, and Young, *Ap. J.* **149**, 179 (1967); reprinted by permission of the University of Chicago Press; copyright 1967 by the University of Chicago.]

d. Sudden Phase Anomaly (SPA). In addition to changes in amplitude, changes in phase of low frequency radio waves are also observed. The cause is similar to that responsible for SEA, i.e., a lowering of the reflection level for low frequency signals by the enhancement of electron concentrations in the *D*-region. A lowering of the reflection height will change the path length of the propagated wave and, hence, will change the phase of the received signal.

e. Sudden Frequency Deviations (SFD). This effect is observed on high frequency (~20 MHz) radio signals. Increased ionization in the *E* and *F*-regions result in changes of refractive index of those regions for radio waves propagated through them, reflected, and propagated back to a receiving station. As the refractive index changes, so does the phase path, and this results in a change of frequency proportional to the rate of change of the ionization. Hence, the more impulsive the event, the more pronounced the SFD.

10.1.3 Balance Equations

The most obvious difference between the steady state conditions of Chapter 8 and the conditions pertaining to impulsive X-ray events is the functional form of $q(z)$ as given by Eq. (8.6). For impulsive events, the time dependent form must be used.

$$q(z, t) = \frac{1}{W} \sum_j n^j(z) \int_\lambda Q_i^j(\lambda)\, F_\lambda(z = \infty; t) \exp\left[- \sum_j Q_a^j(\lambda) \int_z^\infty n^j(z) \sec \chi \, dz \right] d\lambda \tag{10.1}$$

Reference to Figures 8.15 and 10.1 illustrates the fact that during flares, the increased radiation is in the spectral region below 10 Å. Thus, we are concerned with the harder, more penetrating radiation. The X-ray components that can penetrate through the *D*-region increase by several orders of magnitude for a moderate sized flare even though the total X-ray emission (below 10 Å) increases only by factors of 2 to 5. During large flares, the E and F_1 regions may also be affected significantly. Although E and F_1 effects are not yet well-documented, the study of SFD's may provide a better grasp of that problem.

Because we are dealing with ionization of all components of the atmosphere, ionization of minor species such as NO are insignificant and need not be considered. The use of ionization and absorption coefficients for air is adequate for terrestrial applications. Similarly for Mars and Venus, ionization of the major constituents CO_2, N_2, and/or A should be considered; ionization of the minor constituents O, O_2, CO, NO, water vapor, hydrocarbons, and metallic atoms should be unimportant. Hence, for the terrestrial ionosphere:

$$q(z, t) = \frac{n(z)}{W} \int Q_{\text{air}}(\lambda) F_\lambda(t) \exp\left[- \int_z^\infty Q_{\text{air}}(\lambda) n(z)\, dz \sec \chi \right] d\lambda \tag{10.2}$$

where $F_\lambda(t)$ is the flux incident to the atmosphere and is time-dependent. Absorption cross sections Q_{air} for X-rays in air as a function of wavelength are given in Figure 8.2a; for X-ray photons, $Q_i = Q_a$.

The principal ions formed directly in the terrestrial ionosphere by an X-ray pulse will be N_2^+ and O_2^+; the rapid exchange process, given by reaction

(8.7a) ($N_2^+ + O_2 \rightarrow O_2^+ + N_2$), will result in a preponderance of O_2^+. The ion-atom interchange reaction 8.7f is too slow to warrant consideration during a flare. Reactions 8.7b, 8.7c, and 8.7d involve atomic oxygen; these need be considered only for *E* and *F*-region analyses. Therefore, O_2^+ should be the only important positive ion in the *D*-region.† Similar considerations (i.e., 8.7h) should lead to CO_2^+ as the important positive ion in the Mars or Venus atmosphere; however, if O_2 exists in appreciable amounts, even as a minor species, reaction (8.7i) may produce O_2^+ fast enough to make O_2^+ the controlling ion; there is some evidence that O_2^+ may be present at low altitudes.

Under most solar flare conditions, λ, q, and N_e will not increase sufficiently [see (Eq. 8.15b)] to cause $d\lambda/dt$ to be significant; hence, the electron concentration can be described by

$$\frac{dN_e}{dt} = \frac{q(z, t)}{1 + \lambda} - (\alpha_D + \lambda\alpha_i)N_e^2 \tag{8.17a}$$

Inasmuch as solar flares affect only the sunlit hemisphere, photodetachment is probably the dominant detachment process for the terrestrial atmosphere. Hence,

$$\lambda \approx \frac{\beta m}{\rho} \tag{8.18b}$$

The above assumptions must be justified for particular applications and in the light of current knowledge of atmospheric composition. For example, it is conceivable that situations may occur in which either $d\lambda/dt$ or the last term in Eq. 8.15b, or both, cannot be neglected. Also, it is conceivable that associative detachment (reaction 8.13a) could be more important than photodetachment if the atomic oxygen concentration is high.‡ As we noted in Chapter 8, if the lower atmosphere of Mars is composed entirely of CO_2 and either N_2 or A, as is currently thought, no negative ions should be formed; hence

$$\frac{dN_e}{dt} = q(z, t) - \alpha_D N_e^2 \tag{10.3}$$

† The importance of minor species can be illustrated at this point. The identity of the dominant ion in a solar flare-controlled ionosphere actually depends on the rates of competing processes. If the concentration of NO is high when compared with the concentration of free electrons, the dominant ion may become NO^+. The reader should demonstrate this for himself. *Hint:* Use lifetime concepts discussed in last part of Section 10.2.4.

‡ Again, a better perspective of the role of minor constituents can be achieved by the reader if he explores how the importance of associative detachment processes increases with increased atomic oxygen concentrations.

As is also demonstrated in Chapter 8, even small concentrations of O_2 in the Martian atmosphere will produce copious quantities of negative ions.

Using data such as those shown in Figure 10.1, $q(z)$ can be calculated for various values of t. Under most conditions, Eqs. (8.17a) and (8.18b) can then be solved to provide $N_e(z)$ for the same values of t; however, the reader is cautioned to review the implicit and explicit assumptions before applying simplified equations.

10.1.4 Illustrative Examples

In order to illustrate the magnitude of the perturbations caused by solar X-ray flares, three examples are presented. Figure 10.4 shows a composite spectrum of peak X-ray emission from a large flare based on several rocket observations. Figure 10.5 shows the development of a small flare[2]. Ionization rates calculated for the peak fluxes of Figures 10.4 and 10.5 and for an estimated class 3 flare are shown in Figure 10.6. Electron concentrations for the curves of Figure 10.6 are shown in Figure 10.7. These curves should be compared with Figures 8.5, 8.6, 8.15, and 8.16.

10.2 Perturbations by Corpuscular Radiation

Charged particle bombardment of the upper atmosphere at high latitudes causes the very dramatic ionospheric disturbances known as Polar Cap Events and Auroras. In this section we shall consider the characteristics of the charged particle streams and the ionospheric effects. The morphology was discussed in Chapter 6.

10.2.1 Solar Proton Flares

Occasional large solar flares also emit large fluxes of very energetic charged particles. The flares are known as proton flares and the charged particle fluxes are called solar cosmic radiation (SCR). In fact, the SCR is not composed entirely of protons but also contains an appreciable component (perhaps 10 percent) of alpha particles as well as a complement of low energy electrons necessary to provide for charge conservation.

As with the X-ray flare, the generating mechanism of SCR is not known, although some characteristics have been observed. For example, the events are explosive in nature and are associated with bursts of solar radio noise which are considered to be part of the spectrum of synchrotron radiation. These are rather dramatic events; the SCR sometimes arrives at the Earth within a half-hour after the flare is observed optically and continues to bombard the atmosphere for several days. More often, there is a lag of several hours after the optical observation, sometimes as long as 20 hours; and the bombardment is often only a matter of a few hours duration.

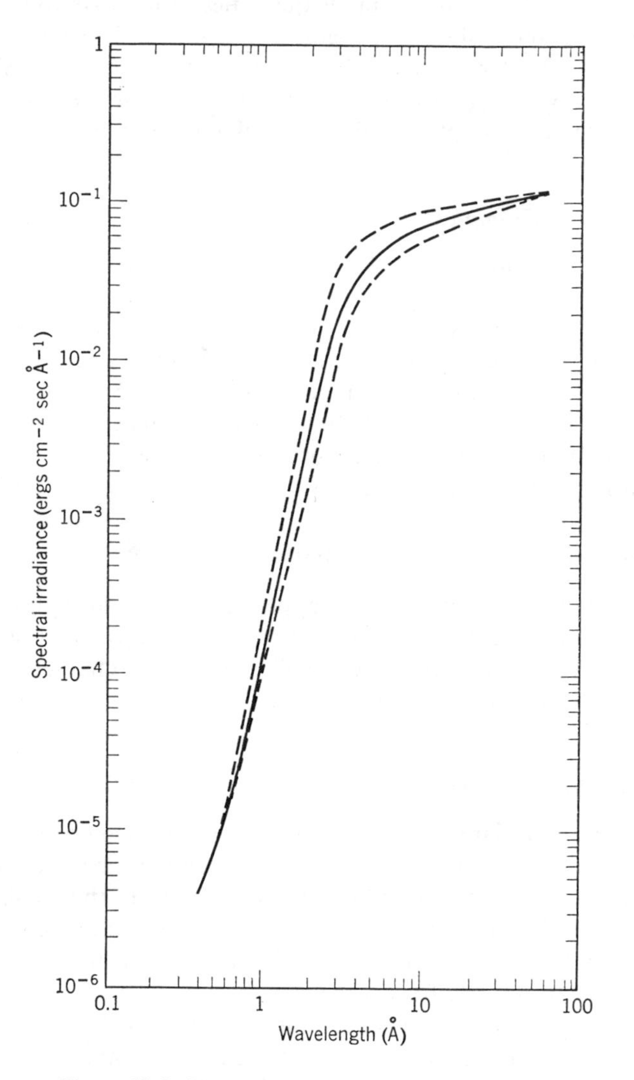

Figure 10.4 X-ray spectrum of the flare observed at 2250 UT on August 31, 1959. The solid curve represents the most probable spectrum based on extrapolated data. The broken lines represent estimated upper and lower limits of intensity. [After Whitten and Poppoff, *JGR* **66,** 2779 (1961); reprinted by permission of the American Geophysical Union.]

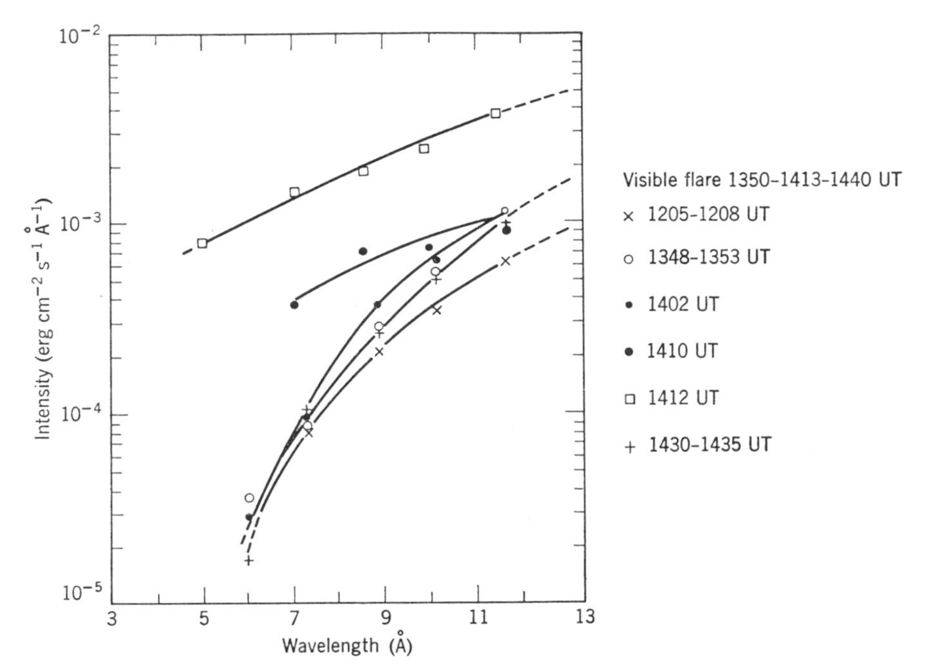

Figure 10.5 Development of a small flare. [After Bowen, Norman, Pounds, Sanford, and Willmore, *Proc. Roy. Soc.* **A281,** 538 (1964); reprinted by permission of the Royal Society.]

Magnetic field lines radiate from the sun. Near the surface of the sun where the magnetic activity originates, the field is radial; but farther away at the distance of the earth's orbit, the angle between the earth-sun line and the magnetic field lines may be almost 45°, and at the distance of Jupiter's orbit the lines may be almost parallel to the orbit. This is because plasma streaming out from the sun has a large electrical conductivity through which the magnetic field lines pass and become "frozen"; a combination of the outward radial flow of the plasma and the rotation of the sun causes the magnetic field lines to be stretched out in Archimedes spirals.[6–8] The pattern can be compared to that of the grooves on a phonograph record; and the pattern of Archimedian spirals of magnetic field lines co-rotating with the sun can be visualized as the phonograph record grooves co-rotating with the record. Furthermore it is observed that the field lines sometimes point toward the sun and sometimes away from the sun; the direction of pointing is characteristic of a sector of the disc containing the spiral-shaped field lines. Within a sector a filamentary structure is sometimes found along which the SCR protons travel. The characteristics of travel of SCR particles are not fully

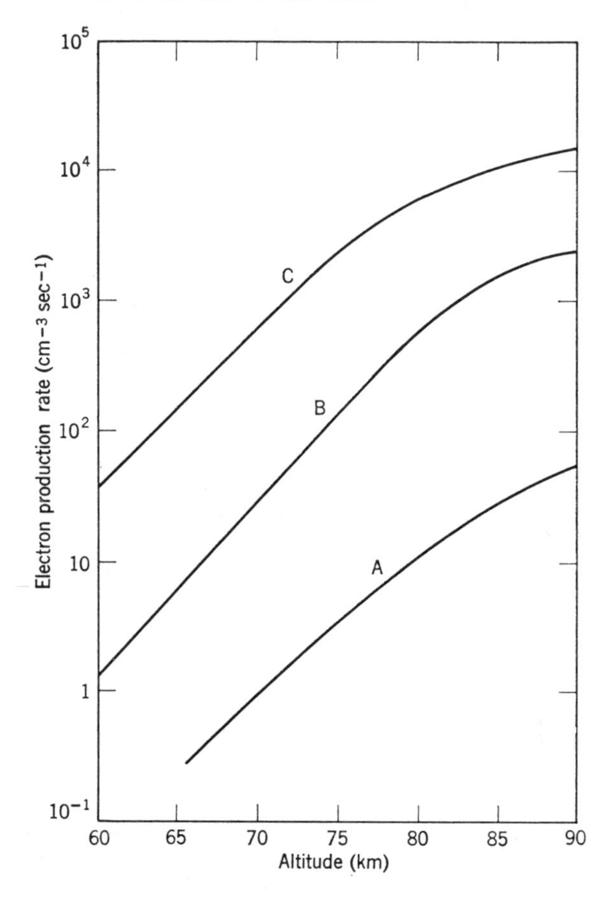

Figure 10.6 Electron production rates during flares. A—Computed from X-ray spectrum shown in Figure 10.5. B—Computed from X-ray spectrum shown in Figure 10.4. C—Possible production rate at peak X-ray intensity of a class 3 or 3+ flare.

understood, but apparently the protons are sometimes trapped for varying periods of time in the filamentary structure and, hence, a considerable lag is sometimes observed between the optical manifestations of a flare and the arrival of SCR protons.

The Earth and Jupiter have strong magnetic fields which effectively shield them from SCR except in the polar zones. In the polar zone, the particles interact with the magnetic field and are funneled into the atmosphere; the resulting trajectories are functions of the particle energy and the magnetic

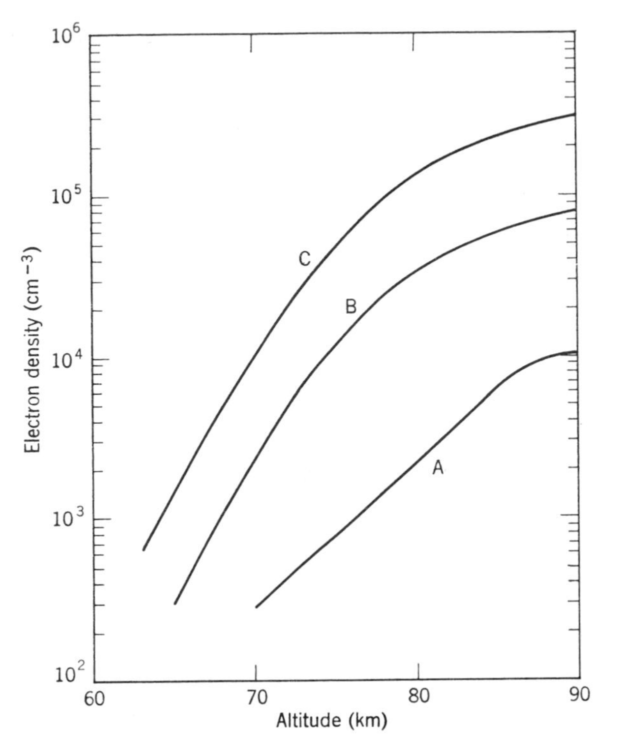

Figure 10.7 A—1415 UT on April 27, 1962. B—2254 UT on August 31, 1959. C—Estimated peak electron density for a 3 or 3+ flare. Electron concentrations produced by the ionization rates of Figure 10.6.

field strength. The orbital theory of charged particles entering the geomagnetic field from interplanetary space was developed many years ago by the Norwegian physicist, Störmer. (See Alpher[9] for a review of the motion of charged particles in the Earth's magnetic field.) While his work did not lead to a valid theory of the aurora as he had originally hoped, it does correctly predict the trajectories of SCR particles within the magnetosphere. The geomagnetic control is pronounced, particularly for the low energy particles.

For a charged particle spiraling along a magnetic line, the magnitude of the centripetal force must equal the magnitude of the Lorentz force or

$$\frac{mv^2}{r} = \frac{eZ}{c} vB \tag{10.4a}$$

where m is the mass of the particle, v is the velocity, r is the gyroradius, e is the electron charge, Z is the number of electron charges carried by the particle, c is the velocity of light and B is the magnetic flux density. We can rearrange Eq. (10.4a) to obtain,

$$\frac{mv}{eZ} = \frac{P}{eZ} = \frac{1}{c} rB \tag{10.4b}$$

or

$$\frac{Pc}{eZ} = rB \tag{10.4c}$$

where P is the momentum of the particle. If Pc/Ze (momentum per unit charge) is expressed in volts, then

$$rB = \frac{1}{300} \frac{Pc}{Ze}$$

is expressed in statvolts and $Pc = 300\ rBeZ$ is in electronvolts. The magnetic rigidity is

$$R = \frac{Pc}{Ze} \tag{10.4d}$$

where $R = 300$ rB is expressed in volts. The particle energy is, of course

$$E = \frac{(Pc)^2}{2mc^2} \tag{10.4e}$$

where mc^2 is the "rest energy" of the particle, e.g., 0.51 MeV for an electron and 938 MeV for a proton. Note that the nonrelativistic form is used because we are dealing with nonrelativistic particles.

Each particle momentum, Pc/Ze, or "rigidity" has associated with it a particular geomagnetic latitude cut-off, λ_c, such that at latitudes below the cut-off, the particles cannot reach the Earth. Conversely, each geomagnetic latitude has associated with it a cut-off rigidity R_c. The two cut-offs are related by the equation

$$R_c = 14.7 \cos^4 \lambda_c \tag{10.5}$$

where R_c is expressed in 10^3 MV. During very large PCA events the geomagnetic field is distorted by the incident plasma cloud and the cut-off latitudes for given momenta are significantly lowered.

Mars and Venus apparently have weak or nonexistent magnetic fields. Obviously magnetic cut-off is not a limitation and SCR could affect large portions of the planet (conceivably, the entire planet) simultaneously. Because of the lack of magnetic shielding it would be expected that "softer" (lower energy) components of SCR would in general be more significant on Mars and Venus than on Earth.

The intensity and spectral distribution of energetic charged particles are often given in the form of a power law of energy E:

$$J = AE^{-n} \tag{10.6a}$$

where J is the particle flux with energy greater than E, and A is a constant. For SCR protons n has been found[10–12] to be about 5 near 1.2 GeV, about 7 between 2 and 15 GeV, and 4 between 85 and 300 MeV. It has been proposed[13] that at least at energies below 1 GeV the spectra follow exponential laws of the type

$$I = I_0 \exp(-P/P_0) \tag{10.6b}$$

where I is the flux per unit momentum and P is particle momentum. In general both I_0, the flux at P_0, and P_0 will vary during the course of the event. Some of the results are shown in Figure 10.8.

For further discussion of SCR generation, characteristics, etc., the reader is referred to the review paper by Bailey[14].

10.2.2 Auroral Electrons

Since McIlwain[15] made the first direct measurement with rocket-borne detectors there has been no doubt that electron bombardments cause most of the auroral zone phenomena called optical or radio aurora; his direct measurements with rocket-borne detectors confirmed earlier ideas based on indirect evidence obtained by analyses of auroral emissions, detection of bremstrahlung, and studies of cosmic noise absorption. In the intervening years a large number of measurements have been made with satellite, balloon, and rocket instrumentation that has confirmed and amplified the earlier observations.

A large body of morphological information has been developed. However, many fundamental questions are still unanswered owing to both experimental and theoretical limitations. For example, we still have not definitely demonstrated a direct correlation between specific electron bursts and specific auroral (optical and radio) features although we assume that it exists. Satisfactory explanations of the origin of auroral electrons and acceleration mechanisms are yet to be proposed. In fact, detailed information on electron energy spectra and temporal characteristics, (particularly in the low energy range) which are needed to check theories are virtually nonexistent. Nonetheless, a number of gross characteristics are known and from these a number of hypotheses have been developed.

Electron energy measurements have been used to support ideas regarding both monoenergetic electron fluxes and fluxes with wide energy ranges. In part, these apparent inconsistencies are easily rationalized as due to instrument

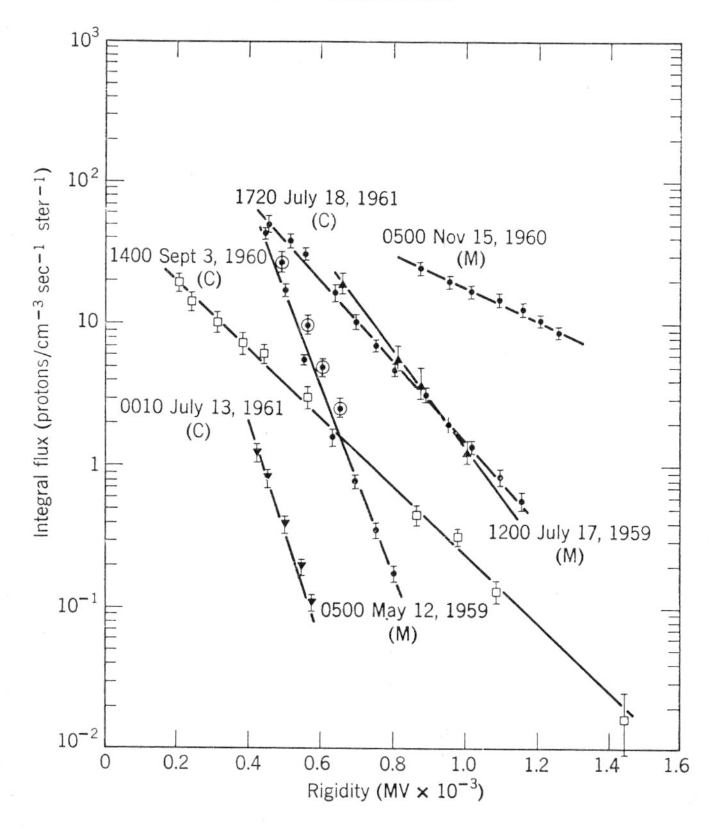

Figure 10.8 SCR proton rigidity spectra. Data points taken from counter ascents are shown as solid symbols; those taken with emulsions shown as open symbols. The letter in parentheses indicates the place of measurement: C—Fort Churchill; M—Minneapolis. Measurements were made by balloon. [After Freier and Webber, *JGR* **68,** 1065 (1963); reprinted by permission of the American Geophysical Union.]

limitations. However, it also seems quite rational to suppose that the measurements also represent wide variations in the natural phenomena known as auroras. It would seem reasonable to assume that the temporal and spectral characteristics of electron fluxes are uniquely associated with the type of auroral event (e.g., low energy monoenergetic beams with rayed auroras and wide energy ranges with diffuse auroras), but not enough data have been obtained to corroborate any such assumptions.

Hence, at present, data are available that indicate the existence of essentially monoenergetic beams[15] of 6-10 keV, or of electron beams with exponential integral energy spectra,

$$N(\geq E) = K \exp(-E/E_0) \text{ electrons cm}^{-2} \text{ sec}^{-1} \tag{10.7}$$

where K is determined by the total integrated flux, and E_0 is the "e-folding" energy which determines the slope of the spectral distribution curve. Values of E_0 have been observed as low[15] as 5 keV and greater[16] than 50 keV. The latter observation, during an auroral breakup revealed E_0 values that increased from 20–35 keV to greater than 50 keV for an energy range of 3.5 to 25 keV. A spectrum deduced from bremsstrahlung measurements[12] had an E_0 of 25 keV; it was found that this spectrum could also be represented by a power law

$$\text{N}(\geq E) = AE^{-4} \text{ electron cm}^{-2} \text{ sec}^{-1} \tag{10.8}$$

Fluxes have been observed in the range 10^5 to 10^9 cm^{-2} sec^{-1} ster^{-1} for energies greater than 25 keV. The observation of monoenergetic electrons by McIlwain[15] is reported as 5×10^{10} cm^{-2} sec^{-1} of 6 keV electrons at peak intensity.

As the electrons penetrate the auroral zone atmosphere, they spiral around the geomagnetic field lines. The angle between the field lines and the velocity vector of the spiraling particle is the pitch angle. Angles from 0-90° indicate downward spiraling particles; angles from 90–180° indicate upward spiraling particles, those that are being scattered back out of the atmosphere. Measurements have been made of pitch angle distributions and they have been found to vary with events and altitudes. Figure 10.9 shows the pitch angle distribution for two altitudes during the same event.[17] At the higher altitude, the distribution is essentially isotropic for the range 0–90° and anisotropic for the range 90–180°. The distribution at lower altitudes is anisotropic for the entire range; the distribution shows a peak at 90° with the same intensity as at the higher altitude. In both cases, the downward flux is much greater than the upward flux. Measurements of the energy distribution of auroral electrons with energies as low as 20 eV showed that the number of electrons continues to increase with decreasing energy;[18] the flux of electrons greater than 1 keV was found to be directional, whereas the flux smaller than 1 keV was isotropic and probably composed of secondaries.

The temporal characteristics of high energy electron bombardments, which may or may not be related to the lower energy electrons that produce optical auroras, have been studied extensively[19] by monitoring the resulting bremsstrahlung radiation at balloon altitudes (~30 km). The duration of electron bursts varies from microbursts (fractions of seconds) to long pulsations (many minutes) with many gradations between. Some of these are

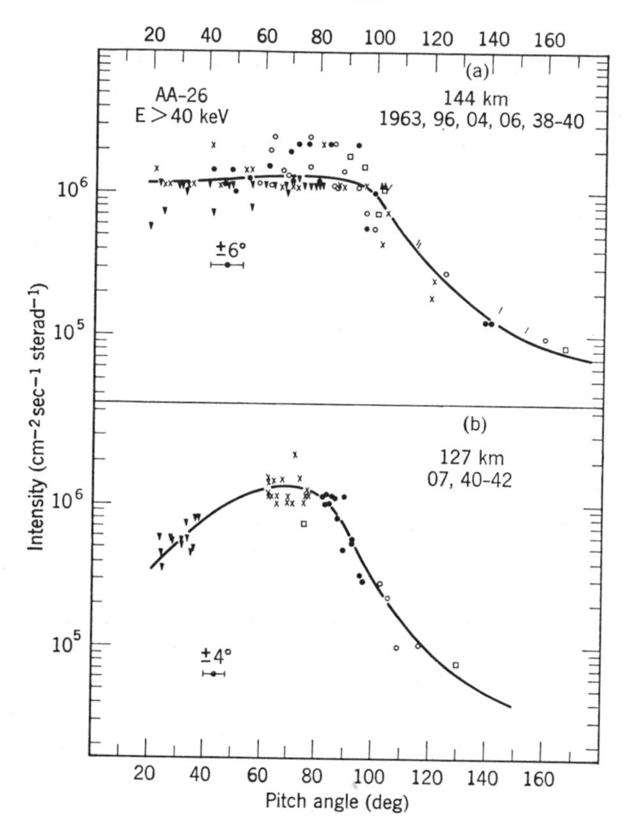

Figure 10.9 Examples of angular distributions of electrons during an auroral event. [After McDiarmid and Budzinski, *Can. J. Phys.* **42,** 2048 (1964) reprinted by permission of the National Research Council of Canada.]

illustrated in Figure 10.10. Correlations have been observed of various temporal characteristics with energy, magnetic pulsations, and auroral types.[19,20] Figure 10.11 illustrates a diurnal pattern of both spectral and temporal characteristics.

10.2.3 Ionospheric Effects

Perturbations caused by corpuscular radiation can be observed by both optical and radio techniques. The optical effects are mainly the excitation and ionization of atmospheric constituents by the impact of low energy electrons, principally secondary and tertiary products of the primary particle bombardment (these are discussed in Chapter 6). Radio effects, however, are

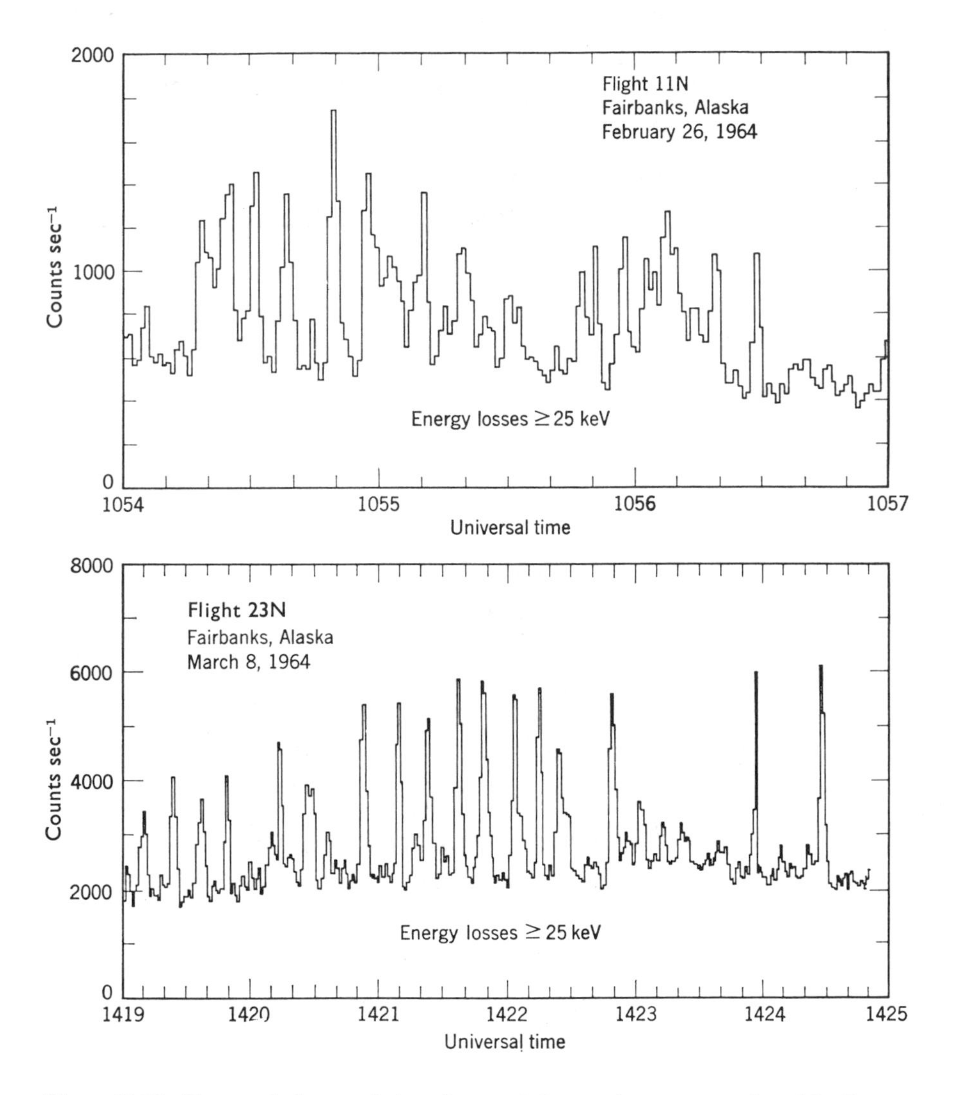

Figure 10.10 Temporal characteristics of auroral electron bursts as monitored by Bremsstrahlung detectors. [After Brown, Barcus, and Parsons, *JGR* **70,** 2599 (1965); reprinted by permission of the American Geophysical Union.] (a) Irregular pulsations in X-ray intensity. (b) Semi-regular pulsations in X-ray intensity.

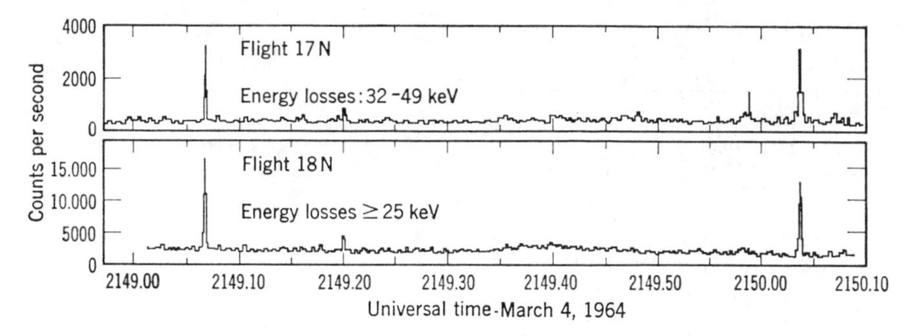

Figure 10.10 continued (c) An example of intense microbursts in time coincidence, recorded by balloon instruments separated by approximately 150 km.

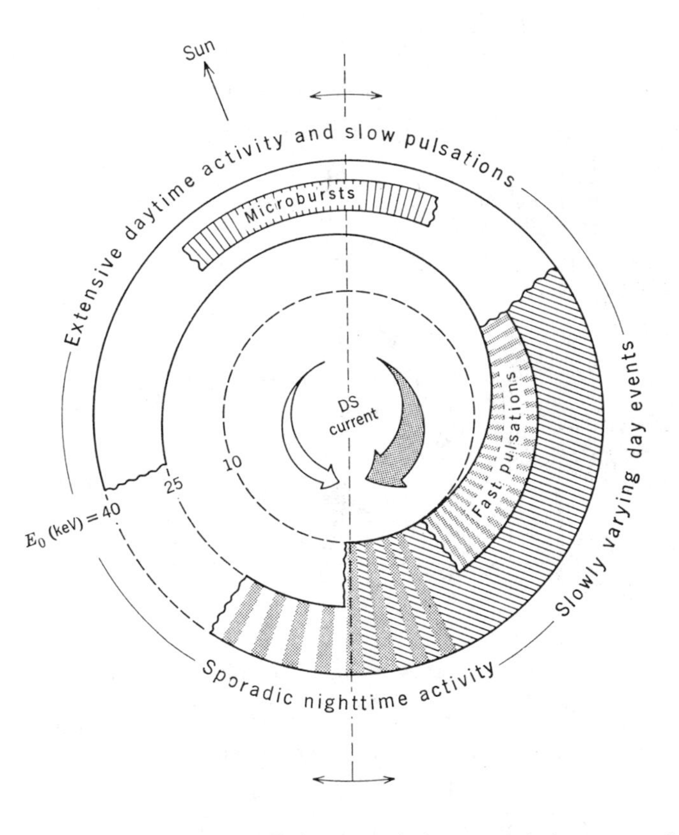

Figure 10.11 Diurnal pattern of the spectral character of electron precipitation, as indicated by X-ray measurements. [After Barcus and Rosenberg *JGR* **71,** 803 (1966); reprinted by permission of the American Geophysical Union.]

mainly the reflection or absorption of radio waves by electrons that have very low energies, and in fact have been thermalized (or are nearly thermalized). Most of the ionospheric effects we shall discuss in this section are concerned with the production and maintenance of electron concentrations that can be detected by radio techniques (and, of course, affect radio communications). Hence, electron energy distributions are not considered to be as important as they were in Chapter 6 because it is assumed that all electrons produced by ionization are eventually (and quite rapidly) thermalized. Although "optical auroras" were known in antiquity, "radio auroras" (and similar phenomena are a modern discovery that by definition could not be observed until the invention of radio frequency techniques. Many of the "radio aurora" phenomena are even more contemporary, dating back no farther than the late 1950's.

Notable among the phenomena detected by radio techniques is the electron enhancement commonly called a Polar Cap Absorption (PCA); the ionizing event itself is called a Polar Cap Event (PCE). The phenomenon, first described by Bailey[21] was soon attributed to solar proton emission. It is now recognized as an important geophysical event and has served to illuminate problems in both low altitude ionospheric processes and solar processes. The term Polar Cap Absorption refers to the effects on radio propagation, which are important in the operation of communication networks. The principal tool for studying PCA's has been the observation of radio propagation effects supplemented with observations by balloon, rocket, and satellite instrumentation. Studies are also being made of the accompanying optical emissions called "Polar Glow" (see Chapter 6).

Compared with Polar Cap Events, auroras are very complex affairs. Geomagnetic pulsations, radio absorption, radio wave scattering, and optical manifestations occur together, in sequence, and, seemingly, in all possible temporal and spatial combinations and mutations. Separation of single effects and correlation of a variety of effects have proven difficult, although intuitively it is felt that correlations must exist.

The radio effects observed during corpuscular bombardments are similar to those observed during X-ray ionization events because the causes are basically the same, *viz.*, increased free electron concentrations. The sources of the electrons differ (particle collision processes instead of photoionization) but the effects differ only in time-scales or penetration altitudes. Cosmic noise absorption, short wave fadeout, and VLF phase changes are all observed, and the observations have contributed to knowledge of high latitude terrestrial ionospheric phenomena.

10.2.4 Balance Equations

Basically, the ionization source term for corpuscular ionization events is given in Chapter 8 [Eq. (8.6b)]. In the case of auroral electrons, an additional ionization source, bremsstrahlung, should be considered. Because of evidence regarding temporal changes in auroral electrons and polar proton spectra, the use of a time-dependent form should be considered

$$q(z, t) = \rho(z) \int_E \int_\Omega \left(\frac{dE}{dx}\right) \frac{1}{W} \frac{dI(z, t)}{dE} \, dE \, d\Omega \tag{10.9}$$

Inasmuch as the temporal characteristics are not known (and are probably quite complex) two limiting cases that seem to be adequate for most analyses are generally used.

(a). For polar proton events and the longer lived auroral electron events, we can assume that for reasonably long periods, no changes occur in the spectral distribution or in the total flux. The periods considered must be longer than those necessary for ion-kinetic processes (i.e., recombination, attachment, detachment, etc.) and shorter than periods over which significant spectral changes are known to occur. This may vary from minutes in the case of auroras to hours in the case of PCE's. Quasi-equilibrium conditions can be assumed.

(b) For micropulsations observed in aurora, we can also assume that no significant changes occur in electron flux or spectrum during the ionization event but the event is a short square wave impulse. In this case, however, the duration of the event is much shorter than the periods necessary for most ion-kinetic processes and is considered smaller than the time necessary for significant spectral changes to occur. The latter condition may be far from correct and should be reconsidered in the light of actual measurements. Time-dependent solutions are probably required.

Ionization by Solar Protons. Brown and Weir[22] assumed that the proton fluxes are isotropic and that the proton paths are rectilinear; they treated the proton ionization rate problem as follows.

The electron production per unit depth at an altitude z, by a proton of kinetic energy E_0, is

$$q' = \frac{1}{W} \left(\frac{dE}{dx}\right)_{E_0} \rho(z) \tag{10.10}$$

Where W is the energy required to form an ion-pair in air, dE/dx is the energy loss (stopping power) of the particle in eV gm^{-1} cm^{-2}, and $\rho(z)$ is the air density at altitude z. The total rate of production is obtained from a knowledge of the number of particles of energy E_0 at this altitude. This is obtained by finding the corresponding energy of a particle E at the top of the atmosphere and calculating the flux of particles with that corresponding energy E from the differential energy spectrum. Then, by weighting the ionization per particle of energy E_0 at altitude z by the number of particles of energy E

incident at the top of the atmosphere, the total electron production rate (in electrons $cm^{-3} sec^{-1}$) is obtained by considering protons incident over the entire upper hemisphere and summing their effects over both direction and energy. Thus, the production rate is

$$q(z) = \frac{\rho(z)}{W} \int_{\theta,\phi} \int_{E_0(\theta,\phi,z)}^{\infty} \left(\frac{dE}{dx}\right)_E N(E)\, dE \sin\theta\, d\theta\, d\phi \tag{10.11}$$

The results are shown for such a calculation in Figure 10.12 for kinetic energy spectra in the power form

$$N(E)\, dE = AE^{-\gamma}\, dE$$

Where $\gamma = 5$ and 7. A geomagnetic cutoff latitude of 65° was assumed. The effect of cutoff energy (See Eq. 10.4) assumptions is illustrated in Figure 10.13.

The foregoing approach used by Brown and Weir is easy to understand as a concept but difficult to use for actual computations. Velinov[33] has developed an analytical expression, which, though very complex, is easier to program for computations of ionization profiles; the derivation is long and will not be reproduced here. However, the following simplified form of Velinov's

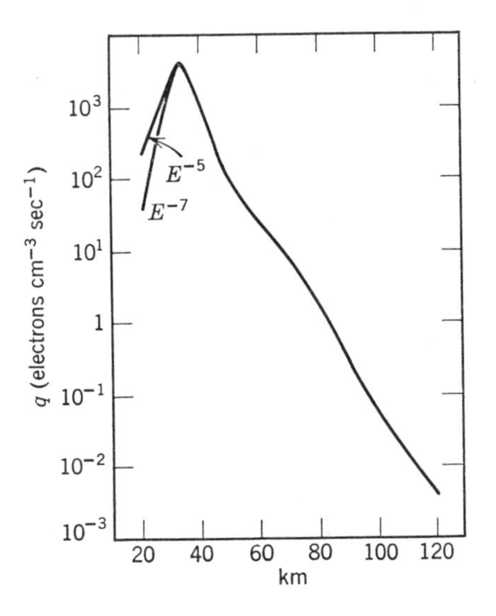

Figure 10.12 Ionization rate profiles for SCR protons. [After Brown and Weir, *Arkiv för Geofysik* **3**, 523 (1962); reprinted by permission of the Royal Swedish Academy of Science.]

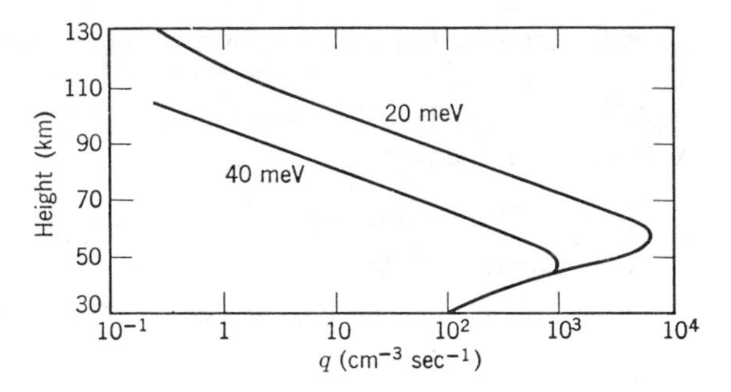

Figure 10.13 Effect of cutoff energy on SCR profile calculations. [After Reid, *JGR* **66,** 4071 (1961); reprinted by permission of the American Geophysical Union.]

equation was derived by Dubach and Barker[34] using the reasonable assumptions that relativistic particles need not be considered in solar flare radiations and that the planetary atmosphere can be considered to be flat.

$$q(h) = \frac{\pi C Z^2}{2W}\,\rho(h)\int_E\int_\theta N(E)\left[\frac{\ln(E_\infty/E_0) + B}{(E_\infty^2 - K\tilde{h}\sec\theta)^{1/2}}\right]\sin\theta\,d\theta\,dE \qquad (10.12)$$

The symbols of Eq. (10.12) have the following definitions:

a. Z, E_0, E_∞, and θ are the atomic number, rest energy, energy of incident particle, and angle of incidence, respectively.
b. $C = 8\pi e^4 Z N_0 m_p/mA$ where e and m are the electron's charge and rest mass, A and Z are the mass and atomic numbers of the stopping atom, m_p is the mass of the proton, and N_0 is Avogadro's number.
c. The density $\rho(h) = n(h)A/N_0$, where $n(h)$ is the stopping atom's number density as a function of altitude h.
d. The atmospheric depth $\tilde{h} = \int_h^\infty \rho(h)\,dh$.
e. $K = CZ^2E_0/2[B + \ln(E_\infty/E_0)]$, where $B = \ln 4mc^2/I$ and I is the mean excitation energy of the stopping atom.
f. The limits of integration for E are from the cutoff energy to infinity for planets with magnetic fields. For planets without magnetic fields, an "atmospheric cutoff" energy, $E_A = (10^3\tilde{h}Z^2)^{1/2}$, is used as the lower limit to permit integration; note that E_A varies with altitude. The limits of integration for θ are 0 to 75° for the flat planet approximation. For $\theta >$ 75°, the curved planet solution requires that sec θ be replaced by the suitable Chapman functions for grazing incidence radiation; the integration limits must also be suitably modified.

Profiles of SCR ionization computed by the different methods presented above are virtually indistinguishable.

Ionization by Auroral Electrons. We shall follow the approach used by Kamiyama[23] which is reasonably realistic, straightforward, and easily applied to the case of isotropic electron bombardment; the reader is also referred to Rees[31] for an alternate treatment.

The incremental energy loss by inelastic collisions of an electron of energy E along its path can be expressed[24] by

$$-\left(\frac{dE}{ds}\right) = nZQ_0 \frac{3\mu^2}{4E}\left(\ln \frac{E}{VZ\sqrt{2}} + \frac{1}{2}\right) \tag{10.13}$$

Where n is the concentration of atmospheric particles, Z is the atomic number of the material through which the electron is passing, Q_0 is the Thomson scattering cross section (6.65×10^{-25} cm^2), μ is the electron rest energy, and V is the ionization energy of the material.

If the electron kinetic energy is relatively high ($>\frac{1}{2}$ keV) the quantity in parentheses on the right hand side of the above equation changes very slowly with E, and can be considered constant over small path lengths. Hence, integration yields (see also Chapter 9)

$$E_{j+1}^2 \approx E_j^2 - \frac{3}{2} ZQ_0\mu^2\left(\ln \frac{E_j}{VZ\sqrt{2}} + \frac{1}{2}\right)\frac{1}{\cos\alpha}\int_x^{x_{j+1}} n(x)\,dx \tag{10.14}$$

where x is measured along the geomagnetic field line, α is the electron pitch angle, and E_j is the energy at x_j. According to Kamiyama, the error in E is less than 1 percent for $E > 400$ eV, if $\Delta x = x_{j+1} - x_j$ is 1 km; the results of his computations are shown in Figure 10.14.

The energy lost per unit path length from a flux† $I(E, x, \alpha)\,dE$ is $I(E, x, \alpha)$ $dE(-dE/dx)$, where $(-dE/dx)$ is given by Eq. 10.13 with the substitution $\cos\alpha\, ds = dx$. If we assume that $E\,dE \approx E_0\,dE_0$,

$$I(E, x, \alpha)\,dE\left(-\frac{dE}{dx}\right) = I_0(E_0, \alpha)\,dE_0 \frac{n(x)}{\cos\alpha}\frac{a}{E}(\ln E - b) \tag{10.15}$$

where

$$a = \tfrac{3}{4}ZQ_0\mu^2; \qquad b = \ln VZ\sqrt{2} - \tfrac{1}{2}$$

The ion production rate is

$$q(E, x, \alpha) = \frac{\cos\alpha}{W} I(E, x, \alpha)\,dE\left(-\frac{dE}{dx}\right) = U(x)I_0(E_0, \alpha)\,dE_0 \tag{10.16}$$

† In this discussion, I rather than dI/dE is used to designate flux per unit energy in order to simplify the notation

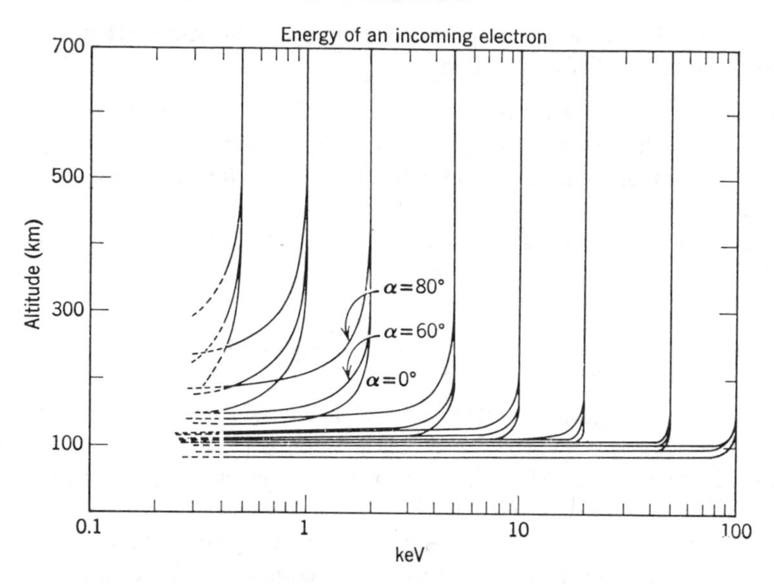

Figure 10.14 Energy of a precipitating electron as a function of initial energy, pitch angle, and altitude. [After Kamiyama, *Report of Ionosphere and Space Res. in Japan* **20,** 171 (1966); reprinted by permission of the Science Council of Japan.]

where

$$U(x) = \frac{n(x)}{W}\frac{a}{E}(\ln E - b)$$

The total rate of ion production by all electrons with energies above $E_{\min}$ is

$$q(x) = 2\pi \int_{E\min}^{\infty} \int_{0}^{\pi/2} q(E, x, \alpha) \sin \alpha \, d\alpha \, dE_0 \tag{10.17}$$

This has been computed by Kamiyama (Figure 10.15) for the spectrum observed by McIlwain, viz.

$$I_0(E) = I_i \exp(-E/5) \text{ electrons cm}^{-2} \text{ sec}^{-1} \text{ kev}^{-1} \text{ ster}^{-1} \tag{10.18}$$

for the case $I_i = 10^3$ electrons cm^{-2} sec^{-1} keV^{-1} ster^{-1} at $E = 5$ keV.

Ionization by Auroral Bremsstrahlung. The principal loss of energy by electrons in gases is through inelastic collisions. However, an additional loss mechanism, the radiation of X-rays (bremsstrahlung) should be considered Thus,

$$-\frac{dE}{dx} = -\left(\frac{dE}{dx}\right)_{\text{coll}} - \left(\frac{dE}{dx}\right)_{\text{brems}} \tag{10.19}$$

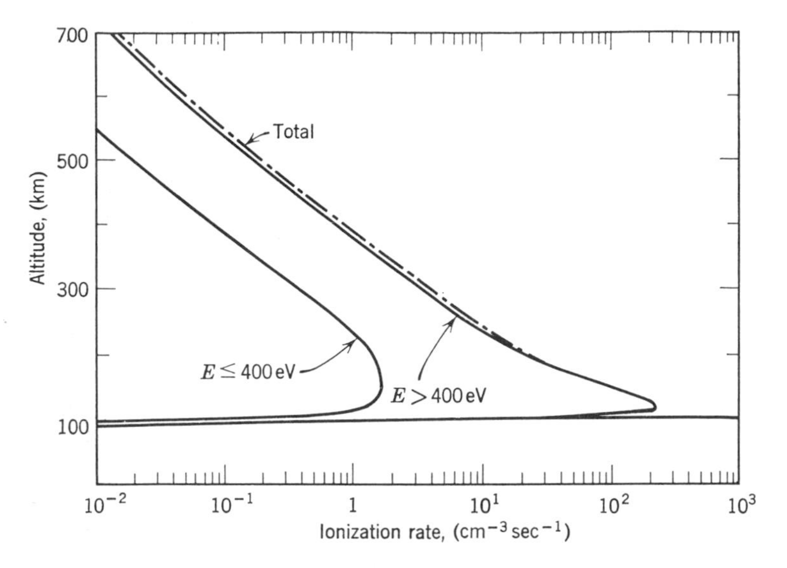

Figure 10.15 Production of ion-electron pairs by the spectrum given by equation 10.18. [After Kamiyama, *Report of Ionosphere and Space Research in Japan* **20,** 171 (1966); reprinted by permission of the Science Council of Japan.]

The ratio of energy loss by radiation to that lost by collision is[25]

$$\frac{(dE/dx)_{\text{brems}}}{(dE/dx)_{\text{coll}}} = \frac{EZ}{8 \times 10^5} \tag{10.20}$$

where E is in keV. Although the ratio is relatively small (about 10^{-3} for 100 keV electrons in air), the emitted X-rays can penetrate much farther into an atmosphere than can the primary electrons. For example, the bremsstrahlung from 25 keV electrons is detected at 30 km in the earth's atmosphere, whereas electrons of that energy are stopped at about 100 km. A rigorous treatment of the production of an ionization by bremsstrahlung is beyond the scope of this book. For a more complete treatment of bremsstrahlung, see the general reference list at the end of the chapter. In order to investigate the role of bremsstrahlung in atmospheric ionization, simplifying assumptions must be made and the resulting inaccuracies accepted. We will consider the treatment by Kamiyama,[26] which is a continuation of the electron-ionization treatment given above; the reader is again referred to Rees[32] for an alternate discussion.

The energy lost by an electron through bremsstrahlung is

$$-\left(\frac{dE}{dx}\right)_{\text{brems}} = nEQ_r \tag{10.21}$$

where Q_r is the total cross section for the radiation of bremsstrahlung. For a nonrelativistic electron with energy E, the cross section for the emission of a photon with energy $h\nu$ is given by

$$Q(E, h\nu)\, d(h\nu) = \frac{8}{3}\, \Phi\, \frac{\mu}{E} \ln \left[\frac{(\sqrt{E} + \sqrt{E - h\nu})^2}{h\nu}\right] \frac{d(h\nu)}{h\nu} \tag{10.22a}$$

where Φ is $r_0Z^2/137$; r_0, the classical electron "radius," is e^2/mc^2, where e is the charge of an electron, m is the mass of an electron, and c is the velocity of light in vacuum.

$$Q_r = \int_0^{h\nu=E} Q(E, h\nu)\, d(h\nu) \tag{10.22b}$$

Inasmuch as the energy lost by bremsstrahlung is small compared to collisional losses, the energy of the electron E_j at x_j [see Eq. (10.14) and Figure 10.14] can be used as the value for E in the above equation (10.22) and x_j can be used as the source altitude. Hence, there is a continuous production of bremsstrahlung with decreasing energies as the electron penetrates the atmosphere. For an electron flux of $I(E, x)$, the bremsstrahlung emission in the energy range $d(h\nu)$ is

$$J_E(x, E, h\nu)\, d(h\nu)\, dE = n(x)Q_r(E, h\nu)I(x, E)\, dE \tag{10.23}$$

For $I_0(E_0, \alpha)\, dE_0$ incident at the top of the atmosphere

$$J_E(x, E_0, h\nu)\, d(h\nu)\, dE = 2\pi n(x) \int_0^{\pi/2} \sin \alpha I_0(E_0, \alpha)\, dE_0 Q_r(E, h\nu)\, d(h\nu)\, d\alpha \tag{10.24}$$

Figure 10.16 shows Kamiyama's calculations of photon emission for a 100 keV electron. Note that the photon flux increases with decreasing photon energy. Although it is not illustrated, it should be noted that the altitude of peak photon production decreases with increasing electron energy.

The photon emitted by a decelerating electron has a preferred direction with respect to the direction of the electron, but we can assume that photon emission from an incremental volume of air is isotropic if the electron velocities are isotropic. Hence, we can expect both an upward and a downward flux of bremsstrahlung photons in the atmosphere.

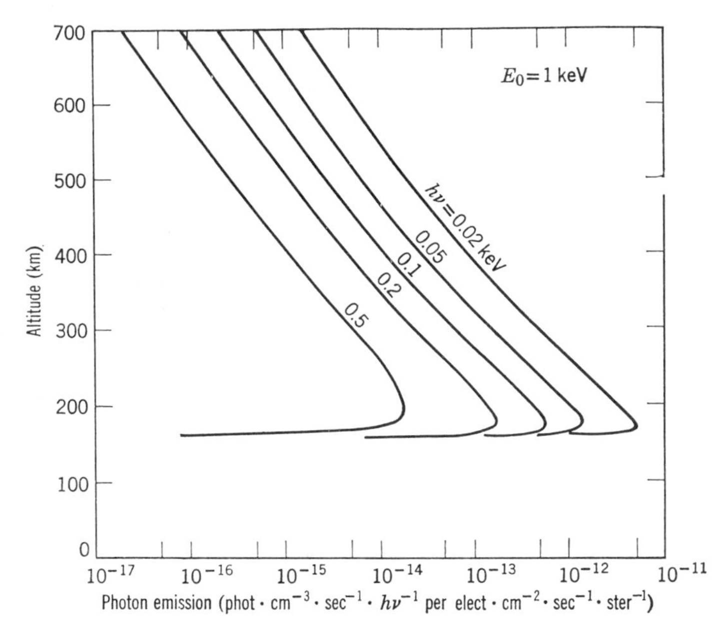

Figure 10.16 Emission of Bremsstrahlung at several photon energies by the deceleration of a 1 keV electron in the terrestrial atmosphere. [After Kamiyama, *Report of Ionosphere and Space Research in Japan* **20,** 374 (1966); reprinted by permission of the Science Council of Japan.]

At any point P at a certain altitude z_P consider the photon flux from a volume element $a\,d\varphi\,da\,dz$ at a distance r as illustrated in Figure 10.17.

$$dJ(E_0, h\nu, z_P, z, \theta) = \frac{1}{4\pi r^2} J_E(E_0, h\nu, z)\,dE_0\,d(h\nu) a\,d\phi\,da\,dz \exp\left(Q_{T(h\nu)} \int_{z_P}^{z} \frac{n(z)\,dz}{\cos\theta}\right) \tag{10.25}$$

Where θ is the angle r makes with the vertical and $Q_{T(h\nu)}$ is the total absorption cross section for photons with energy $h\nu$. Let $r = (z - z_P)/\cos\theta$ and $a = (z - z_P)\tan\theta$; the contribution to the flux at P from photons emitted from the layer dz at level z is

$$\begin{aligned} J(E_0, h\nu, z_P, z)\,dE_0\,d(h\nu)\,dz &= \frac{1}{4\pi} J_E(E_0, h\nu, z)\,dE_0\,d(h\nu)\,dz \int_0^{2\pi}\int_0^{\pi/2} \frac{\sin\theta}{\cos\theta} \exp\left(-\frac{A}{\cos\theta}\right) d\theta\,d\phi \\ &= \tfrac{1}{2} J_E(E_0, h\nu, z)\,dE_0\,d(h\nu)\,dz\,\psi(h\nu, z_P, z) \end{aligned} \tag{10.26}$$

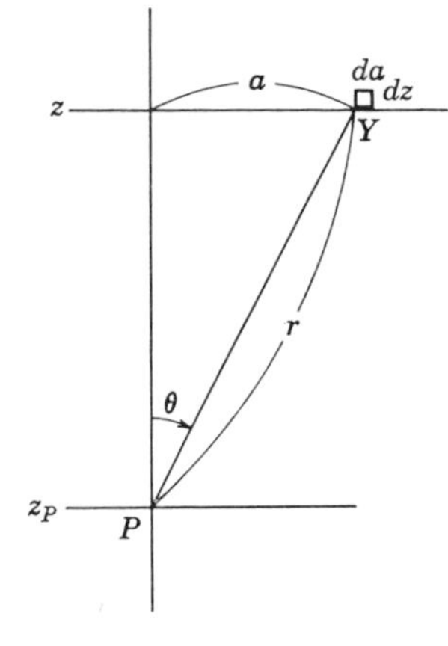

Figure 10.17 Geometry for calculating the contribution to the flux at a point P from a volume element a $d\Phi$ $da\ dz$ at an azimuthal angle Φ. [After Kamiyama, *Report of Ionospheric and Space Research in Japan* **20,** 374 (1966); reprinted by permission of the Science Council of Japan.]

where

$$\psi(h\nu, z_P, z) = \int_1^\infty \frac{e^{-At}}{t}\, dt \quad \text{and} \quad A = Q_{T(h\nu)} \int_{z_P}^{z} n(z)\, dz$$

The downward flux (of photons with energies $h\nu$ to $h\nu + dh\nu$) is obtained by integration with respect to z from z_P to infinity and the upward flux by integrating between 0 and z_P. If the upward flux is J_U and the downward flux J_D, the ion production rate in a unit volume is

$$q(E_0, h\nu, z_P)\, dE_0\, d(h\nu) = Q_i(h\nu) n(z_P)[J_U + J_D]\, dE_0\, d(h\nu) \quad (10.27)$$

Where $Q_i(h\nu)$ is the ionization cross section as a function of photon energy. The total ionization rate is obtained by integrating over the photon energy range and the incident electron energy range.

Kamiyama's calculation for the McIlwain spectrum is shown in Figure 10.18.

Computation of Electron Concentrations Produced by SCR Events. Comparison of Figures 10.12 and 10.6 shows one of the most important aspects of SCR ionization, viz., very high ionization rates at very low altitudes. This aspect makes SCR events especially interesting and, at the same time, difficult to describe in terms of ionic and electronic reactions. The altitudes of maximum ionization are those at which negative ion formation and ion-ion neutralization processes are most important; these are the processes, however, that are least understood for the terrestrial atmosphere. It was, in fact, an attempt to compare radio-absorption values calculated from model electron profiles with radio-absorption measurements made during an SCR event that led Reid[27] to the conclusion that negative molecular oxygen ions could not be the major negative ion; this inference is slowly being confirmed

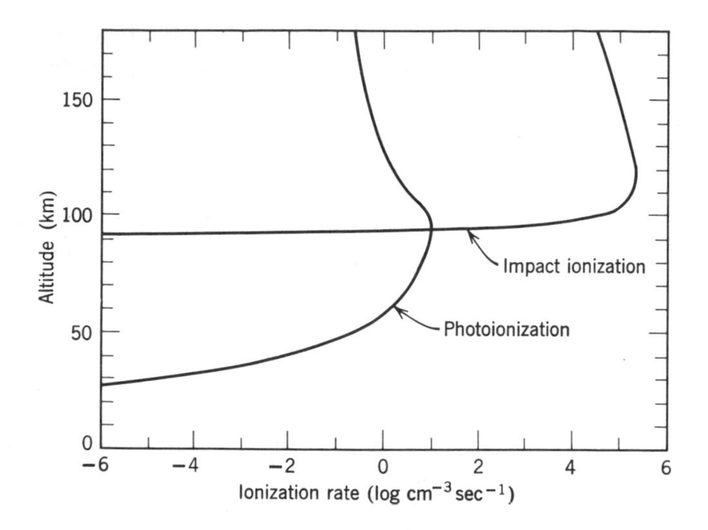

Figure 10.18 Comparison of Bremsstrahlung photoionization and primary electron collisional ionization produced by the spectrum given by equation 10.18. [After Kamiyama, *Report of Ionosphere and Space Res. in Japan* **20,** 374 (1966); reprinted by permission of the Science Council of Japan.]

by both laboratory and ionospheric studies. For atmospheres of Mars and Venus, the role of negative ions is comparatively unknown, although some evidence of electron-attaching gases has been found. There is considerable spectroscopic evidence of electron-attaching gases in the Jupiter atmosphere, however.

Because SCR events are of long duration and decay slowly, equilibrium conditions can be assumed for the solution of the balance equations. Hence, Eqs. (8.18a) and (8.17b) are applicable with the limitation that the calculation of λ is very speculative and may cause large errors in regions where negative ions are important.

$$q(z, t) \cong (1 + \lambda)(\alpha_D + \lambda\alpha_i)N_e(z, t)^2 \tag{8.18a}$$

$$\lambda \cong \frac{\beta m}{\rho + \gamma n} \tag{8.17b}$$

Computation of Electron Concentrations Produced by Auroral Events. As we have discussed earlier, auroral ionization events are characterized by a large range of temporal and altitude conditions. For auroral bursts longer than minutes, the situation can be approximated by the quasi-equilibrium forms of the balance equations, (8.18a, 8.17b), just as for SCR events; but for microbursts the equilibrium forms are clearly not applicable.

Following the treatment by Reid[27] we approximate the buildup of electron concentrations following the abrupt onset of an ionizing pulse by

$$\frac{dN_e(z)}{dt} = \frac{q(z)}{1+\lambda} - (\alpha_D + \lambda\alpha_i)N_e(z)^2 \tag{8.17a}$$

if we assume that λ = constant. If we also assume that q is constant after onset ($t \geq 0$) and that N_e is zero at onset ($t = 0$), Eq. (8.17a) can be integrated to give

$$N_e(z, t) = \left[\frac{q(z)}{(1+\lambda)(\alpha_D + \lambda\alpha_i)}\right]^{1/2} \tanh\left[\left(\frac{q(z)(\alpha_D + \lambda\alpha_i)}{1+\lambda}\right)^{1/2} t\right] \tag{10.28}$$

Figure 10.19 shows the buildup of electron concentrations following the onset of a pulse that produces ionization at the rate of 10^2 cm^{-3} sec^{-1}.

The decay of electron concentrations after the abrupt cessation of ionization can be followed by using the approximation

$$\frac{dN_e(z, t)}{dt} = -(\alpha_D + \lambda\alpha_i)N_e(z, t)^2 \tag{10.29}$$

If we assume that λ is constant, but that $N_e = N_0$ when the pulse ceases, and integrate, we find that

$$N_e(z, t) = \frac{N_0(z)}{1 + (\alpha_D + \lambda\alpha_i)N_0(z)t} \tag{10.30}$$

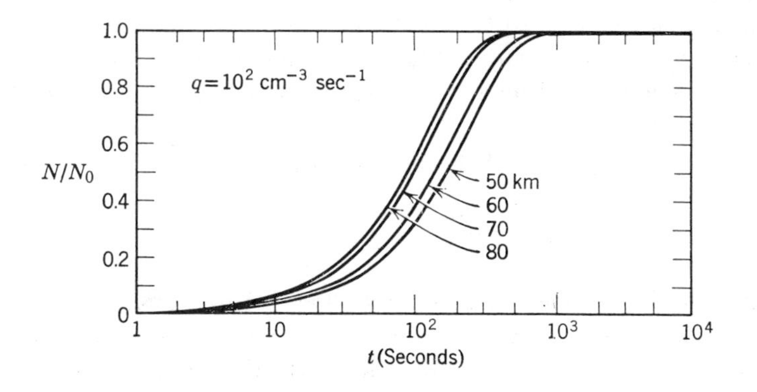

Figure 10.19 Buildup of electron concentrations following the onset of an ionizing pulse. $N_0 = N$ at equilibrium. [After Reid, *Rev. Geophys.* **2,** 311 (1964); reprinted by permission of the American Geophysical Union.]

For auroral bursts of intermediate duration and ionization decay rate, Eq. (8.17a) can be applied just as for solar flare ionization.

Discussion of Approximations. Because of the difficulty of solving the continuity equations exactly—even with the use of numerical techniques and digital computers—tractable approximations must be sought. It should be obvious by now that much of the foregoing discussion (and the discussion in Chapter 8) is concerned with rational simplifications of the continuity equations. Hence, we attempt (see Chapter 8) to divide the ionosphere into regions in which only a few of the terms are important. We attempt to learn about the identities of ionic species so that we can neglect a host of minor but complicating reactions. We look to the time-dependence of ionization events for ways to get around the use of the differential forms of the continuity equations.

The use of the various approximations that have been discussed in this book, as well as the many more that the reader will hopefully derive for specific situations, must be considered carefully. The validity of an approximation should be justified as thoroughly as possible for each specific application. The universal application of an approximation can trap the unwary aeronomer.

Several approaches have been used. One approach is to try to determine the time required to establish equilibrium under specific conditions and then judge the validity of quasi-equilibrium forms. This approach is most pertinent to auroral cases inasmuch as ionizing pulse durations vary so widely. Let us consider a model in which a burst of electrons is suddenly produced by a very short duration delta-function source, and then let us follow the disappearance of electrons and buildup of negative ions. Reid[27] describes such a model for the establishment of equilibrium between electrons and negative ions for a special case in which recombination and mutual neutralization is unimportant. It can be described by (8.14a, b) without the ionization and recombination terms, viz.

$$\frac{dN_e}{dt} = -\beta N_e m + \gamma \lambda N_e n + \rho \lambda N_e \tag{10.31a}$$

$$\lambda\left(\frac{dN_e}{dt}\right) + N_e\left(\frac{d\lambda}{dt}\right) = \beta N_e m - \gamma \lambda N_e n - \rho \lambda N_e \tag{10.31b}$$

Note that the substitution $N^- = \lambda N_e$ was made. Multiply Eq. (10.31a) by λ and subtract from Eq. (10.31b) to obtain

$$\frac{d\lambda}{dt} = \beta m(1 + \lambda) - (\gamma n + \rho)\lambda(1 + \lambda) \tag{10.32}$$

or

$$\int_0^\lambda \frac{d\lambda}{\beta m(1 + \lambda) - (\gamma n + p)\lambda(1 + \lambda)} = \int_0^t dt \tag{10.33}$$

Integration and rearrangement yields

$$\lambda = \frac{2\beta m \tanh(\frac{1}{2}ut)}{u - v \tanh(\frac{1}{2}ut)} \tag{10.34}$$

where $u = \beta m + (\rho + \gamma n)$ and $v = \beta m - (\rho + \gamma n)$. We assume that N^- is zero at $t = 0$, and therefore that $\lambda = 0$ at $t = 0$. At $t \to \infty$, $\lambda \to \lambda_0$ which is the equilibrium value obtained by Eq. (8.17b). Curves showing the establishment of equilibrium at various altitudes are shown in Figure 10.20. Obviously, the validity of the assumption of equilibrium for λ depends not only on the characteristics of the ionizing pulse but also on the altitude for which calculations are made. During daytime at altitudes above 75 km in the terrestrial atmosphere, where $\lambda \ll 1$ and $\gamma n \ll \rho$ are probably valid the integration is simplified to yield the approximation

$$\lambda = \lambda_0(1 - e^{-\rho t}) \tag{10.35}$$

The establishment of equilibrium between negative ions, positive ions, and electrons is given by Eq. (10.28) and illustrated by Figure 10.19 for a case in which q rises as a step function from 0 to q and N_e starts at zero and builds up to an equilibrium value N_0. Obviously, it requires much more time for N_e to attain equilibrium than for λ; on the other hand, the effect of altitude is not as great for N_e as it is for λ.

The effect of varying some of the parameters of the balance equations are shown in Figures 10.21 and 10.22 where the values of the collisional detachment coefficient and the photodetachment rate are varied. Inasmuch as the identities of negative ions are in doubt, the correct values for the collisional detachment coefficient and the photodetachment rate are also in doubt, and

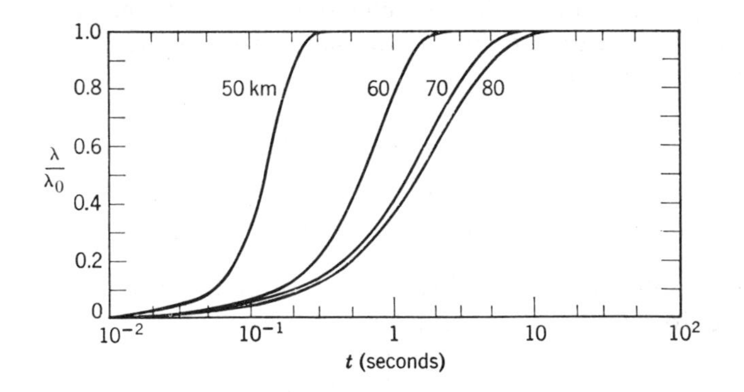

Figure 10.20 Establishment of equilibrium at various altitudes. $\lambda_0 = \lambda$ at equilibrium. [After Reid, *Rev. Geophys.* **2**, 311 (1964); reprinted by permission of the American Geophysical Union.]

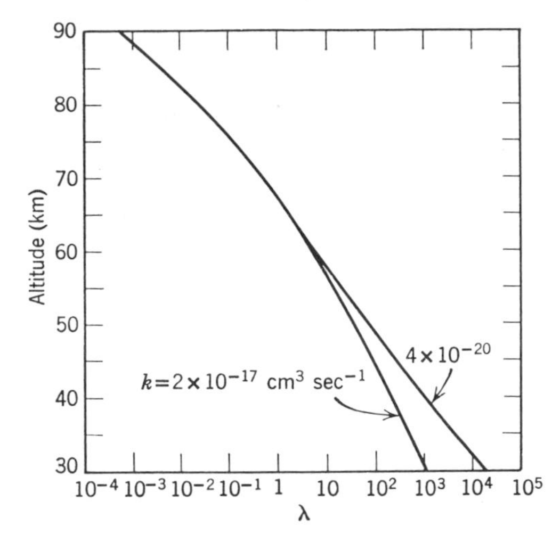

Figure 10.21 Profile of λ for two different values of collisional detachment coefficient. [After Reid, *Rev. Geophys.* **2**, 311 (1964); reprinted by permission of the American Geophysical Union.]

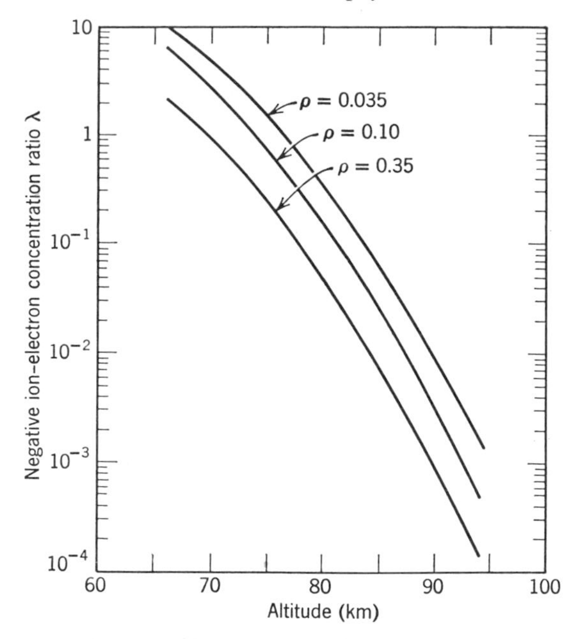

Figure 10.22 Profiles of λ for a range of photodetachment coefficients. [After Whitten and Poppoff, *JGR* **66**, 2779 (1961); reprinted by permission of the American Geophysical Union.]

the values of λ for the terrestrial atmosphere could, conceivably, vary even more than the illustrations indicate.

The use of the approximate Eq. (8.17b) to compute λ can also lead to error for the high values of q that might be attained during auroral events. This is illustrated in Figure 10.23.

In the preceding discussions of approximations it was assumed that each of the recombination, attachment, detachment, or neutralization terms involves only one reaction, e.g., for terrestrial cases, recombination involves only O_2^+ or only NO^+, and attachment involves only O_2, etc. In some regions, this is essentially the case; in others, it is far from true. It is worthwhile, particularly for exploratory calculations, to be able to identify regions in which such assumptions regarding reacting species are valid. A useful technique for identifying these regions involves the concept of species "lifetimes" for the reactions of interest (also discussed in Chapters 2 and 4).

Consider the two-body reaction

$$X^+ + Y \xrightarrow{k} X + Y^+ \tag{10.36}$$

in which $[X^+]$ is decreasing via charge transfer to Y with a rate constant k; but $[Y]$ is so large, it is relatively constant throughout the reaction period (brackets indicate concentrations of species within). Then

$$-\frac{d[X^+]}{dt} = k[X^+][Y] \tag{10.37}$$

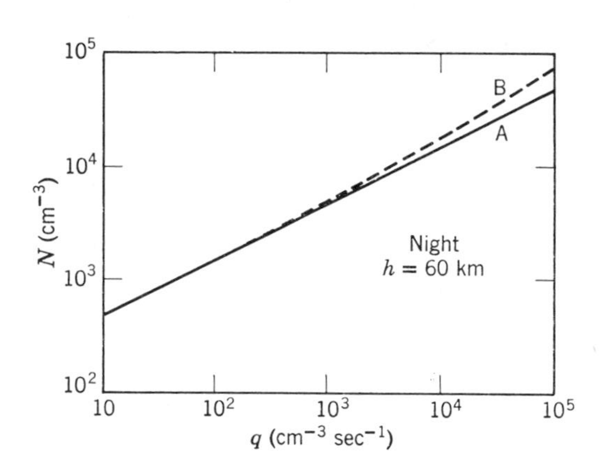

Figure 10.23 Equilibrium electron number density as a function of ionization rate. Curve A was calculated with the approximate equation for λ, whereas curve B was calculated using the full expression. [After Reid, *Rev. Geophys.* **2**, 311 (1964); reprinted by permission of the American Geophysical Union.]

the time τ required for X^+ to decrease by a factor of e is the "lifetime of X^+ against charge transfer" (see Chapter 2).

$$\tau = (k[Y])^{-1} \tag{10.38}$$

The reciprocal of τ might be considered a "reaction probability."

Similarly, for three-body processes such as the attachment reaction,

$$e + Y + M \xrightarrow{k} Y^- + M$$

the lifetime of the electron against attachment is

$$\tau = (k[Y][M])^{-1} \tag{10.39}$$

In the case of reactions in which two components are being depleted significantly, such as electron-ion recombination

$$e + X^+ \xrightarrow{\alpha} X$$

in a plasma composed of electrons and X^+ ions in equal concentrations ($N_e = [X^+]$) it is necessary to use a quadratic form, i.e.,

$$-\frac{dN_e}{dt} = \alpha N_e[X^+] = \alpha N_e^2 \tag{10.40a}$$

$$\begin{aligned} \tau &= (e - 1)/\alpha N_{e0} \\ &= (e - 1)/\alpha[X^+]_0 \end{aligned} \tag{10.40b}$$

in Eq. (10.40a and b), e is the base of natural logarithms. Note that if we consider $[X^+]$ to be constant,

$$\tau = 1/\alpha[X^+] \tag{10.41}$$

and the difference in τ computed by Eqs. (10.40) and (10.41) is the factor $(e - 1)$, which would be unimportant for most applications.

As an illustrative example of the application of species lifetimes, let us assume an ionospheric condition in which N_e is 10^3 cm^{-3}, [NO$^+$] is 10^3 cm^{-3} and [O$_2$] is 10^{15}. Assume also that the rate constant for electron attachment is 10^{-30} cm^6 sec^{-1} (if O$_2$ is the third body) and the ion-electron recombination coefficient is 10^{-6} cm^3 sec^{-1}. The lifetime τ_R of an electron against recombination is $(1 - e)/(10^{-6})(10^3)$ or $\sim 2 \times 10^3$ sec. The lifetime τ_A of the electron against attachment is $(10^{15} \cdot 10^{15} \cdot 10^{-30})^{-1}$ or 1 sec. It is clear that attachment is a much more important process than recombination under these circumstances. If, however, $[O_2]$ were 10^{12} cm^{-3}, τ_A would be 10^3 seconds and neither process could be ruled out (without additional considerations) even for approximate computations.

10.3 Ionospheric Storms

The foregoing discussion of perturbations by corpuscular radiation is, in essence, a description of the better understood details of a class of large scale disturbances called ionospheric storms. An analogy can be made between consideration of the microscopic processes of ionization and deionization to explain electron concentrations in ionospheric storms and consideration of the microscopic processes of nucleation and condensation to explain the presence of rain drops in a hurricane. We understand the production of effects but not the entire phenomenon—and particularly not the origin of the phenomenon. However, a considerable amount of work has been done on the morphology of these storms in the attempt to discover features that can be correlated with solar and magnetospheric perturbations. It has been very definitely established that optical emissions in auroras result from impact by secondary and tertiary electrons, but correlations between observed primary particle bombardments and specific optical emissions are loose even though they are intuitively expected; similarly with magnetic disturbances. The disturbances are obviously large scale, sporadic and nonhomogeneous; on the other hand, observations are necessarily fine scale, and relatively well-confined to particular accessible locations. Hence, storm manifestations are sampled as they pass over a particular spot or intersect the propagation path of a communications link. The reader is referred to the many publications by Chapman and/or Akasofu for discussions of the possible origin and morphology of auroral storms. Similarly, a series of papers by Obayashi and/or Hakura is a good source of material on the morphology of Polar Cap Events.

A class of ionospheric storms is also observed at mid and low latitudes and are also correlated with geomagnetic disturbances. These mid-latitude storms are characterized by rather sudden changes in the F_2 region critical frequency f_0F_2 (frequency above which radio waves penetrate the F_2 region, see Chapter 11) and this, in turn, means that changes are occurring in either the electron concentrations or gradients (or both) in this region. Typically, the f_0F_2 increases for a short period of time ($\sim$5 min) and then slowly decreases during the main phase of the associated magnetic storm; this is known as a negative storm. If the f_0F_2 increases during the main phase, it is known as a positive storm. At mid-latitudes more negative storms are observed than positive storms, whereas at low latitudes the reverse is true. The main phase may last from several hours to days followed by gradual return to normal.

The magnetic storm can be described qualitatively in terms of the solar wind which confines the geomagnetic field within a fairly well-defined cavity called the magnetosphere and of plasma clouds, apparently of solar origin which travel through space. The sudden impingement of a plasma cloud on the magnetosphere produces a hydromagnetic shock that travels on to the earth causing the sudden commencement (SC) of a magnetic storm (actually a series of shocks). After a few minutes, the plasma envelops the earth and the SC ends. The main phase, characterized by a large decrease in the horizontal component of **H**, is due to two effects. First, the plasma cloud feeds charged particles into a belt around the earth, positive particles drifting to the west, negative particles to the east. The result is the establishment of a diamagnetic ring current at several earth radii. Second, auroral effects associated with the storm produce enhanced ionization in the auroral zone at *E*-layer heights. The increased conductivity produced thereby results in greatly increased current flow in the auroral electrojet and, thus, marked changes in the magnetic field, mainly at high latitudes but extending to midlatitudes also. The recovery phase occurs as the currents grow weaker due to the passage of the plasma cloud past the earth.

Unlike the associated magnetic storm, no satisfying explanation of the cause of ionospheric storms is available. One speculation involves the vertical drift of electrons: variations are caused by changes in ionospheric currents which, in turn, are caused by changes in the magnetic field. Another speculation involves the warping of the magnetospheric cavity by the storm which produces a shift in the relative electron contents of the magnetosphere and the ionosphere and thus an increase or decrease in f_0F_2. Recent studies indicate that increased electron concentrations at high altitudes may be due to an expansion of the *F*-region. It is also surmised that the linear ion-depletion rate may increase because of increased neutral densities or excitation of N_2 by electron impact; this may explain the observed decreases of electron densities at lower altitudes. Decreases are observed in f_{min} (this indicates increased electron concentrations) during mid-latitude ionospheric storms; particle bombardment may be responsible.

10.4 Sporadic-*E*

Sporadic-*E* (or E_s) layers are commonly detected at all latitudes. They are thin (usually <1 km) and appear as spikes on normal *E*-region electron profiles, but are not always dense enough to mask the detection of ion-layers above them. They are sometimes long-lived and are observed to settle slowly through the *E*-region during the course of a day. They are occasionally detected as low as 60 km. The cause of E_s may well vary with latitude. It is thought that high latitude E_s is caused by corpuscular bombardment, that mid-latitude E_s is caused by wind-shears in the E-region and that equatorial E_s is caused by plasma instabilities arising from the equatorial electrojet.

Considerable work has been done to establish the wind shear theory of mid-latitude E_s. The detection of strong shears in the E-region by rocket experiments has provided strong impetus to the development of this idea. Some attempts have also been made to correlate E_s with meteoric activity, but the connection is more tenuous than it is with wind shear. Briefly, the wind shear theory is based on the fact that at higher altitudes, charged particles are constrained by magnetic field lines. Therefore, ionospheric winds force the particles to move along the field lines. In case of a wind shear, particles above the shear move in one direction with a downward component, whereas particles below the shear altitude move in the opposite direction with an upward component; under these conditions particles accumulate at the shear altitude, producing E_s. The observations of downward moving strata are also consistent with the theory that shears are caused by atmospheric gravity waves.

Measurements with mass spectrometers in an E_s layer shows that the ions are of metallic elements. The slow recombination rate of atomic ions would explain the persistence of E_s at night. The origin of the atomic ions is not clear, however; inasmuch as the measurements with mass spectrometers were made during a meteor shower, it seems reasonable to attribute the source to meteor activity. On the other hand, the wind shear theory could also explain the existence of large concentrations of otherwise minor ions.

Current ideas are illustrated by the following treatment by Whitehead[28] who assumed a sinusoidal vertical movement of ions and a linear ion-depletion law; his treatment is based on the earlier work of Gleeson[29]. Linear ion-depletion means that the electron concentration is constant and hence only the population of a specific ion changes (see similar assumption for treatment of lifetimes against recombination earlier in this chapter) or that the rate-controlling ion-removal mechanism is not recombination but ion-interchange or chemical reactions. Note also that the production of ions by interchange or transfer reactions is not explicitly noted in the following equation; if it is assumed that the metallic or alkali ion population grows at the expense of oxygen or nitric oxide ions, a complete description must account for all appropriate reactions. Of course, as in most ionospheric problems, the solution of continuity equations requires that many aspects be ignored:

$$\frac{\partial N^+}{\partial t} = q - \beta' N^+ - \operatorname{div}(N^+ W) + D\frac{\partial^2 N^+}{\partial z^2} \tag{10.42}$$

where N^+ is the ion density, q is the ion production rate (in this treatment equal to βN_0^+ where N_0^+ is a constant ambient ion population), β' is the linear ion-depletion coefficient, W is the vertical velocity of ions (in this treatment equal to $W_0 \sin 2\pi z/\lambda$), and D is the diffusion coefficient. The equation is

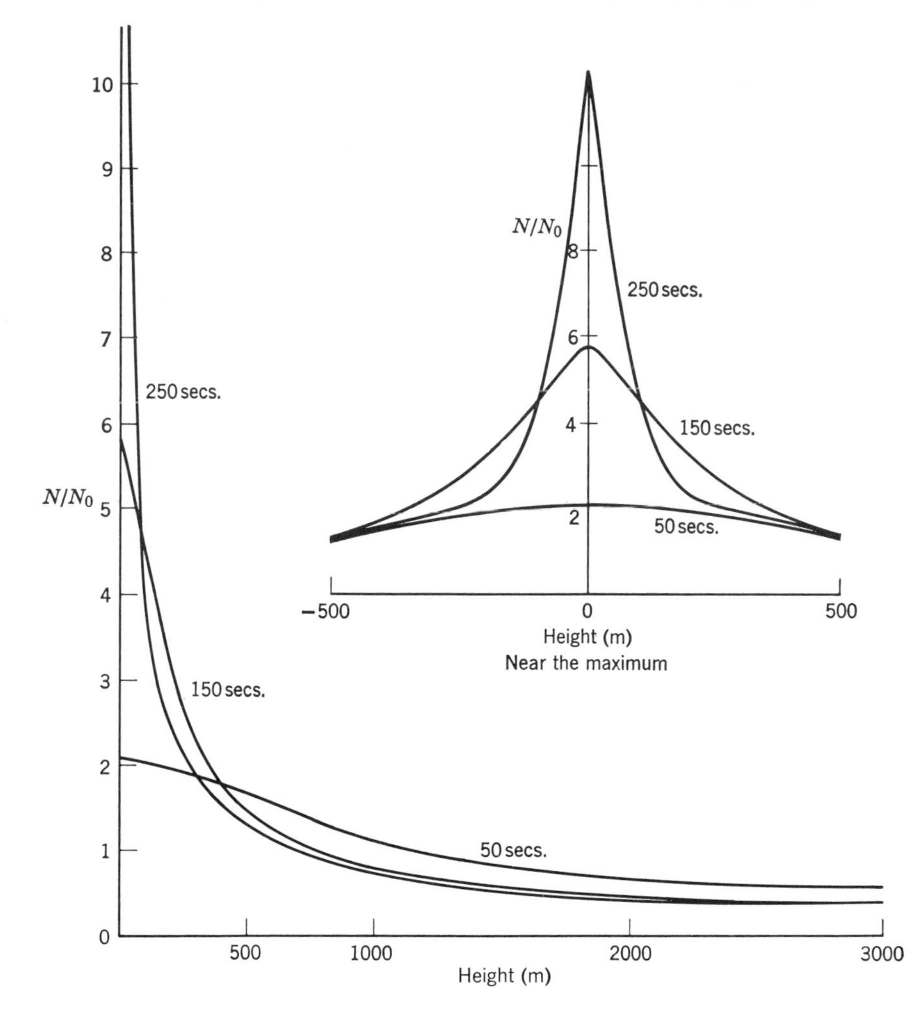

Figure 10.24 Ion layers formed according to the wind shear model for (a) $D = 0$.

solved for the following conditions: $W_0 = 20 \text{ m sec}^{-1}$, $\beta = 0.01 \text{ sec}^{-1}$, $\lambda = 600$ m, and for both $D = 0$ and $D = 100 \text{ m}^2 \text{ sec}^{-1}$. The results are shown in Figures 10.24a and b.

10.5 Solar Eclipses

Solar eclipses afford a unique opportunity to study the interaction of the ionosphere with solar radiation. For ionospheric experiments, eclipses offer several advantages, viz., predictable occurrence and duration, insignificant

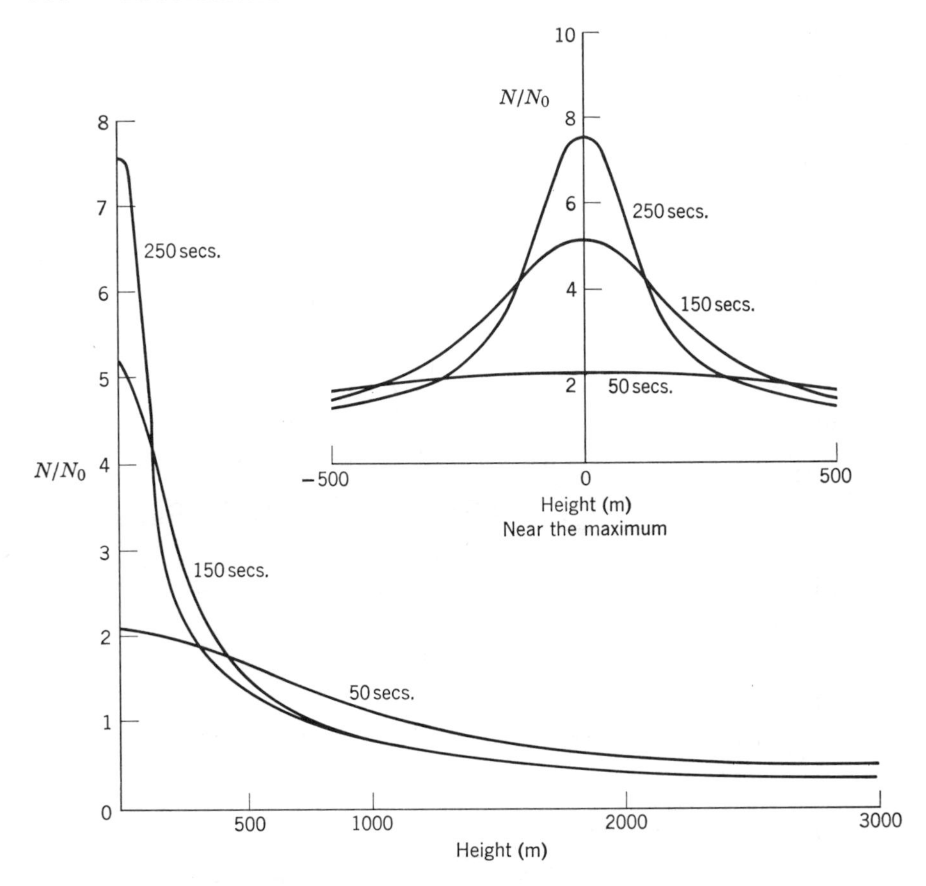

Figure 10.24 continued (b) $D = 100\ m^2\ \text{sec}^{-1}$. [After Whitehead, *Planet. Space Sci.* **15,** 1055 (1967); reprinted by permission of Pergamon Press.]

change in solar zenith angle, and a duration long enough to launch several rockets. A disadvantage not yet overcome is the difficulty of precisely specifying the ionization source at any specific time.

As the solar disc is obscured by the lunar disc, the source q decreases and the electron concentrations decrease correspondingly. The continuity equation for electron concentrations therefore can be written approximately (for the E and upper D-regions):

$$-\frac{d^2N_e}{dt^2} = -\frac{dq}{dt} + 2\alpha_{\text{eff}}N_e\frac{dN_e}{dt} \tag{10.43}$$

During total obscuration, the source may be considered zero, and

$$\alpha_{\text{eff}} = \frac{-(dN_e/dt)}{N_e^2} \tag{10.44}$$

We must caution the reader that it is not at all certain that the source function can be properly specified either during complete obscuration or while the solar disc is being covered or uncovered. It is known from UV and X-ray photographs of the sun that the distribution of ionizing radiation sources is not uniform across the disc. Also, at complete obscuration, it is possible that X-radiation from limb plages or coronal condensations can contribute a significant flux of ionization. Hence, calculated source functions are suspect as is Eq. (10.44). In using solar eclipses to study ionospheric perturbations the unique problem is to determine the source function and use it properly.

PROBLEMS

10.1. a. Find the altitude of unit optical depth for 7 Å X-rays.
b. Plot the data of Figure 10.5 for 6, 7, 8, 9, 10, and 11 Å to show how the intensity of each wavelength varies with time.
c. Using the data in Figure 10.5 for the development of a class 2 flare, plot $q(t)$ produced by 7 Å X-rays at the altitude of unit optical depth during the period 1205 to 1435 UT.
NOTE: For the purposes of Problems 10.1 and 10.2, assume that the number density of air at the altitude of 50 km is 2×10^{16} molecules/cm^3, that air has a constant scale height of 6 km above the altitude of 50 km, and that the sun is at zenith angle of 60°.

10.2. Assume that a pulse of X-rays instantaneously ionizes the atmosphere to the value of q calculated in Problem 10.1c (above) for the peak of the flare. *See* note to Problem 10.1.
a. Plot dN_e/dt and N_e as functions of time assuming no electron attachment occurs. Neglect background (or quiescent) ionization; assume $N_e = 0$ at $t = 0$.
b. Plot dN_e/dt and N_e as in part a assuming attachment does occur.
c. Compute lifetimes of free electrons against attachment to O_2 and against recombination with O_2^+. Compare with the time required for N_e to increase to the value of 0.65 the value it will become at $t \to \infty$. Does this comparison allow one to judge whether or not a quasi-steady state form of continuity equations is justified?

10.3. Plot exponential electron spectra (integral) for fluxes with "e-folding" energies of 5, 25, 50, 100 keV. Assume that the total flux of electrons with energies equal or greater than 5 keV is 10^9 cm^{-2} sec^{-1}.

10.4. Convert Eq. (10.5) to:
a. An equation that relates cut-off energy to cut-off latitude for electrons, and
b. To an equation that relates cut-off energy to cut-off latitude for protons.

10.5. Assume monoenergetic beams of electrons are incident on the earth's atmosphere; for the following electron energies, fluxes, and pitch angles, find stopping power and ionization rate at 100 km altitude.

a. 10^6 electrons $cm^{-2}sec^{-1}$ at 100 keV and 60°.
b. 10^6 electrons $cm^{-2}sec^{-1}$ at 100 keV and 0°.
c. 10^8 electrons $cm^{-2}sec^{-1}$ at 25 keV and 60°.
d. 10^{10} electrons $cm^{-2}sec^{-1}$ at 5 keV and 60°.

10.6. Find the incremental rate of energy emission (eV cm^{-3} sec^{-1}) of bremsstrahlung by 200 keV, 100 keV, 50 keV, 10 keV, and 1 keV electrons at 100 km if the electron pitch angle is 60° and the total flux in each case is 10^8 cm^{-2} sec^{-1}. Also compute the ionization rates at 100 km for each case.

10.7. Derive Eq. (10.28).

10.8. Find an approximate formula for λ [such as Eq. (8.17b)] if the associative detachment process $O_2^- + O \rightarrow O_3 + e$ is also important.

10.9. Derive Eq. (10.30).

10.10. Derive (10.35). How does the time required to attain any specified value of λ depend on the photodetachment rate? Compare the time required for λ to attain the value $\lambda_0(1 - 1/e)$ with the lifetime of an electron against attachment (at 75 km during daytime) and with the lifetime of a negative ion against photodetachment.

10.11. Derive Eqs. (10.40a,b) and (10.41).

10.12. The *D*-region of the ionosphere is strongly ionized by a pulse of radiation. Find an expression for the electron density as a function of time for any altitude in the region. Assume that $\alpha_D = \alpha_i$ and that both electron attachment (*rate* given by β and electron detachment (*rate* given by γ) processes occur. In order to simplify the problem, assume that the initial electron and positive ion number densities are infinite and that the initial negative ion number density is zero.

10.13. Compute both day and night electron density profiles for an aurora in the earth's atmosphere Use the ionization rate profiles computed by Kamiyama (Figure 10.18). Use the λ profiles of Figure 8.8, assume that equilibrium conditions exist, and use any formulas or data in this book.

REFERENCES

1. R. W. Kreplin, *Ann. Geophys.* **17,** 151 (1961).
2. P. J. Bowen, K. Norman, K. A. Pounds, P. W. Sanford, and A. P. Willmore, *Proc. Roy. Soc.* **A281,** 538 (1964).
3. R. W. Kreplin, T. A. Chubb, and H. Friedman, *J. Geophys. Res.* **67,** 2231 (1962).
4. W. M. Neupert, W. Gates, M. Swartz, and R. Young, *Ap. J.* **149,** L79 (1967).
5. R. C. Whitten and I. G. Poppoff, *J. Geophys. Res.* **66,** 2779 (1961); **67,** 3000 (1962) (corrigendum).

6. E. N. Parker, *Astrophys. J.* **128,** 664 (1958).
7. H. S. Ahluwalia and A. J. Dessler, *Planet. Space Sci.* **9,** 195 (1962).
8. John M. Wilcox, *Space Sci. Revs.* **8,** 258 (1968).
9. R. A. Alpher, *J. Geophys. Res.* **55,** 437 (1950).
10. J. A. Van Allen and J. R. Winckler, *Phys. Rev.* **106,** 1072 (1957).
11. P. Meyer, E. N. Parker, and J. A. Simpson, *Phys. Rev.* **104,** 768 (1956).
12. K. A. Anderson and D. C. Enemark, *J. Geophys. Res.* **65,** 2657 (1960).
13. P. S. Frier and W. R. Webber, *J. Geophys. Res.* **68,** 1605 (1963).
14. D. K. Bailey, *Planet. Space Sci.* **12,** 495 (1964).
15. C. E. McIlwain, *J. Geophys. Res.* **65,** 2727 (1960).
16. D. S. Evans, *Goddard Energetic Particles Series*, NASA, 1966.
17. I. B. McDiarmid and E. E. Budzinski, *Can. J. Phys.* **42,** 2048, (1964).
18. W. J. Heikkila and D. L. Matthews, *Nature* **202,** 789 (1964).
19. R. R. Brown, *Space Sci. Revs.* **5,** 311 (1966).
20. J. R. Barcus and T. J. Rosenberg, *J. Geophys. Res.* **71,** 803 (1966).
21. D. K. Bailey, *J. Geophys. Res.* **62,** 431 (1957).
22. R. R. Brown and R. A. Weir, *Arkiv för Geofysik* **3,** 523 (1962).
23. H. Kamiyama, *Reports of Ionosphere and Space Research in Japan*, **20,** 171 (1966).
24. W. Heitler, *The Quantum Theory of Radiation*, 3rd ed., Clarendon Press, Oxford, 1954.
25. H. A. Bethe, and W. Heitler, *Proc. Roy. Soc.* **A146,** 83 (1934).
26. H. Kamiyama, *Reports of Ionosphere and Space Research in Japan*, **20,** 374 (1966).
27. George C. Reid, *Revs. Geophys.* **2,** 311 (1964).
28. J. D. Whitehead, *Planet. Space Sci.* **15,** 1055 (1967).
29. L. J. Gleeson, *Planet. Space Sci.* **15,** 27 (1967).
30. G. C. Reid, *J. Geophys. Res.* **66,** 4071 (1961).
31. M. H. Rees, *Planet Space Sci.* **11,** 209 (1963).
32. M. H. Rees, *Planet. Space Sci.* **12,** 1093 (1964).
33. P. Velinov, *J. Atmos. Terr. Phys.* **30,** 1891 (1968).
34. John Dubach and William A. Barker, *J. Atmos. Terr. Phys.* **32** (in press, 1970)

GENERAL REFERENCES

R. R. Brown, Electron Precipitation in the auroral zone, *Space Science Reviews* **5,** 311–387 (1966). This is an excellent review of auroral electron precipitation and a valuable source of references.

J. W. Chamberlain, *Physics of the Aurora and Airglow*, Academic Press, New York and London 1961. The classic work on aurora and airglow.

W. Heitler, *The quantum theory of radiation*, 3rd ed., Clarendon Press, Oxford, 1954. A basic reference in this field.

H. S. W. Massey and E. H. S. Burhop, *Electronic and ionic impact phenomena*, Oxford University Press, London, 1952. A good introduction to impact processes.

B. Rossi, *Cosmic Rays*, McGraw-Hill, New York, San Francisco, Toronto, London, 1964. A very readable and informative little book.

A. J. Dessler, Solar wind and interplanetary magnetic field, *Reviews of Geophysics* **5,** 1 (1967).

J. M. Wilcox, The interplanetary magnetic field. Solar origin and terrestrial effects, *Space Science Reviews* **8,** 258 (1968).

The articles by Dessler and Wilcox, taken together, comprise a very thorough review of solar wind and magnetic field characteristics.

11

Electromagnetic Waves in the Ionosphere

One of the most important aspects of the ionosphere from a practical point of view is its ability to support the propagation of radio waves from one point on the earth's surface to another. In fact, it was the early investigations of the nature of long distance radio communication which lead directly to the experimental proof of the existence of the ionosphere. Concurrent with the experimental researches, theoretical studies of the propagation of electromagnetic waves in a cold magnetoplasma were carried out by Larmor and Eccles, and later by Appleton and Hartree.

In the present chapter we shall study the propagation of electromagnetic waves in a magnetoionic medium, "hot" as well as "cold." We shall find that under certain conditions oscillations (e.g., acoustic, hydromagnetic) can be excited in the plasma. Next, we discuss the propagation of radio waves of various frequencies, as well as radio wave interaction. The latter is manifested as the modulation of a medium or high frequency wave by a "disturbing wave" of low or medium frequency, the interaction occurring in the *D*-region. Finally we shall consider various techniques for measuring ionospheric parameters such as electron number density.

11.1 Properties of a Cold Magnetoplasma

In this section we shall investigate the physics of electromagnetic waves traveling through a cold plasma which is embedded in a constant magnetic field. The treatment will be restricted at this point to the case where electron number density gradients are small compared to the ratio of the electron number density to the wavelength. Under such conditions the methods of geometrical optics apply, thus simplifying the problem. The more general "full wave" treatment (essentially a boundary value problem) will be discussed briefly later on.

11.1.1 The Fundamental Electrodynamics

The propagation of electromagnetic waves through any medium, including a plasma must obey the basic equations of electrodynamics. These are the well-known equations of Maxwell which were introduced in Section 2.1 but are restated here for easy reference. In differential form they are

$$\text{div}\,\mathbf{B} = 0 \text{ (Gauss' Law for magnetic fields)} \tag{2.1}$$

$$\text{div}\,\mathbf{D} = \rho \text{ (Gauss' Law for electric fields)} \tag{2.2}$$

$$\text{curl}\,\mathbf{H} = \mathbf{j} + \frac{\partial \mathbf{D}}{\partial t} \text{ (Ampere-Maxwell Law)} \tag{2.3}$$

$$\text{curl}\,\mathbf{E} = -\frac{\partial \mathbf{B}}{\partial t} \text{ (Faraday's Law of Induction)} \tag{2.4}$$

In order to insure charge conservation they are supplemented by the equation of continuity.

$$\frac{\partial \rho}{\partial t} + \text{div}\,\mathbf{j} = 0 \tag{2.7}$$

In a dielectric medium we find that the electric displacement $\mathbf{D}$ is related to the field intensity $\mathbf{E}$ by

$$\mathbf{D} = \epsilon \mathbf{E} = \epsilon_0 K \mathbf{E} \tag{11.1}$$

It will be shown later than "K" is really a tensor quantity, thus necessitating the recasting of Eq. (11.1) as

$$\mathbf{D} = \epsilon_0\, \boldsymbol{K} \cdot \mathbf{E} \tag{11.1a}$$

if $\boldsymbol{K}$ is not diagonal.

In order to develop a procedure for determining the effects of the wave-plasma interaction, it is necessary to obtain a relation between the impressed forces (electric and magnetic field intensities) and the motion of the charged particles. Strictly speaking, the methods of (microscopic) plasma kinetic theory should be invoked. This will be done in the appendix. However, in order to develop the physics in a more lucid manner, the macroscopic Langevin equation which governs the motion of a charged particle of mass m and charge e

$$m\frac{d\mathbf{v}}{dt} + m\nu\mathbf{v} = e(\mathbf{E} + \mathbf{v} \times \mathbf{B}) \tag{11.2}$$

will be employed for the present. The symbol $\mathbf{v}$ denotes the mean velocity of the charged particles which is related to the current density $\mathbf{j}$ and charge density ρ by the equality

$$\mathbf{j} = \sum_i \rho_i \mathbf{v}_i \tag{11.3}$$

where the summation is carried out over all species of charged particles. The second term on the left hand side of Eq. (11.3) is a so-called stochastic force (see Section 2.3.2 for definition) arising from momentum transfer at frequency ν due to collision processes. The second term on the right hand side is the well-known Lorentz force discussed in Section 2.1. The charge density is related to the number density n_i of the charged particles by

$$\rho_i = e_i n_i \tag{11.4}$$

11.1.2 The A.C. Conductivity of a Plasma

Before discussing the more complicated case of the conductivity of a multi-component magnetoplasma, it is instructive to first consider the case of an isotropic single component one. Here $\mathbf{B} = 0$ and Eq. (11.2) takes the form

$$m\frac{d\mathbf{v}}{dt} + m\nu\mathbf{v} = e\mathbf{E} \tag{11.2a}$$

If we now specify the electromagnetic wave which is propagating through the plasma to be sinusoidal in nature: $\mathbf{E} = \mathbf{E}_0\, e^{-i\omega t}$, it is easy to show that the solution of Eq. (11.2a) is

$$\mathbf{v} = \frac{ie\mathbf{E}}{m(\omega + i\nu)} \tag{11.5}$$

Substitution into Eq. (11.3) leads immediately to

$$\mathbf{j} = \frac{i\omega_p^2}{(\omega + i\nu)}\,\epsilon_0\mathbf{E} \tag{11.6}$$

where $\omega_p = (Ne^2/\epsilon_0 m)^{1/2}$ is the plasma frequency (see Section 2.1). Comparison with the generalized form of Ohm's Law

$$\mathbf{j} = \underset{\sim}{\boldsymbol{\sigma}} \cdot \mathbf{E} \tag{2.8}$$

which is applicable to any conducting medium yields the following form for the conductivity σ

$$\sigma = \frac{i\epsilon_0\omega_p^2}{\omega + i\nu} = \frac{\epsilon_0\omega_p^2(\nu + i\omega)}{\omega^2 + \nu^2} \tag{11.7}$$

Note that σ is a scalar quantity and that it has an imaginary as well as a real part. The real part is what we usually associate with conductivity: it is the reciprocal of resistivity and is thus associated with energy dissipation. The imaginary part is associated with the dielectric properties of the medium (to be discussed later) and is a purely alternating current parameter (inverse of reactance). If we pass to the (d.c.) case where $\omega = 0$, we find that

$$\sigma_{\text{d.c.}} = \frac{\epsilon_0 \omega_p^2}{\nu} \tag{11.8}$$

which is the one-component analogue of Eq. (7.14c).

The more general anisotropic multi-component case was discussed previously in Section 7.12 for direct currents only. In the alternating current case the acceleration term (term on the left hand side) of Eq. (11.2) must be retained and the driving term $e\mathbf{E}$ is again given a sinusoidal time dependence $e^{-i\omega t}$. Solution of the Langevin equation yields the following expressions for the velocity components of the kth species

$$\begin{aligned}
v_x^k &= [(\omega + i\nu_k)^2 - \omega_{kH}^2]^{-1} \frac{e_k}{m_k} [i(\omega + i\nu_k)E_x - \omega_{kH}E_y] \\
v_y^k &= [(\omega + i\nu_k)^2 - \omega_{kH}^2]^{-1} \frac{e_k}{m_k} [\omega_{kH}E_x + i(\omega + i\nu_k)E_y] \qquad (11.9) \\
v_z^k &= [\omega + i\nu_k]^{-1} i \frac{e_k}{m_k} E_z
\end{aligned}$$

where ω_{kH} is defined as in Eq. (7.10). Substitution into Eq. (11.3) and use of the definitions (11.4) and (2.7) leads to the following form for the components of the conductivity tensor $\underset{\sim}{\sigma}$

$$\begin{aligned}
\sigma_0 &= \sum_k \frac{i\epsilon_0\omega_k^2}{\omega + i\nu_k} \\
\sigma_1 &= \sum_k \frac{i\epsilon_0\omega_k^2(\omega + i\nu_k)}{(\omega + i\nu_k)^2 - \omega_{kH}^2} \qquad (11.10) \\
\sigma_2 &= \sum_k \frac{\epsilon_0\omega_k^2\omega_{kH}}{(\omega + i\nu_k)^2 - \omega_{kH}^2}
\end{aligned}$$

where $\underset{\sim}{\sigma}$ is defined by Eq. (7.13). For most purposes the ionosphere can be satisfactorily treated as a two-component plasma; one component consists of free electrons while the other consists of the dominant species of the positive ions. Under this assumption Eq. (11.10) is approximated by

$$\begin{aligned}
\sigma_0 &= \epsilon_0\left[\frac{i\omega_e^2}{\omega + i\nu_e} + \frac{i\omega_+^2}{\omega + i\nu_+}\right] \\
\sigma_1 &= \epsilon_0\left[\frac{i\omega_e^2(\omega + i\nu_e)}{(\omega + i\nu_e)^2 - \omega_{eH}^2} + \frac{i\omega_+^2(\omega + i\nu_+)}{(\omega + i\nu_+)^2 - \omega_{+H}^2}\right] \\
\sigma_2 &= \epsilon_0\left[\frac{-\omega_e^2\omega_{eH}}{(\omega + i\nu_e)^2 - \omega_{eH}^2} + \frac{\omega_+^2\omega_{+H}}{(\omega + i\nu_+)^2 - \omega_{+H}^2}\right] \qquad (11.11)
\end{aligned}$$

Note the similarity of σ_0 to σ in Eq. (11.7). It was shown in Problem 7.3 that the d.c. conductivity could be "diagonalized" by a suitable transformation. Extension to the alternating current case is left as an exercise at the end of this chapter. We shall return to the conductivity tensor after first discussing the plasma as a polarizable medium.

11.1.3 A Plasma as a Polarizable Medium

In a plasma individual charged particles can move about more or less freely relative to one another. Hence, under suitable conditions one expects the medium to be electrically polarized when an electric field is applied. If a steady field is applied, the positive and negative charges move steadily in opposite directions and the current density is given by Eq. (7.12). Here we have no polarization so long as the charges can move freely. However, consider the effect of an alternating (sinusoidal) electric field $\mathbf{E}_0 e^{-i\omega t}$. Positive and negative charges undergo alternating relative displacements $\mathbf{r}e^{-i\omega t}$ because of the driving effect of the field. If these functional forms for $\mathbf{E}$ and $\mathbf{r}$ are substituted into the Langevin equation Eq. (11.2), one obtains for the kth component of the displacement vector

$$x_k = -[(\omega + i\nu_k)^2 - \omega_{kH}^2]^{-1} \frac{e_k}{m_k\omega} [(\omega + i\nu_k)E_x + i\omega_{kH}E_y]$$

$$y_k = -[(\omega + i\nu_k)^2 - \omega_{kH}^2]^{-1} \frac{e_k}{m_k\omega} [(\omega + i\nu_k)E_y - i\omega_{kH}E_x] \quad (11.12)$$

$$z_k = -[\omega + i\nu_k]^{-1} \frac{e_k}{m_k\omega} E_z$$

where the z axis is chosen as the direction of $\mathbf{B}$. Now, the polarization of any medium is defined as the vector sum of all the dipole moments. Since each such moment is given by the product of the electric charges with their relative displacement (i.e., $\mathbf{P} = e_k\mathbf{r}_k$), we have the relation

$$\mathbf{P} = \sum_k n_k e_k \mathbf{r}_k \quad (11.13)$$

for the polarization produced by all species. This quantity is related to the impressed electric field $\mathbf{E}$, the electric displacement $\mathbf{D}$, and the dielectric tensor $\underset{\sim}{\boldsymbol{K}}$ by the equations

$$\begin{aligned} \mathbf{D} &= \mathbf{P} + \epsilon_0\mathbf{E} \\ &= \epsilon_0 \underset{\sim}{\boldsymbol{K}} \cdot \mathbf{E} \end{aligned} \quad (11.14)$$

If we define the polarizability tensor $\underset{\sim}{\boldsymbol{\alpha}}$ by

$$\mathbf{P} = \epsilon_0 \underset{\sim}{\boldsymbol{\alpha}} \cdot \mathbf{E} \quad (11.15)$$

the components of $\boldsymbol{K}$ are of the form

$$K_{ij} = \delta_{ij} + \alpha_{ij} \tag{11.16}$$

where δ_{ij} is the Kronecker delta. For the two-component plasma discussed in the preceding section,

$$\begin{aligned}
\alpha_0 = \alpha_{zz} &= -\frac{1}{\omega}\left[\frac{\omega_e^2}{\omega + i\nu_e} + \frac{\omega_+^2}{\omega + i\nu_+}\right] \\
\alpha_1 = \alpha_{xx} &= -\frac{1}{\omega}\left[\frac{\omega_e^2(\omega + i\nu_e)}{(\omega + i\nu_e)^2 - \omega_{eH}^2} + \frac{\omega_+^2(\omega + i\nu_e)}{(\omega + i\nu_e)^2 - \omega_{+H}^2}\right] \\
&= \alpha_{yy} \\
\alpha_2 = -\alpha_{xy} &= -\frac{i}{\omega}\left[\frac{\omega_e^2\omega_{eH}}{(\omega + i\nu_e)^2 - \omega_{eH}^2} - \frac{\omega_+^2\omega_{+H}}{(\omega + i\nu_e)^2 - \omega_{+H}^2}\right] \\
&= \alpha_{yx}
\end{aligned} \tag{11.17}$$

where

$$\underset{\sim}{\boldsymbol{\alpha}} = \begin{pmatrix} \alpha_1 & -\alpha_2 & 0 \\ \alpha_2 & \alpha_1 & 0 \\ 0 & 0 & \alpha_0 \end{pmatrix} \tag{11.18}$$

It is evident from comparison of Eq. (11.17) with Eq. (11.11) that

$$\underset{\sim}{\boldsymbol{\alpha}} = \frac{i}{\omega\epsilon_0}\underset{\sim}{\boldsymbol{\sigma}} \tag{11.19}$$

The same result could also have been obtained by noting that

$$\frac{d\mathbf{P}}{dt} = \sum_k n_k e_k(-i\omega)\mathbf{r}_k = \sum_k n_k e_k \mathbf{v}_k = \mathbf{j} \tag{11.20}$$

Referring to Eqs. (7.13) and (11.6) we see immediately that Eq. (11.20) follows.

It is interesting to investigate the behavior of $\boldsymbol{\alpha}$ as ω approaches zero; for example, α_1 is approximated by

$$\alpha_1 \approx \frac{i}{\omega}\left[\frac{\omega_e^2\nu_e}{\nu_e^2 + \omega_{eH}^2} + \frac{\omega_+^2\nu_+}{\nu_+^2 + \omega_{+H}^2}\right] \tag{11.21}$$

a purely imaginary quantity which approaches $i \times \infty$ as ω approaches zero. Under these conditions there is no polarization at all; rather the medium is a good conductor. On the other hand, if ω becomes very large, α_1 is approximated by

$$\alpha_1 \approx -\frac{\omega_e^2}{\omega^2} \tag{11.22}$$

the purely real character indicating that the medium is quite polarizable. This interpretation is in accord with our introductory remarks to this subsection.

11.1.4 The Dispersion Relation for a Cold Magnetoplasma

It is well known that the propagation of electromagnetic waves in any medium is dependent upon its electrical properties, i.e., conductivity and polarizability. If, for example, the real part of the conductivity (imaginary part of the polarizability) is infinite, a high frequency (HF) wave will not propagate at all. Under more favorable conditions the velocity of propagation is strongly influenced by the value of the complex conductivity (or polarizability). Since these quantities are functions of the wave frequency we suspect that a plasma is a *dispersive* medium. Since the behavior of the dielectric properties of the plasma is known from Eqs. (11.11) and (11.17), it is a simple matter to derive an expression for the *index of refraction* (ratio of phase velocity in free space to the phase velocity in the medium). The velocity at which energy flows through the plasma, the *group velocity* (see Section 2.1.4), will be discussed later.

If one wishes to obtain an equation for the index of refraction in terms of the plasma properties, it is necessary to return to Maxwell's equations, (2.1)–(2.4). We assume plane wave solutions of the form $\mathbf{E}_0 \exp i(\mathbf{k} \cdot \mathbf{r} - \omega t)$, $\mathbf{H}_0 \exp i(\mathbf{k} \cdot \mathbf{r} - \omega t)$, $\mathbf{j}_0 \exp i(\mathbf{k} \cdot \mathbf{r} - \omega t)$, and substitute into the equations, obtaining

$$\mathbf{k} \cdot \mathbf{B} = 0 \tag{2.1a}$$

$$\mathbf{k} \cdot \mathbf{D} = 0 \tag{2.2a}$$

$$\mathbf{k} \times \mathbf{E} = \omega \mu_0 \mathbf{H} \tag{2.3a}$$

$$\mathbf{k} \times \mathbf{H} = -i\underset{\sim}{\boldsymbol{\sigma}} \cdot \mathbf{E} - \epsilon_0 \omega \mathbf{E} \tag{2.4a}$$

in the absence of currents and space charge not induced by the passage of the wave. The cross product of the wave vector $\mathbf{k}$ with Eq. (2.3a) yields

$$\begin{aligned} \mathbf{k} \times (\mathbf{k} \times \mathbf{E}) &\equiv \mathbf{k}(\mathbf{k} \cdot \mathbf{E}) - k^2 \mathbf{E} = \mu_0 \omega \mathbf{k} \times \mathbf{H} \\ &= -\epsilon_0 \mu_0 \omega^2 \mathbf{E} - i \mu_0 \omega \underset{\sim}{\boldsymbol{\sigma}} \cdot \mathbf{E} \end{aligned} \tag{11.23}$$

Since $\mathbf{k} \cdot \mathbf{E} \neq 0$ in an anisotropic medium, we obtain the dispersion relation

$$k^2(\underset{\sim}{\mathbf{1}} - \mathbf{kk}) \cdot \mathbf{E} = \frac{\omega^2}{c^2} \left(\underset{\sim}{\mathbf{1}} + \frac{i}{\epsilon_0 \omega} \underset{\sim}{\boldsymbol{\sigma}} \right) \cdot \mathbf{E} \tag{11.24}$$

in which $c = (\epsilon_0 \mu_0)^{-\frac{1}{2}}$ is the velocity of electromagnetic waves in free space and $\underset{\sim}{\mathbf{1}}$ is the unit diadic. If $\boldsymbol{\sigma}$ were not a tensor, the solution of Eq. (11.24) would be quite straightforward. Since it is a tensor, we must solve the determinantal or secular equation

$$\begin{vmatrix} k^2 - \frac{\omega^2}{c^2} - k_1^2 - \frac{i\omega}{\epsilon_0 c^2}\sigma_1 & -k_1k_2 + \frac{i\omega}{\epsilon_0 c^2}\sigma_2 & -k_1k_3 \\ -k_2k_1 - \frac{i\omega}{\epsilon_0 c^2}\sigma_2 & k^2 - \frac{\omega^2}{c^2} - k_2^2 - \frac{i\omega}{\epsilon_0 c^2}\sigma_1 & -k_2k_3 \\ -k_3k_1 & -k_3k_2 & k^2 - \frac{\omega^2}{c^2} - k_3^2 - \frac{i\omega}{\epsilon_0 c^2}\sigma_0 \end{vmatrix} = 0 \tag{11.25}$$

where the components of k,

$$k_1 = 0,\ k_2 = k \sin \theta,\ k_3 = k \cos \theta$$

are defined in Figure 11.1. Defining the index of refraction as $n = kc/\omega$, the secular equation becomes

$$\begin{vmatrix} n^2 - 1 - \frac{i}{\epsilon_0\omega}\sigma_1 & \frac{i\sigma_2}{\epsilon_0\omega} & 0 \\ -\frac{i\sigma_2}{\epsilon_0\omega} & n^2\cos^2\theta^2 - 1 - \frac{i\sigma_1}{\epsilon_0\omega} & -n^2\sin\theta\cos\theta \\ 0 & -n^2\sin\theta\cos\theta & n^2\sin^2\theta - 1 - \frac{i\sigma_0}{\epsilon_0\omega} \end{vmatrix} = 0 \tag{11.26}$$

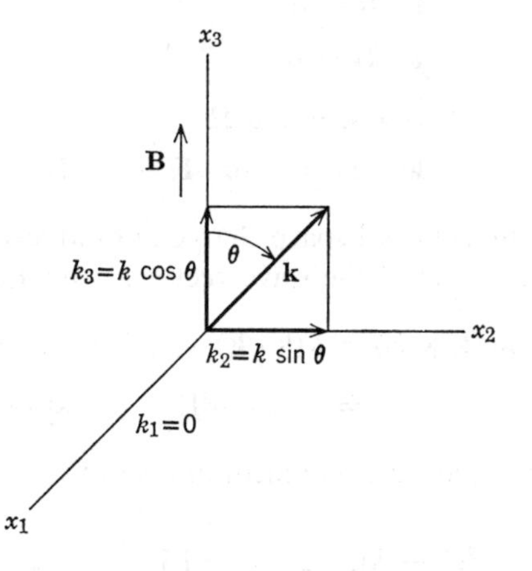

Figure 11.1 The coordinate system used in describing the propagation. The x_3 axis is parallel to the imposed magnetic field **B** and the propagation vector **k** lies in the x_2x_3 plane.

The general solution for n is obviously a complicated expression. It is sufficient for present purposes to investigate the simpler cases where **k** and **B** are (a) parallel and (b) perpendicular.

In case (a) $k_2 = 0$, $k_3 = k$ and Eq. (11.26) simplifies to

$$\begin{vmatrix} n^2 - 1 - \dfrac{i}{\epsilon_0\omega}\sigma_1 & \dfrac{i\sigma_2}{\epsilon_0\omega} & 0 \\ -\dfrac{i\sigma_2}{\epsilon_0\omega} & n^2 - 1 - \dfrac{i\sigma_1}{\epsilon_0\omega} & 0 \\ 0 & 0 & 1 + \dfrac{i\sigma_0}{\epsilon_0\omega} \end{vmatrix} = 0 \tag{11.27}$$

whose solutions are

$$n^2 = 1 + \frac{i}{\epsilon_0\omega}(\sigma_1 \pm i\sigma_2) \tag{11.28a}$$

$$i\omega\epsilon_0 = \sigma_0 \tag{11.28b}$$

The two solutions given by Eq. (11.28a) are the dispersion relations for circularly polarized electromagnetic waves propagating along the magnetic axis; the positive sign represents clockwise (right-handed) rotation of the electric field vector and the negative sign counterclockwise (left-handed) rotation. As we shall see later, Eq. (11.28b) is associated with the occurrence of longitudinally polarized plasma waves. This effect cannot exist in cold plasmas, however, and thus Eq. (11.28b) is not yet a dispersion relation.

In case (b), $k_2 = k$, $k_3 = 0$, and Eq. (11.26) takes the form

$$\begin{vmatrix} n^2 - 1 - \dfrac{i\sigma_1}{\epsilon_0\omega} & \dfrac{i\sigma_2}{\epsilon_0\omega} & 0 \\ -\dfrac{i\sigma_2}{\epsilon_0\omega} & -1 - \dfrac{i\sigma_1}{\epsilon_0\omega} & 0 \\ 0 & 0 & n^2 - 1 - \dfrac{i\sigma_0}{\epsilon_0\omega} \end{vmatrix} = 0 \tag{11.29}$$

whose solutions are

$$n^2 = 1 + \frac{1}{\epsilon_0\omega}\left(\frac{i\epsilon_0\omega\sigma_1 - \sigma_1^2 - \sigma_2^2}{\epsilon_0\omega + i\sigma_1}\right) \tag{11.30a}$$

$$n^2 = 1 + \frac{i\sigma_0}{\epsilon_0\omega} \tag{11.30b}$$

Equation (11.30a) is the dispersion relation for the wave whose electric field vector is perpendicular to **B**. Since the motion of the charged particles is also perpendicular to **B** they will tend to move in orbits about the magnetic field lines. The second solution represents propagation in which the electric field vector is parallel to **B**. Here the motion of the charged particles is not affected by the magnetic field and the plasma behaves like an isotropic one.

11.2 Electromagnetic Waves in a Warm Magnetoplasma

Our treatment in the present section will be more general than in the preceding one in that the thermal effects will be included; as we shall see, these give rise to the traveling plasma oscillations which were associated with Eq. (11.28b). In addition, more complete discussion of ion and electron waves, resonances, and wave polarization will be given.

11.2.1 Dispersion Relations for a Warm Plasma

In treating a warm plasma, we can no longer employ the individual particle approach because of thermal motion. For a rigorous development one must start with the Boltzmann equation [Eq. (2.56b)] and Gauss' Law for electric fields [Eq. (2.2)]. We shall not pursue this line of approach in the present context, however, because it is unnecessary for present purposes. We do point out that the Boltzmann equation without the collision integral predicts "Landau damping", which will be mentioned later. The interested student is referred to the book by Shkarovsky *et al.* listed in the General References for details. The Langevin equation Eq. (11.2) must be replaced by its collective analogue, for each component

$$mn \frac{\partial \bar{\mathbf{v}}}{\partial t} + mN\bar{\mathbf{v}} \cdot \nabla\bar{\mathbf{v}} + mN\nu\mathbf{v} + \nabla p = Ne(\mathbf{E} + \mathbf{v} \times \mathbf{B}) \qquad (11.31)$$

where N is the number density, p is the partial pressure of the component in question, and $\bar{\mathbf{v}}$ is the mean velocity of the component. In order to make the equation tractable, it is necessary to linearize it by writing N and $\bar{\mathbf{v}}$ as the sum of the equilibrium values (N_0, $\bar{\mathbf{v}}_0$) and perturbed parts (N_1, $\bar{\mathbf{v}}_1$), retaining only first order terms. To justify this approach we must assume that $N_1 \ll N_0$ and $\bar{\mathbf{v}}_1 \ll \bar{\mathbf{v}}_0$ which, in fact, proves to be very satisfactory in the ionosphere. For a stationary uniform plasma in its unperturbed condition (here $\bar{\mathbf{v}}_0 = 0$) the second term on the left hand side of Eq. (11.31) vanishes and we are left with

$$mN_0 \frac{\partial \bar{\mathbf{v}}_1}{\partial t} + mN_0\nu\bar{\mathbf{v}}_1 + \nabla p = Ne(\mathbf{E} + \bar{\mathbf{v}}_1 \times \mathbf{B}) \qquad (11.32)$$

It is now necessary to cast the pressure gradient term into a form which is similar to that of its companion terms. In this connection, the question arises as to whether **P** varies isothermally or adiabatically. From our discussion in Section 5.2 we know that the adiabatic approximation is better and so we write the equation of state as

$$\left(\frac{p}{p_0}\right) = \left(\frac{N}{N_0}\right)^{\gamma}; \qquad p = Nk_BT \tag{11.33}$$

where k_B is used to represent Boltzmann's constant in order to distinguish it from the wave number k, and γ is the ratio of the specific heat capacities. The Euler equation thus becomes

$$mN_0\frac{\partial \bar{\mathbf{v}}_1}{\partial t} + mN_0\nu\bar{\mathbf{v}}_1 + \gamma k_BT\nabla N_1 = N_0e(\mathbf{E} + \bar{\mathbf{v}}_1 \times \mathbf{B}) \tag{11.34}$$

which is still not in the desired form. We now assume that $\mathbf{E}(\mathbf{r}, t) \propto \exp i(\mathbf{k} \cdot \mathbf{r} - \omega t)$ and invoke the continuity equation

$$\frac{\partial N}{\partial t} + \operatorname{div}(N\bar{\mathbf{v}}) = 0 \tag{11.35}$$

which becomes

$$\frac{\partial N_1}{\partial t} + N_0 \operatorname{div} \mathbf{v}_1 = 0 \tag{11.36}$$

or

$$\omega N_1 = N_0\mathbf{k} \cdot \mathbf{v}_1 \tag{11.36a}$$

Substitution of Eq. (11.36a) into Eq. (11.34) yields

$$(\underset{\sim}{\mathbf{1}}(\omega + i\nu) - a^2\mathbf{kk}] \cdot \bar{\mathbf{v}}_1 = i\frac{e}{m}[\mathbf{E} + \bar{\mathbf{v}}_1 \times \mathbf{B}) \tag{11.37}$$

where **kk** is the wave vector dyadic with components k_1^2, k_1k_2, k_1k_3, etc., (see Figure 11.1) and a is the acoustic velocity $a = \sqrt{\gamma k_BT/m}$. [cf. Eq. (5.44)].

After some algebraic manipulation and use of Ohm's Law [Eq. (11.7)] we obtain the conductivity tensor whose elements are

$$\begin{aligned}
\sigma_{11} &= iA(\omega + i\nu - a^2k^2/\omega)\\
\sigma_{12} &= -\sigma_{21} = -A\omega_H[1 - a^2k^2\cos^2\theta/\omega(\omega + i\nu)]\\
\sigma_{13} &= -\sigma_{31} = -A\omega_H a^2k^2 \sin\theta\cos\theta/\omega(\omega + i\nu)\\
\sigma_{22} &= iA[\omega + i\nu - a^2k^2(\cos^2\theta)/\omega]\\
\sigma_{23} &= -\sigma_{32} = Aa^2k^2(\sin\theta\cos\theta)/\omega\\
\sigma_{33} &= iA[\omega + i\nu - a^2k^2(\sin^2\theta)/\omega - \omega_H^2/(\omega + i\nu)]\\
A &= \left[\epsilon_0\omega_p^2(\omega + i\nu)^2 - \frac{a^2k^2}{\omega}(\omega + i\nu) - \omega_H^2\left(1 - \frac{a^2k^2\cos^2\theta}{\omega(\omega + i\nu)}\right)\right]^{-1}
\end{aligned} \tag{11.38}$$

By use of the general dispersion relation Eq. (11.23) and the definition of the index of refraction, we obtain the following secular equation for n

$$\begin{vmatrix} n^2 - 1 - \dfrac{i}{\epsilon_0\omega}\sigma_{11} & -\dfrac{i\sigma_{12}}{\epsilon_0\omega} & -\dfrac{i\sigma_{13}}{\epsilon_0\omega} \\ \dfrac{i\sigma_{12}}{\epsilon_0\omega} & n^2\cos^2\theta - 1 - \dfrac{i\sigma_{22}}{\epsilon_0\omega} & -n^2\sin\theta\cos\theta - \dfrac{i\sigma_{23}}{\epsilon_0\omega} \\ \dfrac{i\sigma_{23}}{\epsilon_0\omega} & -n^2\sin\theta\cos\theta + \dfrac{i\sigma_{23}}{\epsilon_0\omega} & n^2\sin^2\theta - 1 - \dfrac{i\sigma_{33}}{\epsilon_0\omega} \end{vmatrix} = 0 \tag{11.39}$$

The solutions are extremely complicated, representing the coupling together of various modes of propagation. In order to simplify the discussion and to make it physically meaningful only certain special cases similar to those treated in Section 11.1.4 will be considered.

11.2.2 Modes of Propagation in a Warm Magnetoplasma

It is convenient to begin with an investigation of the very important case in which $\theta = 0$ (longitudinal propagation). The dispersion relations for a single component plasma (e.g., with infinitely heavy ions) are

$$n^2 = 1 - \left[\frac{\omega + i\nu \pm \omega_H}{(\omega + i\nu)^2 - \omega_H^2}\right]\frac{\omega_p^2}{\omega} \tag{11.40a}$$

and

$$k^2 = \frac{\omega^2}{a^2}\left(1 + i\frac{\nu}{\omega} - \frac{\omega_p^2}{\omega^2}\right) \tag{11.40b}$$

The first equation has the same form as Eq. (11.28a) and describes the longitudinal propagation of two circularly polarized transverse waves whose electric field vectors rotate in opposite directions. The polarization effects can be illustrated by substituting Eq. (11.40b) into the matrix equation which relates n^2 and the conductivity to the electric field vector; the result is

$$\begin{pmatrix} \mp\sigma_{12} & -i\sigma_{12} \\ i\sigma_{12} & \pm\sigma_{12} \end{pmatrix}\begin{pmatrix} E_1 \\ E_2 \end{pmatrix} = 0 \tag{11.41}$$

With the polarization defined as the ratio $\mathscr{R} = E_1/E_2$, we find that $\mathscr{R} = i$ (right-handed polarization) and $\mathscr{R} = -i$ (left-handed polarization) corresponding to the two signs of σ_{12} in Eq. (11.41). By analogy with optical birefringence, the wave with right-handed polarization is called the ordinary wave and the other one the extraordinary wave.

The second equation involves the longitudinal component of the electric field (see Problem 11.4), and thus describes the propagation of a longitudinal wave which has the characteristics of a plasma oscillation. The polarization is longitudinal here because the electric field vector characteristic of this mode (i.e., E_3) is in the direction of the magnetic field and thus of the wave vector **k**.

Now let us consider the case of transverse propagation ($\theta = \pi/2$). The dispersion relations are

$$n^2 = 1 - \frac{\omega_p^2[\omega(\omega + i\nu) - (a^2k^2 + \omega_p^2)]}{\omega^2\left[(\omega + i\nu)^2 - (a^2k^2 + \omega_p^2)\left(1 + \dfrac{i\nu}{\omega}\right) - \omega_H^2\right]} \tag{11.42a}$$

$$n^2 = 1 - \frac{\omega_p^2}{\omega(\omega + i\nu)} \tag{11.42b}$$

The propagation modes corresponding to these dispersion relations are similar to the corresponding modes for a cold plasma [see Eq. (11.30b)]. In fact the second one is identical to Eq. (11.30b). Upon investigation of the polarization of **E** by means of Eq. (11.23) it is seen immediately that for the mode characterized by Eq. (11.42a),

$$\mathscr{R} = -\frac{i}{\sigma_{12}}\left(1 + \frac{i\sigma_{22}}{\epsilon_0\omega}\right) \tag{11.43}$$

Because the transverse components E_1 and E_3 are uncoupled, the wave is linearly polarized in the x direction. Since $\mathscr{R} \neq 0$, there is a component of **E** in the direction of propagation **k**. In the mode characterized by Eq. (11.42b), the electric field vector is linearly polarized in the direction of **B** with the result that the wave propagates as in an isotropic plasma.

So far only single-component plasmas have been considered. However, since every real plasma is macroscopically neutral, there must be both positively and negatively charged particles (i.e., positive ions and electrons) present. In order to simplify the notation we adopt the following (standard) nomenclature.[1]

$$\begin{aligned} X &= \omega_p^2/\omega^2 \\ Y &= \omega_H/\omega \\ Z &= \nu/\omega \\ U &= 1 + iZ \end{aligned} \tag{11.44}$$

Furthermore, we assume for the present that the plasma is collisionless, i.e., $Z = 0$. Under these conditions the dispersion relation for longitudinal propagation takes the form

$$\begin{vmatrix} n^2-1+\dfrac{X_e}{1-Y_e^2}+\dfrac{X_+}{1-Y_+^2} & i\left(\dfrac{X_eY_e}{1-Y_e^2}-\dfrac{X_+Y_+}{1-Y_+^2}\right) & 0 \\ -i\left(\dfrac{X_eY_e}{1-Y_e^2}-\dfrac{X_+Y_+}{1-Y_+^2}\right) & n^2-1+\dfrac{X_e}{1-Y_e^2}+\dfrac{X_+}{1-Y_+^2} & 0 \\ 0 & 0 & 1-\dfrac{X_e}{1-\dfrac{a_e^2k^2}{\omega^2}}-\dfrac{X_+}{1-\dfrac{a_+^2k^2}{\omega^2}} \end{vmatrix} = 0 \tag{11.45}$$

which describes several modes of propagation. It is evident that the 2×2 "block" in the upper left hand corner is associated with the modes with transverse components of $\mathbf{E}$ (i.e., E_1 and E_2) while the remaining element is associated with longitudinal oscillations (i.e., with the component E_3). Let us consider the transverse modes first. The dispersion relations are

$$n^2 = \begin{cases} 1 - \dfrac{X_e}{1 - Y_e} - \dfrac{X_+}{1 + Y_+} & \text{(left handed polarization—extraordinary)} \\ 1 - \dfrac{X_e}{1 + Y_e} - \dfrac{X_+}{1 - Y_+} & \text{(right handed polarization—ordinary)} \end{cases} \tag{11.46}$$

Since $\omega_H = eB/m$ and $\omega_p^2 = N_e e^2/\epsilon_0 m$, we know that $X_+/X_e = m_e/m_+ = \zeta = Y_+/Y_e$ where m is the charged particle mass. Then Eq. (11.46) becomes

$$n^2 = \begin{cases} 1 - \dfrac{X_e}{1 - Y_e} - \dfrac{\zeta X_e}{1 + \zeta Y_e} \approx 1 - \dfrac{X_e}{(1 - Y_e)(1 + Y_+)} \\ 1 - \dfrac{X_e}{1 + Y_e} - \dfrac{\zeta X_e}{1 - \zeta Y_e} \approx 1 - \dfrac{X_e}{(1 + Y_e)(1 - Y_+)} \end{cases} \tag{11.46a}$$

since $\zeta \ll 1$. The index of refraction for these two modes is shown as a function of X_e in Figure 11.2. It is obvious that at high frequencies the ions can be neglected altogether. This is understandable since their large inertia renders them virtually immovable when driven by high frequency waves. It is also evident from Eq. (11.46) that resonances exist at $Y_e = 1$ (extraordinary wave) and $Y_+ = 1$ (ordinary wave). These are the "cyclotron resonances" which result in very strong transfer of energy from wave to the electrons and ions.

At low frequencies, we have $1 \ll Y_+ \ll Y_e$ and

$$n^2 \approx 1 + \frac{X_e}{Y_e Y_+} = 1 + \frac{N m_+}{e_0 B^2} \tag{11.47}$$

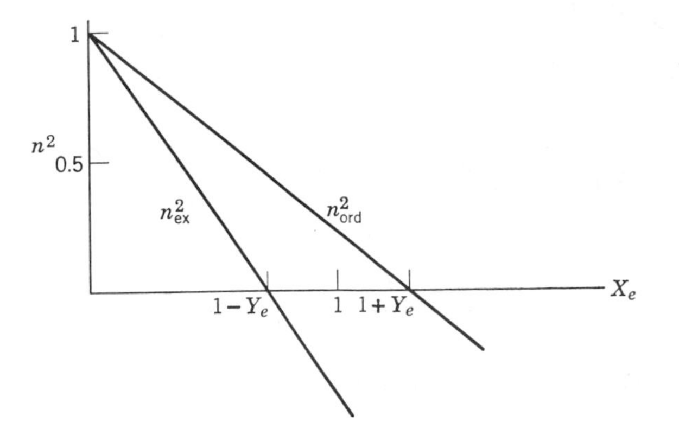

Figure 11.2 Index of refraction as a function of X_e for longitudinal propagation.

for both modes. This corresponds to the propagation of a hydromagnetic wave which will be discussed in more detail later.

As we previously pointed out, the remaining mode for the case of longitudinal propagation is characterized by the dispersion relation

$$1 = \frac{\omega_e^2}{\omega^2 - a_e^2 k^2} + \frac{\omega_+^2}{\omega^2 - a_+^2 k^2} \tag{11.48}$$

or

$$(\omega^2 - a_e^2 k^2 - \omega_e^2)(\omega^2 - a_+^2 k^2 - \omega_+^2) - \omega_e^2 \omega_+^2 = 0 \tag{11.48a}$$

In order to establish the physical significance of this relation, it is most profitable to discuss its properties in particular frequency regimes. We first consider the high frequency case where $\omega_+ \ll \omega$ and $n^2 \ll v^2/a_+^2$ which implies that the phase velocity v is much larger than the ion thermal speed. Under these conditions, the dispersion relation illustrated in Fig. 11.3 becomes

$$k^2 \approx \frac{1}{a_e^2}(\omega^2 - \omega_e^2) \tag{11.49}$$

which describes the propagation of *electron waves*. Such waves are possible only if $\omega_e > \omega$ and if $k < \omega_e/a_e$ (the latter insures that the space charge fields which drive the oscillation are able to develop). If $k_e = \omega_e/a_e$, $\omega = \sqrt{2}\omega_e$ which is the maximum (cut-off) frequency accessible to the wave. Actually this cut-off is not a sharp one because there is no sharp dividing line between individual particle motion and the collective motion of the oscillation. Such a wave is subject to a damping mechanism first suggested by Landau. Most

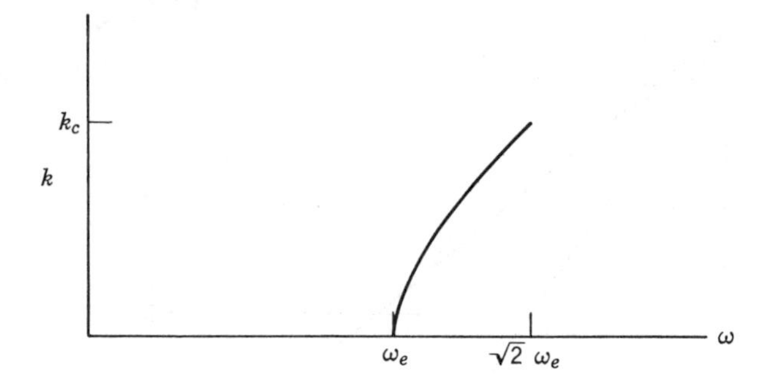

Figure 11.3 Plot of the dispersion relation for electron waves.

of the electrons move rapidly past the wave crests and troughs; however, a few in the "tail" of the energy distribution are approximately in step with the wave and are alternately accelerated and decelerated by the associated space charge oscillation. Landau showed that if the number of electrons with velocities slightly less than the wave speed u is greater than the number with speeds slightly above u, energy is transferred from the wave to the (thermal motion of) the electron gas.

If, on the other hand, one considers the low frequency regime where $\omega \ll \omega_e$ and $\omega \ll ka_e$, the dispersion relation reduces to

$$k^2 = \frac{1}{a_+^2}\left(\omega^2 - \omega_+^2 + \frac{\omega_e^2\omega_+^2}{\omega_e^2 + k^2a_e^2}\right) \tag{11.50}$$

It is convenient to discuss separately two subregimes for this case: $\omega_e \ll ka_e$, and $\omega_e \gg ka_e$. In the first instance, the dispersion relation is approximated by

$$k^2 \approx \frac{1}{a_+^2}(\omega^2 - \omega_+^2) \tag{11.51}$$

which is identical in form to Eq. (11.49) except that we are now dealing with ions instead of electrons. Hence we speak of ion waves. Here the frequency of the space charge variation is sufficiently low that the electrons can move freely to maintain uniform number density. Not so with the ions whose motion is in step with the wave. In the other subregime, the dispersion relation is

$$k^2 \approx \omega^2[a_+^2 + a_e^2\omega_+^2/\omega_e^2]^{-1} \tag{11.52}$$

which can be further simplified, by use of the definition of the plasma frequency and the acoustic velocity $a = \sqrt{\gamma k_B T/m}$, to

$$k^2 \approx (\omega^2/a_+^2)\left(1 + \frac{T_e}{T_+}\right) \tag{11.53}$$

which characterizes a nondispersive wave. In this *pseudosonic* mode the ions and electrons tend to move together. A dispersion diagram for ion and pseudosonic waves (where $T_e = T_+$) is shown in Figure 11.4.

The treatment of the transverse case ($\theta = \pi/2$) is similar to that developed toward the end of Section 11.1.4 but will now be generalized to a hot two-component plasma. The dispersion relation is given by the secular equation

$$\begin{vmatrix} n^2-1+\dfrac{X_eW_e}{W_e-Y_e^2}+\dfrac{X_+W_+}{W_+-Y_+^2} & i\left(\dfrac{X_eY_e}{W_e-Y_e^2}-\dfrac{X_+Y_+}{W_+-Y_+^2}\right) & 0 \\ -i\left(\dfrac{X_eY_e}{W_e-Y_e^2}-\dfrac{X_+Y_+}{W_+-Y_+^2}\right) & -1+\dfrac{X_eW_e}{W_e-Y_e^2}+\dfrac{X_+W_+}{W_+-Y_+^2} & 0 \\ 0 & 0 & n^2-1-X_e+X_+ \end{vmatrix} = 0 \tag{11.54}$$

where

$$W = 1 - \frac{a^2k^2}{\omega^2} = 1 - \frac{a^2}{c^2}n^2$$

The solutions are

$$n^2 = 1 - X_e - X_+ \approx 1 - X_e \tag{11.55}$$

and

$$n^2 = \frac{\left[1 - \dfrac{X_e(W_e+Y_e)}{W_e-Y_e^2} - \dfrac{X_+(W_+-Y_+)}{W_+-Y_+^2}\right]\left[1 - \dfrac{X_e(W_e-Y_e)}{W_e-Y_e^2} - \dfrac{X_+(W_++Y_+)}{W_+-Y_+^2}\right]}{\left[1 - \dfrac{X_eW_e}{W_e-Y_e^2} - \dfrac{X_+W_+}{W_+-Y_+^2}\right]} \tag{11.56}$$

which correspond respectively to the cases where **E** is parallel to, and perpendicular to **B**. The first case is just what we would expect if the plasma

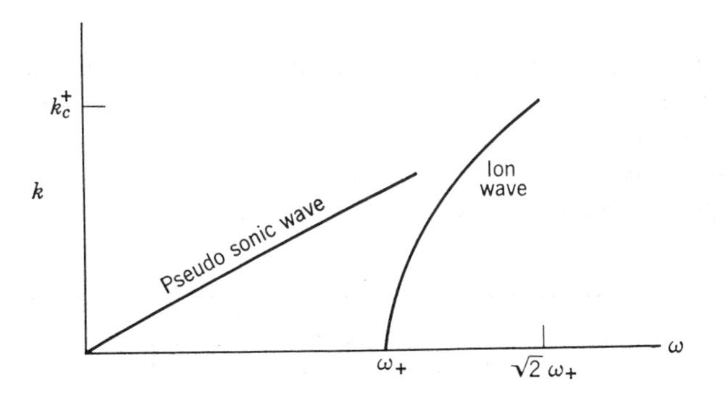

Figure 11.4 Plot of the dispersion relations for ion and pseudosonic waves.

were cold and isotropic. The wave is linearly polarized with **E** parallel to **B** and is the "ordinary wave." It is quite obvious from Eq. (11.55) that the ions produce very little effect. In the form given above, the dispersion relation for the case where **E** is perpendicular to **B** (extraordinary wave) is extremely difficult to interpret physically. We shall therefore consider only certain regimes in which Eq. (11.56) is appropriately simplified.

First we note that in practical situations $a \ll c$. Since $n < 1$ in most cases not near resonances, we have $W \approx 1$. In other words, the plasma behaves as if it were cold. The approximate dispersion relation is

$$n^2 \approx \frac{\left(1 - \dfrac{X_e}{1 - Y_e} - \dfrac{X_+}{1 + Y_+}\right)\left(1 - \dfrac{X_e}{1 + Y_e} - \dfrac{X_+}{1 - Y_+}\right)}{1 - \dfrac{X_e}{1 - Y_e^2} - \dfrac{X_+}{1 - Y_+^2}} \tag{11.56a}$$

In the high frequency regime one can disregard the ions, and Eq. (11.56) simplifies to

$$n_{\text{ex}}^2 \approx 1 - \frac{X_e(1 - X_e)}{1 - X_e - Y_e^2} \tag{11.56b}$$

which has a resonance at $Y_e^2 = 1 - X_e$; the subscript "ex" refers to the extraordinary wave. Again, the ions contribute very little because of their large inertia. The square of the refractive index n^2 plotted against X_e is shown in Figure 11.5; the occurrence of the resonance at $X_e = 1 - Y_e^2$ is the outstanding feature of this diagram.

The polarization of the extraordinary mode is such that the perpendicular component of **E** is in quadrature with the parallel component; that is

$$\mathscr{R} \equiv \frac{E_1}{E_2} = \frac{i(1 - Y^2 - X)}{XY} \tag{11.57}$$

Hence the wave is elliptically polarized in the plane perpendicular to **B**. At resonance we have $\mathscr{R} = 0$ which indicates that the wave is entirely longitudinal (E_2 component only).

Intermediate cases (i.e., $0 < \theta < \pi/2$) are more difficult to treat and we give the dispersion relation only for the high frequency case:

$$n^2 = 1 - \frac{X_e}{U_e - \dfrac{Y_T^4}{2(U_e - X_e)} \pm \left(\dfrac{Y_T^4}{4(U_e - X_e)^2} + Y_L^2\right)^{1/2}} \tag{11.58}$$

where $Y_T = Y_e \sin\theta$ and $Y_L = Y_e \cos\theta$. This is the famous Appleton-Hartree equation developed by these two investigators[2,3] in the 1920's.

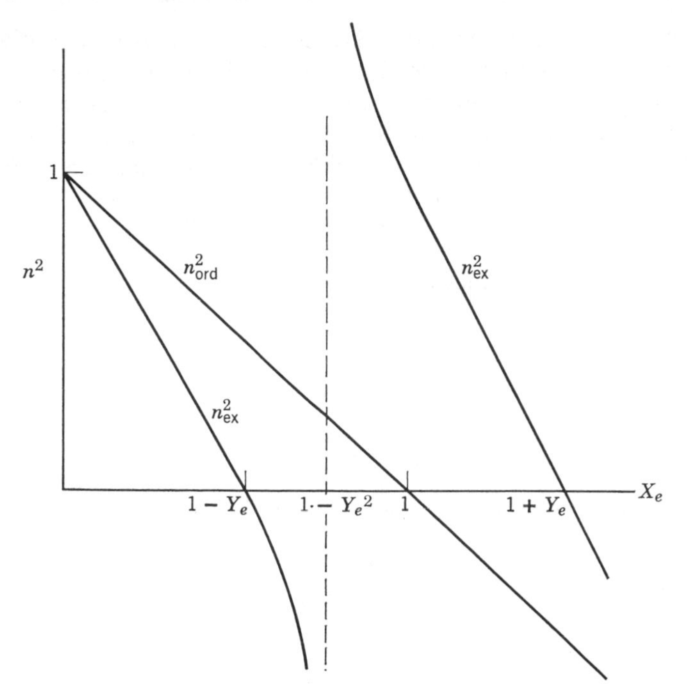

Figure 11.5 Index of refraction as a function of X_e for transverse propagation.

11.2.3 Energy Flow in the Plasma

Investigation of the flow of electromagnetic energy as the wave passes through the plasma is best approached by means of the Poynting vector introduced in Section 2.1

$$\mathbf{\Pi} = \mathbf{E} \times \mathbf{H} \tag{2.19}$$

whose dirction is normal to the wave front (i.e., the plane of **E** and **H** which is not necessarily perpendicular to the direction of propagation). In an isotropic plasma, we find with the aid of Eqs. (2.3), (2.4), and (2.8) that

$$i\mu_0\omega\sigma\mathbf{k} \cdot \mathbf{E} + \frac{\omega^2}{c^2}\mathbf{k} \cdot \mathbf{E} = 0 \tag{11.59}$$

which is equivalent to the statement that $\mathbf{k} \cdot \mathbf{E} = 0$ except where $\sigma = i\omega/\mu_0 c^2$. Hence **E**, **H**, and **k** are mutually perpendicular and the direction of propagation **k** and energy flow **Π** coincide. This is to be expected from considerations of symmetry. Such is not the case with a magnetoplasma. Here

$$\mathbf{k} \cdot \mathbf{E} = -\frac{ic^2}{\omega^2}\mu_0\mathbf{k} \cdot \boldsymbol{\sigma} \cdot \mathbf{E} \tag{11.60}$$

which, in general, is nonvanishing. Hence there is a component of **E** parallel to **k** with the result that **k** and **Π** do not coincide in direction. Further investigation of energy flow in an anisotropic plasma is left as an exercise.

In addition to the direction of power flow, we are interested in its magnitude, in particular the time-averaged power crossing a unit area. In Section 2.1 we stated that

$$|\bar{\mathbf{\Pi}}| = \tfrac{1}{2}\,|R_e(\mathbf{E} \times \mathbf{H}^*)| = \bar{\Pi} \tag{2.20}$$

For an isotropic plasma the equation for $\bar{\Pi}$ is

$$\bar{\Pi} = \frac{1}{2\mu_0 c}\,|\mathbf{E}|^2 \tag{11.61}$$

If **E** is represented by a plane wave

$$\mathbf{E} = \mathbf{E}_0 \exp i(\mathbf{k} \cdot \mathbf{r} - \omega t)$$

with complex wave number $k = k_r + ik_i$, $\bar{\Pi}$ can be expressed as

$$\bar{\Pi} = \frac{e^{-2\mathbf{k}_i \cdot \mathbf{r}}|\mathbf{E}_0|^2}{2c\mu_0} \tag{11.62}$$

which shows that the wave intensity decreases as it passes through the plasma. This effect occurs only if there are collisions between the electrons and the heavy particles (neglecting Landau damping). Otherwise k is real and there is no energy loss to the medium.

In free space the energy flows at the speed limit c, set by special relativity, but in a material medium it flows more slowly. When one first examines a dispersion relation such as Eq. (11.58) it appears that the velocity is greater than c. However, the phase velocity v which is related to c by

$$v = c/n$$

is not the velocity of energy flow. The latter is the so-called group velocity introduced in Section 2.1. It is related to the phase velocity by

$$v_g = c^2/v \tag{11.63}$$

which is less than c. Specifically, for an isotropic plasma

$$v_g = c[1 - X]^{1/2} \tag{11.64}$$

The expression for the group velocity in an anisotropic plasma is more complicated. As a simple example consider the extraordinary ray whose index of refraction is given by Eq. (11.56b). From the definition of the group velocity, we see immediately that

$$v_g = \frac{c^2}{v}\left[1 + \frac{X_e Y_e^2}{(1 - X_e - Y_e^2)^2}\right]^{-1} \tag{11.65}$$

The physical significance of the resonances mentioned previously is easily explained in terms of v_g: at a resonant frequency, e.g., $\omega = \omega_p$ for an isotropic plasma, and $\omega = \sqrt{(\omega_p^2 + \omega_H^2)}$ for the transverse extraordinary ray in a magnetoplasma (the "hybrid" resonance), $v_g = 0$. In other words, the wave does not propagate at these critical frequencies. When an electromagnetic wave with frequency close to a resonant one passes through a plasma such as the ionosphere, its group velocity is markedly decreased. If the plasma is an absorbing one, the wave can lose a great deal of energy in the critical (frequently called the "deviative") region. We shall return to this point in a later section.

11.3 Faraday Rotation in a Plasma

It was found by Faraday over a century ago that a magnetic field rotates the plane of polarization of polarized light. In fact this was probably the most important experimental support for Maxwell in his development of electrodynamics. As one may suspect, a similar effect is experienced by a plane-polarized electromagnetic wave passing through a magnetoplasma. For a wave propagating longitudinally, it can be considered as a superposition of two circularly polarized components, one clockwise, the other counterclockwise. As we have already seen, the speeds of propagation of the two waves are different, and this property results in a relative phase change as the waves propagate. The effect is one of steady rotation of the plane of polarization of the resultant wave. It is instructive to consider briefly the high frequency regime for a collisionless plasma. The relative polarization is given by

$$\mathscr{R} = \frac{E_{\text{ord}}}{E_{\text{ex}}} = \exp\left[\frac{i\omega}{2c}(n_{\text{ord}} - n_{\text{ex}})x\right] = \exp(i\phi) \tag{11.66}$$

where x is the distance over which the polarization change occurs. If we call ϕ the angle of rotation, we obtain the approximate relation

$$\phi \approx \left(\frac{XY}{1 - Y^2}\right)\frac{\omega}{2c}x \tag{11.67}$$

which gives the angle of rotation in terms of the wave and plasma parameters. The Faraday effect affords a very useful method of measuring the electron content (number per unit area column) in the ionosphere (see Section 11.7.6).

11.4 The Hydromagnetic Mode

When we discussed the case of longitudinal propagation in Section 11.2.2, we derived a dispersion relation (ρ is the mass density of the ions)

$$n^2 \approx \frac{Nm_+}{\epsilon_0 B^2} = \frac{\rho\mu_0}{c^2B^2} \tag{11.47a}$$

which was independent of the frequency. The waves which propagate with velocity $A = B/\sqrt{\rho\mu_0}$ are variously called hydromagnetic, magneto hydrodynamic, or Alfvén waves; and A is called the Alfven velocity. The last name stems from the suggestion by the Swedish physicist Alfvén of the occurrence of this mode in connection with a theory of sunspots. Since then they have been associated with many other phenomena as well. The principal interest insofar as the upper atmosphere is concerned is with magnetic storms, particularly the sudden commencement[4] and hydromagnetic heating of the high atmosphere. In the present section we shall investigate hydromagnetic waves somewhat more thoroughly.

We begin by writing down the appropriate equations which govern the motion of ions in the presence of a magnetic field; electrons are not included because they have little effect in the low frequency regime. These equations are the continuity equation

$$\frac{\partial \rho}{\partial t} + \operatorname{div}(\rho \mathbf{v}) = 0 \tag{11.68}$$

where $\mathbf{v}$ is the velocity of the ions and ρ is the mass density; Euler's equation

$$\rho \frac{\partial \mathbf{v}}{\partial t} + \rho \mathbf{v} \cdot \nabla \mathbf{v} = \mathbf{j} \times \mathbf{B} - \nabla p \tag{11.69}$$

where p is the pressure of the ion gas; the Ampere-Maxwell equation

$$\operatorname{curl} \mathbf{B} = \mu_0 \mathbf{j} \tag{2.4a}$$

from which the displacement current term has been omitted because it is very small; the Faraday Induction Law

$$\operatorname{curl} \mathbf{E} = -\frac{\partial \mathbf{B}}{\partial t}\,; \tag{2.4}$$

and Ohm's Law

$$\mathbf{j} = \sigma(\mathbf{E} + \mathbf{v} \times \mathbf{B}) \tag{11.70}$$

By use of the identity $\mathbf{B} \times \operatorname{curl} \mathbf{B} \equiv \frac{1}{2}\nabla B^2 - \mathbf{B} \cdot \nabla \mathbf{B}$, Eq. (11.69) can be transformed into

$$\rho \frac{\partial \mathbf{v}}{\partial t} + \rho \mathbf{v} \cdot \nabla \mathbf{v} = -\nabla\left(p + \frac{1}{2\mu_0} B^2\right) + \frac{1}{\mu_0} \mathbf{B} \cdot \nabla \mathbf{B} \tag{11.69a}$$

We digress slightly at this point to discuss the first term on the right hand side of Eq. (11.69a). Here we have added the quantity $(1/2\mu_0)\,B^2$ to the gas pressure, thus suggesting that the former can be interpreted as magnetic pressure. This is not really surprising because the pressure is in fact the mechanical potential energy density and $(1/2\mu_0)B^2$ is the magnetic potential energy density. Next we substitute Eq. (11.70) into Eq. (2.4a) and then substitute the result into Eq. (2.3), obtaining

$$-\frac{\partial \mathbf{B}}{\partial t} = -\frac{1}{\mu_0 \sigma} \nabla^2 \mathbf{B} - \mathbf{B} \cdot \nabla \mathbf{v} + \mathbf{B} \operatorname{div} \mathbf{v} + \mathbf{v} \cdot \nabla \mathbf{B} \tag{11.71}$$

We have now reached a point beyond which progress is hopeless unless we can simplify the nonlinear equations (11.68), (11.69a), and (11.71). We do this by "linearizing" these equations, realizing that we may have lost sight of some essential properties in doing so.† In such an approximation also employed in Section 5.1, the nonlinear character is treated as a small perturbation (recall the treatment of the second term on the left hand side of Eq. (11.31)):

$$\begin{aligned} \rho &= \rho_0 + \bar{\rho} && (\bar{\rho} \ll \rho_0) \\ p &= p_0 + \bar{p} && (\bar{p} \ll p_0) \\ \mathbf{B} &= \mathbf{B}_0 + \bar{\mathbf{b}} && (|\bar{\mathbf{b}}| \ll |\mathbf{B}_0|) \\ \mathbf{v} &= \bar{\mathbf{v}}, \end{aligned} \tag{11.72}$$

the last equation deriving from the fact that the fluid as a whole is at rest. We also assume that the time-varying quantities are of the form $\mathbf{Q} \exp i[\mathbf{k} \cdot \mathbf{r} - \omega t]$. The basic equations are now the linear forms

$$-i\omega\bar{\rho} + i\rho_0 \mathbf{k} \cdot \bar{\mathbf{v}} = 0 \tag{11.68a}$$

$$-i\omega\rho_0 \bar{\mathbf{v}} = \frac{1}{\mu_0} i(\mathbf{k} \cdot \mathbf{B}_0) \quad - \frac{1}{\mu_0} i\mathbf{k}(\mathbf{B}_0 \cdot \mathbf{b}) - i\mathbf{k}\bar{p} \tag{11.69b}$$

$$-i\omega\bar{\mathbf{b}} = -\frac{1}{\mu_0 \sigma} k^2 \bar{\mathbf{b}} - i\mathbf{B}_0(\mathbf{k} \cdot \bar{\mathbf{v}}) + i(\mathbf{B}_0 \cdot \mathbf{k})\bar{\mathbf{v}} \tag{11.71a}$$

which contain four independent variables. Consequently, we need one more equation in order to obtain a solution. It is assumed that thermal processes in the gas are adiabatic, thus permitting one to write

$$\nabla \bar{\mathrm{p}} = a^2 \nabla \rho \tag{11.73}$$

(See Eq. (11.33).) Thus our equation for the local velocity $\bar{\mathbf{v}}$ is (in diadic form).

$$\left[\underset{\sim}{\mathbf{1}} + \frac{\mathbf{k} \cdot \mathbf{B}_0(\mathbf{B}_0\mathbf{k} + \mathbf{k}\mathbf{B}_0)}{\mu_0\rho_0(\omega + ik^2/\mu_0\sigma)\omega} - \frac{\underset{\sim}{\mathbf{1}}(\mathbf{k} \cdot \mathbf{B}_0)^2}{\mu_0\rho_0(\omega + ik^2/\mu_0\sigma)\omega} - \frac{B_0^2\mathbf{kk}}{\mu_0\rho_0\omega(\omega + ik^2/\mu_0\sigma)} - \frac{a^2}{\omega^2}\mathbf{kk}\right] \cdot \bar{\mathbf{v}} = 0 \tag{11.74}$$

† This is an ever-present danger in dealing with nonlinear phenomena and the investigator must beware of extending this method too far.

In the coordinate system specified by Figure 11.1, the corresponding dispersion relation for a perfectly conducting fluid ($\sigma = \infty$) is

$$\begin{vmatrix} 1 - \frac{k^2A^2}{\omega^2}\cos^2\theta & 0 & 0 \\ 0 & 1 - \frac{k^2}{\omega^2}(A^2 + a^2\sin^2\theta) & -\frac{a^2k^2}{\omega^2}\sin\theta\cos\theta \\ 0 & -\frac{a^2k^2}{\omega^2}\sin\theta\cos\theta & 1 - \frac{a^2k^2}{\omega^2}\cos^2\theta \end{vmatrix} = 0 \quad (11.75)$$

Three modes are associated with this equation. One arises from the element in the upper left hand corner of the determinant:

$$1 - \frac{k^2A^2}{\omega^2}\cos^2\theta = 0 \quad (11.76)$$

for which the phase velocity,

$$v = A\cos\theta \quad (11.76a)$$

is of the same form as Eq. (11.47) except that we now allow propagation in an arbitrary direction. The velocity components are shown in Figure 11.6a. Note that the direction of oscillation is always perpendicular to the plane of $\mathbf{B}_0$ and $\mathbf{k}$.

The remaining two modes have associated phase velocities given by

$$v^2 = \left(\frac{A^2 + a^2}{2}\right)\left[1 \pm \sqrt{1 - \left(\frac{2aA\cos\theta}{A^2 + a^2}\right)^2}\right] \quad (11.77)$$

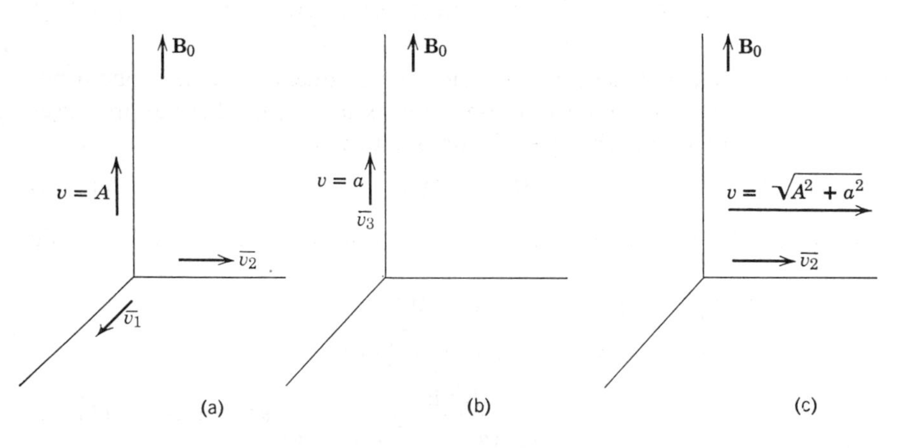

Figure 11.6 Velocity components of various hydromagnetic modes. (a) Alfvén mode. (b) Acoustic mode. (c) Magneto-sonic mode. (After Holt and Haskell, *Foundations of Plasma Physics:* reproduced by permission of the Macmillan Co.)

For illustrative purposes it is useful to consider two cases, $\theta = 0$ and $\theta = \pi/2$; for the former (longitudinal propagation)

$$v = A \text{ or } a \tag{11.78a}$$

Of these, the first solution corresponds to the Alfvén wave already considered while the second one corresponds to an acoustic wave. The velocity components for the latter are shown in Figure 11.6b. For the case $\theta = \pi/2$ (transverse propagation),

$$v = [A^2 + a^2]^{1/2} \tag{11.78b}$$

which can be interpreted as a coupled acoustic-hydromagnetic wave or, more simply a magneto-sonic wave. Its velocity components are shown in Figure 11.6c. For intermediate directions of propagation, there is, of course, no purely acoustic mode; magneto-acoustic coupling is always present.

It is well known that hydromagnetic waves are produced in the magnetosphere by interaction between the fluctuating solar wind and the earth's magnetosphere. These waves propagate through the magnetosphere and eventually reach the ionosphere. The exact character of the HM wave-ionosphere interaction is not known because of the extreme complexity of the HM problem, but wave dissipation does lead to local heating of the plasma and probably the atmosphere also. For the present, therefore, we have to be satisfied with these qualitative (and vague) statements. Hydromagnetic waves are undoubtedly of great significance to the structure of the boundary layer between the solar wind and the ionosphere of a planet without a magnetic field (Mars and Venus). However, the nature of the excitation and propagation of HM waves is even less understood than in the case of the earth.

11.5 The Propagation of Radio Waves

Much of the foregoing development of the theory of electromagnetic wave propagation in a plasma can be applied directly to radio waves in the ionosphere. For example, the treatment of plane waves propagating in a nearly uniform plasma which led to the Appleton-Hartree equation (Eq. 11.58) can be invoked to investigate the behavior of high frequency radio waves ($f >$ 2 MHz) in the ionosphere. At low frequencies ($f <$ 300 kHz) on the other hand, the electron and ion number densities vary greatly within a distance of one wavelength, and a different approach must be employed; this will be done in Section 11.5.2. We shall also discuss radio wave scattering, cross modulation, and partial reflection, all of which assume great importance in the study and observation of ionospheric characteristics.

11.5.1 HF Wave Propagation

The Appleton-Hartree equation which governs the propagation of high frequency electromagnetic waves, given previously in Section 11.2, is

$$n^2 = 1 - \frac{X}{U - \dfrac{Y_T^2}{2(U-X)} \pm \left(\dfrac{Y_T^4}{4(U-X)^2} + Y_L^2\right)^{1/2}};$$

$$Y_L = Y\cos\theta, \quad Y_T = Y\sin\theta \qquad (11.58)$$

where θ is the angle between **B** and **k**.

The two limiting cases which have already been discussed at some length correspond to longitudinal propagation ($\theta = 0$) and transverse propagation ($\theta = 90°$). Now, instead of specifying the limiting cases by θ, suppose we specify them by the inequalities

$$\left|\frac{Y_T^2}{2Y_L}\right| \ll |U - X| \qquad (11.79a)$$

and

$$\left|\frac{Y_T^2}{2Y_L}\right| \gg |U - X| \qquad (11.79b)$$

Then the square of the index of refraction is

$$n^2 \approx 1 - \frac{X}{U \pm Y} \qquad \begin{matrix}(+\text{ ordinary ray})\\(-\text{ extraordinary ray})\end{matrix} \qquad (11.80)$$

for inequality (11.79a) and

$$n^2 \approx 1 - \frac{X}{U} \qquad (\text{ordinary ray}) \qquad (11.81a)$$

$$n^2 \approx 1 - \frac{X}{U - Y_L^2/(U-X)} \qquad (\text{extraordinary ray}) \qquad (11.81b)$$

for inequality (11.79b), which bear striking resemblances to Eqs. (11.46), (11.55), and (11.56b), respectively. Hence the names *quasi-longitudinal* (Q.L.) approximation [Eq. (11.80)] and *quasi-transverse* (Q.T.) approximation [Eq. (11.81a,b)]. Note that the inequalities (11.81a) and (11.81b) can be satisfied even when θ is far from 0 or 90°. For example, at a frequency of 10 MHz, the Q.L. approximation is valid for a value of θ as small as a few degrees. The behavior of n as a function of X is shown in Figures 11.2 and 11.5. The corresponding polarizations are (see Section 11.2.2)

$$\begin{aligned}\mathscr{R} &= i \text{ (extraordinary ray)}\\ &= -i \text{ (ordinary ray)};\end{aligned} \qquad (11.82)$$

that is, the X and Y components of the electric field have a relative phase of 90°. The resultant field vector rotates clockwise for the ordinary ray, counterclockwise for the extraordinary one. In the Q.T. approximation the electric field vector is linearly polarized with

$$\begin{aligned} \mathscr{R} &= 0 \text{ (ordinary ray)} \\ \mathscr{R} &= \infty \text{ (extraordinary ray)} \end{aligned} \tag{11.83}$$

At vertical incidence "reflection" of a high frequency radio-wave occurs when the group velocity vanishes or, equivalently when the group-index of refraction n_g is infinite. Since

$$n_g = \frac{\partial(\omega n)}{\partial n} = \frac{\partial}{\partial\omega}(\sqrt{\omega^2 - \omega_p^2}) = \frac{\omega}{\sqrt{\omega^2 - \omega_p^2}} \tag{11.84}$$

in an isotropic collisionless plasma, reflection occurs when $\omega = \omega_p$ (resonance), or when the phase index of refraction vanishes. It has already been shown that in a plasma in which random electron-heavy particle collisions occur, energy is absorbed from the wave by the plasma. Two cases are of interest in this respect, one in which the phase index of refraction n is not much different from unity, and a second in which n may be nearly zero. Absorption of the first type is called nondeviative since the ray direction does not change appreciably during the absorption process while the second one is called deviative (recall our discussion at the end of Section 11.2.3). It is quite easy to show that in the general case the respective absorption coefficients (i.e., product of the imaginary part of the phase index of refraction with the wave number) for $Z \ll 1$ are

$$K = \frac{\omega}{2c}\frac{\beta}{\alpha^2 + \beta^2} \qquad \text{(nondeviative)} \tag{11.85a}$$

$$K = \frac{\omega}{2c}\frac{\beta}{\alpha}(\mu^{-1} - \mu) \qquad \text{(deviative)} \tag{11.85b}$$

where μ is the real part of n and α and β are defined in Table 11.1.

The spatial dependence of a field component in a slowly varying plasma can be approximated by

$$A_i \approx A_i^0 \exp\left[-ik\int_0^z n(z')\,dz'\right] \tag{11.86}$$

where the argument appearing on the right hand side of Eq. (11.86) is called the phase integral. Hence at vertical incidence E_y of the incident wave, for example, is

$$E_y^{\uparrow} \approx E_y^0 \exp\left[-ik\int_0^z n\,dz'\right] \tag{11.87a}$$

TABLE 11.1. DEFINITION OF QUANTITIES α AND β

Ray	Q.T. Case	Q.L. Case
Ordinary	$\alpha = 1/X;\ \beta = Z/X$	$\alpha = (1 + \lvert Y_L \rvert)/X;\ \beta = Z/X$
Extraordinary	$\alpha = \dfrac{1}{X}\left[- \dfrac{Y_T^2(1 - X)}{(1 - X)^2 + Z^2} \right]$	$\alpha = (1 - \lvert Y_L \rvert)/X$
	$\beta = \dfrac{Z}{X}\left[1 + \dfrac{Y_T^2}{(1 - X)^2 + Z^2} \right]$	$\beta = Z/X$

while the reflected E_y is

$$E_y^{\downarrow} \approx E_y^0 R \exp\left[ik \int_0^z n\, dz' \right]; \tag{11.87b}$$

R is the *reflection coefficient* which we can now define as

$$R = \exp\left[-2ik \int_0^z n\, dz' \right] \tag{11.88}$$

in the phase integral approximation. Reflection coefficients are very useful in quantitative treatments of radio wave propagation and will be referred to again in the discussion of low frequencies.

Nondeviative absorption of high frequency waves occurs mainly in the ionospheric D-layer, during the daytime or during ionospheric disturbances and in the E and F_2 regions at night. Hence the electron collision frequency at these altitudes is of crucial importance to signal propagation. The dependence of the conductivity of an isotropic plasma on the electron velocity distribution is developed in the appendix. It is quite obvious from the form of Eqs. (11.A19) and (11.58) that if the collision frequency ν for momentum transfer is independent of velocity, the index of refraction is given by the isotropic form of the Appleton-Hartree equation [Eq. (11.58) with $Y_e = 0$]. However, ν is in general a function of electron energy $\varepsilon = k_B T u$; in particular $\nu \propto u$ for elastic collisions of electrons with nitrogen molecules as we saw in Section 2.5.4. Under these conditions

$$\begin{aligned} n^2 &= 1 - \frac{4}{3\sqrt{\pi}} \frac{\omega_p^2}{\nu_M^2} \left[\int_0^\infty \frac{u^{3/2} e^{-u}\, du}{\dfrac{\omega^2}{\nu_M^2} + u^2} - \frac{i\nu_M}{\omega} \int_0^\infty \frac{u^{5/2} e^{-u}\, du}{\dfrac{\omega^2}{\nu_M^2} + u^2} \right] \\ &= 1 - \frac{\omega_p^2}{\nu_M^2} \left[C_{3/2}\left(\frac{\omega}{\nu_M}\right) - \frac{5i}{2} \frac{\nu_M}{\omega} C_{5/2}\left(\frac{\omega}{\nu_M}\right) \right] \end{aligned} \tag{11.89}$$

where ν_M is an average collision frequency and $C_p(x)$ is a "Dingle" function

$$C_p(x) = \frac{1}{\Gamma(p+1)} \int_0^\infty \frac{u^p e^{-u}}{x^2 + u^2}\, du \tag{11.90}$$

This form of the dispersion relation, frequently called the Generalized Appleton-Hartree equation was first derived by Sen and Wyller[5] who carried out the computation with the effect of the geomagnetic field included. The latter is left to the reader as an exercise.

At oblique incidence the conditions for reflection must be modified in such a way that the vertical component of the group velocity vanishes. This is equivalent to requiring that the quantity

$$q = n \cos \theta \tag{11.91}$$

vanishes; θ is the angle between the wave front normal in the ionosphere and the vertical. Using Snell's Law

$$n \sin \theta = \sin \theta_{\mathrm{I}} \tag{11.92}$$

where θ_{I} is the angle between the wave front normal and vertical *below* the ionosphere, we can rewrite Eq. (11.91) as

$$n^2 = q^2 + \sin^2 \theta_{\mathrm{I}} = q^2 + S^2 \tag{11.93}$$

or

$$\sin \theta = \frac{S}{\sqrt{q^2 + S^2}}; \qquad \cos \theta = \frac{q}{\sqrt{q^2 + S^2}} \tag{11.94a}$$

$$\sin \theta_I = S; \qquad \cos \theta_I = C \tag{11.94b}$$

Before obtaining the dispersion relation it is necessary to find an expression for the longitudinal component of Y_e, i.e., Y_L, in terms of S, q, and the direction cosines of **B** with respect to **k**. In an obvious notation in which the propagation path is assumed to lie in the xz plane, the direction cosines are

$$\cos(\mathbf{B}, x) = \alpha$$

$$\cos(\mathbf{B}, y) = \beta$$

and thus

$$Y_L = Y_e \left[\frac{\alpha S}{\sqrt{q^2 + S^2}} + \frac{\beta q}{\sqrt{q^2 + S^2}} \right] \tag{11.95}$$

It is immediately evident from Eqs. (11.58) and (11.93) that

$$q^2 = C^2 - \frac{X}{U - \dfrac{Y_T^2}{2(U - X)} \pm \sqrt{\dfrac{Y_T^4}{4(U - X)^2} + Y_L^2}}$$

or

$$U - \frac{Y_T^2}{2(U - X)} + \frac{X}{q^2 - C^2} = \pm \sqrt{\frac{Y_T^4}{4(U - X)^2} + Y_L^2} \tag{11.96}$$

Squaring, subtracting $Y_T^2/4(U - X)$ from both sides, and multiplying by $U - X$ yields

$$(U - X)\left(U + \frac{X}{q^2 - C^2}\right)^2 - Y^2\left(U + \frac{X}{q^2 - C^2}\right) + XY^2\left(\frac{\alpha S + \beta C}{q^2 - C^2}\right)^2 = 0 \tag{11.97}$$

This is a quartic equation in q originally discovered by Booker[6]; it is important not only to high frequency propagation but to low frequencies as well. The four independent solutions correspond to the upgoing ordinary, upgoing extraordinary, downcoming ordinary, and downcoming extraordinary rays.

We conclude this subsection by briefly investigating two limiting cases of the Booker quartic. In the first, we let $\alpha = 0$ which implies that the plane of propagation is perpendicular to $\mathbf{B}$. Then

$$q^2 = C^2 - \frac{X}{U - \dfrac{Y^2(1 - C^2\beta^2)}{2(U - X)} \pm \sqrt{\dfrac{Y^4(1 - C^2\beta^2)^2}{4(U - X)^2} + \dfrac{Y^2\beta^2(C^2U - X)}{U - X}}} \tag{11.98a}$$

In the second case $\mathbf{B}$ is horizontal so that $\beta = 0$. The equation again reduces to quadratic form

$$q^2 = C^2 - \frac{X}{U - \dfrac{Y^2(1 - S^2\alpha^2)}{2(U - X)} \pm \sqrt{\dfrac{Y^4(1 - S^2\alpha^2)^2}{4(U - X)^2} + \dfrac{Y^2S^2\alpha^2U}{U - X}}} \tag{11.98b}$$

Both equations are very reminiscent of the Appleton-Hartree equation. This is hardly surprising since they must both reduce to the A-H form when C is unity (vertical incidence). The signs of the real and imaginary parts of q are as yet undetermined. Conventions in general use require that $\operatorname{Im} q$ be negative for upgoing waves; positive for downward traveling ones. This ensures that $\operatorname{Im} q$ corresponds to absorption of energy from the wave.

Reflection at oblique incidence can occur at higher frequencies than the penetration frequency for vertical incidence. This is illustrated in Figure 11.7. Ray a is refracted by the ionosphere, but not far enough to turn it back to the earth. Ray b on the other hand enters the ionosphere and is refracted back toward the earth. In the ground interval between points 1 and 2, sometimes called the "skip distance," no signal reflected from the ionosphere can be received. The angle of incidence of ray c is even larger, and the ray is, of course, reflected.

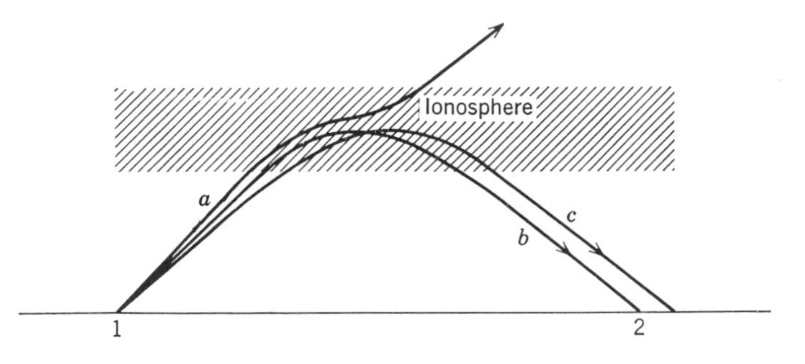

Figure 11.7 High frequency rays obliquely incident on the ionosphere.

11.5.2 LF and VLF Wave Propagation

In the present section we shall restrict our treatment to the case of a sharply bounded isotropic ionosphere and a flat earth for the sake of simplicity. For the more realistic cases of anisotropy and inhomogeneity as well as for a curved earth the reader is referred to the treatises by Budden listed at the end of the chapter.

We begin by considering a plane wave which is incident on the bottom side of the ionosphere. Part of the wave is transmitted through the boundary and part is reflected. The field components of each can be expressed as

$$F^{(\mathrm{I})} = F_1 \exp\,[-ik(Cz + Sx)] \tag{11.99a}$$

$$F^{(\mathrm{R})} = F_2 \exp\,[ik(Cz - Sx)] \tag{11.99b}$$

$$F^{(\mathrm{T})} = F_3 \exp\,[-ik(qz + Sx)] \tag{11.99c}$$

where the geometry is illustrated in Figure 11.8; and C, S, and q have the same meanings as in the preceding section. At a given height z_0, the reflection and transmission coefficients R and T, can be defined as

$$R \equiv \frac{F^{(\mathrm{R})}}{F^{(\mathrm{I})}} = \frac{F_2}{F_1} \exp\,(2ikCz_0) \tag{11.100a}$$

$$T = \frac{F^{(\mathrm{T})}}{F^{(\mathrm{I})}} = \frac{F_3}{F_1} \exp\,(-ik(q - C)z_0) \tag{11.100b}$$

where the region above the boundary is assumed to be a homogeneous ionosphere. It is worth mentioning at this point that there are four reflection

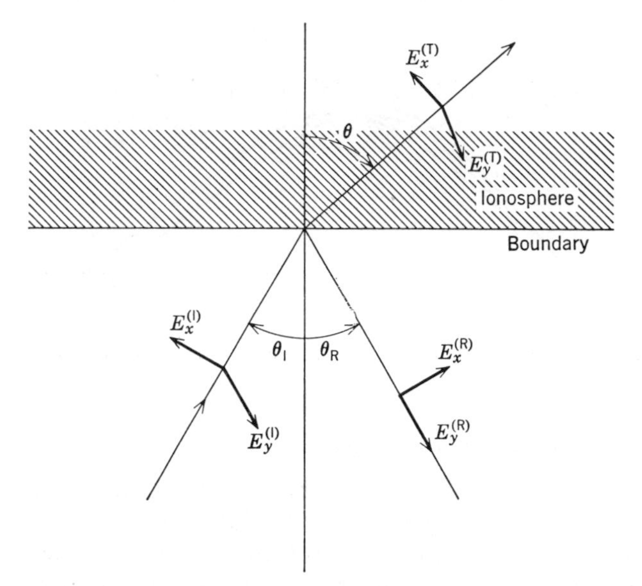

Figure 11.8 The electric fields at reflection by a sharp boundary. The y axis is perpendicular to the page, the x axis is in the page and normal to the propagation vectors. The superscripts (I), (R), and (T) denote the incident, reflected, and transmitted waves, respectively.

coefficients and four transmission coefficients when the ionosphere is anisotropic:

${}_{\|}R_{\|}$ – $E^{(I)}$ and $E^{(R)}$ are both in the plane of incidence.

${}_{\perp}R_{\perp}$ – $E^{(I)}$ and $E^{(R)}$ are both perpendicular to the plane of incidence.

${}_{\|}R_{\perp}$ – $E^{(I)}$ is in the plane of incidence; $E^{(R)}$ is perpendicular to the plane of incidence.

${}_{\perp}R_{\|}$ – $E^{(I)}$ is perpendicular to the plane of incidence; $E^{(R)}$ is in the plane of incidence.

with similar definitions for the T's. The relation between incident and reflected fields is now a matrix equation

$$\begin{pmatrix} E_{\|}^{(R)} \\ E_{\perp}^{(R)} \end{pmatrix} = \begin{pmatrix} {}_{\|}R_{\|} & {}_{\|}R_{\perp} \\ {}_{\perp}R_{\|} & {}_{\perp}R_{\perp} \end{pmatrix} \begin{pmatrix} E_{\|}^{(I)} \\ E_{\perp}^{(I)} \end{pmatrix} \tag{11.101}$$

It is evident from Eq. (11.101) that the field components of the wave are rotated by the anisotropy; this is not unexpected since we have already seen that a superposed magnetic field will rotate the plane of polarization. There are two limiting cases to be considered here. In case (a) the incident electric field $\mathbf{E}^{(I)}$ is in the plane of incidence and the magnetic field $\mathbf{H}^{(I)}$ is perpendicular to it ($E_y^{(I)} = H_x^{(I)} = 0$). Then

$$_{\|}R_{\|} = \frac{H_y^{(R)}}{H_y^{(I)}}; \qquad _{\|}R_{\perp} = \sqrt{\epsilon_0/\mu_0}\,\frac{E_y^{(R)}}{H_y^{(I)}}; \qquad _{\perp}R_{\|} = 0 = {}_{\perp}R_{\perp} \tag{11.102a}$$

In case (b) $\mathbf{H}^{(I)}$ is in the plane of incidence and $\mathbf{E}^{(I)}$ is perpendicular to it; then

$$_{\|}R_{\|} = 0 = {}_{\|}R_{\perp}; \qquad _{\perp}R_{\|} = \sqrt{\mu_0/\epsilon_0}\,\frac{H_y^{(R)}}{E_y^{(I)}}; \qquad _{\perp}R_{\perp} = \frac{E_y^{(R)}}{E_y^{(I)}} \tag{11.102b}$$

In the general case $\mathbf{E}^{(I)}$ is somewhere between the limits of (a) and (b); then all four reflection (and transmission) coefficients are nonvanishing. Of course, for an isotropic ionosphere we must have $_{\perp}R_{\|} = {}_{\|}R_{\perp} = 0$.

The coefficients can be evaluated by imposing the usual boundary conditions which must be satisfied as the wave enters the ionosphere, namely continuity of the tangential components of $\mathbf{E}$ and $\mathbf{H}$. It is evident from Faraday's induction law (Eq. (2.3)) and Snell's law (Eq. (11.92)) that the following relations must be obeyed:

$$E_x^{(I)} = \sqrt{\mu_0/\epsilon_0}\, H_y^{(I)} \cos\theta_I \tag{11.103a}$$

$$E_x^{(R)} = \sqrt{\mu_0/\epsilon_0}\, H_y^{(I)} \cos\theta_R = -\sqrt{\mu_0/\epsilon_0}\, H_y^{(R)} \cos\theta_I \tag{11.103b}$$

$$nE_x^{(T)} = \sqrt{\mu_0/\epsilon_0}\, H_y^{(T)} \cos\theta_T \tag{11.103c}$$

$$\sqrt{\mu_0/\epsilon_0}\,[H_y^{(I)} - H_y^{(R)}] \cos\theta_I = \frac{1}{n} H_y^{(T)} \cos\theta_T \tag{11.104a}$$

$$H_y^{(I)} + H_y^{(R)} = H_y^{(T)} \tag{11.104b}$$

where the index of refraction below the ionosphere is assumed to be unity. Solving for the appropriate field components and substituting the results into Eqs. (11.102) yields expressions for the reflection coefficients

$$_{\|}R_{\|} = \frac{n^2 \cos\theta_I - \sqrt{n^2 - \sin^2\theta_I}}{n^2 \cos\theta_I + \sqrt{n^2 - \sin^2\theta_I}} \tag{11.105a}$$

$$_{\perp}R_{\perp} = \frac{\cos\theta_I - \sqrt{n^2 - \sin^2\theta_I}}{\cos\theta_I + \sqrt{n^2 - \sin^2\theta_I}} \tag{11.105b}$$

with similar expressions for the transmission coefficients. These are the well-known Fresnel formulas which appear in physical optics as well as in ionospheric theory.

If $n = \sin \theta_I$ we find that both reflection coefficients are unity in the absence of collisions. In this case there is total reflection and the transmitted field components are of the form

$$F = F_0 \exp(-k|n|z) \tag{11.106}$$

which describes an *evanescent* "wave." Such a "wave" does not really propagate because **E** and **H** are in quadrature, with the result that the time-averaged Poynting vector vanishes. This is strongly reminiscent of the current I and potential difference V in a purely reactive a.c. circuit; no net energy is transmitted by the circuit because $\overline{IV} = 0$. On the other hand, if n is complex (nonvanishing electron collision frequency), the transmitted wave is not evanescent, some of the energy is lost by transmission and $|R| < 1$ at angles of incidence greater than critical. The ionosphere is said to be "lossy" in this case.

When the wavelength is of the same order as the height of the ionosphere above the earth, the cavity between them acts as a wave guide. Such is the case for VLF waves (3 to 30 kHz). There are two modes which can be excited in the cavity, one with **E** perpendicular to the plane of propagation (TE mode), the other with **H** perpendicular (TM mode). Practical restrictions on VLF antenna configurations usually result in excitation of the latter.

The basic concept behind the wave guide theory is the occurrence of constructive interference between waves which are traveling down the guide, having been reflected at the boundaries. For example, consider a plane wave incident on the ionosphere at angle θ_I; a field component is given by Eq. (11.99b). The reflected wave for this case can be expressed as

$$F^{(R)} = F_1 R(\theta_I) \exp[ik(Cz - Sx)] \tag{11.107}$$

where the last exponential factor is included to ensure that $F^{(R)}/F^{(I)} = R$ when $z = h$. The wave is again reflected at the lower boundary with a reflection coefficient $R_g(\theta_I)$, and a field component $F^{(RR)}$ which is of the form

$$F^{(RR)} = F_1 R(\theta_I) R_g(\theta_I) \exp(-2ikhC) \exp[-ik(Cz + Sx)] \tag{11.108}$$

In order that the two waves $F^{(I)}$ and $F^{(RR)}$ interfere constructively to form a propagating mode, it is necessary that $F^{(I)} = F^{(RR)}$ or

$$R(\theta_I) R_g(\theta_I) \exp[-2ikhC] = 1 \tag{11.109}$$

In the very simple (and artificial!) case where $R(\theta_I) = R_g(\theta_I) = 1$, Eq. (11.109) has the solutions

$$khC = n\pi; \qquad (n = 1, 2, 3, \ldots) \tag{11.110}$$

in which n specifies the mode and h is the duct width. It is evident that the greater the mode number, the larger the number of reflections suffered by the waves. If the boundaries are "lossy," one would thus expect the higher modes to be attenuated more strongly. This is the case, and it is in fact the strong attenuation of all but the first mode or two which makes the wave guide mode approach so useful at great distances from the transmitting antenna. One can show from Eq. (11.110) that there is a certain cut-off wavelength for each mode above which the wave will not propagate. In order to prove this statement, we rewrite Eq. (11.110) as

$$\frac{C}{\lambda} = \frac{n}{2h} \tag{11.111a}$$

or

$$\frac{1}{\lambda^2} = \left(\frac{n}{2h}\right)^2 + \frac{1}{\lambda_g^2} \tag{11.111b}$$

where λ is the free space wavelength, λ_g is the wave guide wavelength, and $\lambda_c = 2h/n$ is the maximum (cut-off) wavelength which will propagate. If $\lambda > \lambda_c$, the wave is evanescent and does not propagate.

11.5.3 Scattering of HF and VHF Waves

In Chapter 10 the scattering of VHF electromagnetic waves in aurorally disturbed regions was briefly mentioned but the mechanism was not considered. It is the purpose of this section to discuss in a very simple manner two theories which have been proposed for the scattering phenomena. In the first, it is assumed that electron number density gradients are large over the distance of a wavelength. The result is the occurrence of partial reflections from the ionized region. From Maxwell's equations and the generalized Ohm's law, one obtains the usual wave equation in one dimension

$$\frac{d^2E}{dz^2} + \frac{\omega^2}{c^2}\left(1 - \frac{\omega_p^2}{\omega^2}\right)E = 0 \tag{11.112}$$

for which the solution is (in integral form)

$$E(z) = E_0 e^{i\omega z/c} + \frac{1}{2i\omega c}\int_{-\infty}^{z} e^{i\omega(z-z')/c}E(z')\omega_p^2(z')\,dz' + \frac{1}{2i\omega c}\int_{z}^{\infty} e^{i\omega(z'-z)/c}E(z')\omega_p^2(z')\,dz' \tag{11.113}$$

where the first term represents the incident wave at $z = -\infty$. We now expand $\omega_p^2(z')$ in a Taylor series about z' and employ the "Born approximation" (in which the scattered wave is assumed to differ only slightly from a plane wave). Then the electric field at $z \to -\infty$ is

$$E(-\infty) \sim E_0 e^{i\omega z/c} + \frac{E_0 e^{-i\omega z/c}}{2i\omega c} \omega_p^2(z) \int_{-\infty}^{\infty} e^{2i\omega z'/c} \frac{l}{N_e} \frac{dN_e}{dz'} dz' \quad (11.114)$$

if ω is well below the critical frequency and the reflection coefficient is

$$R = \frac{\omega_p^2}{2\omega} \int_{-\infty}^{\infty} e^{2i\omega z'/c} \frac{l}{N_e} \frac{dN_e}{dz'} dz' \quad (11.115)$$

where l is the characteristic dimension of the scattering layer. Of course, the foregoing is grossly oversimplified—the problem is three dimensional and **E** is a vector, not a scalar field. Nevertheless, it does offer a revealing qualitative description of the reflection process.

According to the alternative proposal, the waves are scattered by small scale irregularities in the region. Now, it is known from classical electron theory that the cross section for scattering of radiation by electrons (differential Thomson cross section) is

$$Q = \tfrac{1}{2}(1 + \cos^2 \Theta)\left(\frac{e^2}{mc^2}\right)^2 \quad (11.116)$$

where Θ is the scattering angle. If it is assumed that the electrons in the region contribute to the scattering process with appropriate phase differences S for scattered waves from different electrons, the scattered intensity at a mean distance r from the scattering region with volume V is

$$I^{(R)}(\Theta) = \frac{V^2(1 + \cos^2 \Theta)}{r^2} \left(\frac{e^2}{mc^2}\right)^2 N_e^2 \, |\mathscr{R}|^2 \, I_0; \quad (11.117)$$

I_0 is the incident intensity and

$$\mathscr{R} = \frac{1}{V} \int e^{i\delta dV} \quad (11.118)$$

is an interference factor. Note that $I^{(R)}$ is proportional to the square of the electron concentration *gradient* in the first case and to the square of the electron number density in the second. Which mechanism best describes radio auroral effects is still unsettled.

If one integrates both sides of Eq. (11.115) over the azimuthal angle and computes Q for $\Theta = \pi$, one obtains the "backscatter" Thomson cross section

$$Q_T = \pi\left(\frac{e^2}{mc^2}\right)^2 \quad (11.119)$$

In 1958 Gordon[7] suggested that if a very high power radar beam is directed at the ionosphere from the ground, a small but measurable amount of radiation will be incoherently backscattered. Because of the incoherence, which results from the random positions of the electrons, the scattered power is directly proportional to the electron number density. If the situation were one of coherence, the scattered power would be proportional to the *square* of the electron number density.

Random motion of the ions and electrons complicates the scattering because of Doppler shifts in the frequency of the scattered power, and (if the wavelength is large compared to the Debye length) excitation of acoustic oscillations. If we regard the scattered radiation as being caused by local fluctuations in the electron density, ΔN_e, the scattered electric field $\mathbf{E}_S$ in the "farfield" zone can be expressed as[8]

$$\mathbf{E}_S(t) = \sqrt{Q_T}\,\frac{\mathbf{E}_0 e^{i\omega_0 t}}{R_0}\left[\frac{1}{V_S}\int_{VS} \Delta N_e e^{-2ik_0 z}\, dV\right] \tag{11.120}$$

where E_0 is the magnitude of the incident electric field, V_S is the volume element in which the scattering occurs, k_0 is the wave number of the incident radiation, and ω_0 is its frequency. If we Fourier analyze ΔN_e in $\mathbf{k}$ space, we can express the expectation value of the square of the Fourier transform of $\mathbf{E}_S$ (power spectrum) as

$$\overline{|E_S(\omega_0 \pm \omega)|^2}\, d\omega = \frac{Q_T\,|E_0|^2}{R_0^2}\,\overline{|\Delta N_e(2k_0, \omega)|^2}\, d\omega \tag{11.121}$$

If thermal fluctuations only are important, it can be shown that

$$\overline{|\Delta N_e(\mathbf{k}, \omega)|^2} = \frac{N_0}{k}\left(\frac{m_e}{2\pi k_B T}\right)\exp - \ [(\omega^2/k^2)(m_e/2k_B T)] \tag{11.122}$$

where N_0 is the mean electron number density, m_e is the mass of the electron, T is the temperature of the electron gas, and k_B is Boltzmann's constant.

The foregoing approach is satisfactory if the wavelength of the radiation is small compared to the Debye length of the plasma. However, if λ is large, the long range Coulomb interaction between electrons and ions permits the excitation of collective phenomena and the approach must be modified.[8] The scattered power is still proportional to the mean electron number density, but the dependence of the scattered power spectrum on electron and ion temperatures becomes quite complex. The effect of wavelength on the power spectrum is shown in Figure 11.9. In a later section we shall briefly discuss the use of the Thomson scatter technique in ionospheric measurements.

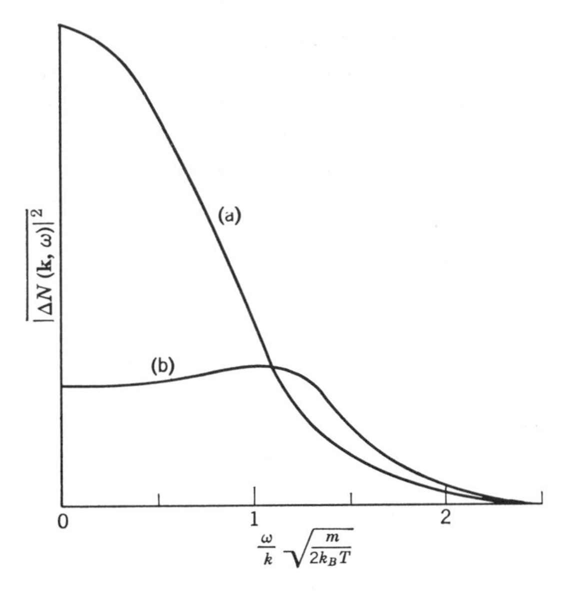

Figure 11.9 Spectrum of the thermal density fluctuations for (a) the case when $\lambda \ll$ Debye length, (b) $\lambda >$ Debye length.

11.5.4 Radio Wave Interaction (Luxembourg Effect)

The Luxembourg effect was discovered in the 1920's when a powerful radio station in Luxembourg was observed to modulate the signals transmitted by other stations. In 1934 V. A. Bailey and D. F. Martyn[9] correctly proposed that free electrons at *D*-layer heights were heated by the "disturbing wave"; the time constant for cooling was assumed to be sufficiently short that the electron temperature could "follow" the modulation pattern. As the wanted wave passed through the same region, it was absorbed in a time-varying manner due to the energy dependence of the electron collision frequency. The time variation of the absorption was thus very close to the modulation pattern of the disturbing signal. It is interesting to note that the Appleton-Hartree theory in which the collision frequency is assumed to be energy-independent does not predict radio wave interaction.

In order to describe the Luxembourg effect on the microscopic scale, we refer to the appendix at the end of the chapter where the distribution of electron velocities in a plasma is investigated. There it is assumed that the distribution function can be written as the sum of an isotropic part f° and a directional perturbation f'. In addition f° is assumed to be time-independent and the plasma to be homogeneous (f is not a function of position). In the case of radio wave interaction the entire distribution function is dependent upon both time and position, since the electron energy spectrum must "follow" the modulation frequency of the disturbing wave. In order to obtain an equation for the absorption coefficient of the plasma it is necessary to solve simultaneously Eqs. (11.A14) and (11.A20), (i.e., the perturbed Boltzmann equation). A rigorous solution would be an horrendous task, but if the amplitude of the disturbing wave is not too great, the solution expressed by Eq. (11.A23) can be employed.

In order to obtain the rate of energy transfer to the electron gas, we multiply the left hand side of Eq. (11.A22) by $\frac{1}{2}mv^2$ and integrate over $\mathbf{v}$. Because the relaxation time for electron cooling is much greater than the period of the disturbing pulse, we average over the time, obtaining for the rate of energy gain by the electron gas

$$\frac{dU_e}{dt} = \int_0^\infty \frac{m_e v^2}{2} \left\{ \frac{1}{3v^2} \frac{\partial}{\partial v} \left[\nu \left(\frac{e^2 E_{\mathrm{rms}}^2}{m_e^2(\nu^2 + \omega^2)} + G \frac{3}{2} \frac{kT_N}{m_e} \right) v^2 \frac{\partial f^\circ}{\partial v} \right] + \frac{1}{3v^2} \frac{\partial}{\partial v} \left(\frac{3}{2} \nu G v^3 f^\circ \right) \right\} 4\pi v^2 \, dv \quad (11.123a)$$

or

$$\frac{dU_e}{dt} = \frac{e^2}{m_e} E_{\mathrm{rms}}^2 \int_0^\infty \frac{f^\circ}{3v^2} \frac{\partial}{\partial v} \left(\frac{\nu v^3}{\nu^2 + \omega^2} \right) 4\pi v^2 \, dv + \frac{3}{2} kT_N \int_0^\infty \frac{f^\circ}{3v^2} \frac{\partial}{\partial v} (G\nu v^3) 4\pi v^2 \, dv - \frac{m_e}{2} \int_0^\infty \nu G v^2 f^\circ 4\pi v^2 \, dv \quad (11.123b)$$

after integrating by parts; E_{rms} is the root mean square electric field, and k is the Boltzmann constant from which the subscript B is now omitted. This equation can be written in terms of electron and neutral gas energies, U_e and U_N, respectively, as

$$\frac{dU_e}{dt} = \epsilon_0 \omega_p^2 E_{\mathrm{rms}}^2 \overline{\frac{1}{3v^2} \frac{\partial}{\partial v} \left(\frac{\nu v^3}{\nu^2 + \omega^2} \right)} + \frac{N_e}{N_N} \overline{(G\nu)_N} U_N - \overline{(G\nu)_e} U_e \quad (11.124)$$

where

$$\overline{(G\nu)_N} = \overline{\left[\frac{1}{3v^2} \frac{\partial}{\partial v} (G\nu v^3) \right]} \quad (11.125a)$$

$$\overline{(G\nu)_e} = \overline{G\nu v^2} \quad (11.125b)$$

and the horizontal bars indicate averages over the distribution function; N_N and N_e are neutral and electron gas number densities. If f° is very close to Maxwellian as is usually the case, $(G\nu)_N \approx (G\nu)_e$, and, in terms of electron and neutral gas temperatures, (T_e and T_N)

$$\frac{\partial T_e}{\partial t} + \overline{(G\nu)}(T_e - T_N) = \overline{\frac{2E_{\text{rms}}^2}{3m_e k} \frac{1}{3v^2} \frac{\partial}{\partial v} \left(\frac{\nu v^2}{\nu^2 + \omega^2} \right)} \tag{11.126}$$

Because of the energy (temperature) dependence of the electron collision frequency, it is obvious that an electromagnetic wave passing through the plasma can be modulated by varying T_e in a periodic manner.

The fractional decrease in amplitude of the "wanted" signal can be expressed as

$$R = \int_0^{h_0} \frac{\partial K_2}{\partial T} \Delta T \tau \, dz \tag{11.127}$$

where K_2 is the absorption coefficient for the "wanted" pulse, ΔT is the change in electron temperature caused by the disturbing pulse and τ is the relaxation factor:

$$\Delta T = \frac{E_0^2 K_1}{6\pi h^2 N_e k_B} \exp \left(- \int_0^h 2K_1 \, dz \right) \tag{11.128}$$

$$\tau = \exp \left[- \frac{2G\bar{\nu}}{c} (h - h_0) \right] \tag{11.129}$$

Here, K_1 is the absorption coefficient of the "disturbing" pulse and h is the height of the interaction region. By computing R for various electron number density and collision frequency profiles and comparing the results with the actual observations, it is possible to obtain quite reliable estimates of both of these ionospheric properties in the region of interaction. We shall discuss the application of the cross modulation technique to the ionosphere in a later section.

11.5.5 Partial Reflection of Radio Waves in the Lower Ionosphere

It was shown some years ago by Gardner and Pawsey[10] that medium frequency waves (e.g., 2 MHz) partially reflected by steep electron number density gradients in the *D*-layer could be used to determine the electron number density profile there. These investigators (wrongly) assumed ordinary Fresnel reflection by a uniform gradient; it was shown later that local fluctuations in the electron number density rather than a uniform gradient were responsible for the reflection phenomena. Nevertheless, the results of Gardner and Pawsey were not invalidated; the theoretical basis was merely changed slightly.

One can easily show (see Section 11.5.2) that the reflection coefficient R for the wave is approximated by

$$R \approx \frac{\delta n}{2n} \tag{11.130}$$

where n is the mean index of refraction of the medium and δn is the change in n which causes the reflection. For a wave propagating in a magneto-ionic medium, there will be two reflection coefficients, one for the ordinary ray and one for the extraordinary ray. Since the medium is a "lossy" one, the coefficient R must be multiplied by the factor $\exp\left(-2ik\int n\,dz\right)$ where k is the wave number and n is the (complex) refractive index. [cf. Eq. (11.88)].

In practice one measures the relative intensities of the reflected ordinary and extraordinary components, A_x and A_0, which are related to the respective indices of refraction by[11]

$$\frac{A_x}{A_0} = \frac{R_x}{R_0} \exp\left[-2\int (K_x - K_0)\,dz\right] \tag{11.131}$$

Here K_x, K_0 are the respective absorption coefficients introduced in Eq. (11.85a). The application of the partial reflection technique to D-layer measurements will be discussed in a later section. It is interesting to note that the precise nature of the reflection mechanism (i.e., whether Fresnel reflection or reflection due to local fluctuations) does not affect the relationship expressed by Eq. (11.131).

11.6 The Two-Stream Instability

It was shown in Section 11.2.2 that the dispersion relation describing acoustic oscillations in a hot plasma is

$$1 = \frac{\omega_e^2}{\omega^2 - a_e^2 k^2} + \frac{\omega_+^2}{\omega^2 - a_+^2 k^2} \tag{11.48}$$

where ω is the wave frequency, k the wave number, ω_e and ω_+ the electron and ion plasma frequencies, and a_e and a_+ the phase velocities of the oscillations. By using the same approach, one can show (see Problem 11.21) that the corresponding relation for a plasma in which the electrons have a mean velocity $2V$ relative to the ions is

$$1 = \frac{\omega_e^2}{(\omega - Vk)^2 - a_e^2 k^2} + \frac{\omega_+^2}{(\omega + Vk)^2 - a_+^2 k^2} \tag{11.132}$$

for disturbances propagating in the same direction as V. Equation (11.132) is a quartic equation in ω whose solution usually must be obtained numerically. If such a program is carried out, it is found that for certain ranges of the relative velocity V, the roots are imaginary. An imaginary root means that the wave amplitude must grow or decay with time; if it grows, the plasma is "unstable." However, it does not grow indefinitely as our simple linear theory seems to indicate. The nonlinearity of the plasma oscillation eventually arrests the growth by means of feedback, and the initial instability becomes more or less "stable."

One can see the way in which imaginary roots arise by studying the simple case of two cold streams of electrons which move with relative velocity $2V$. Equation (11.132) then simplifies to

$$1 = \omega_e^2 \left(\frac{1}{(\omega - Vk)^2} + \frac{1}{(\omega + Vk)^2} \right) \tag{11.133}$$

for which the solution is

$$\omega^2 = V^2k^2 + \omega_e^2 \pm \sqrt{\omega_e^4 + 4\omega_e^2 V^2 k^2} \tag{11.134}$$

Thus we have imaginary roots for all $V > \sqrt{2}(\omega_e/k)$, and the flow is unstable under such a condition. In the case where the two streams are composed of dissimilar particles (i.e., electrons and ions) the onset of instability occurs at a different value of V.

The two-stream instability which we have just discussed in an admittedly oversimplified manner is widely believed to be the mechanism responsible for equatorial sporadic-E; the instability leads to the formation of thin layers of abnormally high electron concentration. Compare with the theory of mid-latitude E_s discussed in Section 10.4.

11.7 Techniques for Ionospheric Measurements

There are numerous methods for carrying out measurement of ionospheric properties. Some of these are ground based radio propagation experiments, some involve propagation of radio waves between ground and rockets or ground and satellites, and some are *in situ* experiments in which the vehicle (rocket or satellite) performs measurements of local properties. The oldest of these techniques is radio sounding in which the time delay between pulse transmission and echo reception is used to determine the height of the reflecting region. We shall consider this method as well as some others in general use in some detail.

11.7.1 Radio Wave Sounding

A radio wave vertical sounder consists basically of a transmitter which sweeps in frequency from about 1 to 25 Hz, an antenna array which emits vertically, a receiver which measures time delay between transmission and echo reception, and a recorder which produces a trace of time delay versus frequency.

As we saw in Section 11.5.1, the index of refraction in the Q.L. approximation (which is usually valid in "sounder" work) is (neglecting collisions)

$$n^2 \approx 1 - \frac{X}{1 \pm Y} \tag{11.80a}$$

The condition for reflection is that $n^2 = 0$. Hence at the reflection point $X = 1 + Y$ or $X = 1 - Y$. If the radio wave traveled through free space except while in contact with a sharp boundary, it would propagate with velocity c and the height of the reflecting layer would be $h' = ct$ where t is the time delay discussed earlier. In actual fact the wave travels over a long path through the ionosphere before reflection occurs; here the group refractive index of the wave is n_g. The apparent or actual height h' is thus related to the true height of reflection h by the integral equation

$$h' = \int_0^h n_g[\omega, \omega_p(h'')]\, dh'' \tag{11.135}$$

A typical record of the h' recorded during the frequency sweep is shown in Figure 11.10. The cusps occur when h' is very large as it would be if $dN_e/dz = 0$ (in fact it would be infinite in this case). It was the appearance of the

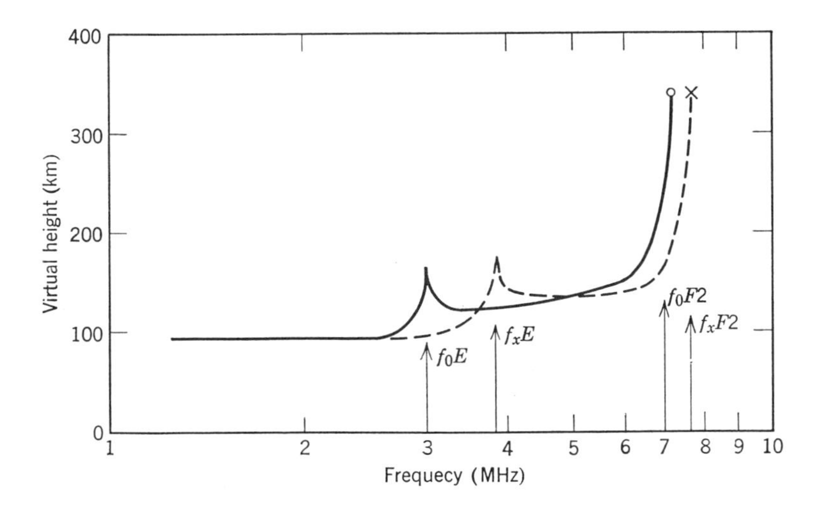

Figure 11.10 Typical idealized ionogram. The solid and broken lines refer to the ordinary and extraordinary modes, respectively.

cusps which led the early investigators to conclude that distinct layers exist in the ionosphere. Actually a point of inflection in the N_e profile will appear to give infinite virtual height.

It is possible in principle to unfold the height h from the integral equation (11.135), thereby yielding a precise electron number density profile. In practice, the precision is not as good as may be desired because it is quite sensitive to small uncertainties in h'. Also it does not resolve the question of "valleys" in the profile. Although these do not occur in the daytime, it is known from rocket experiments that they are present at night.

We begin the analysis[12] by dividing the ionosphere into slabs in which the lowest slab is at the bottom of the ionosphere. We also make a change of independent variable in the integral so that Eq. (11.135) becomes

$$h'(\omega) = h(0) + \int_0^{\omega} n_g(\omega, \omega_p(h''))] \frac{dh''}{d\omega_p} \, d\omega_p \tag{11.136}$$

Then in the slab approximation, we can replace the integral equation by the set of equations (for k slabs)

$$h'(0) = h(0)$$

$$h'(\omega_1) = h(0) + \frac{h(\omega_1) - h(0)}{\omega_1 - 0} \int_0^{\omega_1} n_g(\omega_1, \omega_p) \, d\omega_p$$

$$h'(\omega_2) = h(0) + \frac{h(\omega_1) - h(0)}{\omega_1 - 0} \int_0^{\omega_1} n_g(\omega_2, \omega_p) \, d\omega_p + \frac{h(\omega_2) - h(\omega_1)}{\omega_2 - \omega_1} \int_{\omega_1}^{\omega_2} n_g(\omega_2, \omega_p) \, d\omega_p$$

$$h'(\omega_K) = h(0) + \sum_{i=1}^{k} \frac{h(\omega_i) - h(\omega_{i-1})}{\omega_i - \omega_{i-1}} \int_{\omega_{i-1}}^{\omega_i} n_g(\omega_k, \omega_p) \, d\omega_p \tag{11.137}$$

which are solved for h. It is convenient to re-express Eq. (11.137) in matrix form

$$\Delta h_k' = \sum_{i-1}^{k} M_{ki} \, \Delta h_i \tag{11.138}$$

where

$$\Delta h_k' = h'(\omega_k) - h(0)$$

$$\Delta h_i = h(\omega_i) - h(\omega_{i-1})$$

$$M_{ki} = [\omega_i - \omega_{i-1}]^{-1} \int_{\omega_{i-1}}^{\omega_i} n_g(\omega_k, \omega_p) \, d\omega_p$$

Upon inversion of the square matrix **M**, one can solve for the Δh_i

$$\Delta h_i = \sum_j (\mathbf{M}^{-1})_{ij} \Delta h'_j \tag{11.139}$$

and thus for h

$$h = h(0) + \sum_{i=1}^{k} \Delta h_i \tag{11.140}$$

Actual unfolding of the electron density profile from the ionogram is a tedious affair and is subject to the uncertainty discussed previously. Nevertheless, it is still one of the most useful methods for probing the ionosphere. Since the advent of sophisticated satellite systems, the same approach has been used to sound the topside of the ionosphere,[13] thus yielding information about this region that is not otherwise available. An example of this is the discovery of the spread-F condition on the topside by the Canadian Alouette sounder.

11.7.2 Cosmic Noise Absorption

Many objects in the universe (e.g., the center of our galaxy, the Crab nebula, quasars, etc.) are powerful emitters of radiation in the high frequency radio wave band. Most of the radiation, however, comes from synchrotron emission in the interstellar medium. Observations of the intensity of such cosmic noise by means of a suitable antenna and receiver show that it varies diurnally in a more or less regular manner, with stronger noise levels occurring at night. Part of the variation is due to the antenna's looking at different parts of the sky as the earth rotates, but most is due to ionospheric absorption. Observations of cosmic noise offer a very useful method studying the ionosphere in the region where the absorption occurs (D and F_2 regions).

Little and Leinbach[14] were the first to develop a receiver which was sufficiently stable to obtain meaningful measurements of cosmic noise intensity over long periods of time. Their instrument, called a *riometer* (relative ionospheric opacity meter) employed such a stable receiver and a nearly-dipole antenna directed at the zenith.

The intensity of radiation at a radiofrequency is usually expressed in terms of an equivalent black body temperature T. Typically, T is fairly uniform over a riometer antenna, and the received power P per H_z of band width is

$$P = \frac{kT}{2\pi} \int_0^{2\pi} \int_0^{\pi/2} G(\theta, B) \exp\left(-\frac{A_0}{4.3} \sec\theta\right) \sin\theta \, d\theta \, d\phi \tag{11.141}$$

where T is the source temperature, $G(\theta, \phi)$ is the power gain of the antenna expressed as a function of the zenith and azimuthal angles θ and ϕ, and A_0

is the zenithal absorption in decibels. In terms of the integral of the absorption coefficient K over the path length as given in Eq. (11.88),

$$A_0 = 8.6\int K\,dz \tag{11.142}$$

In all measurements made by riometers, one is interested only in the received power *relative* to that if there were no ionospheric absorption. Denoting the power without absorption as P_0 $(= kT)$, we have

$$P = \frac{P_0}{2\pi}\int_0^{2\pi}\int_0^{\pi/2} G(\theta, \phi)\exp\left(-\frac{A_0}{4.3}\sec\theta\right)\sin\theta\,d\theta\,d\phi \tag{11.143}$$

Unfortunately, $G(\theta, \phi)$ is typically a rather "messy" function and is difficult to determine. The best technique available is to use a helicopter-borne receiver and measure the received power from the antenna when used for transmission. However, if one is satisfied with values of K which are correct to within 15 percent, the measured absorption A_0, given by

$$P = P_0\exp\left(-\frac{A_0}{4.3}\right)$$

can be used to obtain an approximate $A_0 \approx (0.7 \pm 0.1)A$.

A riometer operated at a single frequency can tell us very little about the electron number density in the absorbing region (D or F_2). However, if we make absorption measurements at several frequencies, it is possible to unfold the electron concentration provided we have a profile of collision frequency. According to Eq. (11.85a) the absorption coefficient has the following

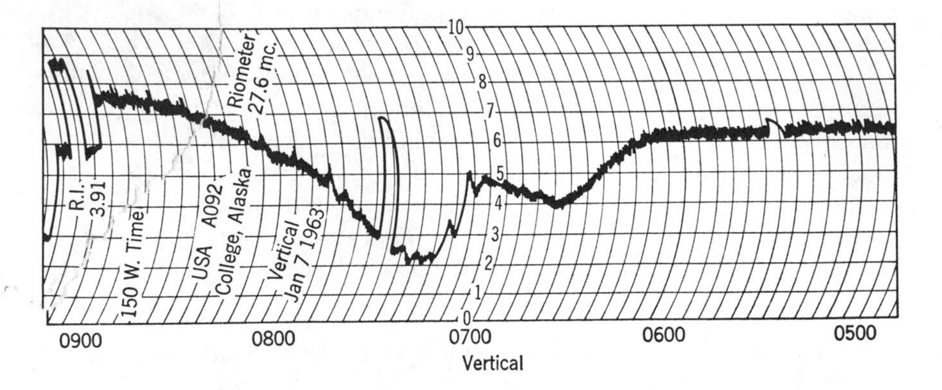

Figure 11.11 Radio wave absorption at 27.6 MHz measured at College, Alaska. [After Parthasarathy and Hessler, *J. Geophys. Res.* **69,** 2867 (1964); reprinted by permission of the American Geophysical Union.]

dependence upon the frequency ω, electron number density N_e, and collision frequency ν

$$K = \frac{N_e e^2 \nu}{2mc(\omega^2 + \nu^2)\epsilon_0} \tag{11.144}$$

where e is the electronic charge, m the electronic mass, c the velocity of light and ϵ_0 the susceptibility of free space. Measurements of P thus lead, by a rather circuitous route, to knowledge of N as a function of height.

A typical riometer record is shown in Figure 11.11.

11.7.3 Thomson Scatter Radar

The Thomson scatter radar technique utilizes the incoherent scattering (see Section 11.5.3) of high power radar waves by ionospheric electrons. As we mentioned earlier, the method can be used to determine ion and electron temperature as well as electron number density and ion identity as a function of altitude, altitude being obtained from the time lapse between transmission and receipt of echo. A number of Thomson scatter radars are now in service in various parts of the world, two of the best known are the facilities at Millstone Hill in Massachusetts, and the Arecibo radar in Puerto Rico.

The analysis of the radar data is an extremely complicated affair which we shall not pursue here because of space limitations. Figure 9.14c shows ion and electron temperature profiles determined by Thomson scatter measurements. As we stated in Chapter 9, these results do not agree with measurements made by rocket and satellite probes carried out under similar ionospheric conditions. The reasons for the discrepancy are not known at present.

11.7.4 Cross Modulation

The pulse interaction technique which was originated by Fejer[16] has been employed by many investigators (in various modifications) to measure electron number density, collision frequency, and energy transfer coefficient G (see Section 11.5.4 and the Appendix) in the D-region. Basically, the method consists of transmitting "wanted" (W) and "disturbing" (D) pulses, and observing the interaction of the pulses by means of a suitable receiver. Figure 11.12 shows two schemes for conducting the experiment. In Type I,† the pulse D precedes the pulse W and the interaction is measured at several time delays between the trailing edge of D and the leading edge of W. This technique has the advantage that observation of the decay of the collision frequency perturbation yields the parameter G; however, there is no height determination. On the other hand, Type II in which the pulse W precedes

† The mode of classification given here is that of R. A. Smith, University of New England, Australia.

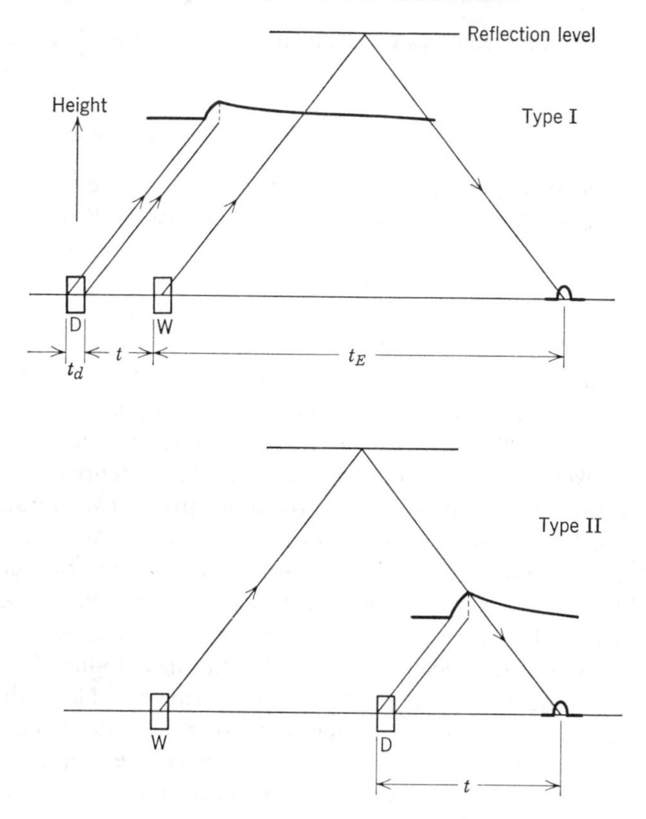

Figure 11.12 Types I and II pulse wave interaction experiments.

pulse D, permits one to determine the interaction height h, but not the parameter G. Type III combines Types I and II and thus yields both G and h. Electron number density and collision frequency profiles between 65 and 85 km altitude are obtainable in all three types by application of the theory outlined in Section 11.5.4.

11.7.5 Partial Reflection

The partial reflection of medium frequency radio waves by D-region inhomogeneities was discussed in Section 11.5.5. This technique has been applied with great success by Belrose and coworkers.[11] Unfortunately, the method is strongly dependent upon the collision frequency profile, which must be assumed in order to deduce electron number densities from the data.

Belrose's apparatus, located at Ottawa, Canada, consists of a transmitting section and receviing section operating at 2.66 or 6.275 MHz. The transmitter, which operates at 1000 kw, is coupled to a square antenna array, opposite sides of which are fed in phase, and perpendicular sides in quadrature so as to provide circular polarization. The receiver is coupled to a 128-dipole circularly polarized antenna array in order to resolve the ordinary and extraordinary echoes. Figure 11.13 shows typical profiles of measured A_x/A_0 and computed R_x/R_0.

11.7.6 Faraday Rotation

The techniques for measuring ionospheric parameters which were discussed in the preceding sections employed ground based instrumentation (with the exception of the satellite-borne sounder) to measure characteristics of the earth's ionosphere. In this and the following sections we shall discuss techniques for use with rockets and satellites, one method has been successfully employed to investigate the ionospheres of Mars and Venus.

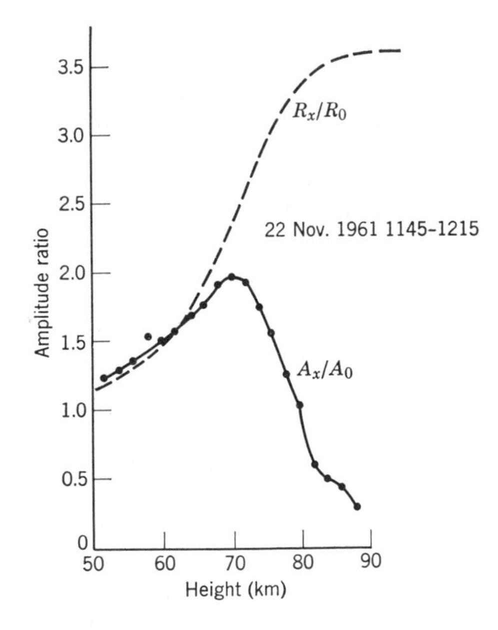

Figure 11.13 Function R_x/R_0 (calculated) and A_x/A_0 (measured) with height for 2.66 MHz for a quiet day, November 22, 1961. The amplitude ratios are averages for 1145–1215 local time. [After Belrose and Burke, *J. Geophys. Res.* **69**, 2799 (1964); reprinted by permission of the American Geophysical Union.]

Of these, the application of Faraday rotation (see Section 11.3) has been widely employed. The wave transmitted from the rocket or satellite is linearly polarized; the elliptically polarized ordinary and extraordinary waves propagate independently. The E_{ex} and E_{ord} fields are added within the receiver, and the change in angle ϕ [see Eq. (11.66)] measured. Because the indices of refraction change with altitude, ϕ must be expressed as an integral

$$\phi = \frac{\omega}{2c} \int (n_{\text{ord}} - n_{\text{ex}})\, dx$$

$$\approx \frac{\omega}{2c} \int \frac{XY}{1 - Y^2}\, dx \tag{11.145}$$

where in general the path is at an angle to the vertical. The electron number density which is contained in X is readily unfolded from the measured changes in ϕ.

11.7.7 Differential Doppler-Effect

If a rocket or satellite is transmitting to or receiving a radio signal from ground, the motion of the satellite causes the observed frequency, ω_D, to be Doppler shifted.

$$\omega_D = \omega - k \frac{d}{dt} \int_{\mathbf{r}_S}^{\mathbf{r}_T} \mathbf{n} \cdot d\mathbf{r}$$

$$= \omega - k\left[\mathbf{n}_T \cdot \frac{d\mathbf{r}_T}{dt} - \mathbf{n}_S \frac{d\mathbf{r}_S}{dt} + \int_{\mathbf{r}_S}^{\mathbf{r}_T} \frac{d}{dt} \mathbf{n} \cdot d\mathbf{r}\right] \tag{11.146}$$

Here, ω is the true frequency, k is the wave number in free space, $\mathbf{n}$ is the refractive index *vector* (in a magnetoionic medium, $\mathbf{n}$ and $\mathbf{k}$ will not usually have the same direction), $\mathbf{r}_S$ is the position of the transmitter, and $\mathbf{r}_T$ the position of the receiver. The first term in brackets on the right hand side of Eq. (11.146) represents the motion of the receiver, the second term the motion of the transmitter, while the third term represents variations in the ray path by virtue of the motion or of temporal changes in the medium.

One of the most successful techniques is the differential Doppler method in which the Doppler frequency is observed on two coherent frequencies, one of which is a harmonic of the other. After reception the lower frequency is multiplied by the frequency ratio so that the difference in the Doppler frequencies would be zero if no dispersive medium were present. Then

$$\Delta\omega_D = \omega_{D_2} - \omega_{D_1} = -k \frac{d}{dt} \int_{\mathbf{r}_S}^{\mathbf{r}_T} (\mathbf{n}_2 - \mathbf{n}_1) \cdot d\mathbf{r} \tag{11.147}$$

where ω_{D_1} and ω_{D_2} are the Doppler frequencies after multiplication of the lower one.

The determination of the electron density profile from observations of $\Delta\omega_D$ (the Doppler "beat" frequency) is usually quite difficult even when simplifying assumptions (e.g., neglect of electron collisions and the geomagnetic field) are made. One ordinarily uses a ray-tracing technique to compute ray paths by iterative means, starting at the lowest level at which the ionosphere has a detectable effect and "bootstrapping" to higher and higher altitudes.[17]

11.7.8 Bistatic Radar

The bistatic radar technique, which is closely related to the differential Doppler method, has been successfully employed to investigate the atmosphere and ionosphere of Mars and Venus[18] (see Sections 3.7.3, 3.8.3, 8.4, and 9.8.1) for discussion of results.

For a more detailed discussion the reader is referred to the original papers by Fjeldbo and Eshleman;[18] however, the following outline should suffice to acquaint him with the principles involved. As the signals from the transmitter successively pass through denser and denser parts of the atmosphere and ionosphere, they are refracted with resultant phase shifts in the received waves given by

$$\phi = \lambda^{-1} \int_0^{\Delta 4} (n - 1)\, dy' \tag{11.148}$$

where λ is the free space wavelength and integration is taken through the planetary atmosphere. Assuming a specific shape for the atmosphere (and ionosphere) such as a spherically symmetric one, it is quite straightforward to compute the phase shifts from Eq. (11.148) if the atmospheric densities, electron number densities, and limb distances are known. Since the phase shifts are in fact the known quantities and the indices of refraction the desired ones, it is necessary to invert the equation. Then neutral particle and electron number densities are easily obtained from the computed indices of refraction.

11.7.9 Differential Absorption

The differential absorption of a pair of radio waves, each at a different frequency, has been successfully employed to measure electron number density or collision frequency in the D-region. Using the form of the nondeviative absorption coefficient given by Eq. (11.85a), we find that the wave amplitudes E are related in the Q.L. approximation to the electron number density by

$$\ln \frac{E_1}{E_2} = \frac{e^2}{2mc\epsilon_0} \int_p N\nu\{[\nu^2 + (\omega_2 \pm \omega_H)^2]^{-1} - [\nu^2 + (\omega_1 \pm \omega_H)^2]^{-1}\}\, dz \tag{11.150}$$

where N is the electron number density and the other symbols have the same meanings as earlier in the chapter; the path of integration P is the ray path. In order to obtain the electron number density profile by this means, one must know the profile of the collision frequency ν. Conversely, independent measurement of N would enable one to determine ν by the differential absorption technique. The method is particularly useful for measuring D-region electron number density during a disturbance such as a PCA event or an SID.

By using a single frequency not too far removed from the gyro frequency $\omega_{H'}$, one can measure N by observing the differential absorption of the extraordinary and ordinary components of a plane polarized wave. The amplitudes are then given by

$$\ln\left(\frac{E_{\text{ex}}}{E_{\text{ord}}}\right) = \frac{e^2}{2mc\epsilon_0}\int N\nu\{[\nu^2 + (\omega - \omega_H)^2]^{-1} - [\nu^2 + (\omega + \omega_H)^2]^{-1}\}\, dz \qquad (11.151)$$

At altitudes where $\omega - \omega_H \sim \nu$, one must use the Sen-Wyller formula[5] (see Section 11.5 and Problem 11.17) rather than Eq. (11.151). Measurement of differential absorption of the extraordinary and ordinary waves is most useful in the quiet D-layer.

11.7.10 Langmuir Probe

The Langmuir probe is a direct current instrument, and thus its discussion does not strictly belong in the present chapter. We include it here in order to group together all of the ionospheric measuring systems.

This probe, invented many years ago by Irving Langmuir, determines the current-voltage characteristic I of a weak direct current discharge in a plasma. By operating the charged particle collector of area A a few volts negative, it collects all positive ions, but repels electrons. The probe current thus provides a direct measure of the ion number density. If the collector potential V is changed to several volts positive, all electrons are collected, permitting one to determine (in principle) the electron number density N_e. Actually, for a space probe surrounded by an ion sheath, the limitations of probe theory do not permit one to determine reliable values of N_e.

When the collector potential V is very slightly negative (i.e., retarding for electrons), the probe current I can be expressed as

$$I = AeN_e\sqrt{\frac{kT}{2\pi m}}\exp\left(\frac{eV}{kT}\right); \qquad V < 0 \qquad (11.152)$$

where e and m are the electronic charge and mass, k is Boltzmann's constant, and T is the electron temperature. We have assumed that the electron velocity distribution is Maxwellian (see Section 2.3.1). One then obtains T from the slope of the V-I characteristic:

$$T = \frac{e}{k} I \frac{\partial V}{\partial I} \tag{11.153}$$

With this very brief discussion of the theory of the Langmuir probe, we bring to a close our treatment of the science of aeronomy. We sincerely hope that it has been of assistance to the reader, and that he has a better grasp of the field than when he began to read page one.

PROBLEMS

11.1. Construct electric circuit analogues of the conduction phenomena discussed in Sections 11.1.2–3; compare analogous quantities.

11.2. Find a transformation which diagonalizes the conductivity tensors whose elements are given by Eq. (11.11).

11.3. What are the "eigenmodes" or "eigenvectors" of the conductivity tensor in Problem 2; i.e., what are the components of $\mathbf{E}$ and $\mathbf{j}$ when $\underset{\sim}{\boldsymbol{\sigma}}$ is in diagonal form?

11.4. Obtain Eqs. (11.40b) from Eq. (11.39). Find an expression for the group velocity and discuss the propagation characteristics of this mode (e.g., is it dispersive?).

11.5. Obtain Eq. (11.52) from Eq. (11.51).

11.6. Find an approximate expression for the phase speed of the longitudinal plasma oscillations as the wave frequency tends to approach zero. Assume that the ion and electron gas temperatures are equal.

11.7. Find the low frequency limit of the index of refraction of the extraordinary mode as given by Eq. (11.56a).

11.8. Discuss the reasons for the coupling of the transverse and longitudinal modes in a hot magnetoplasma.

11.9. Prove relations (11.59) and (11.60).

11.10. Find an expression for the instantaneous Poynting vector of an electromagnetic wave propagating in a magnetoplasma in the direction of the magnetic field. Illustrate the behavior of $\boldsymbol{\Pi}$ with the aid of a diagram and show what happens to it when it is averaged over a cycle.

11.11. Prove that the phase velocity u and group velocity v_g for an electromagnetic wave propagating in an isotropic collisionless plasma are related by $uv_g = c^2$ [Eq. (11.63)]. Prove Eq. (11.64).

11.12. Derive the "WKB solution" $E(z) = (A/n^{1/2}) \exp [ik \int n\,dz - \omega t]$ for an electromagnetic wave in a slowly varying plasma (i.e., the fractional change in the index of refraction, $(1/n)(dn/dz)$, is much less than unity). *Hint:* Assume a solution $E \propto e^{i\phi(z)}$ and expand $\phi(z)$ in inverse powers of the wave number k.

11.13. Compute the Alfvén velocity for the ionosphere. Assume the electron number density is 10^5 cm^{-3}.

11.14. Derive the analogue of Eq. (11.A17) for a magnetoplasma.

11.15. Derive Eqs. (11.A3). *Hint:* Use the identity $\partial/\partial v_i = [(\partial v/\partial v_i)(\partial/\partial v)]$.

11.16. Derive Eqs. (11.85) for absorption of electromagnetic waves in a plasma in the quasi-longitudinal approximation.

11.17. Obtain the Sen-Wyller formula for the index of refraction of a magnetoplasma in the quasi-longitudinal approximation. Obtain an expression for the absorption coefficient in the high frequency limit. How does this compare with the Appleton-Hartree result?

11.18. Compute and plot the real and imaginary parts of the index of refraction for the case of longitudinal propagation. Assume that $Y = 0.5$, $Z = 0.1$.

11.19. This is a continuation of Problem 10.12. Obtain in closed form an expression for the decrease in intensity of HF cosmic noise due to absorption in the D-layer. Assume the following: wave frequency $\omega \gg$ collision frequency ν in the absorbing layer; that $\nu = \nu_0 \exp [-z/H]$ where H is scale height; that the attachment process is three-body and the molecular oxygen number density varies as $\exp [-z/H]$, and that the detachment coefficient γ vanishes.

11.20. Find expressions for the reflection coefficients ${}_{\parallel}R_{\perp}$ when the electric field vector makes an angle θ with the plane of incidence of a wave incident on an anisotropic ionosphere.

11.21. Using the methods developed in Section 11.2, derive Eq. (11.132).

11.22. Find the errors in this book and send them to the authors.

REFERENCES

1. J. A. Ratcliffe, *Magnetoionic Theory*, Cambridge University Press, Cambridge, 1959.
2. E. V. Appleton, *Proc. Phys. Soc.*, **37,** 16D (1925).
3. D. R. Hartree, *Proc. Cambridge Phil. Soc.*, **25,** 97 (1929).
4. A. J. Dessler and F. C. Michel, *Particles and Fields in Space*, John Wiley & Sons, New York, to be published.
5. H. K. Sen and A. A. Wyller, *Phys. Rev. Lett.*, **4,** 355 (1960).
6. H. G. Booker, *Proc. Roy. Soc.* **A155,** 235 (1936).
7. W. E. Gordon, *Proc. Inst. Radio Engrs.* **46,** 1824 (1958).

8. J. P. Dougherty and D. T. Farley, *Proc. Roy. Soc.* **A259,** 79 (1960).
9. V. A. Bailey and D. F. Martyn, *Nature*, **133,** 218 (1934).
10. F. F. Gardner and J. L. Pawsey, *J. Atmos. Terr. Phys.* **3,** 321 (1953).
11. e.g., J. S. Belrose and M. J. Burke, *J. Geophys. Res.* **69,** 2799 (1964).
12. K. G. Budden, in *Physics of the Ionosphere*, The Physical Society, London, 1955.
13. *See*, for example, G. L. Nelms in *Space Research IV*, Ed. by P. Müller, North-Holland Pub. Co., Amsterdam, 1964.
14. C. G. Little and H. Leinbach, *Proc. IRE* **47,** 315 (1959).
15. R. Parthasarathy, G. M. Lerfald, and C. G. Little, *J. Geophys. Res.* **68,** 3581 (1963); G. M. Lerfald, C. G. Little, and R. Parthasarathy, *J. Geophys. Res.* **69,** 2857 (1964).
16. J. A. Fejer, *J. Atmos. Terr. Phys.* **7,** 322 (1955).
17. *See*, e.g., R. C. Whitten, I. G. Poppoff, R. S. Edmonds, and W. W. Berning, *J. Geophys. Res.* **70,** 1737 (1965).
18. G. Fjeldbo and V. R. Eshleman, *J. Geophys. Res.*, **70,** 3217 (1965).

GENERAL REFERENCES

K. G. Budden, *Radio-Waves in the Ionosphere*, Cambridge University Press, Cambridge, 1961.

K. G. Budden, *The Wave Guide Mode Theory of Wave Propagation*, Prentice-Hall, Englewood Cliffs, N.J., 1961.

J. A. Ratcliffe, *Magnetoionic Theory*, Cambridge University Press, Cambridge, 1959.

K. Rawer and K. Suchy, Radio Observations of the Ionosphere, in *Handbuch der Physik*, Vol. XLIX/2 (Geophysik II/2), Ed. by S. Flügge, Springer-Verlag; Berlin, Heidelberg, New York, 1967.

These four are very good general references on radio wave propagation; much of the theory involved has been developed by these authors.

J. L. Delcroix, *Introduction to the Theory of Ionized Gases*, Interscience (John Wiley), New York, 1960.

J. F. Denisse and J. L. Delcroix, *Plasma Waves*, Interscience (John Wiley) New York, 1963.

E. H. Holt and R. E. Haskell, *Foundations of Plasma Dynamics*, Macmillan, New York, 1965.

I. P. Shkarovsky, T. W. Johnston, and M. P. Bachynski, *The Particle Kinetics of Plasmas*, Addison Wesley, Reading, Mass., 1966.

These are very good references for the general principles of plasma physics. The development of the subject by Holt and Haskell was in fact followed quite closely in several instances in the present chapter.

APPENDIX

The velocities of the molecules of a gas in equilibrium are distributed according to the Maxwellian law (see Section 2.3):

$$f(v)\,dv = 4\pi N\left(\frac{m}{2\pi kT}\right)^{\frac{3}{2}} v^2 \exp\left(-\frac{mv^2}{2kT}\right) dv \tag{11.A1}$$

which is equal to the number of molecules per unit volume with speeds between v and $v + dv$; here N is the molecular number density, m is the molecular mass, and T is the absolute temperature. Since a plasma is in fact a gas, we should expect a similar result, modified by the effects of collisions of electrons with neutral particles and ions, as well as by applied electric and magnetic fields. Just as with a neutral gas, the Boltzmann equation

$$\frac{\partial f}{\partial t} + \mathbf{v}\cdot\nabla f - \frac{e}{m}(\mathbf{E}\cdot\nabla_v f + \mathbf{v}\times\mathbf{B}\cdot\nabla_v f) = \left(\frac{\partial f}{\partial t}\right)_{\text{coll}} \tag{11.A2}$$

governs the spatial and temporal form of f. The operators ∇ and ∇_v are the gradient operators in configuration and velocity space, respectively, and $(\partial f/\partial t)_{\text{coll}}$ is the collision integral which will be discussed later.

Suppose that the electric and magnetic fields are sufficiently weak that f is perturbed only slightly from its equilibrium value f^0; it can be approximated by the expansion

$$f = f^\circ + \hat{\mathbf{v}}\cdot\mathbf{f}'(\mathbf{v}); \qquad |\mathbf{f}'| \ll f^\circ \tag{2.59}$$

Upon substitution, the various terms on the left hand side of Eq. (11.A2) become

$$\frac{\partial f}{\partial t} = \frac{\partial f^\circ}{\partial t} + \frac{\mathbf{v}}{v}\cdot\frac{\partial \mathbf{f}'}{\partial t} \tag{11.A3a}$$

$$\mathbf{v}\cdot\nabla f = \mathbf{v}\cdot\nabla f^\circ + \sum_{ij}\frac{v_j v_i}{v}\frac{\partial f'_j}{\partial x_i} \tag{11.A3b}$$

$$\mathbf{E}\cdot\nabla_v f = \frac{\mathbf{E}\cdot\mathbf{v}}{v}\frac{\partial f^\circ}{\partial v} + \frac{\mathbf{E}\cdot\mathbf{v}}{v}\nabla_v\cdot\left(\frac{\mathbf{f}'}{v}\right) + \mathbf{E}\cdot\frac{\mathbf{f}'}{v} \tag{11.A3c}$$

$$\mathbf{v}\times\mathbf{B}\cdot\nabla_v f = (\mathbf{v}\times\mathbf{B})\frac{\mathbf{f}'}{v} \tag{11.A3d}$$

yielding the following form of the Boltzmann equation

$$\frac{\partial f^\circ}{\partial t} + \frac{\mathbf{v}}{v} \cdot \frac{\partial \mathbf{f}'}{\partial t} + \mathbf{v} \cdot \nabla f^\circ + \sum_{ij} \frac{v_i v_j}{v} \frac{\partial f'_j}{\partial x_i}$$

$$- \frac{e}{m}\left[\mathbf{E} \cdot \mathbf{v} \frac{\partial f^\circ}{\partial v} + \mathbf{E} \cdot \mathbf{v} \frac{\mathbf{v}}{v} \cdot \frac{\partial}{\partial v}\left(\frac{\mathbf{f}'}{v}\right) + \mathbf{E} \cdot \frac{\mathbf{f}'}{v} + \mathbf{v} \times \mathbf{B} \cdot \frac{\mathbf{f}'}{v}\right] = \left(\frac{\partial f}{\partial t}\right)_{\text{coll}} \quad (11.\text{A}4)$$

We now integrate both sides of Eq. (11.A4) over a complete solid angle whose element is $d\Omega = \sin\theta\, d\theta\, d\phi$. After dividing by 4π we obtain†

$$\frac{\partial f^\circ}{\partial t} + \frac{v}{3} \operatorname{div} \mathbf{f}' - \frac{e}{3m} \frac{\mathbf{E}}{v^2} \cdot \nabla_v (v^2 \mathbf{f}') = \frac{1}{4\pi} \int_0^{2\pi} \int_0^{\pi} \left(\frac{\partial f}{\partial t}\right)_{\text{coll}} \sin\theta\, d\theta\, d\phi \quad (11.\text{A}5)$$

Eq. (11.A4) is next multiplied by v_l and again integrated over a full solid angle. The result is, after dividing by $4\pi v/3$

$$\frac{\partial f'_l}{\partial t} + v \frac{\partial f^\circ}{\partial x_l} - \frac{e}{m}\left[E_l \frac{\partial f^\circ}{\partial v} + (\mathbf{B} \times \mathbf{f}')_l\right]$$

$$= \frac{3}{4\pi v} \int_0^{\pi} \int_0^{2\pi} v_l \left(\frac{\partial f}{\partial t}\right)_{\text{coll}} \sin\theta\, d\theta\, d\phi \quad (11.\text{A}6)$$

It remains to consider the collision term which is of the form (see section 2.3)

$$\left(\frac{\partial f}{\partial t}\right)_{\text{coll}} = \int_0^{\infty} \int_{\Omega} (\bar{f} \bar{f}^N - f f^N) v_{\text{rel}}\, Q(\theta)\, d\Omega\, dc^N \quad (11.\text{A}7)$$

where f, f^N and $\bar{f}$, $\bar{f}^N$ are the distribution functions for the electrons and neutral particles before and after collision, respectively; $Q(\theta)$ is the momentum transfer cross section for scattering through angle θ, c is the molecular speed and v_{rel} is the speed of an electron relative to a neutral particle. Substitution of the expansion for f given by Eq. (2.59) into $(\bar{f}\bar{f}^N - ff^N)$, yields

$$(\bar{f} \bar{f}^N - f f^N) = \bar{f}^\circ \bar{f}^N - f^\circ f^N + \frac{\bar{\mathbf{v}}}{v} \cdot \mathbf{f}' f^N - \frac{\mathbf{v}}{v} \cdot \mathbf{f}' f^N \quad (11.\text{A}8)$$

If it is assumed that the neutral particles are very heavy, we have $f^\circ = f^N$ and $\bar{v} \approx v$. In the notation used in Figure 11.A, Eq. (11.A8) becomes

$$(\bar{f} \bar{f}^N - f f^N) = f' f^N (\cos \bar{\Theta} - \cos \Theta) \quad (11.\text{A}9)$$

† The following integrals are employed in the computation:

$$\int_0^{2\pi} \int_0^{\pi} \sin\theta\, d\theta\, d\phi = 4\pi; \qquad \int_0^{2\pi} \int_0^{\pi} v_i \sin\theta\, d\theta\, d\phi = 0$$

$$\int_0^{2\pi} \int_0^{\pi} v_i v_j \sin\theta\, d\theta\, d\phi = \frac{4\pi}{3} \delta_{ij} v^2; \qquad \int_2^{2\pi} \int_0^{\pi} v_i v_j v_k \sin\theta\, d\theta\, d\phi = 0$$

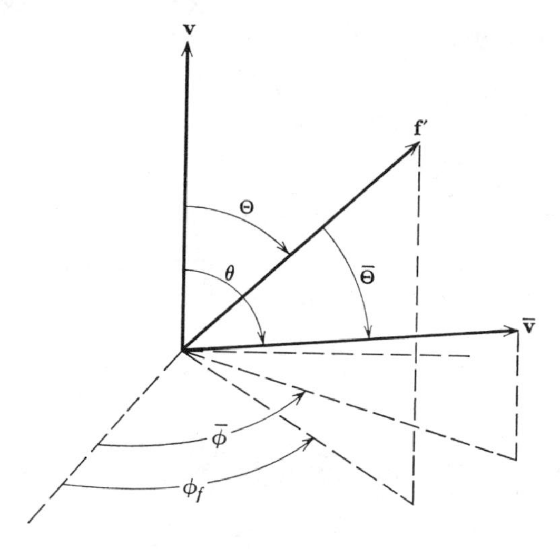

Figure 11.A Coordinate system used to describe the perturbed part of the distribution function **f**′.

The identity

$$\cos \bar{\Theta} = \cos \Theta \cos \theta + \sin \Theta \sin \theta \cos \phi$$

is used in the collision integral

$$\left(\frac{\partial f}{\partial t}\right)_{\text{coll}} = \int_0^\infty \int_0^{2\pi} \int_0^\pi f^N f' \cos \Theta(\cos \theta - 1) v_{\text{rel}} Q(\theta) \sin \theta \, d\theta \, d\phi \, dc^N$$

$$+ \int_0^\infty \int_0^{2\pi} \int_0^\pi f^N f' \sin \Theta \sin \theta Q(\theta) \sin \theta \, d\theta \cos \phi \, d\phi \, dc^N \quad (11.\text{A}10)$$

of which the last term vanishes because

$$\int_0^{2\pi} \cos \phi \, d\phi = 0$$

Hence

$$\left(\frac{\partial f}{\partial t}\right)_{\text{coll}} = -2\pi \int_0^\infty \int_0^\pi f^N \frac{\mathbf{v}}{v} \cdot \mathbf{f}'(1 - \cos \theta) v_{\text{rel}} Q(\theta) \sin \theta \, d\theta \, dc^N \quad (11.\text{A}11)$$

For the case of infinitely heavy neutral particles, moreover, $v_{\text{rel}} = v$ and

$$N_N = \int_0^\infty f^N \, dc^N, \quad (11.\text{A}12)$$

the neutral particle number density. The collision integral in Eq. (11.A6) becomes

$$\left(\frac{\partial f_l}{\partial t}\right)_{\text{coll}} = -2\pi N_N v f'_l \int_0^\pi (1 - \cos\theta) Q(\theta) \sin\theta \, d\theta = -\nu_{eN} f'_l \quad (11.\text{A}13)$$

where ν_{en} is the electron-neutral particle collision frequency. Using this form for the collision integral, one can now write the equation for f'_l [Eq. (11.A6)] as

$$\frac{\partial f'_l}{\partial t} + v \frac{\partial f^\circ}{\partial x_l} - \frac{e}{m}\left[E_l \frac{\partial f^\circ}{\partial v} + (\mathbf{B} \times \mathbf{f}')_l\right] = -\nu_{\text{en}} f'_l \quad (11.\text{A}14)$$

As an example of the application of this equation, consider the conductivity of a homogeneous isotropic plasma in the presence of a time-varying electric field $\mathbf{E}_0 e^{-i\omega t}$. The equation for the perturbed distribution function is

$$(\omega + i\nu) f'_l = \frac{ie}{m} E_l \frac{\partial f^\circ}{\partial v} \quad (11.\text{A}15)$$

Since f° is isotropic, the current density is

$$\begin{aligned} j_l &= -Ne \sum_k \iint v_l \frac{v_k}{v} f'_k \, d\Omega v^2 \, dv \\ &= -i \frac{Ne^2}{m} \sum_k E_k \iint \frac{v_l v_k}{\omega + i\nu} \frac{\partial f^\circ}{\partial v} \, d\Omega v \, dv \end{aligned} \quad (11.\text{A}16)$$

From Ohm's Law we see that the conductivity is

$$\sigma = -\frac{4\pi}{3} i\epsilon_0 \omega_p^2 \left[\omega \int_0^\infty \frac{v^3}{\omega^2 + \nu^2} \frac{\partial f^\circ}{\partial v} \, dv - i \int_0^\infty \frac{\nu v^3}{\omega^2 + \nu^2} \frac{\partial f^\circ}{\partial v} \, dv\right] \quad (11.\text{A}17)$$

If f° is Maxwellian, i.e.,

$$f^\circ = N \left(\frac{m}{2\pi k_B T}\right)^{3/2} \exp\left(-\frac{mv^2}{2k_B T}\right) \quad (11.\text{A}18)$$

the conductivity is obviously

$$\begin{aligned} \sigma &= \frac{8}{3\pi^{1/2}} \epsilon_0 \omega_p^2 \left(\frac{m}{2k_B T}\right)^{5/2} \\ &\times \left[\int_0^\infty \frac{\nu v^4}{\omega^2 + \nu^2} e^{-(mv^2/2k_B T)} \, dv + i\omega \int_0^\infty \frac{v^4}{\omega^2 + \nu^2} e^{-(mv^2/2k_B T)} \, dv\right] \end{aligned} \quad (11.\text{A}19)$$

The result given in Eq. (11.A19) was used to establish the equation for the index of refraction in a nitrogen atmosphere [Eq. (11.89)].

It remains to discuss the case where the neutral particles are not infinitely heavy, and some of the electrons' momentum is transferred to a neutral particle at each electron-neutral collision. Under these conditions the zeroth order electron distribution function is non-Maxwellian. Momentum can also be transferred by inelastic as well as by elastic collisions, as for example, by rotational excitation of nitrogen molecules. This mechanism is of particular importance in the *D*-region of the ionosphere where it is responsible for radio wave interaction (Luxembourg effect)—*see* Section 11.5.4.

It can be shown by some rather tedious but straightforward arguments that when momentum loss by electrons is included, Boltzmann's equation for the zeroth order term in the distribution function expansion is

$$\frac{\partial f^\circ}{\partial t} + \frac{v}{3}\operatorname{div}\mathbf{f}' - \frac{e}{3mv^2}\mathbf{E}\cdot\frac{\partial}{\partial v}(v^2\mathbf{f}') = \left(\frac{\partial f^\circ}{\partial t}\right)_{\text{coll}} \tag{11.A20}$$

in the absence of magnetic fields. Substituting the form of $\mathbf{f}'$ given by Eq. (11.A15) into Eq. (11.A17), and assuming a steady state, we obtain

$$\frac{ie^2E^2}{3m_e^2v^2}\frac{\partial}{\partial v}\left(\frac{v^2\,\partial f^\circ/\partial v}{\omega + i\nu}\right) + \frac{1}{2v^2}\frac{\partial}{\partial v}\left(G\nu v^3 f^\circ + G\nu v^2\frac{k_BT_N}{me}\frac{\partial f^\circ}{\partial v}\right) = 0 \tag{11.A21}$$

where G is the energy transfer coefficient ($G = 2m_e/m_N$ for elastic collisions). When a time average of this equation is taken, the imaginary terms vanish and we are left with

$$\frac{e^2\nu E_{\text{rms}}^2}{3m_e^2v^2}\frac{\partial}{\partial v}\left[\frac{v^2}{\omega^2 + \nu^2}\left(\frac{\partial f^\circ}{\partial v}\right)\right] + \frac{1}{v^2}\frac{\partial}{\partial v}\left[G\nu v^3 f^\circ + G\nu v^2\frac{k_BT_N}{m_e}\left(\frac{\partial f^\circ}{\partial v}\right)\right] = 0 \tag{11.A22}$$

which integrates to

$$f^\circ = C\exp\left[-\int_0^\infty \frac{m_e}{k_BT_N}\frac{v\,dv}{(1 + eE^2/3Gm_ek_BT_N(\nu^2 + \omega^2))}\right] \tag{11.A23}$$

The velocity distribution represented by Eq. (11.A23) is sometimes called a *Margenau* distribution.

List of Symbols

a	acoustic velocity (plasmas)
$\mathbf{A} \cdot \mathbf{B}$	scalar product of vectors A and B
$\mathbf{A} \times \mathbf{B}$	vector product of vectors A and B
$A_{n \to n_1}$	Einstein spontaneous emission coefficient
A	exchange coefficients (fluids); Alfvén velocity; mass number
b	impact parameter
$\mathbf{B}$	magnetic field induction
$B_{n \to n_1}$	Einstein absorption coefficient
c	velocity of light; acoustic velocity (fluids) ratio of electron to neutral particle temperature
C_D	ballistic coefficient
C_p	specific heat capacity at constant pressure
C_v	specific heat capacity at constant volume
D	molecular diffusion coefficient
$\mathbf{D}$	electric field displacement
$đ$	inexact differential
D/Dt	total derivative with respect to an inertial reference frame
D_{ij}	elements of structure tensor (eddy diffusion)
e	electron charge; electron
$\hat{\mathbf{e}}$	symbol for a unit vector
$e\mathbf{r}$	electric dipole operator
E	energy
$\mathbf{E}$	electric field intensity
(dE/dx)	stopping power for charged particles
$e\langle n' \vert \mathbf{r} \vert n \rangle$	electric dipole operator connecting states n and n'
f	distribution function (kinetic theory of gases)
$f(\Theta)$	scattering amplitude
$\mathbf{f}$	frictional (viscous) deceleration (fluids)
F	free energy of activation
$\mathbf{F}$	force
F_λ	spectral irradiance (energy per unit area-unit time-wavelength interval)
$\mathbf{F}_r$	viscous force on a single charged particle
$F(\theta, Z)$	Van Rhijn function

$\mathbf{g}$	gravitational acceleration
g_n	statistical weight of state n
G	energy transfer coefficient
h	Debye shielding length; Planck's constant, equivalent depth; height
$\hbar$	Planck's constant/2π
H	enthalpy; Hamiltonian operator; scale height
$\mathbf{H}$	magnetic field intensity
i	imaginary number ($\sqrt{-1}$)
I	total current; magnetic dip angle; intensity of photons from a column of emitters (photons per unit area-unit time-unit solid angle); particle flux per unit area
I_λ	intensity (spectral irradiance per unit solid angle)
dI/dE	particle differential flux per unit area
IP	ionization potential
j	total angular momentum quantum number
$\mathbf{j}$	current density
J	total angular momentum operator; source function
$\mathbf{J}$	linear current density
$J_E(X, E, h\nu)\, d(h\nu)\, dE$	bremsstrahlung production rate (per unit volume-unit time) at photon energy $h\nu$, electron energy E
k	wave number; Boltzmann's constant; rate coefficient
$\mathbf{k}$	wave propagation vector
k_B	Boltzmann's constant
K	kinetic energy; thermal conductivity; number of excitations per recombination; magnetic activity index; dielectric susceptibility scalar; absorption coefficient; thermal conductivity
K_h	coefficient of eddy diffusion (heat transport)
K_m	coefficient of eddy diffusion (mass transport)
K_p	world-wide weighted average of magnetic activity index
$\mathbf{K}$	dielectric susceptibility tensor
l	orbital angular momentum quantum number; mean free path; mixing length; charge rearrangement coefficient for negative ions
L	orbital angular momentum operator; particle loss term; thermal energy loss term
m	mass of a particle
$\mathbf{M}$	magnetization
n	number density of a species; number of degrees of freedom; principal quantum number; phase index of refraction
n_g	group index of refraction
N	number density of a species; Brunt-Väisälä frequency

N_c	number of particles in a column of unit cross section
N_T	total number of particles in a system
p	pressure
P	momentum magnitude; power; production rate per unit volume of excited species
$\mathbf{P}$	electric polarization; momentum
q	total charge; ionization rate per unit volume; vertical component of refractive index
q_T	thermal energy deposition rate per unit volume
Q	cross section; heat energy
Q_a	absorption cross section
Q_{air}	absorption cross section for X-rays in air
Q_i	ionization cross section
Q_m	momentum transfer cross section
$Q(\theta)$	differential scattering cross section
Q	total cross section for radiation of bremsstrahlung
$Q_T(h\nu)$	total absorption cross section for photons of energy $h\nu$
Q_{T}	Thomson cross section
R	gas constant; electronic transition operator; reflection coefficient; particle rigidity in magnetic field; planetary radius
Re	real part of $(\cdots)$; Reynold's number
R_i	Richardson's number
R_{rad}	radiative loss term (in heat conduction equation) wave polarization; interference factor
s	spin angular momentum quantum number
S	entropy; spin angular momentum operator
S_{27}	27-day average sunspot number
T	absolute temperature; transmission coefficient
U	internal energy; wave characteristic in plasma [Eq. (11.44)]
v	velocity; phase velocity
v_g	group velocity
v_T	thermal wind
V	potential energy; photon production rate per unit volume; ionization energy; electric potential
w	vertical drift velocity
w_D	diffusion velocity
w_{DE}	convective ion drift velocity
W	mechanical work
X	wave characteristic in plasma [Eq. (11.44)]
Y	wave characteristic in plasma [Eq. (11.44)]
Z	atomic number; wave characteristic in plasma [Eq. (11.44)]
$\partial f/\partial x$	partial derivative of f with respect to x

$\mathbf{\nabla}$	gradient operator
$\mathbf{\nabla}\times$ (or curl)	curl operator
$\mathbf{\nabla}\cdot$ (or div)	divergence operator
∇^2	Laplacian operator
$[X]$	number density of species X
X^*	excited state
$\underset{\sim}{\mathbf{1}}$	unit tensor
$\oint_c$	line integral over a closed path c
α	specific volume; polarizability; wave characteristic (Table 11.1); recombination coefficient
$\underset{\sim}{\boldsymbol{\alpha}}$	polarizability tensor
α_D	dissociative recombination coefficient
α_{eff}	effective recombination coefficient
α_i	ion-ion recombination coefficient
α_T	coefficient of thermal diffusion
β	electron attachment coefficient; linear loss rate (positive ions); wave characteristic (Table 11.1)
γ	ratio C_p/C_v; collisional detachment coefficient
δ	variation symbol
ϵ	electric permittivity; energy; heating efficiency
ζ	ratio of mass of electron to mass of ion
$\hat{\boldsymbol{\zeta}}$	vertical unit vector
θ	angle; potential temperature
$\Theta(\theta)e^{im\phi}$	Hough function
λ	wavelength; ratio of negative ion to electron number density; ionospheric drag coefficient
Λ	molecular orbital angular momentum operator
μ	magnetic permeability; reduced mass; electron rest energy
ν	wave frequency; collision frequency; also used for ν_m where meaning is clear; kinematic viscosity of a fluid
ν_m	momentum transfer collision frequency
ξ	vertical displacement
π	parity of atomic state; fluid stream function
$\mathbf{\Pi}$	Poynting vector
ρ	charge, mass density;
σ	scalar electrical conductivity
	electrical conductivity tensor
$\tilde{\Sigma}$	electronic spin operator (molecules)
$\underset{\sim}{\mathbf{\Sigma}}$	conductivity tensor integrated over altitude
τ	lifetime of a state; optical depth; relaxation factor
ϕ	angle, electrical potential function, velocity potential function

Φ	heat energy or mass flux per unit area
χ	zenith angle; divergence of fluid velocity; gravitational potential (in units of length)
ψ	quantum mechanical state function
ω	angular wave frequency
ω_b	Brunt-Väisälä frequency
ω_c	cut-off frequency
ω_k	plasma frequency of species k
ω_p	plasma frequency
$\tilde{\omega}$	vorticity (fluid)
ω_{kH}	gyro frequency of species k
Ω	total angular momentum operator (molecules)
$\mathbf{\Omega}$	planetary angular velocity

Physical Constants

Planck's constant	$h = 6.6257 \times 10^{-34}$ joule sec
Rationalized Planck's constant	$\hbar = h/2\pi = 1.0544 \times 10^{-34}$ joule sec
Velocity of light in vacuum	$c = 2.99793 \times 10^{8}$ m sec^{-1}
Charge of the electron	$e = 4.8029 \times 10^{-10}$ esu
	$= 1.6021 \times 10^{-19}$ Coulomb
Rest mass of electron	$m_e = 9.1091 \times 10^{-31}$ kg
Ratio of proton mass to m_e	$(m_p/m_e) = 1836.1$
Boltzmann's constant	$k = 1.3806 \times 10^{-23}$ joule $(°K)^{-1}$
Avogadro's number	$N_0 = 6.0224 \times 10^{26}$ (kg mole)$^{-1}$
Universal gas constant	$R = 8.314 \times 10^{3}$ joule (kg mole °K)$^{-1}$
Permittivity of free space	$\epsilon_0 = 8.8542 \times 10^{-12}$ farad m^{-1}
Permeability of free space	$\mu_0 = 4\pi \times 10^{-7}$ henry m^{-1}
Acceleration of gravity (at surface of Earth)	$g = 9.8$ m sec^{-1}

Index

Absorption, ionospheric, 399, 418, 419
Acoustic cut-off frequency, 164
Acoustic-gravity waves, *see* Gravity waves
Acoustic velocity, 163, 164, 388
Adiabatic processes, 28
Aeronomy, 2, 5, 7, 14, 117
Airglow, 3, 4, 143, 189-202
Alfven waves, *see* Hydromagnetic waves
Alpha particles, 215, 335
Angular momentum, 41-3, 46, 51
  addition of, 44
  orbital, 41-3
  spin, 42-3
Anomalies (F_2 region), "December," 299
  equatorial, 297-8
  nocturnal ionization, 296
  Ottawa depression, 299
  seasonal, 299
Appleton-Hartree equation, 390, 397, 400, 402, 410
Archimedes spirals, 337
Ashen light, 200
Association, *see* Recombination
Atmospheric drag, 87, 88, 90, 154, 294
Atmospheric infrared bands, *see* Infrared atmospheric oxygen band
Atmospheric oxygen band system, 194, 200, 210-11
Atmospheric regions, 18
Atomic spectra, 48, 79
Attachment, Electron, 63, 246, 248, 251, 253, 257, 260, 263, 273, 357
Aurora, 2, 3, 75, 202-15, 327, 341-6, 364-5
Aurora excitation, 203-7
Aurora morphology and classification, 202
Aurora, polar glow, 215, 347
Auroral bremsstrahlung ionization, 352-6
Auroral spectra, 207-13

Balmer Hα line (atomic hydrogen), 11-2, 200, 213, 215, 328
  Lα line, 15, 123, 194, 200, 207, 241, 254, 273, 276, 328-9
  Lβ line, 15, 123, 263, 265, 273
Band spectra, 49, 78
Barbier's equation, 199
Barometric equation, 70, 80, 157
Bistatic radar occultation experiments, 145, 267, 320, 423
Blackbody spectrum, 37, 38, 78
Bohr model of H atom, 39, 40
Bohr radius, 39
Boltzmann equation, 32-4, 36, 73, 382, 411, 428-9
Boltzmann transport equations, 33-4, 221, 285
Booker quartic, 402
Born approximation, 53, 407, 408
Boundary conditions (electric and magnetic), 24, 405

Bremsstrahlung, 55, 206, 244, 341, 343, 345-6, 352-6
Brunt-Väisälä frequency, 157-9, 163-4

C-layer, 17, 253
Cameron bands (CO), 201
Carnot cycle, 27
Chapman function, 17, 74, 242, 284
Chapman profile, 16, 289
Chapman reaction, 134, 197, 199
Chapman theory, 242-4, 256
Chappius bands, 126
Charge density, 22, 23, 374
Chemical loss rates, 118, 132-6, 138, 140, 142, 144, 146-8
Chemiluminescence, 186-7, 202
Chromosphere (sun), 10
Circulation, atmospheric, 153, 170, 172-3
Collision frequency, 33, 35, 222-3, 227, 286, 288, 410, 418-9, 451
Collision integral, 33, 382, 428, 430
Collision processes, charge transfer, 63, 136, 213, 245, 256, 282, 307, 314, 333-4
  chemical reaction, 49, 117, 186
  elastic, 302-3, 306, 400
  excitation, 50, 54, 187-8, 206, 211, 213, 301, 302, 306
  ionic reaction, 49, 245-7, 252-3
  ionization, 49, 56-8, 187, 213, 347-56, 351-6, 358
  momentum transfer, 49, 375, 429
  rearrangement, 50, 63, 64, 136, 193, 214, 251, 253, 259-60, 271-2, 366
Collision theory, 50-4
Conductivity, electrical, 22, 219, 227-32, 234, 375-6, 381, 431
Conductivity, thermal, 35
  electron-ion gas, 300, 303, 307
  neutral atmosphere, 75-6, 100, 107
Conductivity tensor, electrical, 22, 221, 223-4, 226, 375-6, 378-9
  Cowling, 227, 232
  Hall, 225
Conductivity tensor *(Continued)*
  longitudinal, 225, 227
  Pedersen, 225
Conservation of energy, 26, 34, 72, 89, 155, 178
Conservation of momentum, 34, 154
Continuity equation, 22, 34, 81, 155, 192, 248-50, 263, 271-2, 287-8, 293, 296, 333-5, 347-60, 366, 374, 383
Convection, 81, 83, 159, 181
Cooling of electron-ion gas, 293, 296, 302, 306-9, 323, 411
Coriolis force, 166, 175
Corona (sun), 10
Cosmic noise, 329, 341, 417
Cosmic rays, galactic, 14, 15, 17, 241, 245, 253-4, 259, 273, 278
  solar, 15, 215, 335-41
Critical height, 108, 109, 110
Cross modulation, 412, 419
Cross section, collision, 50-2
  differential, 50, 53, 54
  momentum transfer, 51, 60, 222, 429
  total, 50
Cross section, photoionization, 54, 301
Cross section, photoabsorption, 54-57, 122-27, 129-31, 133, 254, 255
Current density electric, 22, 224 ff, 229, 375, 377, 431
  Hall current, 226-7
  Pedersen, 226-7
Cut-off, atmospheric, 350
  geomagnetic latitude, 340, 349-50
  rigidity, 340
Cyclogenesis, 161
Cyclotron frequency, 224, 376
Cytherean, atmosphere, 96-7, 111, 181, 221
  ionosphere, 253, 259, 267, 269-74, 276, 320-3, 333-4, 357

D-region, 17, 137, 253-9, 273-4, 277-8, 328-9, 332-4, 400, 410, 412, 417, 419-20, 432

Dayglow, 144, 189-195
Debye shielding, 36, 37
Debye length, 37-409
De-excitation, 135-6, 186-8, 191-4, 211, 214, 302
Density, atmospheric mass, 71, 73, 83, 87-91, 156, 181
Density, atmospheric number, 71, 90, 92, 101, 108, 112, 140-2, 144, 146-7, 174, 181
Density, charged particle number, 223-4, 228, 235, 257-8, 262, 268-71, 274, 277, 285-94, 335, 339, 358, 362, 367-8
Detachment, associative, 247, 252, 259, 260, 263
  collisional, 247-8, 252, 260, 263, 360
  photo, 192, 247-8, 252, 260, 263, 273, 334, 360
Diameters, planetary, 8
Dielectric tensor, 374, 377
Differential absorption, 423
Differential Doppler effect, 422-3
Diffusion, ambipolar, 236, 247, 253, 282, 285, 287, 314
Diffusion coefficient, 34, 81, 287
Diffusion, eddy, 72, 80, 110, 175-182, 137
  molecular, 34-5, 75, 80, 94-5, 137, 153, 180, 181
Diffusion equation, 82, 287
Diffusion velocity, 81-2, 286, 296
Diffusive separation, 80
Dispersion relation, 24, 163, 381-2, 386-7, 390, 396, 401, 413
Dissociation, molecular, 15, 96, 97, 118, 187
Distribution function, 30-3, 36, 75, 109-10, 223, 283, 400, 410, 412, 428-32
Doppler shift, 206, 211, 213
Drift velocity, 235, 287, 296
Drifts, electrodynamic, 234-5, 282, 287, 293-4, 298
Dynamo theory, 229, 232
E-region, 4, 137, 174, 226, 232, 234-5, 253, 256, 259-71, 273, 283, 294, 319, 332-3, 365-7, 400
Eddy viscosity, 176
Eigenfunctions, 168
Eigenvalues, 168
Einstein coefficients, 37, 38, 100, 191, 193
Electric currents, ionospheric, 266, 219-27
Electric field displacement, **D**, 22, 24, 374, 377
Electric field intensity, **E**, 22-5, 221-3, 226, 229, 234, 374-5, 377, 379, 385, 405
Electric polarization, **P**, 22, 24, 377
Electric potential, 36, 229
Electrodynamics, 21-6, 39, 374
Electrojets, 232, 234, 365-6
Electromagnetic energy flow, 25, 391-3
Electromagnetic waves, 373-424
  polarization of, 381, 384-5, 390, 393, 398-9
Electrons, auroral, 341-6
  secondary, 206, 210, 215, 344
  tertiary, 207, 344
Electron energy levels, 42, 45-7
Electron loss, 284
  rates, 247, 282, 289, 294, 299, 322
Emission rate, 188-9, 193
Emissivity, 77
Energy deposition, 72-3, 75, 155, 168
  nonlocal, 300
Energy flux, 156
Energy loss mechanisms (electron), 301
  rates, 304-5
Energy transfer, 187-8, 191
Enthalpy, 28, 64
Entropy, 27-30, 64, 99, 155-9, 163
Equilibrium, chemical, 314, 247-52
  diffusive, 80, 82-3, 271, 288, 291, 293

Equilibrium *(Continued)*
hydrostatic, 70, 156, 176
mixing, 80
photochemical, 80, 82, 118, 137, 145-7, 263, 277, 282
Equivalent depth, 168, 170
Escape, atmospheric, 109-12
Exchange coefficients, 176, 181
Excitation functions, 209-12
Exosphere, 18, 70, 107-12
Extraordinary ray, 398, 400
Extraordinary wave, 384, 390

F region, 17, 118, 191, 193, 199, 214, 253, 267, 276, 278, 332, 365
F_1 region, 17, 253, 256, 271-3, 283, 333
F_2 region, 17, 174, 199, 226, 234, 235, 253, 271, 282-4, 286-8, 290, 292-300, 315-20, 364, 400, 417
F region drifts, 234
Faraday rotation, 393, 421-2
Filamentary structure, 337-8
First negative bands, (N_2) 194, 197, 208-9, 215
First negative system, 211
First positive bands, (N_2) 208, 211
Fluid dynamics, 153-182
Fluorescence, 186, 195
Fourth positive system (CO), 201
Fox-Duffendack-Barker bands (CO_2), 201
Free energy of activation, 64

Geocorona, 18, 200, 273
Geomagnetic disturbances, 219, 232, 234
Geomagnetic field, 86, 166, 201, 213, 215, 219, 240-1, 244-5, 253, 282, 285, 327, 337, 339-40, 343, 364, 366
Geomagnetic rigidity, 340, 342
Geomagnetic variations, 86, 166, 219, 229
Geostrophic approximation, 160
Geostrophic wind, 160
Gravitational potential, 8, 154, 168, 170
Gravitational separation, 80
Gravity waves (internal), 153, 162-5, 173, 175
Green line (atomic oxygen), 134, 190, 199, 214-5
Group velocity, 25, 164, 379, 392, 401

Harmonic oscillator, 46
Hartley bands (ozone), 125, 194
Heat conduction, 34, 75-6, 79, 99, 100, 300, 303, 307, 322
Heat equation, 76, 84, 99, 308, 322
Heat flux, 72, 75, 76, 297, 309, 311
Heating, chemical, 84
Heating efficiency, 303, 306
Heating of electron gas, 296, 300
Heisenberg uncertainty principle, 40
Herzberg bands (O_2), 121-4, 200
Heterosphere, 17
Homosphere, 17
Hough functions, 169
Huggins bands (O_3), 126
Hydromagnetic mode, 393-7
Hydromagnetic waves, 87, 207, 228, 387
Hydrostatics, 6, 70, 243

Ideal gases, 28, 35, 71
equation of state, 31, 32
IGY, 229
Index of refraction, 379-80, 386, 390, 398-9, 413, 415
Infrared atmospheric bands (O_2), 194, 197, 210
Instability, baroclinic, 171
two stream, 232, 413, 414
International brightness coefficient, 189
Ionization potential, 54
Ionogram, 315-8, 415
Ionosphere, 4, 5, 15, 106, 219-237, 247, 282-323, 367, 374, 397-424

Ionosphere-solar wind screening, 236-7, 323
Ionospheric currents, 229-37
Ionospheric disturbances, 400
Ionospheric drag, 174
Ionospheric storms, 327, 364-5
Ions, 15, 36, 51, 52, 58, 62, 248, 366
Irregularities, ionospheric, 283, 315-20

Jovian atmosphere, 106, 107, 357
  ionosphere, 320
Jupiter, 7, 8, 215, 221, 283, 337-8

K index, 219
Kennelly-Heaviside layer, 4, 15
Kinetic theory of gases, 21, 30-37, 80, 374, 428-32

L current system, 219
Landau damping, 382, 387-8, 392
Langevin equation, 374, 376-7, 382
Langmuir probe, 310, 311, 424-5
Layer formation, 16, 17
Lifetime, 61, 62, 120-1, 135, 137-8, 146-7, 188, 362
Lorentz force, 23, 222, 339
Luxemburg effect, 410, 432

M-arcs, 213-4
Magnetic, disturbance 234
Magnetic field induction, **B**,22, 24, 222-3, 226, 229, 374-5, 385, 405
Magnetic field intensity, **H**, 22-5, 229, 379
Magnetic storms, 87, 234
Magnetic storm time variation, 234
Magneto-acoustic coupling, 397
Magnetoplasma, 373-97
Magnetosphere, 220, 234, 319
Margenau distribution, 432
Mariner spacecraft, 6, 145
Mars, 6-8, 15, 145-8, 201-2, 215, 221, 235, 236, 241, 244, 283, 320, 340, 397
Mars airglow, 201-2
Martian atmosphere, 87, 96, 97, 181
  ionosphere, 253, 259, 261-2, 267-8, 272-3, 333-5, 357
Mass, planetary, 8
Maxwell's equations, 21, 236, 374
Maxwell-Boltzmann distribution, 31
Measurements, of atmospheric properties, 87-98
  of ionospheric properties, 414-25
  balloon, 206, 329, 341, 343, 347
  rocket, 92-5, 143-4, 198, 206, 207, 292, 329, 341, 347, 348
  satellite, 87-91, 207, 292, 312, 329, 341, 342
Meinel bands (N_2+), 209
Mercury, 7, 8
Mesopause, 69, 80, 84, 107, 159
Mesophere, 69, 78-9, 84, 153, 165, 170-1, 173
Metallic ions, 254, 263, 333, 366
Metastable states, 48, 121-4, 134-6, 139, 188, 191-3, 197, 199, 208, 210-11, 214, 254
Mixing, 138, 147
Mixing length, 176, 180
Mode condition (wave guide), 407
Model atmosphere, 76
  CIRA, 98, 101, 105, 142-3
  Harris and Priester, 99, 100, 102-4
  Mars, 105
  US Standard, 98, 101, 105
  Venus, 105
Mögel-Dellinger effect, 328
Molecular spectra, 37, 48-9

Negative ions, 62, 63, 246-8, 250, 254, 259-60, 262-3, 357, 359
Negative ion-to-electron ration, 248-50, 257-8, 334, 357-60
Nightglow, 134, 197-201
Noctilucent clouds, 159

Occultation experiment *see* Bistatic radar
Ohm's Law, 22, 221, 375, 383, 394, 407
Optical depth, 73, 100, 242

Ordinary ray, 398, 400
Ordinary wave, 384, 390
Ozone, 170, 171
Ozonosphere, 17

Parity, 42
Partial reflection, 407, 412-13, 420-21
Pauli exclusion principle, 43
Permeability (magnetic), 22
Permittivity (electric), 22
Perturbation theory, 156, 383, 395
Phase velocity, 23, 392, 396
Photodissociation, 54, 72, 107, 118, 121-32, 135, 146, 175, 181-2, 190-2, 197, 271
Photoelectron excitation, 194
Photoexcitation, 54, 72
Photoionization, 54, 63, 72, 241-5, 247, 253, 256, 261, 264-7, 273, 276, 283-4, 300, 333-5, 338
Photon, 24, 38, 54, 187
Photosphere, 10
Physical constants, 438
Pitch angle, 343
Planetary probes, 6, 145
Plasma, 2, 24, 36-7, 373-7, 400, 413
Plasma frequency, 24-5, 224, 375, 388
Plasma waves, 381
Polar cap absorption (PCA), 215, 340, 347, 424
Polar cap aurora, 335
Polar cap event, 335, 347
Polarizability, 51, 377-8
Potential energy curves, 47, 54-5, 122-3, 125-6, 129, 133
Poynting vector, 25, 391-2, 406
Predissociation, 121
Pressure, atmospheric, 6-8, 71, 83, 87, 90, 98-9
  plasma, 286, 382, 394
  fluid, 155 ff
Proton, 2, 56, 58, 213, 215
Protonosphere, 18, 297, 311, 314-16, 319
Pseudosonic mode, 389

Quantum of action, 37
Quantum mechanics, 40-1
Quantum numbers, 39, 41-3
Quasilongitudinal approximation, 398, 415, 423
Quasisteady state, 249, 274, 359
Quasitransverse approximation, 398
Quenching, *see* De-excitation

Radiation, black body, 37, 38
  corpuscular, 75, 101, 240, 243-5, 327, 335-65
  electric dipole, 48, 54
  electric quadrupole, 48
  electromagnetic, 240-1, 327
  magnetic dipole, 48
  selection rules, 48-9
  solar, 5-7, 10, 14-16, 69, 73, 84-5, 97, 100, 111, 113, 117, 197, 245, 254, 263, 267, 272, 314
Radiative cooling, 77, 79, 100
Radiative transfer, 77-9, 99, 100
Radio blackout, 329
Radio effects, 329-33, 344-7
Radio wave interaction, 373, 410-12
Radio wave propagation, high frequency, 398-402
  low frequency, 403-7
Radio wave scattering, 407
Radio wave sounding, 4, 253, 414-7
Radioactive decay, 220
Rayleigh, 288
Reactions, chemical, 64, 118, 153, 247-52
  ion-molecule, 63, 64, 107, 284
Reaction rate constants, 61, 63, 64, 123, 127, 136, 251-2, 272
Recombination, ionic, 118, 132-6, 146, 187, 197, 201, 211, 214, 246, 248, 251, 253-4, 271-2, 359, 366
  dissociative, 62, 284, 322
  effective, 250, 252, 369
  ion-ion, 62, 246, 251, 257, 259, 263, 359
  radiative, 246, 251, 272
Red arc, 199

Red line (atomic oxygen), 190, 199, 214
Reflection coefficients, 400, 403-6, 413
Reflection height, 415
Relaxation time, 61-2
Resonances, 390, 393
Reynolds number, 179
Reynolds stresses, 176-7
Richardson number, 179
Ring current, 234
Riometer, 329, 417-8
Rotational energy levels, 47, 60

Satellites, artificial, 88-90
Saturn, 7, 8
Scale height, 71, 73, 80-2, 90, 94, 101, 108-9, 112, 168, 181, 242, 278, 288-9, 293
Scattering, atmosphere, 77
Scattering, resonance, 144, 186, 189, 193-5, 197, 200, 202, 209
Scattering theory, *see* Collision theory
Schrödinger equation, 40, 53
Schumann-Runge bands, 55
  continuum, 55, 121-24, 191, 197
Second positive system (N_2), 208
Secular equation, 163, 380, 384
Sen-Wyller formula, 424
Shortwave fadeout (SWF), 329
Solar activity, 84, 86
Solar control, 76
Solar cycle, 14, 84, 86, 90, 220, 275-6, 278, 327
Solar eclipse, 328, 367-9
Solar flare, 327-41
Solar plage, 328, 369
Solar plasma, 337, 340
Solar radio flux, 84-5, 335
Solar spectrum, 9, 10, 276
Solar wind, 84, 236, 320, 322, 397
Specific heat capacity, 26, 28-9, 72, 76, 99, 157
Specific volume, 156-7
Spectral irradiance, 73, 77, 118, 243, 329
Spectral line broadening, 77, 93
Spectrum, electron energy, 300-1
Spin conservation rule (Wigner), 65
Spin-orbit coupling, 42, 44
Spherical harmonics, 48, 169
Sporadic-E, 173, 232, 328, 365-7, 414
Spread-F, 283, 315-7, 417
Sq current system, 219-20, 232-4
Stability, atmospheric, 159, 179
Steady state equations, 118, 333
Stopping power, 58, 59, 75
Stratopause, 69, 79
Stratosphere, 69, 84, 159, 162, 170-2
Structure tensor, 176, 302
Sudden commencement, 365
Sudden cosmic noise absorption (SCNA), 329
Sudden enhancement of atmospherics (SEA), 329, 332
Sudden frequency deviations (SFD), 332-3
Sudden ionospheric disturbance (SID), 329-33, 424
Sudden phase anomaly (SPA), 332
Sudden polar warming, 84
Sun, 6, 7
Sunspot number, 84, 85, 241, 275, 278, 299, 327-8
Symbols, list of, 433
Symmetry, molecular, 46
Synchrotron radiation, 335

Temperature, neutral atmosphere, 26, 69, 71-2, 81, 85-6, 92, 97-8, 112, 155
  plasma, 283, 286, 294, 300, 308-13
  surface, 6-8
Thermal efficiency, 73
Thermal energy, internal, 26, 72, 117, 155
Thermal wind, 160
Thermodynamics, 2, 21, 26-30, 153, 155
Thermosphere, 70, 76, 85-6, 94, 153, 159, 165, 170, 173-4

Thomson cross section, 408
Thomson scatter radar, 310-11, 313, 409, 419
Tides, atmospheric, 153, 160, 162, 165-71, 174-5, 219-21
Transmission coefficient, 403-4, 406
Traveling ionospheric disturbance (TID), 320
Triangulation, 198
Tropopause, 2, 69
Troposphere, 69, 165
Turbopause, 17, 80, 176, 180
Turbulence, 80, 81, 107, 175-182
Turbulent heat transport, 180
Turbulent mixing, 81, 153, 180-2
Twilight glow, 144, 195-7, 366
  illumination, 195-7

Van Rhij formula, 195
  variations, atmospheric structure, 83-6
  diurnal, 84, 273-7
  seasonal, 84, 277
  solar cycle, 83-6
Vector model of atom, 40-5
Vegard-Kaplan band system (N_2), 208
Velocity, fluid, 254 ff
  moments, 31
Venera spacecraft, 94, 145
Venus, 6-8, 15, 145-8, 200-1, 215, 221, 235-6, 241, 244, 283, 320-3, 340, 397
Vibrational energy levels, 46, 49
Virtual height, 199
Viscous interaction, 158, 174-5, 179, 222, 294
Vortex, 162, 171
Vorticity, 161-2

Water vapor, 170
Waves, electromagnetic, *see* Electromagnetic waves
  electron, 382, 387
  ion, 382, 388
  plasma, *see* Plasma waves
Wave equation, 23
Wave function, 40, 53
Wigner spin rule, 126
Winds, 81, 92, 101, 152, 160, 169, 170-5, 219, 229-33, 297
Wind shears, 153, 160-1, 165, 171, 173, 179, 366
Winter anomaly, 277

X-rays, 7, 10-15, 241, 243, 253-4, 259, 263, 265, 267, 272-3, 329, 369
  solar flare, 328-39
  spectra, 276, 330-32, 336-7